AF794802

Pollutant Fate and Transport in Environmental Multimedia

Pollutant Fate and Transport in Environmental Multimedia

Frank M. Dunnivant
Elliot Anders

WILEY

This edition first published 2019
© 2019 John Wiley & Sons Inc.,

All rights reserved. No part of this publication may be reproduced, stored in a retrieval system, or transmitted, in any form or by any means, electronic, mechanical, photocopying, recording or otherwise, except as permitted by law. Advice on how to obtain permission to reuse material from this title is available at http://www.wiley.com/go/permissions.

The right of Frank M. Dunnivant and Elliot Anders to be identified as the authors of this work has been asserted in accordance with law.

Registered Office
John Wiley & Sons, Inc., 111 River Street, Hoboken, NJ 07030, USA

Editorial Office
111 River Street, Hoboken, NJ 07030, USA

For details of our global editorial offices, customer services, and more information about Wiley products visit us at www.wiley.com.

Wiley also publishes its books in a variety of electronic formats and by print-on-demand. Some content that appears in standard print versions of this book may not be available in other formats.

Limit of Liability/Disclaimer of Warranty
In view of ongoing research, equipment modifications, changes in governmental regulations, and the constant flow of information relating to the use of experimental reagents, equipment, and devices, the reader is urged to review and evaluate the information provided in the package insert or instructions for each chemical, piece of equipment, reagent, or device for, among other things, any changes in the instructions or indication of usage and for added warnings and precautions. While the publisher and authors have used their best efforts in preparing this work, they make no representations or warranties with respect to the accuracy or completeness of the contents of this work and specifically disclaim all warranties, including without limitation any implied warranties of merchantability or fitness for a particular purpose. No warranty may be created or extended by sales representatives, written sales materials or promotional statements for this work. The fact that an organization, website, or product is referred to in this work as a citation and/or potential source of further information does not mean that the publisher and authors endorse the information or services the organization, website, or product may provide or recommendations it may make. This work is sold with the understanding that the publisher is not engaged in rendering professional services. The advice and strategies contained herein may not be suitable for your situation. You should consult with a specialist where appropriate. Further, readers should be aware that websites listed in this work may have changed or disappeared between when this work was written and when it is read. Neither the publisher nor authors shall be liable for any loss of profit or any other commercial damages, including but not limited to special, incidental, consequential, or other damages.

Library of Congress Cataloging-in-Publication Data applied for
ISBN: 9781119414629

Cover design by Wiley
Cover images: Background: Courtesy of Karen Bauschlicher, Whitman College, Class of 2022.
Overlay: Courtesy of Frank M. Dunnivant

Set in 10/12pt WarnockPro by SPi Global, Chennai, India

Printed in the United States of America

V10008837_031419

To the chemists who have had a positive influence on my career: Loretta McLean, John Coates, Alan Elzerman, Philip Jardine, and Rene Schwarzenbach. And to my family, Marion, Lukas, and Marley. Thanks to all of you,

Frank M. Dunnivant

To my family, Tessa, Jude, and Elsa, thank you for supporting me throughout.

Elliot Anders

Contents

Preface

While there are many of textbooks on environmental chemistry and fate and transport phenomena, and they each have their place in education, this is the first book to truly integrate the subjects chemistry, fate and transport modeling, assessment, laws, and environmental laboratory experiments. The entire text was edited to ensure a consistent modeling theme that is vital to the regulatory applications of environmental fate and transport of contaminant science. These applications are truly multimedia in scope and encompass modern-day regulations in successful EPA approaches in the Clean Air Water and Drinking Water Acts. The related RCRA and CERCLA (Superfund) programs address hazardous and toxic materials and ensure that environmental and human health are equally protected. These laws use mathematical models to assess contaminant exposure that contribute to multiple adverse effects in the biosphere. Most books in this genre are conceptual in nature or require a working knowledge of differential calculus to effectively derive and use new models. This prerequisite effectively eliminates many people from working in fate and transport.

We approach each of these topics initially from a conceptual perspective, and then, we explain the concepts in terms of the math necessary to model the problem. The only prerequisites for understanding the concepts covered in this book are a basic knowledge of algebra and first-year college chemistry. Mathematic enthusiasts will find the extensive, step-by-step, end-of-chapter derivations useful. This last feature is unique to all other textbooks. For the fate and transport modeling chapters (Chapters 4–9), we have included a simple, user-friendly web-based, online simulator, Fate®, which uses basic step and pulse models to predict the fate and transport of pollutants in lake, river, groundwater, and atmospheric systems. Fate® can be an effective teaching and learning tool, as discussed in the "How to Use Fate" section of the introductory materials.

This book is the result of a challenge I made to one of my senior chemistry students. I challenged Elliot Anders, the coauthor of this book, to create a new version of EnviroLand, the precursor to Fate®. If he did so, I told him, I would write a book to accompany it. To my surprise, Elliot finished the software in a few months and to meet my end of the bet, I had to write the 2006 textbook around the software. We feel that this textbook provides a very unique instructional tool for students and environmental professions who lack the rigorous mathematical backgrounds to be able to derive the governing fate and transport equations, but nonetheless require an understanding of the subject. This book can be used to teach a variety of classes, from a new type of environmental chemistry course to new fate and transport courses for support personnel who want to work in the environmental arena. I use the book to teach environmental chemistry to undergraduate students majoring in chemistry, geology, and biology, to mostly prepare them for work in the environmental consulting arena and government. These students usually have sufficient background to work in environmental remediation but lack the basic engineering knowledge to be truly effective in this area. I have had great success in expanding the academic horizons of science students to areas of chemistry, modeling, risk assessment, and environmental legislation. In addition, there is no reason that this book cannot be used in a graduate course in fate and transport, since it provides an especially extensive and complete development of fate and transport models. In this case, the professor can use the book as a conceptual guide while teaching the derivation portion of the course in the classroom.

New additions to this book include free software available at https://sites.google.com/a/whitman.edu/frank-dunnivant-webpage/environmentalsoftware:

- an Web-based Pollutant Fate and Transport Simulator, (Fate®)
- a pC-pH Simulator
- Water and Wastewater Plant Tour Video
- EnviroLab
- More end-of-chapter problems with answers to the more complicated problems

- New environmental lab chapters on the United States, European Union, and China
- Student lab results for suggested experiments
- A supporting discussion-based book entitled *Environmental Success Stories* with Columbia University Press

We hope you enjoy our approach to environmental chemistry and pollutant fate and transport modeling.

Frank M. Dunnivant
September 2018

Acknowledgments

A book of this detail and 20-year effort has involved many people. First, I thank my coauthor, Elliot Anders, for challenging me to make the software better than the original Visual Basic version. Next, I thank Raymond Whittemore for writing, editing, and making many, many excellent suggestions. Note Dr. Whittemore's experience with, and knowledge of regulatory modeling were essential to the revised theme of this edition. Ray participated in water quality modeling training workshops for more than a decade for EPA's Exposure Assessment Laboratory in Athens, GA. Late in his career, he participated in the EPA Council for Regulatory Environmental Modeling (CREM) development in a leadership role in the surface water quality modeling group discussions.

Then, I thank Dr. Nate Boland and his students for testing this book and materials in his Environmental Chemistry and Engineering class at Whitman College. Certainly, a great thank you goes to my former student John Brooksbank for derivations of the many pollutant fate and transport equations that do not appear in any textbook in this detail. Many students can earn an A in differential equations but few can apply the knowledge like John has. The end-of-chapter problems, answers, and laboratory experiment results have been provided by 19 years of students taking my classes. I thank all of you for a job well done.

Acronyms

α	alpha type of radiation
ADD	average daily doses
AF	Risk: soil to skin adherence factor (mg/cm^2)
AID	maximum acceptable daily intake
Al_2O_3	aluminum oxide
A_s	mean lake or system surface area, m^2
AT	Risk: average time period of exposure (days)
AMD	acid mine drainage
β	beta type of radiation
BAP	best available practice
BFC	bioconcentration factors
BOD	biochemical oxygen demand
Bq	Becquerel
BW	Risk: body weight (kg)
c	the speed of light in a vacuum (2.99×10^8 m/s)
C	the pollutant concentration in the water (mass/length^3 or volume; mg/l, etc.) or molar charge
Ci	curies, a measure of radiation
C_i	the pollutant concentration in the main inlet river or system, kg/m^3
$C_6H_{12}O_6$	chemical formula for glucose
CA	Risk: pollutant concentration in the air (mg/m^3)
CAS	Chemical Abstract Number
CCA	Air Quality Act, The Clean Air Act
CCRs	coal combustion residuals
CDC	Center for Disease Control and Prevention
CEAM	EPA Center for Exposure Assessment Modeling
CEC	cation exchange capacity
CERCLA	Comprehensive Environmental Response, Compensation, and Liability Act
CF	Risk: volumetric conversion factor for water (1.00 l/1000. cm^3)
CF	Risk: pollutant concentration in the food (mg/kg)
CF	Risk: pollutant concentration in the fish or shellfish (mg/kg)
CF	Risk: pollutant concentration in the fish (mg/kg d)
CFL	compact fluorescent lamp
CH_4	chemical formula for methane (gas)
CO	carbon monoxide; Colorado, a state in the United States
CO_2	carbon dioxide
CO_3^{2-}	carbonate
COC	contaminants of concern
COE	Army Corps of Engineers
C_p	the net pollutant concentration in precipitation, kg/m^3
CR	Risk: contact rate (0.050 l/h)
CREM	EPA Council for Regulatory Modeling
C_s	the average pollutant release from suspended lake sediments, kg/m^3-time
CS	Risk: pollutant concentration in the soil (mg/kg)
CSTR	continuous stirred tank reactor
C_T	the total concentration of all chemical species for this chemical
C_{trib}	the net pollutant concentration of the tributaries, kg/m^3
C_w	the pollutant concentration in the wastewater, kg/m^3
CW	Risk: pollutant concentration in the drinking water (mg/L)
CWRA	Clean Water Restoration Act
d	average stream depth
DBMS	Database Management System
DDT	1,1′-(2,2,2-trichloroethylidene) bis[4-chloro-benzene]
DEM	Dynamic Estuary Model
DNAPL	dense nonaqueous phase liquid(s)
DO	dissolved oxygen
DOD	Department of Defense
DOE	Department of Energy
DWEL	drinking water equivalent levels
e^-	an electron

E	energy; longitudinal eddy diffusion term
ED	Risk: exposure duration (number of years)
EDC	endocrine disrupting compound
EDTA	ethylenediaminetetraacetic acid
EF	Risk: exposure frequency (d/yr)
EFDC	Environmental Fluid Dynamics Code
E_H	oxidation–reduction potential
EIS	environmental impact statement
EPA	Environmental Protection Agency
EPA LUST	Environmental Protection Agency Leaking Underground Storage Tank
EPCRA	Emergency Planning and Community Right to Know Act
EPTT	extraction procedure toxicity test
ESA	Endangered Species Act
EXAMs	Exposure Analysis Modeling System
F	Faraday constant (96 485 C/mol of electrons)
FIFRA	Federal Insecticide, Fungicide, and Rodenticide Act
Fe_2O_3	iron oxide
$FeCO_3$	iron carbonate
FI	Risk: fraction of soil ingested from the polluted site
FWPCA	Federal Water Pollution Control Act
γ	gamma type of radiation
g	the acceleration due to gravity (length/time^2)
g	grams
GEOCHEM	a numerical modeling program for determining chemical speciation
GLBTS	Great Lakes Binational Toxics Strategy
η	kinematic viscosity of the fluid (length^2/time)
h	Planck's constant (6.63×10^{-34} J s)
H	the water depth (length)
H^+ or H_3O^+	hydronium ion, acidic hydrogen
H_2O	chemical formula for water
HRS	hazardous ranking system
HCl	hydrochloric acid
HCO_3^-	hydrogen carbonate, bicarbonate
H_2CO_3	carbonic acid
HQ	hazard quotient
H_2S	hydrogen sulfide gas
HS^-	chemical formula for hydrogen sulfide ion
INEEL	Idaho National Engineering and Environmental Laboratory
IR	Risk: inhalation rate (m^3/d)
IR	Risk: ingestion rate of water (l/d)
ISC	industrial source complex
k	the first-order removal rate for the pollutant, 1/time
k_d	the first-order BOD decay rate
k_r	the DO re-aeration rate
k_{NOM}	the rate constant in the presence of Natural Organic Matter (NOM)
K_a	the acidity equilibrium constant
K_b	the basicity equilibrium constant
K_d	distribution coefficient
K_H	air-water Henry's law constant
K_p	partition coefficient
K_{sp}	solubility product constant
K_w	the water ionization constant; 1.00×10^{-14} (at 25 °C)
λ	the wavelength in meters
l	liters, a measure of volume
LCA	life cycle assessment
LCOE	levelized cost of energy
LED	light-emitting diode
ln	natural logarithm
LNAPL	light nonaqueous phase liquid
LOAEL	lowest-observed-adverse-effect level
LOEL	lowest-observed-effect level
LUST	leaking underground storage tank
J	Joules, a measure of energy
μ	ion strength
M	molarity, mol/l
M_o	the total mass of pollutant in the river
meq	milliequalivance
NEPA	National Environmental Policy Act
MIC	methyl isocyanate
MINTECH	a numerical modeling program for determining chemical speciation
MINEQL+	a numerical modeling program for determining chemical speciation
NIOSH	National Institute for Occupational Safety and Health
MnO_2	chemical formula for manganese dioxide
MPRSA	Marine Protection, Research, and Sanctuaries Act
MTBE	methyl *tert*-butyl ether
NAAQS	national ambient air quality standard
NEPA	National Environmental Policy Act
N_2	chemical formula for nitrogen gas
NOM	natural organic matter
NO_x	combinations of nitrogen oxide and nitrogen dioxide
NO_3^-	chemical formula for nitrate
NOAEL	no-observed-adverse-effect level
NOEL	no-observed-effect level
NPL	National Priorities List
NTA	nitrilotriacetic acid
O*	radical gaseous oxygen

OH^-	hydroxide ion
OH^*	radical gaseous hydroxide
O_3	ozone gas
OPA	Oil Pollution Act
OSHA	Occupational Safety and Health Act or Administration
p	as in pC or p(anything); represents the negative log of the quantity; example: pCO_2 represent the negative log of the carbon dioxide molar concentration in pC–pH diagrams
pH	negative log of the hydrogen ion concentration
pH_{ZPC}	the pH where there is no surface charge
P	annual precipitation, m/time; phosphorus
Pb	lead
PBDE	polybrominated diphenyl ethers
PCBs	polychlorinated biphenyls
PFOA	perfluorooctanoate
PFOS	perfluoroocotanesulfonate
PPA	Pollution Prevention Act
ppb	parts per billion
ppm	parts per million
PSD	prevention of significant deterioration
PZC	point of zero charge in clay titrations
PZSE	point of zero salt effect
Q_e	the outlet or effluent flow of water from the lake or system (m^3/time)
Q_i	the inflow rate of the main river water or system, m^3/time
Q_{trib}	the net inflow rate of water from all other tributaries, m^3/time
Q_w	the inflow rate of water or wastewater, m^3/time
r	the spherical particle radius (length)
R	ideal gas law constant, with a variety of units; ion radius
r_A	the rate of decrease in pollutant A concentration per unit volume of water (mass/length^3-time)
RCRA	Resource Conservation and Recovery Act
r_f	the density of the fluid (mass/length3)
RfD	reference dose
RI-FS	remedial investigation-feasibility study
r_s	the density of the spherical particle (mass/length^3)
s	the slope or gradient of the stream bed
S	the suspended solids concentration (mass/length^3)
SA	(cross) sectional area
SA	Risk: skin surface area available for contact (cm^3)
SAP	sampling and analysis plan
SARA	Superfund Amendments and Reauthorization Act
SCRAM	support center for regulatory atmospheric modeling
SDWA	Safe Drinking Water Act
SIP	state implementation program(s)
SPARC	chemical parameter estimator software from EPA
SOD	sediment oxygen demand
SO_x	a combination of sulfur dioxide and sulfur trioxide gases
SO_4^{2-}	chemical formula for sulfate
SWDA	Solid Waste Disposal Act
T	temperature
t_o	detention time of water or chemical in a system
TAN	Test Area North
TAN-05	injection well (Test Area North)
TCE	trichloroethylene
TCLP	toxicity characteristic leaching procedure
TDS	total dissolved solids
TESS	threatened and endangered species system
TEA	terminal electron acceptor
THM	trihalomethane
TMDL	total maximum daily load
TRI	toxic release inventory
TSCA	Toxic Substances Control Act
TSS	total suspended solids
US EPA	United States Environmental Protection Agency
USGS	United States Geologic Survey
USTs	underground storage tanks
v	the frequency of light (1/s)
V	volume, in length^3 or l
ω	settling velocity of a particle in natural water
w	average stream width
W	the mass input of pollutant to the lake (or system) per unit time, kg/time
WASP	water quality analysis simulation program
WES	Waterways Experiment Station of the Corps of Engineers
Z	charge of an ion
z/R	ionic potential (charge/radius)

Glossary

Adsorption	The process of accumulating an excess of a chemical on a surface. As used in this text, it refers to the concentrating of pollutant on a mineral or NOM-coated mineral surface.
Advection	The transport of pollution in the direction of flow.
Anaerobic	An aquatic or atmospheric system that does not contain oxygen. This term is the same as anoxic.
Anoxic	Devoid of oxygen.
Abiotic reaction	A chemical reaction that takes place without the aid or in the complete absence of microorganisms.
Basel convention	An international treaty regulating the reporting, disposal, and transport of hazardous waste. The United States is currently not a member of this treaty.
Biota	Any living organism in an ecosystem.
Biotic reaction	A chemical reaction that occurs due to a microbial process (enzymes in the microbe cell).
BOD	Biochemical oxygen demand, the amount of dissolved oxygen required by microorganisms to oxidize organic matter present in the water.
Cation exchange capacity (CEC)	The concentration of sorbed cations that can be readily exchanged for other cations.
Confined aquifer	An underground body of flowing water that is located below the layer of strata that is impermeable to water.
Conservative tracer	A chemical tracer that does not degrade and is not sorbed by anything in the water. The chemical moves freely with the water.
Contaminant	A chemical that is out of its proper place. In this text, we will use the term contaminant and pollutant interchangeable.
Dispersion	A mixing process resulting from advection which always dilutes the concentration of pollutant.
DNAPL	Dense nonaqueous phase liquid. An example is carbon tetrachloride.
DO	Dissolved oxygen; the concentration of dissolved oxygen in water, usually from 0 to 14 mg/l.
DOM	Dissolved organic matter. These are components of NOM that are soluble is water.
E	The longitudinal dispersion (eddy) coefficient for streams.
Environmental impact statement	A study required by the National Environmental Policy Act to attempt to determine if any adverse effects will occur from governmental actions such as the building of a building, plant, or even a remediation effort.
Empirical	A relationship based on experiment data.
Epilimnion	The upper region of a stratified lake.
Eutrophication	An overproductive aquatic system. Excess alga growth occurs during the day due to the presence of excess nutrients, but during nighttime hours, the oxygen is depleted below levels that can support aerobic life.

Explanative modeling	A model that attempts to explain how something happened. For example modelers can use mathematical relationships to explain how a pollutant moved to where it is, where it originally came from, or how much pollutant was originally released.
Head	The height of a water column.
Hypolimnion	The lower or bottom region of a stratified lake.
LNAPL	Light nonaqueous phase liquid. An example is gasoline.
Longitudinal dispersion	The mixing of pollution in water or air in the direction of flow.
Modeling	An attempt to explain a process in a simpler form. Models can take on several forms including physical models that are usually small-scale versions of the real thing. We will limit our discussions in the text to mathematical models, which are very simple mathematical relations of more complicated processes. The more the terms we include in our models, the more the processes we account for in our model and theoretically, the more accurate our model will predict the real system.
NAPL	Nonaqueous phase liquid. An example would be oil or gasoline.
Nitrobenzene	A benzene molecule with an attached nitro (NO_2) group. Nitrobenzenes can also have other functional groups such as methyl (CH_3), chloro (Cl), and many other chemical groups attached to the benzene ring.
NOM	Natural organic matter. NOM results from the accumulation of degradation products from plants and animals in soil and water.
Nonpoint pollutant source	A source of pollutant that cannot be identified as a specific location. Examples would be runoff from a large area of land such as a parking lot or agricultural setting and atmospheric inputs.
Partitioning	Partitioning is very similar to adsorption but does not involve a site-specific reaction. It is more of a solvation or dissolving of a pollutant into NOM.
pC	A way of representing concentration units on the log scale. "p" stands for the negative log of anything. "C" stands for the concentration (in any units) of any chemical species.
pH	The negative low of the hydrogen ion concentration.
Point pollutant source	A source of pollution that can be pinpointed to a specific location. For example the output pipe from an industrial process or sewage treatment plant would be a point source.
Predictive modeling	A model that attempts to predict what will happen at a future time. For example, in this text, we are concerned with predicting the concentration of pollutant at some point (location or time) in the future.
Pulse or instantaneous release of pollutant	A release of pollutant that occurs over a very short time scale and contains a finite volume or mass of pollutant. This type of release is in contrast to a step or continuous pollutant release.
Refractory pollutant	A pollutant that does not readily degrade (degrades slowly or not at all).
Remediate	To cleanup a waste site to acceptable pollutant concentrations.
Residence time	The average time a chemical spends in an environmental compartment. This can be obtained by dividing the volume of the compartment or mass of a chemical in a compartment by the outflow from the system (volume / volume per time = average residence time).
Sensitivity analysis	An iterative process of testing a model where one parameter at a time (volume, rate constant, etc.) is systematically increased or decreased while the result (pollutant concentration) is recorded.

Sorption	A generic term referring to the combination of adsorption and partitioning.
Stratification	A process that results in two distinct layers of water in a lake system. Stratification results due to heating of the surface water and cooling of lower waters by the Earth that sets up a density difference in the two bodies of water. The cool water settles to the bottom of the lake, while the warmer water is present at the surface.
Surface aquifer	The ground water closest to the land surface.
Step or continuous pollutant release	A release of a pollutant that occurs over a long-time scale. Examples include the constant release of sulfide from a pulp mill and the release of nitrate from a sewage treatment plant.
Vadose zone	The portion of the ground that is unsaturated with respect to water.
Variable	A symbol representing a mathematical term or the term of interest in a mathematical expression. For example, velocity is an important term in fate and transport, and it is represented by the variable, v.
Watershed	The drainage area of land surrounding and feeding water into a lake or river basin.

About the Companion Website

This book has a companion website hosted by Wiley:

www.wiley.com/go/Dunnivant/pollutant_fate&Transp_EnvironMultimedia

To the Instructor

The material in this textbook has been used for 19 years to teach an undergraduate course in environmental chemistry to chemistry, biology, and geology students who have not completed linear algebra or differential equations. Students from these disciplines tend to have excellent skills and knowledge in environmental studies but lack an understanding of how to apply this knowledge to environmental applications such as fate and transport modeling and risk assessment. Students feel a strong sense of empowerment as they come to understand how to approach an environmental pollution event and gain an appreciation of the many scientific and political factors controlling the remediation of contaminated sites. Software packages are integrated into the textbook, and we have suggested important technical papers for class discussion for many of the chapters. In addition, we have designed laboratory exercises specific to the topics covered in the lecture material. This textbook would be ideal for an introductory course in environmental chemistry, pollution science, environmental science, environmental engineering, or pollutant fate and transport. One particular useful discussion book is my popular science book, entitled *Environmental Success Stories: Solving Major Ecological Problems & Confronting Climate Change.*

A set of laboratory exercises have been designed for this textbook. As with all laboratory experiments, it will be instrumental that the instructor test the procedures prior to using them. This will ensure the success of your laboratory class. A complete set of instructions for solution preparations, suggested level of difficulty, suggested lab time requirements, apparatus construction guidelines, hints for success, and *detailed example student results* are included in this book.

To the Student

This is a new type of textbook, in the sense that it combines many disciplines to craft a practical discussion of environmental pollution and its remediation. We first start with an overview of the chemistry of fate and transport processes. Next, we introduce you to how professionals model a pollution event in the real world and then present you with a conceptual understanding of the models that these professionals use. At the end of each fate and transport chapter, we incorporate our conceptual ideas into two simple and common models that can be used to predict the pollutant concentration in most environmental systems. Next, we use the results from fate and transport modeling as a basis for our risk assessment calculations to determine whether a hazardous situation is present. The final chapters include environmental legal aspects and case studies of common pollutants that have been spread around the world.

To the Environmental Professional

While most environmental engineers will find this book to be a basic review, many of other environmental professionals who do not understand the derivation of complex environmental models will find this book to be a valuable asset for developing a conceptual and mathematical understanding of the problem. Our goal is to bridge the knowledge gap between the engineers and mathematicians and the many important workers associated with a Superfund site. The only prerequisites for understanding the subjects covered in this book are a working knowledge of algebra and college-level general chemistry. We provide the background and explain everything else. If you are an engineer or modeler working in the environmental arena, you will find this book to be a must for your colleagues.

How to Use the Book with Fate® and Associated Software

This book comes with several online software packages, Fate®, the pC-pH Simulator®, EnviroLab®, and The Water and Wastewater Tutorial®available at https://sites.google.com/a/whitman.edu/frank-dunnivant-webpage/environmentalsoftware. Fate is certainly the most important tool, since it can be used in a variety of ways (https://educationalsolutions.github.io/fate/). First, Fate can be used as an in-class tool to illustrate how each of the fate and transport models work. It enables the instructor to quickly and easily show how changing model input parameters affect the resulting pollutant concentration. Students and instructors will find Fate to be an invaluable resource in working homework problems. All these problems require lengthy, multistep calculations, and each step can be checked with Fate. This decreases the need for students to rely on a tutor or their instructors, who in turn would have to manually examine through the students' work to find errors. We suggest the following approach. First, work the homework problem manually, consulting Fate only when you do not understand which calculation step to complete next. Check your manually calculated answers at each step in the process against Fate to see if you are correct. If you do not understand why you are doing a calculation, then certainly consult your instructor. The pC-pH Diagram Simulator (https://educationalsolutions.github.io/pCpH//open/) works in a similar manner to Fate but is used only in Chapters 2 and 3 to understand acid–base equilibrium and buffers. The Water and Wastewater Tutorial (https://sites.google.com/a/whitman.edu/frank-dunnivant-webpage/environmentalsoftware) can be viewed on your own time or in class to illustrate how modern water and wastewater treatment facilities work. Components of EnviroLab (https://sites.google.com/a/whitman.edu/frank-dunnivant-webpage/environmentalsoftware) are useful in understanding details of several laboratory experiments. Readers will find the Internet sites of EPA, USGS, and the Army Corps of Engineers useful in acquiring source code, documentation manuals, and training for most of those models used in the regulatory domains. Historically, these agencies have supported these models and developed new ones when the science dictates. More importantly, personnel at the state and national levels may assist in their use by the public.

Instructor/Student Resources

The following computer resources are available for this textbook from https://sites.google.com/a/whitman.edu/frankdunnivant-webpage/environmentalsoftware:

Computational Simulators for Chemistry and Fate and Transport Calculations:

Fate-2016®
pC-pH®
Dispersion Calculator for groundwater experiments

Videos:
Global Issues
Water

Laboratory Experiment Simulators:
Alkalinity
Hardness
Dissolved Oxygen – Biochemical Oxygen Demand
Total Suspended Solids

In addition, all lab experiments presented in Chapter 16 have example student results.

Part I

Introduction

Through the history of literature, the guy who poisons the well has been the worst of all villains

Author unknown

Pollutant Fate and Transport in Environmental Multimedia, First Edition. Frank M. Dunnivant and Elliot Anders.
© 2019 John Wiley & Sons, Inc. Published 2019 by John Wiley & Sons, Inc.
Companion website: www.wiley.com/go/Dunnivant/pollutant_fate&Transp_EnvironMultimedia

1

Sources and Types of Pollutants, Why We Need Modeling, and the Need to Study Historical Pollution Events

1.1 Introduction

A good starting point for our discussions and this book is to ask, "Is there a common sequence of events leading to the identification, characterization, and remediation of an old hazardous waste site?" There are a variety of answers to this question, but a general order of events often occurs as follows:

- First, a pollutant is observed to be present or the potential of a pollutant release from a proposed industrial site is identified. This can result from routine monitoring of a pollutant's concentration at the site, through the known manufacturing of the pollutant at the site, through research identifying the cause of an illness or cancer cluster in the community of a site, or during an environmental planning assessment (also called an environmental impact statement, EIS).
- Second, the source of the pollutant is identified at the hazardous waste site or a theoretical release can be simulated.
- Third, a remedial investigation is conducted to determine the mass/volume of pollutant released and responsible parties.
- Fourth, pollutant fate and transport modeling is completed to determine what pollutant concentrations will result at specific points at the site over time (referred to as receptor sites where humans are the receptors of the pollutants).
- Fifth, the results of the pollutant fate and transport modeling are used in risk assessment calculations to estimate health or environmental/ecological risks. A common use of modeling in the twentieth and twenty-first centuries is in air and water quality environmental management and related to compliance with ambient air or water quality standards (EPA 2017a). Essentially, the model purports the allowable discharge or emission rates to maintain acceptable air or water quality.
- And finally, a decision or plan of remediation is negotiated between the local citizens, local and federal governments, and the party responsible for the hazardous waste site.

Today, new pollutant releases from regulated industries take on a slightly different approach. In the Global North, previously referred to as developed countries, most industries act responsibly, but all human-designed, engineered, and operated systems are prone to failures and accidents. In general, a slightly modified sequence of events occur: a spill or accident occurs, emergency response systems respond, and remediation (when possible) occurs. Responsible parties usually pay the incurred expenses, and appropriate fines if regulations have been broken.

These steps also outline our approach in this book. In this chapter, we will look at several historical hazardous waste release events, as well as recent ones, and describe types and sources of potential pollutant releases. In Chapters 2 and 3, we will look in-depth at the chemistry associated with fate and transport phenomena in environmental media. In Chapters 4–10, we will develop and learn to use pulse and step fate and transport models for rivers, lakes, groundwater, and the atmosphere. In Chapter 11, we will introduce basic chemical toxicology and use pollutant concentrations from the fate and transport modeling to perform basic risk assessment calculations to estimate risk to human health, look briefly at cost–benefit analysis, and the relatively new field of environmental life cycle assessment. In Chapters 12–14, we will look at the development and evolution of environmental laws in the United States and Europe. In Chapter 15, we will look at the history of several "world class" pollutants that are present at undesirable concentrations across the globe. In the last Chapter 16, laboratory experiments that support key chemical and modeling concepts will be presented, with actual student results.

Pollutant Fate and Transport in Environmental Multimedia, First Edition. Frank M. Dunnivant and Elliot Anders.
© 2019 John Wiley & Sons, Inc. Published 2019 by John Wiley & Sons, Inc.
Companion website: www.wiley.com/go/Dunnivant/pollutant_fate&Transp_EnvironMultimedia

1.2 Need for Modeling of Pollutants in Environmental Media

Many pollutants have been found to be ubiquitous in nature; that is every environmental compartment that has been tested has shown some level of contamination. Note that presently ambient monitoring of water bodies and the atmosphere are both part of the missions of both state and federal government agencies. Historically, two of the chemicals that fall into this category and receiving much public attention are PCBs (polychlorinated biphenyls) and DDT (1,1′-(2,2,2-trichloroethylidene)bis[4-chloro-benzene]). These compounds accumulate especially well in the environment due to their refractory behavior (they are not easily chemically or biologically degraded). More importantly, their chemical and physical properties are such that they bioaccumulate in fatty tissue in many organisms. Modelers like to divide nature into more easily mathematically described boxes, such as the atmosphere, rivers, lakes, groundwater, and biota (living organisms). In this book, we simplify this further by only considering one section of a river, a small portion of a groundwater aquifer or a small portion of the atmosphere.

There are two basic goals in pollutant modeling: to explain how a pollutant arrived where it is (a form of thermodynamic equilibrium) and to predict how fast a pollutant will move through an environmental compartment in the future (a form of kinetics). Some engineers call this all-compassing field "chemodynamics" to reflect its transport, phase transfer, and equilibrium components. There are several excellent textbooks based on these principles (Thibodeaux 1996; Hemond and Fechner 2000). Later in this chapter, we will look at several examples of major pollutant release events; the approach taken to these reflects the first goal of modeling (an explanation of how a pollutant arrived where it is). Although this approach is very important to understand how pollutants move in the environment, a more common use of modeling today is to predict if and/or how fast a pollutant will move once it is accidentally released. In the United States, this is included in an EIS. Most Global North countries require a variety of industries to perform some type of environmental impact modeling in the unlikely and unfortunate event of an accidental pollutant release, which again is inevitable with most human-engineered systems. This book will concentrate on predictive modeling, but by understanding the processes involved in the predictive modeling approach, it is relatively easy to appreciate how the explanation type of modeling is completed.

In order to predict where a pollutant will go we use a set of fate equations that describe the chemical and physical processes occurring in the environmental compartment under evaluation, such as mixing, outflow through the effluent of the system, evaporation, volatilization, and chemical or biological degradation. Transport of chemicals is part of its fate. The same modeling equations and concepts are used for the explanation type of modeling (where pollution already exists) but a reverse process is used. In explanation modeling, the modeling equation is fit to pollutant concentration data from field measurements in order to obtain estimates of mixing and degradation rates in the specific system. Both modeling approaches will become clearer when we reach the modeling chapters in this text, especially beginning in Chapter 4.

1.3 Pollution versus Contamination; Pollutant versus Contaminant

Is there really a difference between the terms pollutant and contaminant? Some would argue no, while others will argue vehemently yes. In the United States, environmentalists and those working for US Environmental Protection Agency (US EPA) prefer the terms pollution and pollutant. But what makes a chemical a pollutant? Basically this distinction is determined by where the chemical is located and how much of the chemical is present. For example, a bottle of mercury chloride is typically not considered a pollutant when it is sitting on the shelf in a chemical storeroom. But pour that same chemical down the drain, and it immediately becomes a pollutant, even at extremely low concentrations. This is the logic behind usage preferred by other branches of the US government, such as the Department of Energy (DOE). The DOE prefers to use the terms contamination and contaminant. If a chemical is in its proper place it is not considered a problem and is thus not a contaminant. Likewise, if a chemical is present in an environmental media below the concentration deemed to be a problem (which is subject to a variety of views and laws), then the chemical is not a problem and contamination is not necessary present. Thus, the legal definition usually specifies two factors, where the pollutant is and its concentration. For example, the first author of this book (Dunnivant) worked on a project for the Idaho National Engineering and Environmental Laboratory (INEEL), one of many US DOE sites, in which we intentionally placed a rapidly degrading radioactive substance in a water-filled lagoon in order to characterize the movement of water and radionuclides in the subsurface media. As long as the chemicals stayed within the controlled boundaries of the lagoon (the approved experimental area) they were not considered a contaminant. But one day a violent storm blew a small amount of the water and sediment a few feet out of the lagoon. The DOE safety officials overreacted and suspended all operations at the site for a time period that jeopardized a multimillion dollar field experiment! Although the vast majority (we would speculate over 99.99%) of the chemical was still present in the lagoon, the extremely small amount of chemical "released" (amounting to a few specks of dirt) was officially classified as a contamination event. Similarly, in the 1980s chlorinated dioxins and furans were attributed to wood pulp bleaching to achieve acceptable brightness attributes. In these cases, concentrations in the parts per trillion (10^{-12}) and parts per quadrillion (10^{-15}) were routinely part of the technical conversation, and they were termed "pollution" in most cases by technical and lay personnel (EPA 2017b). Today, wood pulp and paper processing plants have replace chlorine with peroxide to achieve the product with less pollution. So, when you use the terms pollution versus contaminant be aware who you talking to. In this book, we will use the more common terms pollutant and pollution.

1.4 Pollution Classifications

There are perhaps as many ways to categorize types of pollution as there are pollutants. One broad way to by physical phase: solid, liquid, or gaseous. This categorization is useful from a standpoint of treatment and disposal. Many treatment technologies are based on the physical phase of the pollutant, such as bag filtration houses for atmospheric particulates or filtration of particles in liquids. Another general but more chemical means of categorization is the inorganic, organic, and radioactive nature of the waste. Mixed waste, a very difficult, if not impossible waste to treat, usually refers to organic and radioactive waste being present in the same waste product. Inorganic waste can be further broken down into non-toxic and toxic metals, metalloids, and non-metals. Metal waste can be subdivided into heavy metals and transition metals. Toxic wastes can be grouped as carcinogens, teratogens, and mutagens (discussed in Chapters 11 and 12). Compounds that are subjects of heightened public awareness, such as PCBs, are divided out even further. Other wastes are listed as hazardous based solely on their origin from a specific industrial process (for example, metal plating wastes) or when a specific chemical is present in the waste stream (for example, the presence of PCBs). As you see there are many ways to categorize or list a waste and each country will have their own system for classifying pollutants. In the modern era, some compounds are known to accumulate in organism tissue with high lipid content that further increases the human health risk associated with their presence.

Another interesting way to categorize waste is by the risk it poses. For example, say you have a hazardous waste site that needs to be remediated and there are 20 pollutants of interest. But, from a risk standpoint, 5 of the 20 pose a significantly higher hazard based on risk assessment calculations (Chapter 11). In some countries, it is customary to focus the remediation effort on these 5 chemicals and disregard the other 15 "less hazardous pollutants." Such an approach is used by the US EPA to calculate health risk based on source and an overview of the approach is shown in Figure 1.1.

1.5 Sources of Pollution

As with types of waste, there are many ways of categorizing sources of waste (or potential pollutants). We will only mention a few of the more useful approaches and give examples of generated volumes from major industrialized countries. First, we will discuss point and nonpoint sources.

Sources of pollutants are commonly divided into one of two types, point or nonpoint sources. Point sources are well-defined sources, such as the end of a pipe, smoke stack, or drain. Nonpoint sources are less well defined, but contain all sources where you cannot directly pinpoint the emission. The distinction becomes a bit more vague depending on physical scale. For example, if you are studying a large watershed (the land surface that drains into a stream or lake), a cattle holding lot could be considered a point source. Obviously, the farm is the source of the cattle waste. But on a smaller scale, what is the point source? The answer is possibly each cow, or the entire holding lot, since the pollution will be spread over its entirety. But, in general the terms point and nonpoint are used to describe sources.

Table 1.1 lists sources of waste by 10 categories: agricultural, chemical industry, mining industry, energy industry, municipal and hazardous landfills, medical industry, food processing, domestic waste, municipal government, and federal government. General wastes for each source are also listed in the right-hand column. This list is certainly not complete, but contains the major sources of waste generated and provides a starting point for discussion of waste sources.

The terminology of wet and dry deposition applies to ambient air and the mode of transport in the atmosphere. Rain and snow can act to "scrub" some contaminants from the air and deposit them on trees, plants, and the soil where erosion and runoff infiltration further distribute them across the surface and subsurface environments.

Since this book deals mostly with the transport of toxins, we will be especially concerned with hazardous

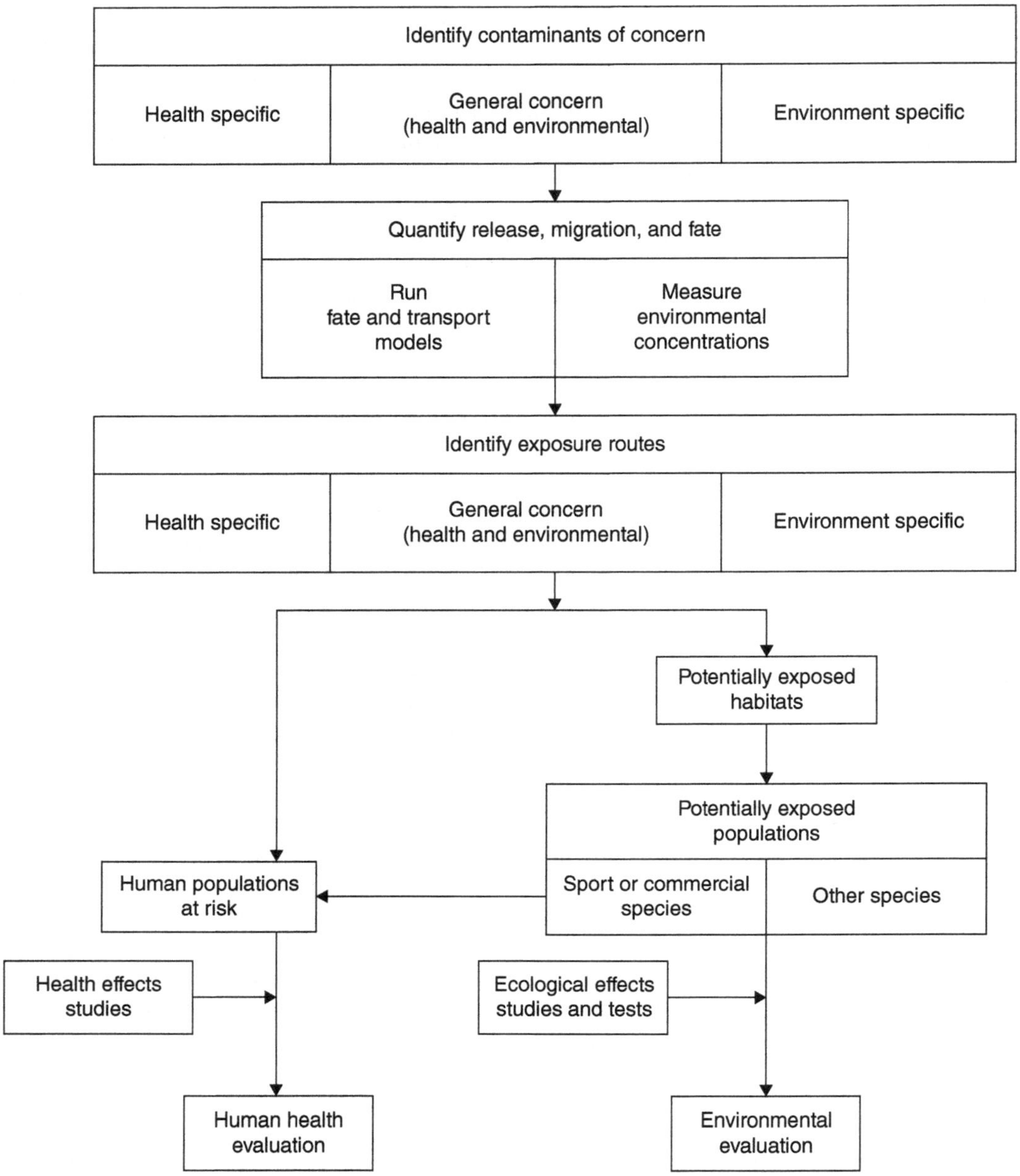

Figure 1.1 An overview of the approach to access human risk used in the United States. Source: DOE.

waste. A summary of the amounts of hazardous waste generated by country members of the Basel Convention (an international treaty regulating the reporting, disposal, and transport of hazardous waste) is given in Table 1.2 for the year 2000. Unfortunately this dataset is no longer available on the Internet, nor is updated total generation data available since data are now reported on an import–export basis. One would only speculate that the magnitudes of these values have increased and, in general, the more developed the country, still the more hazardous waste it generates. This correlation of waste to economical level is a homework exercise that you should conduct and, by the way, it would make an excellent examination question. You will undoubtedly note the lack of data on two highly industrialized countries: Japan and Germany. Even without these two countries the total amount of hazardous waste annually generated is an obscene 93 992 999 metric tons in year 2000. Another country you will note absence from the list is the United States, since it has not signed the Basel Convention (UN 2000). For comparison purposes, the United States generated 40 821 481 tons in the year 2001 (EPA 2001),

Table 1.1 Potential sources of chemical emissions (point and nonpoint sources).

Source	General waste (representative but not exhaustive)
Agricultural	Field and chemical wastes
	Nutrients (fertilizers)
	Pesticides/herbicides
	Petroleum fuels
	Feedlot waste
	Dairy waste
Chemical industry	Metal products
	Metal sludges
	Non-metal waste
	Electrical equipment waste
	Detergents/soaps/cleaners
	Petroleum-related waste
	Metal plating
	Cooling-tower additives
	Film-processing waste
	Solvents
	Wastewaters and off-specification wastes
	Pesticides
	Insecticides
	Chlorinated hydrocarbons
	Organophosphates
	Carbamates
	Microbial insecticides
	Pyrethroids
	Rodenticides
	Herbicides
	Petroleum oils
	Carbamates
	Triazines
	Phenoxy herbicides
	Amide herbicides
	Fungicides
	Precursors to smog (NO_x, hydrocarbons)
Mining industry	Mine tailings
	Mineral leaching (cyanide)
	Acid mine drainage
	Coal
	Smelting waste
	Atmospheric particulates
Energy industry	Petroleum-based waste
	Solvents
	Gas and vapor emissions
	Coal tars
	Boiler waste
	Nuclear waste
	Petroleum stored in underground storage tanks (gasoline stations)
	Precursors to acid rain and smog
Municipal and hazardous landfills	All chemicals disposed in the landfill
Incinerators	Incomplete combustion of feedstocks
	Combustion by-products
	Metals
	Particulates
Medical industry	Biodegradable waste (biohazards)
	Pharmaceutical waste
	Solvents
Food processing	Waste food products
	Rinsing waste
	Slaughter house waste
Domestic waste	Detergents/cleaners
	Pesticides, etc.
	Fertilizers
	Compose materials
	Paints/solvents
	Gasoline
Municipal governments	Chemicals associated with water and wastewater treatment
	Sewage (biochemical oxygen demand)
Federal governments	Weapons-related waste (conventional and nuclear)
	Nuclear waste
	Petroleum-based waste

which is 43.4% of the world total shown in Table 1.2. The US data delineated by geography and industry are reported under TSCA (EPA 2017c) and reported in public Internet sites. More recently, data for year 2011 show that the United States generated 28 million tons of RCRA (landfill and nonliquid) defined waste (US EPA 2017b). A summary of hazardous waste generated in 2015 for the United States is given in Table 1.3 by state and territory

Table 1.2 Total hazardous waste generated in year 2000 by the Parties to the Basel Convention.

Country	Total hazardous waste generated (metric tons)
Austria	980 558
Bulgaria	755 677
Canada	No data
China	8 300 000
Croatia	25 999
Cuba	1 023 638
Czech Republic	2 603 337
Denmark	287 491
Ecuador	85 859
Egypt	170 000
Estonia	5 965 750
Finland	1 203 000
France	9 150 000
Georgia	93 000
Germany	No data
Hungary	3 392 628
Iceland	12 550
Israel	279 987
Italy	No data
Japan	No data
Jordan	17 390
Kyrgyzstan	6 087 869
Luxembourg	96 526
Malaysia	344 550
Morocco	987 000
Netherlands	2 722 828
New Zealand	No data
Norway	650 000
Oman	242 098
Poland	1 627 143
Portugal	194 724
Qatar	280
Republic of Korea	2 756 984
Republic of Moldova	7 122
Romania	860 892
Russian Federation	12 800 000
Singapore	121 500
Slovakia	1 600 000
Slovenia	128 395
Spain	3 293 705
Sri Lanka	40 617
Sweden	1 100 000
Ukraine	2 613 400
United Kingdom	6 296 043
Uzbekistan	15 074 459
Total	93 992 999

Source: www.basel.int/natreporting/index.html (accessed 8 December 2003).

Table 1.3 Number of hazardous waste generators and amount of waste produced by the States of the United States.

Location name	Number of generators	Waste generated (tons)
National	Total	
	26 794	33 646 771
Alabama	432	709 666
Alaska	45	3 965
American Samoa	0	0
Arizona	320	38 696
Arkansas	109	338 703
California	3 395	309 109
Colorado	209	40 440
Connecticut	458	35 939
Delaware	101	12 577
District of Columbia	74	1 078
Florida	520	90 847
Georgia	821	172 752
Guam	19	140
Hawaii	100	484 603
Idaho	72	7 049
Illinois	1 083	601 542
Indiana	822	930 379
Iowa	209	56 946
Kansas	257	1 280 076
Kentucky	326	135 106
Louisiana	508	5 010 182
Maine	138	2 606
Maryland	342	125 807
Massachusetts	813	39 108
Michigan	535	407 754
Minnesota	362	120 862
Mississippi	147	1 946 349
Missouri	482	291 067
Montana	72	6 773
Navajo Nation	1	45
Nebraska	104	26 839
Nevada	141	13 141
New Hampshire	232	3 714
New Jersey	948	333 804
NEW MEXICO	89	5 631
New York	3 636	281 259
North Carolina	611	81 153
North Dakota	35	265 051
Northern Marianas	1	2
Ohio	1 324	1 711 110
Oklahoma	242	105 306
Oregon	252	52 830
Pennsylvania	1 422	293 947
Puerto Rico	97	20 312
Rhode Island	184	7 252

Table 1.3 (Continued)

Location name	Number of generators	Waste generated (tons)
South Carolina	519	154 595
South Dakota	43	1 917
Tennessee	290	47 756
Texas	1 686	16 288 728
Trust Territories	1	0
Utah	167	33 927
Vermont	62	288 019
Virgin Islands	5	299
Virginia	567	62 413
Washington	512	135 451
West Virginia	259	52 163
Wisconsin	559	172 716
Wyoming	34	7 250

Source: EPA (2015).

(EPA 2015a), and in Table 1.4 by type of industry (EPA 2015b). An excellent annual source of information is available from the US EPA's Toxic Release Inventory website at https://www.epa.gov/trinationalanalysis (EPA 2017c) but this will be used as a homework exercise.

From the standpoint of modeling, the main subject of this text, we would like to be able to express our waste emission or "sources" in mathematical terms. Two mathematical source terms we will use throughout this text will be the "pulse" and the "step" inputs of pollutant. The pulse and step inputs represent both extremes of the duration of pollutant inputs to a system. You should familiarize yourself with these terms. These two terms can be slightly confusing, mostly because of the use of multiple names for them. A pulse release, also commonly referred to as an instantaneous release, is one that occurs over a small time scale (an instant in time, hence instantaneous), but may contain large or small amounts of pollutant. Examples of pulse releases would be the immediate release of 1 gallon of acetone or antifreeze into a river, or the same into a groundwater well, or the flushing of one toilet of human waste into a stream. We use the terms pulse and instantaneous to stress that the release occurs immediately, travels through the system as a pulse of pollution (a narrow band of high pollutant concentration), and is not a prolonged release. Some prefer to characterize environmental releases as buoyant or nonbuoyant plumes. A plume is characterized by a conical release emanating from a discharge pipe or stack. An air emission appears white or gray when condensing moisture is present in the cooling discharge. The emission appears to spread because it is entrains (draws in) ambient air or water. Picture an aquarium tank with

Table 1.4 Top hazardous waste generators in the United States (greater than 10 000 tons).

Description	Generated (tons)
Basic chemical manufacturing	19 725 340
Petroleum and coal products manufacturing	4 827 640
Waste treatment and disposal	3 183 389
Iron and steel mills and ferroalloy manufacturing	1 333 956
Nonferrous metal (except aluminum) production and processing	914 990
Semiconductor and other electronic component manufacturing	429 610
Pharmaceutical and medicine manufacturing	268 950
Pesticide, fertilizer, and other agricultural chemical manufacturing	262 367
Coating, engraving, heat treating, and allied activities	246 545
Resin, synthetic rubber, and artificial synthetic fibers and filaments manufacturing	234 919
Remediation and other waste management services	212 438
Warehousing and storage	162 140
Waste collection	150 780
Other chemical product and preparation manufacturing	121 401
Other electrical equipment and component manufacturing	115 746
Paint, coating, and adhesive manufacturing	104 333
Alumina and aluminum production and processing	103 599
Administration of economic program	91 224
Steel product manufacturing from purchased steel	70 711
Motor vehicle parts manufacturing	58 445
Plastics product manufacturing	56 018
Electric power generation, transmission and distribution	55 986
Highway, street, and bridge construction	49 015
Cement and concrete product manufacturing	45 180
Aerospace product and parts manufacturing	43 889
Forging and stamping	43 651
National security and international affairs	43 326
Architectural and structural metals manufacturing	34 093
Miscellaneous durable goods merchant wholesalers	31 094
Motor vehicle manufacturing	29 908
Other fabricated metal product manufacturing	28 192
Foundries	27 672
Printing and related support activities	19 806

Table 1.4 (Continued)

Description	Generated (tons)
Soap, cleaning compound, and toilet preparation manufacturing	19 610
Glass and glass product manufacturing	19 079
Agriculture, construction, and mining machinery manufacturing	16 725
Chemical and allied products merchant wholesalers	16 292
Other personal services	15 726
Medical equipment and supplies manufacturing	14 860
Scientific research and development services	14 170
Administration of environmental quality programs	14 048
Furniture stores	12 663
Support activities for water transportation	12 534
Lessors of real estate	12 267
Sawmills and wood preservation	12 188
Household and institutional furniture and kitchen cabinet manufacturing	11 930
Beverage manufacturing	11 671
Other miscellaneous manufacturing	10 865
Rubber product manufacturing	10 507
Electrical and electronic goods merchant wholesalers	10 453
Colleges, universities, and professional schools	10 300
Residential building construction	10 179

Source: EPA (2015b).

a diffuser–aerator that appears to widen or spread as the bubbles rise. We also call a subsurface discharge a plume but clearly not visible but its transport has this plume characteristic as the leak moves by potential energy–driven diffusion from higher to lower concentration. We often call this motion one characterized by Fick's law of diffusion (Fick's law 2017) and is a useful process mathematical model in nature.

In contrast, a step or continuous release occurs over longer periods of time. For modeling purposes, we like to assume that the release occurs indefinitely. Examples of step releases would be the constant release of a pollutant from an industry at "acceptable" or "non-acceptable" levels, the continuous release of leachate from a landfill into the groundwater system, and the 24-hour outflow from a sewage treatment plant. Pollutant release concentrations can be high or low, but as long as they are constant, we can model them using relatively simple transport equations. When a discharge or release occurs at a constant rate, engineers and scientists refer to steady-state flow and its rate of change with time is zero and is a convenient mathematical assumption one makes to avoid complexity.

Many pollutant sources fit into these two simple categories, but you might think of them as end-members of a continuum because there are also other pollutant sources that do not fit neatly into these categories. For example, suppose an industry legally emits a pollutant for five years into a lake, but installs better waste treatment technology or shuts down the process completely. How could we treat this? One approximation would be to use a step model for the first five years and then switch to a pulse model. To be more accurate, we would derive a specific model to fit the exact emission scenario. As noted earlier, the pulse and step inputs are very common, and since this is an introductory text on pollutant fate and transport we will only use models based on these two types of pollutant inputs. As long as there is an equation describing the pollution input, mathematicians can modify the final fate and transport equation to model the pollution as it moves through an environmental compartment.

1.6 Historic Examples of Where Fate and Transport Modeling Are Useful

There is no lack of unfortunate and disastrous pollution events that fate and transport models have been applied to in an effort to better understand and be able to predict pollutant movement in the environment. These are divided into surface water, groundwater, and atmospheric events. You may recall many of the events, discussed below, from news reports or case studies that you have covered in other classes.

Surface water: Both the Great Lakes and Chesapeake Bay in the United States were subjected to extensive modeling in the later decades of the twentieth century and became by default the field laboratories of present-day state of the art for eutrophication and toxic fate modeling practice.

The Water Quality Analysis Simulation Program (WASP 1983 discussed in Chapter 10) was first used on the Great Lakes and the detail in benthos nutrient-water column exchange and kinetics evolved from Chesapeake Bay research. Further, load allocations based on water quality models, many of which resembled the classic Streeter–Phelps analysis for the Ohio River after World War I, were responsible for the attainment of water quality standards for dissolved oxygen (DO) in thousands of stream and river miles. This maturation of modeling science for water quality control is still evolving for nutrient-rich areas and further water quality improvements are very likely using tools in EPA's model toolbox.

Groundwater: The tragedy of Love Canal that arose during the late 1970s resulted in the statute we now call Superfund and resulted in the migration of toxic chemicals into a residential neighborhood causing disease, birth, and early childhood development defects. The EPA LUST (leaking underground storage tank) program and Resource Conservation and Recovery Act (RCRA) and Superfund Corrective Action decrees have hundreds of similar episodes, albeit with less tragic results. Detection of leaks from industrial landfills are monitored very closely with detection wells and modeled as an aid to corrective actions. Models are used to delineate source water drinking water protection.

Atmospheric events: The national ambient air quality standards (NAAQSs) were established early by the Clean Air Act to address deteriorating and unhealthy air throughout the country. State implementation programs (SIPs) were established in hundreds of areas, where these standards for CO, HCl, SO_x, NO_x, ozone (O_3), and particulates were not achieved. State SIPs were created to tell EPA and the public in turn how attainment would be achieved to the extent possible. Near-field and long-range dispersion models was part of these assessments.

1.6.1 Preenvironmental Movement and Legislation

Prior to the 1960s environmental movement and the 1970s rush of environmental legislation, environmental disasters were common place. As we have seen, regulations do not prevent disastrous environmental events from occurring; think Three-Mile Island, Deepwater Horizon, and Bhopal India. Probably, the first environmental concern was the same as today in many Global South (undeveloped) countries, mammalian waste and associated diseases. Since the evolution of mammals, animals have excreted their waste into waterways that resulted in highly degradable organic waste and microbial contamination. Although the concentrations were minor, disease vectors such as *Giardia lamblia* and *Escherichia coli* were common in many species. As humans evolved and populations concentrated into sedentary civilizations, management of human waste became a problem. This problem has only been addressed in recent centuries, but traced their origins to Ancient Rome and China. One of the most profound events shaping environmental movements in the United States was the open fires of the Cuyahoga River in 1969. This is less than 100-mi river flows from Akron to Cleveland, Ohio, into Lake Erie, and at the time one of the most polluted rivers and lakes in the United States. To attribute to the historic level of pollution, 13 fires were recorded between 1868 and 1969. The 1969 fire (from surficial oil slicks) certainly spurred the US environmental movement. But environmental disasters did not end after the passage of comprehensive legislation. Superfund sites were located, requiring extensive remediation that continues today, and accidents continue to happen. As noted several times in this book, human-engineered and human-maintained systems are prone to accidents or misguided national policy.

The release of biochemical oxygen demand (BOD) waste to streams; timeline – since the dawn of civilization up until today: One of society's greatest accomplishments, and possibly most extensive over time, has been the treatment and disposal of our sewage waste. However, you may be shocked to learn that as late as the 1980s, several major coastal (East and West) cites in the United States only partially treated their sewage before dumping it into estuaries, bays and oceans. Many major cities across the world are still modernizing their treatment facilities, and huge strides have been made in this effort. Since the installation of sewage treatment plants, then Global North has eradicated many diseases associated with human waste. But from the standpoint of nature, human pathogens are of little concern and the main problem is the large amount of biodegradable organic waste, known as BOD, that is released into natural systems. As we will discuss in detail in Chapter 7, water contains relatively small amounts of DO and when untreated sewage is released into a stream, the oxygen needed to oxidize these organic wastes far exceeds the oxygen present in the stream water. This causes the stream to become anaerobic (devoid of DO) and the biological organisms dependent on this DO quickly die. In addition, the stream smells like an open sewer, which is essentially what it is.

The modeling of the consumption of DO by organisms oxidizing BOD is relatively simple. The governing equation, known as the Streeter–Phelps equation, will be discussed in Chapter 7. For now, we will only concern ourselves with a plot of DO downstream from a sewage treatment plant. Figure 1.2 shows the DO profile of a stream starting at the input of untreated sewage and downstream from the inlet point. Note how the DO immediately plummets as waste enters the stream and microbes consume the organic matter. In fact, the stream modeled here becomes anaerobic from 180 to 430 km downstream from the waste input (an incredibly long distance) and only slowly recovers as the stream undergoes natural reaeration and as the BOD is oxidized by microorganisms.

Now note the effect of adding a sewage treatment plant to the system where sewage waste is treated prior to release to the stream (Figure 1.3). In this modeling effort, we assume that 95% of the BOD is removed,

$$DO = \frac{\%\ \text{Saturation}}{100} * \text{Saturation concentration} - \left(\frac{k'\ BOD_L}{k_2' - k'}\left(e^{k'(x/v)} - e^{k_2'(x/v)}\right) + D_0 * e^{k_2'(x/v)}\right)$$

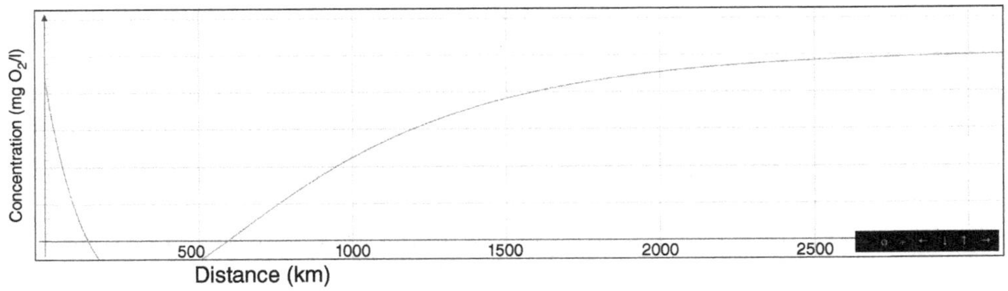

Figure 1.2 The dissolved oxygen concentration profile of a stream receiving untreated sewage waste (Results from Fate® 2018).

$$DO = \frac{\%\ \text{Saturation}}{100} * \text{Saturation concentration} - \left(\frac{k'\ BOD_L}{k_2' - k'}\left(e^{k'(x/v)} - e^{k_2'(x/v)}\right) + D_0 * e^{k_2'(x/v)}\right)$$

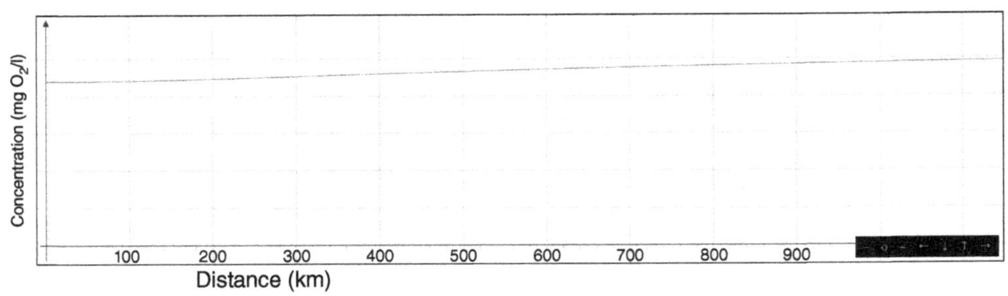

Figure 1.3 The dissolved oxygen concentration profile of a stream receiving 95% treated sewage waste (Results from Fate® 2018).

which is easily achievable in modern treatment facilities. The DO of the stream and its associated wildlife are not noticeably affected by the oxidized organic matter (treated waste). Thus, we can not only predict the effects of adding sewage to a stream but also evaluate the effects on the stream by adding treatment systems.

1.6.2 Legacy Waste Sites

Many legacy future Superfund sites existed across the United States (and world-equivalent sites) prior to the US and Europe environmental movements of the 1960s and prior to the series of environmental laws passed in the 1980s. Of course, industry was responsible for many of these environmental disasters such as the Love Canal site and others, but a surprising number of the most egregious acts were conducted by each country's governments prior to, during, and after the Cold War in the name of self-preservation (nuclear weapon programs). Examples of such sites in the United States are the Hanford site in Washington State, the Savanna River site in South Carolina, the Oak Ridge National Laboratory in Tennessee, and Idaho National laboratories site in Idaho, with each location having multiple Superfund designations. As an example, we will look at one of the Superfund sites at the INEEL. Then, a similar pre-World War II (WWII) site in Europe.

1.6.2.1 INEEL Test Area North Deep Well Injection; Timeline – Pre- and During World War II

The deep well injection project that the authors are most familiar with was located on the INEEL at the Test Area North (TAN) site (Dunnivant et al. 1994). This US government owned and operated hazardous waste site was placed on the US EPA's National Priorities List (NPL) on 21 November 1989. Injection well TAN-05 was drilled in 1953 to a depth of 93 m (305 ft) with a diameter of 30.5 cm (12 in). The groundwater surface at the site is approximately 63 m (206 ft) below the land surface. During its years of operation, the well received approximately 133 000 l (35 000 gallons; 193 000 kg) of liquid and dissolved trichloroethylene (TCE), organic sludges, treated sanitary sewage, metal filing process waters, and low-level radioactive waste streams. Basically, the well was the means of disposal for any liquid or semiliquid waste (domestic and hazardous) produced at the site. Although several pollutants of concern were disposed of and detected in the groundwater at TAN, we will only

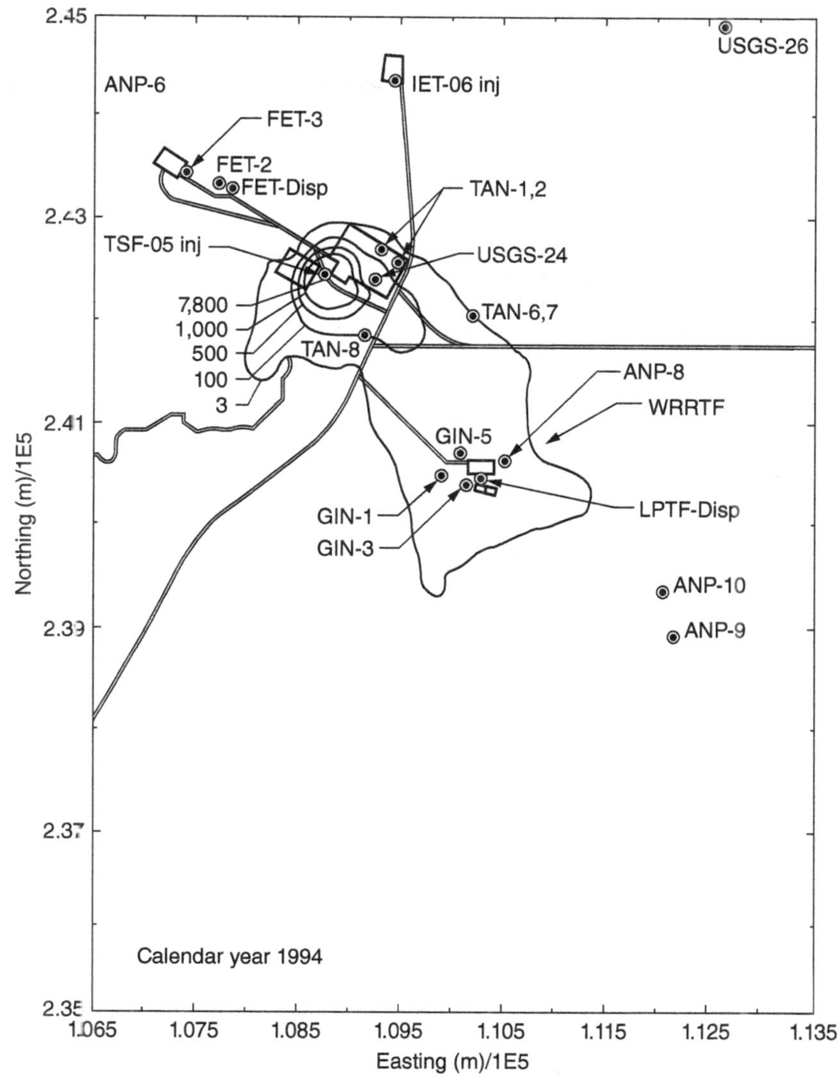

Figure 1.4 Predicted isoconcentration lines for TCE at the INEEL Tan injection site for the year 1994. Source: INEEL, Department of Energy, Dunnivant et al. (1994), and public domain.

show the results for TCE. It should be noted that the site is located in the mostly uninhabitable region of the Snake River Plain.

Figure 1.4 shows isoconcentration circles (isopleths; lines of equal TCE concentration) for the TAN that were obtained by fitting a step model to field measurements of TCE concentrations in the groundwater (explanatory modeling). Since TCE is suspected of being present in the aquifer as a dense nonaqueous phase liquid (DNAPL), it solubilizes slowly and the input of TCE can be considered a continuous source (step input over an extended period of time). You will note that near TSN-05 injection well, the isopleth is for 1000 parts per billion (ppb) while the lowest concentration shown in this figure is 3 ppb (the maximum allowed drinking water concentration). If no remedial action was to be taken in 1994 (the proposed year of remediation if any was to be attempted) and the DNAPL continued to release TCE to the groundwater, another step model predicted that the TCE plume would expand to the one shown in Figure 1.5 for the year 2044 if no remediation was taken.

So, how do we decide what is to be attempted in the way of remediation? Pollutant fate and transport modeling can be one of many tools to evaluate this question. For example, what if, through extensive and expensive remediation of the site, the source of TCE could be removed. How would this affect the TCE concentrations for year 2044 (a date selected by DOE and EPA)? These results are shown in Figure 1.6. As you can see, there is significant improvement in the groundwater quality. In this scenario, the source of the contamination has been removed and the plume of contamination has migrated down gradient and been significantly diluted. Still, pollutant fate and transport modeling cannot be the only tool in the decision process. Other factors, such as the cost of this magical cleanup and the associated health

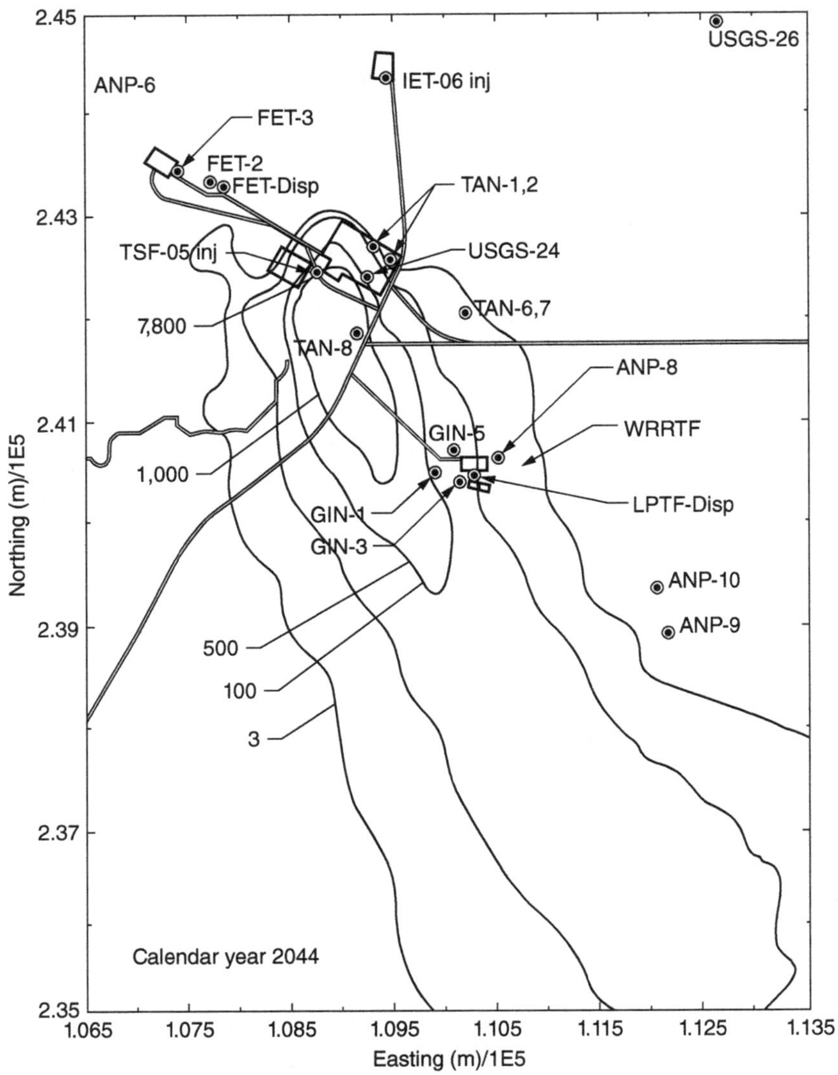

Figure 1.5 Predicted isoconcentration lines for TCE at the INEEL Tan injection site for the year 2044 without remediation. Source: INEEL, Department of Energy, Dunnivant et al. (1994), and public domain.

risk of drinking the contaminated groundwater and of remediation practices must also be considered. And note again, the contaminated groundwater is inaccessible to any human through restrictions on the installation of groundwater wells.

1.6.2.2 The Release of Nitrobenzene-Based Solvents at the Sondermulldeponie Landfill in Kolliken, Switzerland; Timeline – Pre- and During World War II

This pre-WWII, Superfund-type site is a bit more complicated since it involves a lot of chemistry to understand the pollution scenario. Around 1978, Sondermull disposed of nitroaromatic liquid wastes in metal drums in a shallow pit (Dunnivant et al. 1992; Schwarzenbach et al. 1990; Colombi 1986). The disposal pit was lined with sawdust in the event that the drums would someday leak. Due to flooding from changes in the groundwater table, the metal drums rusted and eventually leaked their contents. In 1985, groundwater wells located downgradient ("downhill" in a groundwater system) from the disposal area were found to contain reduced aromatics, such as chloroanilines, but no nitrobenzene compounds. Anyone who knows the Swiss knows that they are meticulous recordkeepers and always have been. The company that disposed of the waste produced records showing that they had not disposed of any aniline compounds, and thus asserted they were not the ones responsible for the contamination. So, an investigation was conducted to attempt to explain this dilemma.

Groundwater modeling could have easily been conducted using a step model if you were only modeling the transport of nitrobenzene compounds. But this was not the issue since the monitoring well downgradient from the point source (the landfill) only contained anilines. Thus, the importance of considering the chemistry occurring in the landfill and groundwater became apparent. Schwarzenbach et al. (1990) proposed that as the water table changed it could have allowed water

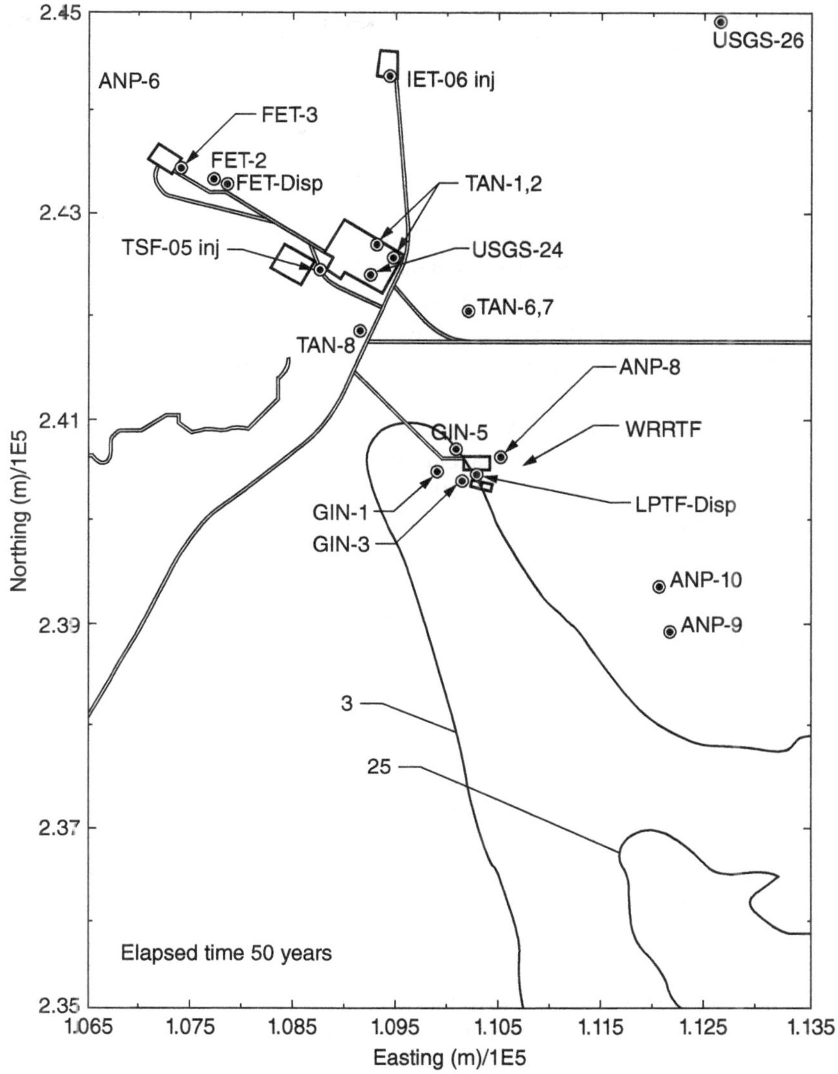

Figure 1.6 Predicted isoconcentration lines for TCE at the INEEL Tan injection site for the year 2044 after remediation of the immediate area around the injection well. Source: INEEL, Department of Energy, Dunnivant et al. (1994), and public domain.

to enter the landfill and allowed oxidation of the metal drums. After the drums ruptured, the leaking nitrobenzene compounds were subjected to reducing conditions from the presence of reduced humic acid (degraded and dissolved sawdust), and the nitrobenzene compounds were reduced to their respective anilines. This case demonstrates the importance of recordkeeping, and the integration of chemistry and modeling in assessing the history of and in predicting the effects of a pollution event. This is why we will be discussing the role of chemistry as part of the modeling approaches in this book.

1.6.3 Postenvironmental Movement Era Accidents

Even with extensive environmental regulations, accidents still continue to happen around the world. These are expected in countries with poor or no environmental regulations or enforcement, but citizens of the Global North are shocked with these disasters happen in their country. The following examples are in chronological order and result from a variety of causes from lax maintenance, poor inspections, to simply no fault accidents.

1.6.3.1 Methyl Isocyanate Release in Bhopal, India; Timeline 1984

In the early hours of 23 December 1984, a pesticide manufacturing plant in Bhopal, India, accidentally released methyl isocyanate (MIC). The event was a pulse release (short term) resulting in estimated concentrations that range from 13 (Dave 1985) to 100 ppm (Varma 1986). Death estimates range from 2000 (Dave 1985) to 10 000 (Shrivastava 1987) while approximately 300 000 injuries were reported (Shrivastava 1987). Although records indicate that concerns were raised before the accident, pollutant fate and transport modeling, as well as risk assessments should have been conducted before the plant was even built. If the citizens of Bhopal had known of such modeling results, they may have been willing to surrender the small economic benefits from the factory in order to avoid the deaths and injuries caused by the accident. We will look at the Bhopal accident in more detail at the beginning of Chapter 9.

1.6.3.2 Accident at the Nuclear Power Plant at Chernobyl, Ukraine (Chernobyl Accident 2003; Chernobyl Disaster 1986); Timeline 1986

A number of radioactive releases have occurred since our development of nuclear power plants. England experienced a major release at the Windscale site as early as 1957 (Bailey et al. 2002), and the United States' nuclear industry suffered the accident at Three Mile Island, Pennsylvania, in March 1979. The world was shaken by the serious accident at the Chernobyl Unit 4 in the Soviet Union (present day Ukraine) in 1986. Radionuclides were spread across both Eastern and Western Europe contaminating dairies in Austria, Hungary, Poland, and Sweden, not to mention the hazards posed by eating fresh fruit and fish from the immediate area. Released radionuclides were subsequently detected in the atmosphere over Canada, Japan, and the United States (Bailey et al. 2002). Thirty workers at the plant were directly killed from accident (20 from radiation exposure), another 209 were treated for acute radiation poisoning, and countless others were exposed to lower doses of radiation with unknown effects (Chernobyl Accident 2017). In the end, 210 000 people had to be resettled. Modeling of such a large, widespread area, such as that affected by the Chernobyl accident is also possible but less accurate given the scale of the system and dynamic nature of weather systems. Still, you can rest assure that worst-case scenarios have now been modeled by governing agencies of nuclear power plants in the Global North.

1.6.3.3 A Chemical Spill on the Rhine River (Capel et al. 1988); Timeline 1986

The Rhine River system in Western Europe is lined with chemical and pharmaceutical plants. It should come as no surprise that there are occasional chemical spills into this aquatic system. A major release occurred on 1 November 1986, when a storehouse owned by Sandoz Ltd. near Basel, Switzerland, caught fire and released pesticides, solvents, dyes, and various raw and intermediate chemicals (Capel et al. 1988). Figure 1.7 shows a map of the route that the Rhine River takes through this part of Europe. This release has been referred to as "one of the worst chemical spills ever" (Anonymous 1987). For modeling purposes, this was treated as a pulse (instantaneous, short duration) release. Environmental and academic groups rapidly formed monitoring points for the various chemicals at the stations labeled in Figure 1.8 (Capel et al. 1988). One of the most abundant pesticides released was disulfoton, a thiophosphoric acid ester insecticide. Figure 1.8 shows the movement and flushing of disulfoton through the Rhine (Capel et al. 1988). Note the bell or Gaussian shape of the concentration profile, which is characteristic of a pulse release. As the pulse of disulfoton moved downstream, it was diluted and degraded as indicated by the broader peaks and lower concentrations shown in Figure 1.8. This is a case where the model can be fit to the data to better understand how pollutants move through the system and thus can be used to predict downstream concentrations of future releases of pollutants. Of course, this does not help the wildlife affected by a disaster, but it can help to minimize the impact on drinking water systems that can be shut down as the peak concentrations of pollution pass. As we will see in Chapter 6, most rivers have a built-in cleansing system – flow downstream – eventually into a lake or the ocean. We will discuss this further in Chapter 6.

1.6.3.4 A Chemical Spill into the Tisza River Bordering Hungary and Romania (UNEP 2002); Timeline 2000

On 30 January 2000 a mine tailing dam, owned by the Aurul SA Baia Mare Company, broke and released a cyanide-rich waste into the Săsar River, an already highly contaminated river, which subsequently drained into the Tisza and Danube before entering the Black Sea (UNEP 2002). It has been estimated that nearly 50 000 and 100 000 m^3 of aqueous and suspended cyanide and heavy metals were released that resulted in the death of 500 tons of fish. This pollution event, like the one of the Rhine discussed above, illustrates the international effort needed in controlling release events and the politics of such a release. While the company in Switzerland bore the full responsibility (liability) for the Rhine incident, Romania took the stand that they had no international treaties with Hungary, thus, they were not responsible for damages to the fishing industry (Schaefer 2000). Liable or not, if an environmental impact study (utilizing fate and transport modeling) had been conducted showing the possible extent of damage to the affected rivers, more stringent safety measures would certainly have been in place.

1.6.3.5 The Kingston Fossil Plant Incident (Coal Ash Spill, Tennessee 2008; Dewan 2008; EPA 2017d; Kingston Fossil Plant Coal Fly Ash Slurry Spill 2008; Lisenby 2014; Sludge Graphic 2008; Southeast Coal Ash site 2017), Tennessee (USA); Timeline 2008

On 22 December 2008, an ash dike ruptured at an 84-acre (19.4 ha) solid waste containment area at the Kingston Fossil Plant in Roane County, Tennessee. The plant

Figure 1.7 The route of the Rhine River in Western Europe. Source: Capel et al. (1988). Reprinted with permission of American Chemical Society.

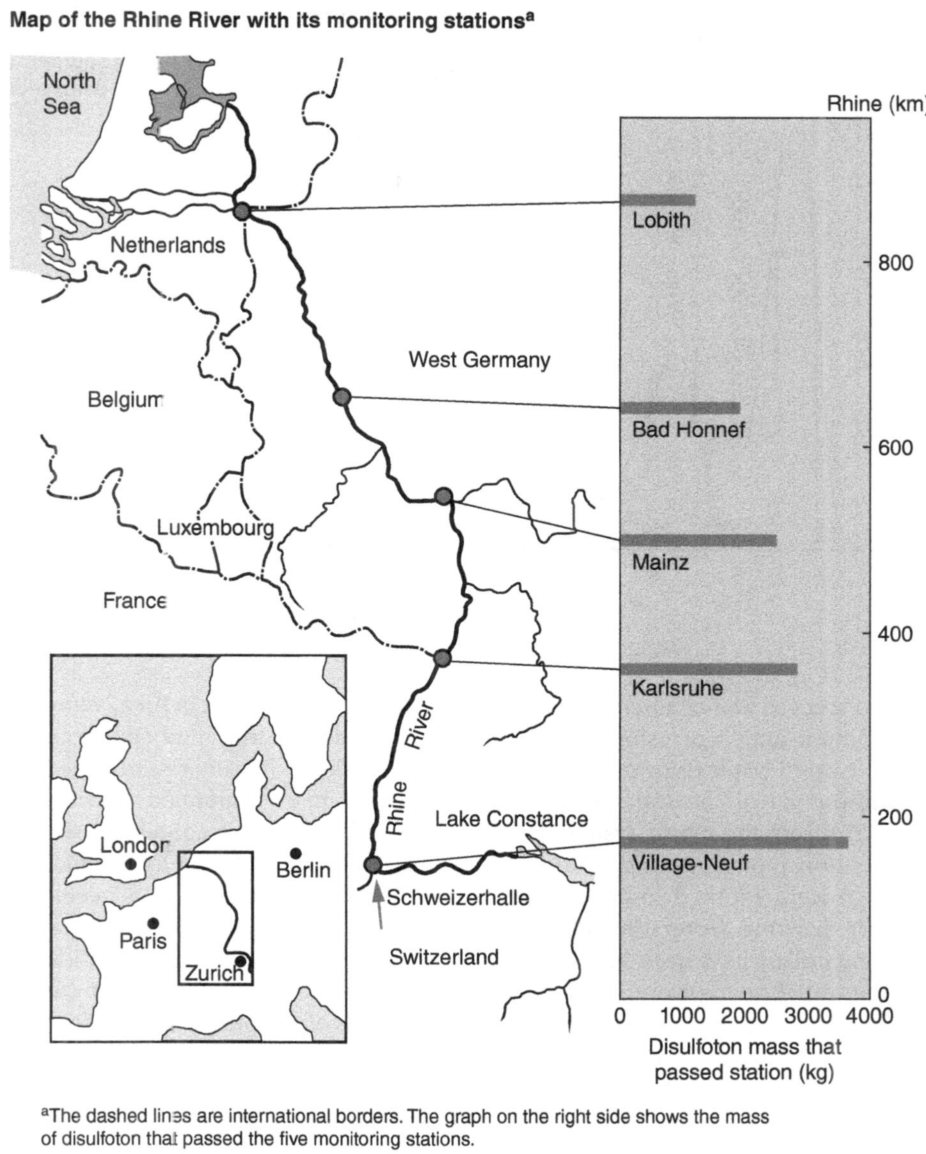

released over 2.2 million pounds of coal fly ash slurry into a holding pond. The slurry then traveled across the Emory River and its Swan Pond embayment onto the opposite shore, covering about 300 acres (121 ha) of surrounding land. The incident damaged homes and polluted the water systems nearby (the Emory River and Clinch River).

According to one source (Kingston Fossil Plant 2017), the spill dumped around 45 000 lb of arsenic, 49 000 lb of lead, 1.4 million pounds of barium, 91 000 lb of chromium, and 140 000 lb of manganese. As a result, a lot of fish were killed and a test on the river water was taken on 1 January 2009 which indicated that there were significantly elevated levels of toxic metals (including arsenic, copper, barium, cadmium, chromium, lead, mercury, nickel, and thallium). These pollutants are well known to be factors in causing cancer, liver damage and other neurological problems. The type of release was a pulse release that continues to affect the water systems as pollutants become more available depending on environmental conditions.

1.6.3.6 Lead in the Domestic Drinking Water of Flint River, Michigan (Flint Water Crisis 2014); Timeline 2014

The City of Flint Michigan normally received its drinking water from Lake Huron, but in an effort to decrease operating cost, switched sources to the Flint River, a river known for its exceptionally polluted water. The Flint River is polluted from a couple of main sources before processing, in the form of natural biological waster, as well as treated and untreated human and industrial waste. This leads to a higher level of bacteria

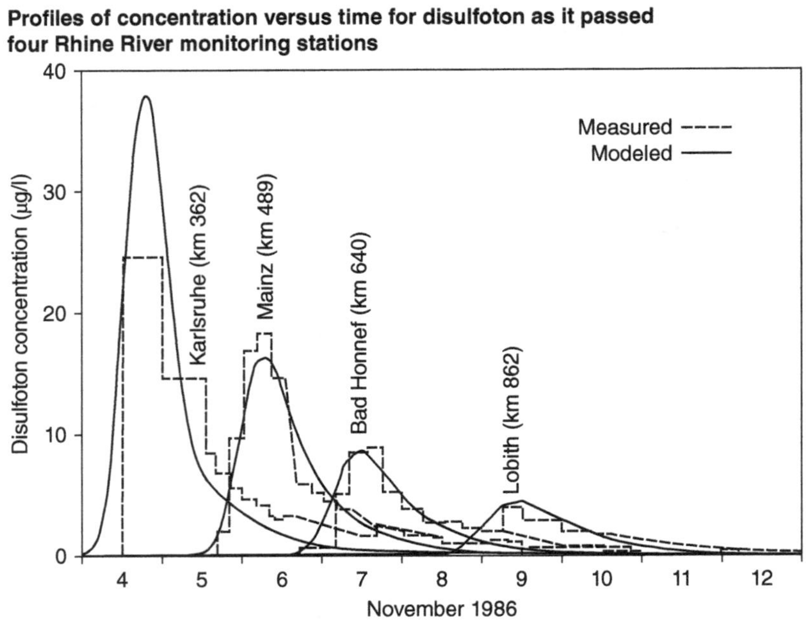

Figure 1.8 The concentration of disulfoton at each monitoring station on the Rhine river. Source: Capel et al. (1988). Reprinted with permission of American Chemical Society.

and organics as well as potential inorganic contaminants. Most importantly, corrosion inhibitors were not added to the treated water prior to distribution. The corrosion and lower pH of the water resulted in the release to lead (Pb) from the piping system and also damaged the distribution piping network. As a result, 6000–12 000 children were exposed to elevated lead concentrations in their drinking. Several hundreds of millions of government dollars have been spent to address the problem, yet the initial saving from switching water sources has been estimated to be between a few tens of thousands of dollars to 1 million dollars. Governmental officials were blamed for the oversight and several have resigned due to this likely avoidable event. As of 2017, residents continue to drink bottled water and will do so until a permanent remediation is expected in 2020. The different chemistry of the source waters was the main cause of the problem, and while this problematic chemistry was noted by some, others choose to ignore it. Chemical speciation of pollutants will be covered in Chapters 3 and 4. One problematic result of the Flint crisis is the current lack of trust of domestic water treatment plants that is woefully misguided. Treatment plants in the Global North are routinely tested and shown to be perfectly safe; piping in residential homes on the other hand can still be a source of toxic lead (Pb) and are not the fault of the city's treatment system.

1.6.3.7 Ajka Alumina Plant Accident (Hungary) (Ajka alumina plant accident 2010); Timeline 2010

On 4 October 2010, the dam of a reservoir collapsed containing 1 million m^3 of caustic waste from an alumina plant in Ajka, Veszprem County, in western Hungary. The red mud spill flooded the town of Devecser and eventually reached the Danube River. The red mud primarily contained transition metal oxides (i.e. Fe_2O_3, Al_2O_3) but also contained 660 mg/kg (parts per million) chromium, 110 mg/kg arsenic, and 1.2 mg/kg mercury. Ten people died and 150 were injured. The liquid–solid waste killed all life in the Marcal river, and finally reached the Danube on 7 October. Cities and countries downriver developed emergency plans for their drinking water as the pollutant plume flowed downgradient. Initial estimate placed the cleanup timeframe and costs to be at least a year and tens of millions of dollars. The plant was found guilty of several violations and a second dam was installed to prevent another disaster. As the modeling standpoint, the release could be treated as a pulse input into a river and the soluble pollutants were relatively quickly washed downstream. Toxic metals adsorbed to oxide particles sank to the river bottom and were more slowly released.

1.6.3.8 Talvivaara Mine Leak (Finland) (Talvivaara Mine Release 2012); Timeline: 2012. (Author: Aaron Lieberman Class Assignment, 2016)

The Talvivaara mine primarily mined nickel metal, but other toxic metals such as uranium were found. All the excess metals were dumped in the Gypsum Pond to be contained in a single source. In November 2012, the pond leaked ~250 000 m^3 of contaminated liquid as a pulse release. The leak was quickly capped, but later the following year in April 2013 the pond began to release wastewater 7000 m^3 an hour. This step release caused ~250 000 m^3 wastewater to be released into the environment. The major environmental issues that were brought up were that the pollutant was a mixture of both metal–inorganic compounds as well as radioactive waste caused by the uranium. This caused fish downstream to be affected with abnormally high cadmium, aluminum, and nickel. The overall fate of the compounds would eventually end up in human consuming the fish, but the government made fishing in the nearby river systems illegal. Furthermore, nothing has been done by the company to attempt to fix the large amount of contamination in the local rivers, as evident by the nearby River Lumijoki turning red and orange from contamination. These two contamination events allow both a pulse and step model to be used.

1.6.3.9 Huangpu River (China) (Shanghai River's Dead Pig Total Approaches 15,000 2013; Rivers of Blood 2013); Timeline: 2013. (Author: Teddy Pierce, Class Assignment, 2016)

In March 2013, more than 16 000 young pigs were fished out of the Huangpu River and its tributaries in China. The river is the major source of tap water for Shanghai. The pigs were dumped into the river by pig farmers up river in the Zhejiang province. Some analysis on the pigs showed that they were infected with porcine circovirus, which is not known to infect humans. The waters of the river were described as "inky black, covered in a slick of lime green algae, and it smells like a blocked drain." This incident is believed to be a sign of the collapse of the pig farming industry that boomed in the region starting in the 1980s.

This unique, but specific release of thousands of pigs was a pulse release. This release was likely caused by farmers going out of business or suffering an outbreak of disease disposing of their pigs in the cheapest way possible, tossing them in the river. The high amount of organic matter being added to the river likely resulted in an anaerobic aquatic system since thousands of rotting carcasses would further increase the BOD to the point that the river DO levels could not be restored, killing of fish and other life forms.

Fishermen along the river have been affected by these lower-level step releases and this pulse release because they are finding it harder and harder to make a living fishing on the river as animal life in the river is choked out due to high BOD and nitrogen levels. Many of these fishermen, including the ones in the village of Xinfeng, make most of their money being paid by the government to help clean the river.

The water was reported to meet national standards by Shanghai's municipal water department but no test results were released. This opacity combined with the whole situation caused demonstrations for reform in Shanghai that were quickly shut down by police. Fortunately for the people who live along the river and depend on it for water, this specific event had a relatively short effect. A real cleanup effort would need to monitor and control and/or treat the amount of sewage and manure being added to the river by the pig farms. The river recovered rapidly as it flushed excess organic matter out and the DO increased to healthier levels.

1.6.3.10 Tianjin Chemical Plant Explosion (China) (Tianjin Explosions 2015); Timeline: 2015. (Author: Adrian Morton, Class Assignment, 2016)

The 2015 Tianjin explosions were a series of explosions from chemical containers in the port of Tianjin, China. Because this explosion occurred at a transit facility, the chemicals actually involved in the explosion were difficult to distinguish. The people storing these chemicals were less practiced at their safekeeping, recordkeeping because these chemicals were spotty at best, and the amount being stored was beyond legal limitations. This hindered the ability to deal with these disasters after they had occurred and also made it extremely difficult to assess the amount of risk the public was at to contamination and harm.

The first explosion was caused by an overheated container of nitrocellulose (also known as flash paper), which created a large fire. A smaller fire was the most likely cause of this overheating, but once this initial explosion occurs, the situation turned much more serious. A second explosion occurred within 30 seconds of the first and was much larger. This explosion involved the detonation of over 800 tons of ammonium nitrate. This was the largest and most damaging explosion of the event, but fires continued throughout the week, setting off a series of other explosions in the facility. These explosions included calcium carbide being heated and mixed with water to create acetylene gas. The most harmful species involved in this disaster was sodium cyanide. Seven hundred tons were stored at this facility, much of which was dispersed around the surrounding area of Tianjin and its port waters through the explosions.

This disaster was a series of pulse releases of various pollutants into the environment. The main pollutant, sodium cyanide dust, was sprinkled around the area as well as in the water and atmosphere. This cyanide salt has severe effects on humans and animals. When the first rains came after the event, the streets foamed white, and people who came in contact with the rain complained of chemical burns and rashes on their skin. Sodium cyanide is very mobile because it is rather soluble in water, and because it was turned to dust and thrown into the atmosphere, it has plenty of time to react with rain and become the much more harmful gaseous form of cyanide. Cyanide reacts under acidic conditions to form hydrogen cyanide gas, which is extremely harmful to most organisms including humans. Carbon dioxide in the air is acidic enough to complete this transition, and since the rainwater was more acidic than usual in this area, this transformation to hydrogen cyanide gas was further assisted.

Thousands of dead stickleback fish washed up on shore soon after this event. The Chinese government attempted to calm fears by blaming these fish deaths on lowered oxygen levels in the water due to a large chemical oxygen demand created by new pollutants in the water. While anoxic waters are preferable to large amounts of cyanide, other types of pollutants likely occurred as sodium cyanide was detected in the water around the explosion site. The main factors affecting the fate and transport of sodium cyanide in this case is the pH of the water, affecting its reaction into hydrogen cyanide, and the solubility of the species allowing it to be easily incorporated into rainwater and the surrounding bay. The reduction potential, E_H, is also related to the chemical oxygen demand and its effect, and in this case the E_H became very low (reducing conditions) as a result of the pollution from this event.

1.6.3.11 Animas River Mine Tailing Spill (Animas River Spill 2015; Brumfield 2015; Schlanger 2015); Timeline 2015 (Author: Dylan Price, Class Project, 2016)

The Animas River in Colorado (USA) is surrounded by abandoned gold mines, some dating back to the late nineteenth century and have been closed for years while others that have only been closed for only 10 years. These gold mines, especially the old ones, often contain large tailings ponds ridden with heavy metals. Ironically, a team of EPA workers were sent to one of these mines to reinforce the walls of a tailings pond and add a tap that would clean and slowly release the tailings over time and ended up breaching the tailings and causing a massive release of toxic mine waste into Cement Creek, a tributary to the Animas river. An estimated 3 million gallons (11 400 m^3) were released over about a week starting on 5 August, while efforts were taken to stop the spill. The waste drained into the Animas and quickly made its way to the San Juan River by 10 August. The San Juan river flows into the Colorado and then into Lake Powell on 14 August – at which point the waste was diluted to a point where it was no longer dangerous according to the fate and transport model constructed for the disaster.

The main toxins of concern in the spill were lead, iron, cadmium, and arsenic. Measurements were taken 15 mi upstream from Durango (CO) and lead was found in levels 100 times larger than the limit. Iron was found to be 326 times the limit. Local residents, relying on water from wells fed by the Animas, were told not to drink well water and that it had to be analyzed for purity. The river was also closed from recreation for weeks. The upper Animas River was devoid of fish prior to this spill – from previous mine leakage in the area. Had there been fish to kill this spill certainly would have.

This release was treated a step pollutant release at a well-defined point source, containing inorganic heavy metals. The heavy metals were both suspended and aqueous. There was not a prolonged release of tailings at this rate and the spill was controlled with emergency tailings ponds. The sediment in the river is expected to dilute the released heavy metals to acceptable levels.

The spill required Superfund money that had, again ironically, been offered to this site years earlier and declined due to expected decrease in tourism with the major cleanup operation. The US EPA did not notify locals until almost 24 hours after the spill, and overlooked warning New Mexico of this spill for even longer. The governor of New Mexico sought legal action against the EPA and the governor of Colorado declared the spill an official disaster. This spill is a prime example of the politics associated with major pollution and the importance of fate and transport modeling.

1.6.3.12 Duke Power Coal Ash Spill; Timeline 2016 (Author: Ellie Finnegan, Class Project, 2016)

Coal ash is found in multiple forms. Produced from coal burning factories and power plants, coal combustion residuals (CCRs) range anywhere from small fly ash particles to boiler slag glassy pellets (EPA 2017e); Southeast Coal Ash site 2017. Plants that generate coal are required under regulations such as water discharge permits to properly treat and manage coal ash waste; however, sometimes regulations of these standards are not kept. In early February 2014, 39 000 tons of coal ash and 24 million gallons of wastewater were released as a pulse into the Dan River of North Carolina near the town of Eden. Duke Energy and the state agency responsible for protecting North Carolina waters were held responsible for the disaster.

Duke Energy, and other coal-fired power plants, primarily stores coal ash waste in unlined lagoons that are notoriously difficult to manage. This mechanism of storage is typically considered a poor choice because heavy metals are still able to leach into the soil and thus groundwater. In the scenario of the Duke Energy spill, the infrastructure was dated. Despite multiple voices of concern from Duke employees, and at least two requests from Duke engineers for funding to better monitor the lagoons, the company headquarters chose not update the outmoded waste treatment system. Finally, drainage pipes collapsed under the lagoons and caused the major spill in 2014 contaminating the Dan River with 70 mi of sludge.

Coal ash is considered a hazardous waste when not treated properly because it holds contaminants such as mercury, cadmium, and arsenic. Pollutants such as these can leach into groundwater systems and, as seen in the Duke Energy spill, contaminate downriver ecosystems. Because toxicity of heavy metals is concentration dependent, the spill caused the most harm at the point source. Though over time the Dan River diluted the system as the pollutant traveled downstream, heavy metal transport can rely on the mass of contaminant and the conditions of the river water system. If the metal-associated sediment is heavy, it may sink to the bottom of the river and thus take more time to be "flushed out" through resuspension events. If the river has high levels of dissolved organic matter, the metal ion may attach to the chemical sites on the soluble and mobile organic molecules thus increasing mobility. The oxygen levels of the water system also may play a large role in the fate and transport of associated metal pollutants. Generally, a metal in oxygenated water (metal oxide) will stay in its precipitated form thus decreasing the solubility and potential toxicity. If, however, the Dan River had low oxygen levels at the time of the spill, some metals would be reduced and may have become more soluble, traveled further, and created more damage. Heavy metal ions are more mobile under acidic conditions and/or anoxic conditions. This is the type of chemistry that will be discussed in Chapters 3 and 4.

In May 2015, Duke Energy was sued under the Clean Water Act for 64 million dollars in fines and 36 million dollars to go toward environmental projects and land conservation of rivers and wetlands in North Carolina and Virginia. During prosecution, Duke Energy pleaded guilty for a step pollution release of coal ash wastewater into the Dan River dating back to 2010.

The events above are only the major terrestrial disasters, and not ocean shipping accidents or the Fukushima nuclear plant disaster which are beyond the scale of any modeling scenario. It is likely that hundreds to thousands of accidents have happened since the year 2000 but were not reportable news events. Again, pollution events will continue to happen but strict enforcement of environmental regulations will deter these from becoming common place.

The news savy reader will note the absence of several large-scale estuary and oceanic incidents, such as the Deepwater Horizon disaster that released tons of dispersed petroleum in the Caribbean, the mercury contamination in Minamata Japan (Minamata Disease 1956), the many zones of hypoxia across the globe (USGS 2017; discussed at the end of Chapter 7), and the piles of plastics accumulating in our oceans, with microspheres documented at every beach in the world. These events are beyond the modeling scale of this text but are still noteworthy human-caused disasters.

1.7 Environmental Laws

Each country has developed a set of environmental laws governing the adequate premanufacture testing, use, and appropriate disposal or treatment of hazardous chemicals. Some countries have extensive rules, while others are still in the process of writing enforceable laws as their chemical industry develops. Environmental movements and the major environmental laws for the United States, Europe, and China will be discussed, in detail, in Chapters 12, 13, and 14, respectively. If the reader is from a different country, you may wish to research the laws for your country and compare them to those for the United States, Europe, and China. The main point here is most countries have environmental laws in place but compliance with these laws varies from the national, to state, to local level. In many countries, the creation of jobs overrides enforcement of laws, both environmental and safety.

Internationally, we are all affected by a process known as "Environmental Colonialism" in which developed countries prey on less-developed nations for disposal of hazardous wastes. This egregious act that exports hazardous chemical risks to unknowing nations for badly needed cash was addressed by the United Nations in its Basel Convention of 1989.

In the United States, the Environmental Protection Agency, along with other agencies, has made great strides since the 1970s to improve environmental compliance. Our rivers and lakes are not no longer open sewers, nor do they catch on fire any longer, but closer to the swimmable–fishable vision of the pre-1960s. The leakage of underground storage tanks has mostly been solved. And our metropolitan areas are much cleaner. Hazardous waste are strictly controlled and disposed. Recent revisions in the Toxic Substances Control Act (TSCA 2017) corrected long overdue expansion of Federal authority regarding toxic chemical management. But we do still have accidents, some that could have been prevented and others that are a result of inevitably flawed human-engineered and political systems. Several good sources of this good news is available including Protecting the Environment: A Half Century of Progress from EPA (http://www.epaalumni.org/hcp/Overview.pdf), Success Stories in Human Health (https://www.cdc.gov/features/sharingsuccessstories/index.html), the recognition of Seven Spectacular Places Saved by the Environmental Movement (http://www.slate.com/articles/health_and_science/science/2013/04/environmental_success_stories_on_earth_day_visit_places_saved_by_conservationists.html), and the popular science book entitled *Environmental Success Stories* by Frank Dunnivant (2017). Our successes are remarkable and will only continue to progress as we educate our citizens.

Concepts

1.1 What is the difference between explanatory and predictive modeling?

1.2 Explain the difference between a pollutant and pollution.

1.3 What is meant by chemodynamics in the concept of this course?

1.4 Define and give three examples of nonpoint source pollution from the Internet.

1.5 Define and give three examples of point source pollution from the Internet.

1.6 Define and give three examples of a step pollution event.

1.7 Define and give three examples of a pulse pollution event.

1.8 What does steady-state imply regarding environmental discharges or emissions?

1.9 Give three examples of environmental plumes, and why are they significant?

1.10 Using the Internet provide two examples of why the UN created the Basel Convention in 1989. Why do you think the United States did not sign this Treaty?

Exercises

1.1 Outline one of the pollution events given in Section 1.6.2 and include date, location, and main pollutant.

1.2 The US government, as with most Cold War governments around the globe, is the most serious of polluters. For your country (in the United States, this will be current DOE and DOD installations), research your weapon programs sites and write a two-page summary for one Superfund-type site.

1.3 Research a pollution event on the Internet. Write a one- to two-page summary including as many of the following as possible: the location, the company responsible for the site, the pollutants, the concentration of pollutants, the extent (surface area or volume) of pollution, the health effects of the pollutants, and the planned remediation. Include any retrospective (what did we learn and/or what's different because of the incident?)

1.4 Become socially active and research one of the chemical companies given in Table 1.4 using the Internet or by telephone or email. What chemicals do they use? What are their wastes? How do they dispose or destroy their wastes? How has their reported hazardous waste amounts changed in recent years?

1.5 Research the term "Pollution Prevention" on the Internet and for three industries. How has pollution prevention affected these manufacturing processes?

1.6 Become socially active and research one of the chemical companies in your home state using the Internet or by telephone or email. What chemicals do they use? What are their wastes? How do they dispose or destroy their wastes?

1.7 Visit the US EPA's Toxic Release Inventory website at https://www.epa.gov/trinationalanalysis and research a topic of your choice based on your zip code.

Bibliography

Ajka alumina plant accident (2010). https://en.wikipedia.org/wiki/Ajka_alumina_plant_accident (accessed 21 September 2017).

Animas River Spill (2015). Department of public health and environment. https://www.colorado.gov/pacific/cdphe/animas-river-spill (accessed 21 September 2017).

Anonymous (1987). *Environ. Sci. Technol.* **21** (March): 5.

Associated Press (AP) (2015). http://www.foxnews.com/us/2015/05/15/duke-energy-pleads-guilty-fined-102m-in-connection-with-2014-coal-ash-spill.html (accessed 21 September 2017).

Bailey, R.A., Clark, H.M., Ferris, J.P. et al. (2002). *Chemistry of the Environment*, 2e. New York: Academia Press (Chapter 14).

Basel Convention (1989). Table 1.2. www.basel.int/natreporting/index.html (accessed 8 December 2003)

Bioaccumulation (2017). Wikipedia. https://en.wikipedia.org/wiki/Bioaccumulation (accessed 3 October 2017).

Brumfield, B. (2015). Animas River spill: the massive toll by the numbers, CNN. http://www.cnn.com/2015/08/13/us/animas-river-spill-by-the-numbers/index.html (accessed 24 September 2017).

Capel, P.D., Giger, W., Reichert, R., and Warner, O. (1988). Accidental input of pesticides into the Rhine River. *Environ. Sci. Technol.* **22** (9): 992–997.

Coal Ash Spill, TN (2008). http://earthobservatory.nasa.gov/NaturalHazards/view.php?id=36352 (accessed 21 September 2017).

Chernobyl Accident (2003). http://www.basel.int/ (accessed 1 November 2003).

Chernobyl Disaster (1986). https://en.wikipedia.org/wiki/Chernobyl_disaster (accessed 25 September 2017).

Colombi, C. (1986). Sondermulldeponie Kolliken, wie weiter? *Phoenix Int.* **1**: 10–15.

Dave, J.M. (1985). The Bhopal methyl isocyanate (MIC) incident: an overview. In: *Proceedings of an International Symposium, Highly Toxic Chemicals: Detection and Protection Methods* (ed. H.D. Schiefer), 1–38. Saskatoon, Saskatchewan, Canada: Springer.

Dewan, S. (2008). At plant in coal ash spill, toxic deposits by the ton. http://www.nytimes.com/2008/12/30/us/30sludge.html (accessed 21 September 2017).

Dimdam, E. (2014). Discovery magazine. http://list25.com/25-biggest-man-made-environmental-disasters-in-history/ (accessed 3 October 2017).

Dunnivant, F.M. (2017). *Environmental Success Stories: Solving Major Ecological Problems and Confronting Climate Change.* Columbia University Press 9780231179195.

Dunnivant, F.M., Stromberg, G.J., Wiley, A.H. et al. (1994). Feasibility Study Report for Test Area North Groundwater Operable Unit 1-07B at the Idaho National Engineering Laboratory (INEEL), EFFF-ER-10802, January, 1994, Idaho Falls, Idaho, USA.

Dunnivant, F.M., Schwarzenbach, R.P., and Macalady, D.L. (1992). Reduction of substituted nitrobenzenes in aqueous solutions containing natural organic matter. *Environ. Sci. Technol.* **26** (11): 2133–2141.

EPA (1989). *Risk Assessment Guidance for Superfund Volume II, Environmental Evaluation Manual, Interim Final*, EPA/540/1-89/001, 4. Washington, DC: Office of Emergency and Remedial Response.

EPA (2001). National Analysis, The National Biennial RCRA Hazardous Waste Report (based on 2001 data). https://www.epa.gov/hwgenerators/biennial-hazardous-waste-report (accessed 14 December 2018).

EPA (2013). https://www.epa.gov/sites/production/files/2015-09/documents/rcras_critical_mission_and_the_path_forward.pdf (accessed 25 July 2018).

EPA (2015a). U.S. hazardous waste by state. https://rcrapublic.epa.gov/rcrainfoweb/action/modules/br/summary/view (accessed 25 September 2017).

EPA (2015b). U.S. type of waste by industry. https://rcrainfo.epa.gov/rcrainfoweb/action/modules/br/naics) (accessed 25 September 2017).

EPA (2016). https://blog.epa.gov/blog/2016/06/tsca-reform-a-bipartisan-milestone-to-protect-our-health-from-dangerous-chemicals/ (accessed 3 October 2017).

EPA (2017a). https://www.epa.gov/caa-permitting (accessed 3 October 2017).

EPA (2017b). http://www.ecologydictionary.org/EPA-Terms-of-Environment-Dictionary/Waste_Load_Allocation (accessed 3 October 2017).

EPA (2017c). https://www.epa.gov/toxics-release-inventory-tri-program (accessed 3 October 2017).

EPA (2017d). www.epa.gov/aboutepa/our-mission-and-what-we-do (accessed 3 October 2017).

EPA (2017e). www.epa.gov/coalash/coal-ash-basics (accessed 21 September 2017).

Schlanger, A. (2015). EPA Causes Massive Spill of Mining Waste Water in Colorado, Turns Animas River Bright Orange. Newsweek, 7 Aug., 2015. www.newsweek.com/epa-causes-massive-colorado-spill-1-million-gallons-mining-waste-turns-river-361019 (accessed 21 September 2017).

Fick's Law (2017). en.wikipedia.org/wiki/Fick%27s_laws_of_diffusion (accessed 3 October 2017).

Flint Water Crisis (2014). en.wikipedia.org/wiki/Flint_water_crisis (accessed 21 September 2017).

Hemond, H. and Fechner, E.J. (2000). *Chemical Fate and Transport in the Environment*. Amsterdam: Elsevier, Academic Press.

Kingston Fossil Plant (2017). https://www.tva.gov/Energy/Our-Power-System/Coal/Kingston-Fossil-Plant (accessed 22 September 2017).

Kingston Fossil Plant Coal Fly Ash Slurry Spill (2008). en.wikipedia.org/wiki/Kingston_Fossil_Plant_coal_fly_ash_slurry_spill (accessed 21 September 2017).

Lisenby, D. (2014). Duke energy coal ash spill pollutes river and threatens drinking water. ecowatch.com/2014/02/04/duke-energy-coal-ash-spill/ (accessed 21 September 2017).

Minamata Disease (1956). en.wikipedia.org/wiki/Minamata_disease (accessed 3 October 2017).

NPR (2015). What caused the warehouse explosions in Tianjin, China? Interview. www.npr.org/2015/08/14/432280627/what-caused-the-warehouse-explosions-in-tianjin-china (accessed 21 September 2017).

Rivers of Blood (2013). The dead pigs rotting in China's water supply. www.theguardian.com/world/2013/mar/29/dead-pigs-china-water-supply (accessed 21 September 2017).

Schaefer, A. (2000). Disastrous cyanide spill could spawn liability reforms. *Environ. Sci. Technol.* **34**: A-203.

Schwarzenbach, R.P., Stierli, R., Lanz, K., and Zeyer, J. (1990). Quinone and iron porphyrin mediated reduction of nitroaromatic compounds in homogeneous aqueous solution. *Environ. Sci. Technol.* **24**: 1566–1574.

Shanghai River's Dead Pig Total Approaches 15,000 (2013). phys.org/news/2013-03-shanghai-river-dead-pig-total .html (accessed 21 September 2017).

Shrivastava, P. (1987). *Bhopal: Anatomy of a Crisis.* Cambridge, Massachusetts: Ballinger Publishing Company.

Sludge Graphic (2008). http://www.nytimes.com/imagepages/2008/12/25/us/20081225_SLUDGE_GRAPHIC.html (accessed 21 September 2017).

Southeast Coal Ash site (2017). http://www .southeastcoalash.org/ (accessed 21 September 2017).

Tianjin Explosions (2015). https://en.wikipedia.org/wiki/2015_Tianjin_explosions (accessed 21 September 2017).

Talvivaara Mine Release (2012). http://www.nuclear-heritage.net/index.php/Talvivaara_mine:_environmental_disaster_in_Finland (accessed 21 September 2017).

Thibodeauz, L.J. (1996). *Environmental Chemodynamics: Movement of Chemicals in Air, Water, and Soil*, 2e. New York: Wiley.

TSCA (2017). https://en.wikipedia.org/wiki/Toxic_Substances_Control_Act_of_1976 (accessed 3 October 2017).

United Nations Environment Programme, UNEP (2002). Office for the co-ordination of Humanitarian Affairs, OCHA. *Spill of Liquid and Suspended Waste at the Aurul S.A. Retreatment Plant in Baia Mare.* 23 February – 6 March 2000, Geneva, March.

United Nations (U.N.) (2000). Secretariat of the Basel Convention, United Nations Environment Programme, Country Fact Sheets. http://www.basel.int/natreporting/index.html (accessed 21 July 2018).

USGS (2017). https://toxics.usgs.gov/hypoxia/hypoxic_zone.html (accessed 3 October 2017).

Varma, D.R. (1986). Anatomy of the methyl isocyanate leak on Bhopal. In: *Hazard Assessment of Chemicals* (ed. J. Saxena), 233–299. Hemisphere Publishing Corp.: New York.

WASP (1983). Water quality analysis simulation program. https://www.epa.gov/exposure-assessment-models/water-quality-analysis-simulation-program-wasp (accessed 2 October 2017).

Part II

Chemistry of Fate and Transport Modeling

The noblest of the elements is water.

Pindar, 476 BC

Pollutant Fate and Transport in Environmental Multimedia, First Edition. Frank M. Dunnivant and Elliot Anders.
© 2019 John Wiley & Sons, Inc. Published 2019 by John Wiley & Sons, Inc.
Companion website: www.wiley.com/go/Dunnivant/pollutant_fate&Transp_EnvironMultimedia

2

Basic Chemical Processes in Pollutant Fate and Transport Modeling

2.1 The Liquid Medium: Water and the Water Cycle

There are several factors responsible for life on Earth, including the Earth's ideal distance from the Sun and a stable planetary rotation which result in hospitable temperature conditions for species suitable in Earth's past and present habitats. A view of the Earth from space almost tells the complete story as to why life as we know it thrives; the planet's surface consists of approximately 71% water. Water vapor rising from the oceans replenishes the landmasses with essential freshwater and, as a greenhouse gas, warms the earth. As we will see in this textbook, water acts not only as the medium of life but also as a transport medium for chemicals that can threaten life. Whether we are discussing water in rivers, lakes, groundwater, or as it leaves the oceans and passes through and cleanses the atmosphere, the unique properties of water will play an important role in understanding transport phenomena.

Water occurs on Earth in distinct settings that we generally refer to as compartments: atmosphere, land, groundwater, rivers, lakes, and oceans. The water cycle is the exchange between these compartments and is a highly dynamic system. Figure 2.1 is a simplified representation of the global water cycle, showing each major compartment with its respective mass of water, water's average residence time in each compartment, and the flux between compartments. Residence time refers to the average amount of time a molecule of water (in this case) spends in the compartment of interest. Of course, the largest reservoir of water is the oceans, which are also associated with the longest residence time (40 000 years). Some molecules of water falling on the surface of the oceans are almost immediately volatilized (evaporation) and return to the atmosphere to fall across landmasses. Other molecules are taken deep into the ocean, not to see the light of day for tens of thousands of years. Water that is volatilized from the ocean surface usually only spends a short time in the atmosphere, about nine days on average. Most of the water returns to the oceans, but some also enters the terrestrial water cycle. As shown in Figure 2.1, water falling on the surface of land makes up glaciers, lakes, rivers, and land runoff. Surface water also percolates into the ground and forms two types of water: water in the unsaturated zone (vadose zone) and water that makes up aquifers or groundwater. Only a small fraction of the freshwater is actually available to terrestrial life, since glaciers and deep groundwater are difficult to access and water in these compartments has relatively long residence times.

It is an understatement to say that water is not distributed evenly across the planet. Table 2.1 shows the total renewable water resources, water withdrawals, percent of renewable resources, and average percent used, by sector, for major geographic regions of the world. Perhaps, the most important column is the percent of renewable resources used, which is an indicator of future crisis areas of the world. As this percentage approaches 100, the population in the area is using water faster than it can be supplied. The most important areas of concern are Pakistan, Turkmenistan, and Uzbekistan in Asia and most of the countries of the Middle East and Northern Africa. Several countries in the Middle East and Northern Africa are supplementing their freshwater with desalination plants fueled by their reserves of fossil fuels, but this cannot last forever. It is also interesting to note the use of freshwater by agriculture and industry, given in the last two columns in Table 2.1. Water from these sectors is almost always subject to contamination, and the fate and transport of these polluted waters is the subject of this textbook. Water scarcity will be an increasingly important issue in the future and has been listed as a major security issue by world leaders, and agencies concerned with climate change. The United Nations perceives that access to clean water resources is one of the major challenges for human life to exist beyond the twenty-first century.

Pollutant Fate and Transport in Environmental Multimedia, First Edition. Frank M. Dunnivant and Elliot Anders.
© 2019 John Wiley & Sons, Inc. Published 2019 by John Wiley & Sons, Inc.
Companion website: www.wiley.com/go/Dunnivant/pollutant_fate&Transp_EnvironMultimedia

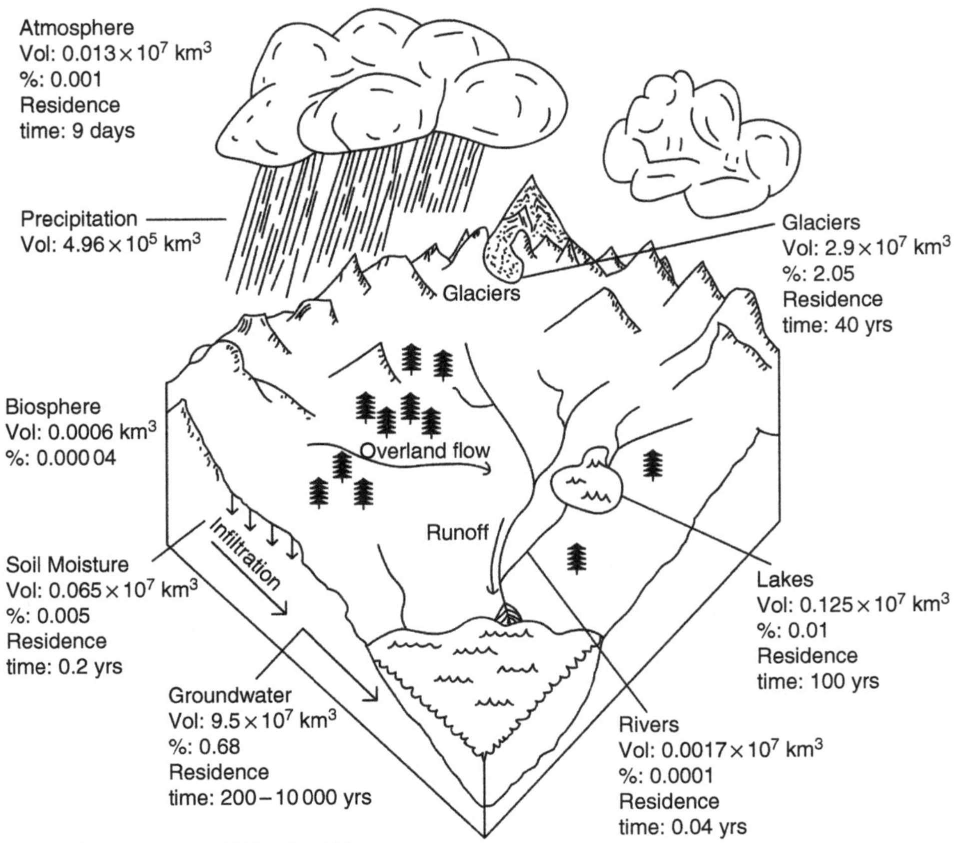

Flux:	Fluxes km^3/yr
Evaporation	496 000
Ocean	425 000
Land	71 000
Precipitation	496 000
Ocean	385 000
Land	111 000
Runoff to oceans	39 700
Streams	27 000
Ground Feed	12 000
Glacial Ice	2 500

Figure 2.1 The global water cycle. Source: Data from: http://www.geog.ouc.bc.ca; http://web.sfc.keio.ac.jp; and Stumm and Morgan 1996.

2.2 Unique Properties of Water

Most people, even most college science students, do not realize how incredibly unique water is. In fact, as we will see later, if water behaved as a chemist would predict based solely on chemical structure, life as we know it would not exist on Earth. But first, let us discuss the general nature of water. Table 2.2 lists 38 unique properties of water that you might not predict if you just looked at water as a typical molecule. So, why is water so unique? All of the unique properties listed in Table 2.2 can be traced back to hydrogen bonding between individual water molecules. You should recall from first-year chemistry that hydrogen bonding is a special form of dipole–dipole interaction, reserved for intermolecular bonding between hydrogen and the nonbonding electrons of N, F, and O (the lone pair of electrons in the Lewis dot structure). Hydrogen bonding in water is especially important, since each water molecule has two bonded hydrogen atoms and two lone pairs of electrons. This allows for a complete alignment and bonding between each lone pair of electrons and hydrogen. In ice, virtually all of the nonbonding electrons on the oxygen atoms are bonded to adjacent hydrogens of another water molecule (see Figure 2.2). As ice melts to liquid water, approximately 15% of these hydrogen bonds between

Table 2.1 The global water budget by region (Food and Agriculture Organization of the United Nations 2017).

Region/country	Total renewable water resources (km^3/yr)	Total water withdrawals (m^3/yr)	Per capita ($m^3/person$)	Average percentage of renewable resources	Average percentage used by agriculture	Average percentage used by industry
World	43 219	3 414 000	650	—	71	20
Asia (excluding Middle East)	11 321	1 516 247	1 028	29	79	10
Cambodia	121	520	60	0	94	1
Pakistan	52	155 600	1 382	100	97	2
Turkmenistan	1.4	23 779	5 801	116	98	1
Uzbekistan	16	58 051	2 598	132	94	2
Europe	6590	367 449	503	9	25	48
Bulgaria	21	13 900	1 573	58	22	75
Middle East and North Africa	518	303 977	754	423	80	5
Egypt	1.8	66 000	1 055	127	82	25
Israel	0.8	1 620	287	108	54	7
Jordan	0.7	984	255	151	75	3
Kuwait	<1	538	306	3 097	60	2
Libyan Arab Jamahiriya	0.6	4 500	870	801	84	3
Oman	1	1 223	658	181	94	2
Saudi Arabia	2.4	17 018	1 056	955	90	1
United Arab Emirates	0.2	2 108	896	1 614	67	9
Yemen	4.1	2 932	253	123	92	1
Sub-Saharan Africa	3 901	72 556	147	5	66	7
North America	4 850	512 440	1 720	14	27	58
Central America and Caribbean	1 186	105 741	513	9	65	14
South America	12 246	156 948	614	1	76	7
Oceania	1 693	16 730	318	1	45	15

water molecules are broken; the net result is a shrinking of the volume of water (an increase in density). This accounts for the fact that water shrinks on melting and that ice floats on liquid water, the first unique property of water listed in Table 2.2. All other inorganic and organic liquids have the reverse relationship; their solid phase is more dense and sinks in their liquid.

Two other unique properties that result from hydrogen bonding in water are its unusually high melting and boiling points. You will recall from general chemistry that elements in the Periodic Table are arranged based on periodicity, which is the term used for the relationships between atomic structure and behavior. If we take the hydrides formed by the elements in Group VI, the group containing oxygen and sulfur (hence H_2O, H_2S, etc.), and plot the melting and boiling points as a function of molecular mass of these hydrides, we observe a directly proportional relationship (refer to Figures 2.3 and 2.4). As the molecular weight of the hydride increases, so do the melting and boiling points. Thus, based on this trend and theoretical calculations, water should boil at −73 °C and freeze at −98 °C. As you know, fortunately this does not occur; water boils at 100 °C and freezes at 0 °C. The reason for this discrepancy is again due to hydrogen bonding. First, let us look at the observed melting point of water ice. Since water can undergo hydrogen bonding, there is a stronger than expected attraction between individual water molecules; in order to melt solid ice into gaseous water, you must break 15% of these stronger intermolecular bonds. To do this, we must put in more heat, manifested by an increase in temperature. Thus, the melting point of water rises from the predicted −98 to 0 °C. In order to volatilize water, in turn, we must break the remaining intermolecular hydrogen bonds. Again, the strength of these bonds raises the melting point of water from the expected −73 °C to the observed 100 °C. Also note that not only have the absolute boiling and melting points of water increased but also the range between the two has increased from a difference of 25 °C to a difference of 100 °C. This also allows liquid water, and therefore aquatic life, to exist over a broader range of temperatures on Earth.

Table 2.2 Unique properties of water.

Properties discussed in text:
Water shrinks on melting (ice floats on water)
Unusually high melting point
Unusually high boiling point
Unusually high surface tension
Unusually high viscosity
Unusually high heat of vaporization
Unusually high specific heat capacity
Other properties dependent on hydrogen bonding:
The solubilities of nonpolar gases in water decrease with temperature to a minimum and then rise
Specific heat capacity has a minimum at 36 °C
A high density that increases on heating
The number of nearest neighbors increases on melting
The number of nearest neighbors increases with temperature
Pressure reduces water's melting point
Pressure reduces the temperature of maximum density
D_2O and T_2O differ from H_2O in their physical properties much more than might be expected
Unusually large viscosity increase as the temperature is lowered
Viscosity decreases with pressure
Unusually low compressibility
Compressibility drops as temperature increases down to a minimum
Low coefficient of expansion
Thermal expansivity decreases increasingly at low temperatures
The speed of sound in liquid water increases with temperature
Liquid water has over twice the specific heat of steam or ice
NMR spin–lattice relaxation are very small at low temperatures
Solute have varying effects on properties such as density and viscosity
Nonaqueous solutions even approach thermodynamic ideality
X-ray diffraction shows unusually detailed structure
Supercooled water has two phases and a second critical point
Liquid water may be supercooled, in tiny droplets, down to about −70 °C
Solid water exists in a wider variety of stable crystal and amorphous structures than do other materials
Hot water may freeze faster than cold water
The refractive index of water has a maximum value at just below 0 °C
At low temperatures, the self-diffusion of water increases as the density of pressure increases
The thermal conductivity of water rises to the maximum at about 130 °C and then falls
Proton and hydroxide ion mobilities are anomalously fast in an electric field
The heat of fusion of water with temperature exhibits a maximum at −17 °C
The dielectric constant is high
Unusually high critical point

http://www.sbu.ac.uk/water/anmlies.html; Internet site no longer available, last accessed in 2006.

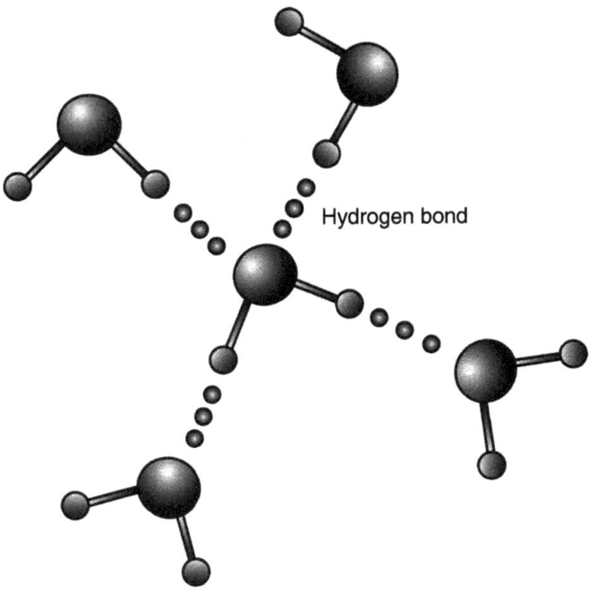

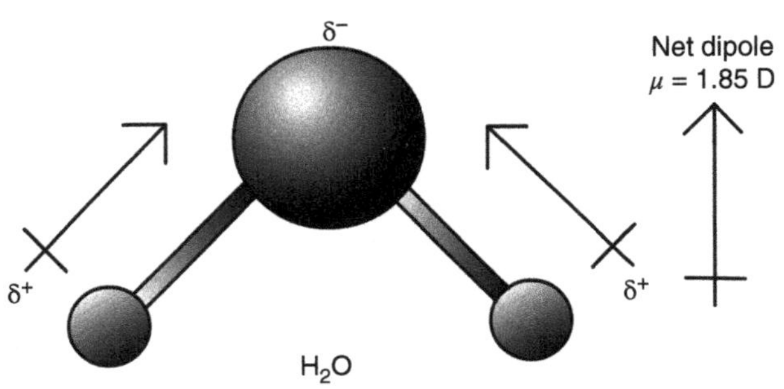

Figure 2.2 Electronic structure of water.

Hydrogen bonding between water molecules is also responsible for the unusually high viscosity and high surface tension of water. In order to understand surface tension, you must picture in your mind the three-dimensional nature of water. Each water molecule is completely surrounded by other water molecules, and each is hydrogen bonded to four other water molecules (minus the 15% of bonds not present in liquid water). In order for water to flow and to have relatively low viscosity, this three-dimensional bonding must be elastic and allow the relatively free movement of adjacent water molecules. Since the hydrogen bonding is stronger than expected, the water molecules are less fluid than expected, resulting in a high viscosity. Now take the three-dimensional structure of water and remove the upper layer so that you have an exposed isolated layer of water. Water does not like this, and the surface molecules cling very strongly to the water molecules below the surface. This gives rise to a very strong surface tension in water. Surface tension is related to the force it takes to break the surface of the water. An illustration of this high surface tension can be seen when you gently place a metal

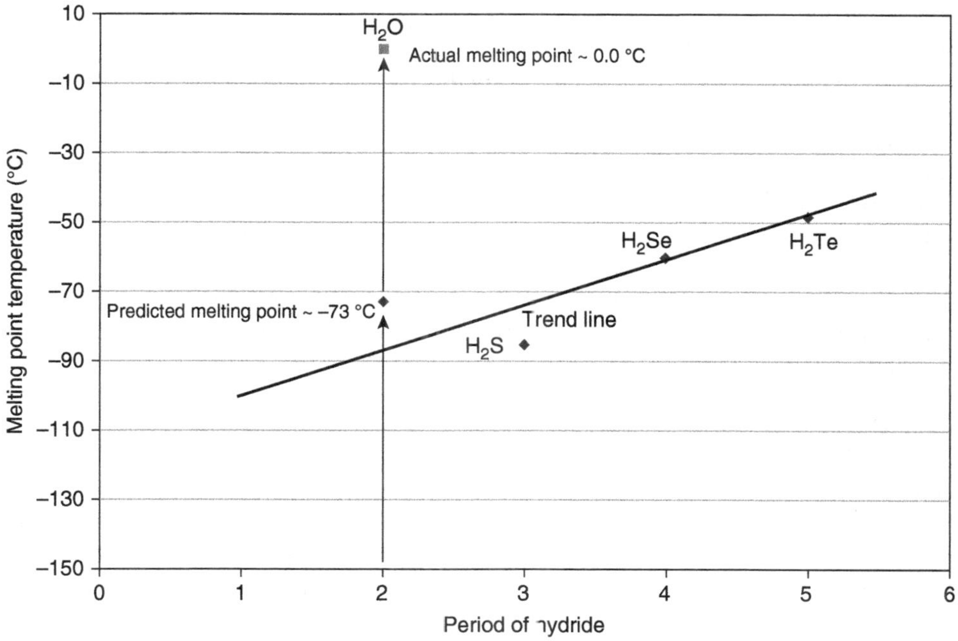

Figure 2.3 Melting points of Group VI hydrides. Source: Data from Weast and Astle 1980.

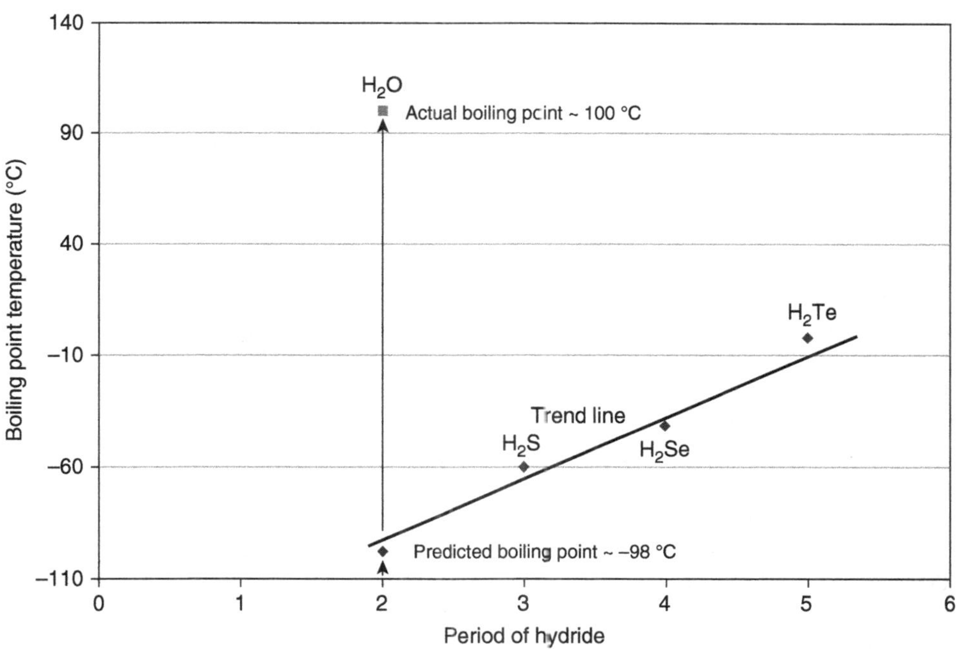

Figure 2.4 Boiling points of Group VI hydrides. Source: Data from Weast and Astle 1980.

paper clip on the surface of water. It will float even though the metal is more dense than water. However, if you add a surface active agent, a surfactant or soap, the surfactant molecules will align themselves between the surface water molecules, break the surface tension (hydrogen bonds), and allow the paper clip to sink. Surface tension and viscosity will be important when we study water flow and pollutant transport in groundwater systems.

Two final unique properties of water are its unusually high heat of vaporization and high specific heat capacity. The high heat of vaporization is directly related to water's unusually high boiling point. In the absence of hydrogen bonding, water would more freely evaporate and go from a liquid to a gas. However, hydrogen bonding is a strong force holding water molecules together, and significant heat must be put into the system in order to separate liquid water molecules into gaseous molecules. Likewise, the strength of the bonds makes a lot of heat necessary to increase the average kinetic energy, or temperature, of water (its relatively high specific heat is 4.184 J/g °C). This ability to absorb large amounts of heat without significantly changing temperature is important to life of Earth. Water, in the form of lakes, rivers, and oceans, serves to store heat on Earth's surface and greatly affects the weather. The climates of landmasses near large water bodies are significantly affected by the temperature of these water bodies. You have most likely heard of microclimates and the lake effect. In colder climates, cities located downwind from a large body of water receive large amounts of snow in the winter. Dry cold air moves across the relatively warm water body, evaporating water, and on reaching the shore and elevated landmasses, the moist air mass rises with the topography and snow forms. However, when the source of the water vapor is closed off, after the water body cools and freezes later in winter, the snowfall greatly diminishes or completely ends. Water is a very important thermal regulator on a global scale, and its ability to only slightly change temperature with absorption of heat is a property directly related to hydrogen bonding.

2.3 Concentration Units

In general chemistry, a variety of concentration units are normally introduced, including g/l, mol/l, mole fraction, molarity, and molality. Some general chemistry textbooks even briefly mention mg/l (parts per million, ppm). Chemists tend to use molarity (mol/l) more often, while engineers, hydrologists, and other researchers dealing with pollutant fate and transport prefer mg/l (ppm) or μg/l (parts per billion, ppb). We will generally use ppm and ppb in this textbook, since we are mostly dealing with pollutant modeling, except in Chapters 2 and 3, where we discuss chemistry. But where do the terms ppm and ppb come from? It is assumed that pure water contains 1.00 million parts of water per million parts of water, or 100%. We typically use a density of 1.00 g/ml for water and assume (incorrectly sometimes) that pollutants have a similar or identical density. So, for example, how many ppm is a pollutant with a concentration of 1000 mg/l?

$$1000\ \text{mg/l of } H_2O \left(\frac{1.00\ \text{g pollutant}}{1000\ \text{mg}}\right) \times \left(\frac{1.00\ \text{l}\ H_2O}{1000\ \text{g}\ H_2O}\right) = \frac{1000\ \text{g pollutant}}{10^6\ \text{g}\ H_2O} = \frac{0.001\ \text{g}}{\text{g}} = 1000\ \text{ppm}$$

Parts per billion (ppb) is defined as μg/l, and parts per trillion (ppt) is defined as ng/l. These are the three basic units we use in pollutant monitoring. As shown above, in water we use a mass per mass calculation, while for gas we use volume per volume calculations. When we are dealing with radioactivity, we will use the milli, micro, and nano prefixes with the radioactivity unit of Curies (Ci). A Curie is the radioactivity of 1.00 g of pure ^{226}Ra, which is equal to 3.7×10^{10} disintegrations/s. Another commonly used unit (the SI equivalent) in the study of radioactivity is the Becquerel (Bq). One Becquerel is equal to 2.703×10^{-11} Ci.

You should recall from general chemistry that molarity (M) is another concentration unit, meaning moles of chemical per liter of solution (mol/l). This unit will be used when we are dealing with problems or situations that involve stoichiometry. If you are unfamiliar with this unit, you should review a general chemistry textbook. The molarity of a chemical species is often represented using brackets: [A] denotes the mol/l of species A.

$$\text{Molarity (M)} = \frac{\dfrac{\text{mass (g) of substance}}{\text{Molar mass of substance (g/mol)}}}{\text{Volume of solution (l)}} = \frac{\text{Moles of solute}}{\text{Liter of solution}} \tag{2.1}$$

2.4 Chemical Aspects of Environmental Systems

There are several general properties that are vitally important to chemical fate in all phases of planet Earth's environment and will be reviewed here.

2.4.1 pH

You learned in general chemistry that pH is the negative log of the molar hydrogen ion concentration:

$$\text{pH} = -\log[H^+] \tag{2.2}$$

Ions in water are always surrounded by water molecules that partially cancel out the charge of the ion. The hydrogen ion, like all ions in solution, is surrounded by waters of hydration. For ease of writing, we normally represent the hydrogen ion as H_3O^+, suggesting one water of hydration, although in reality it can have 5–9. In fact, it is a cluster of water molecules with one extra proton.

The pH of a water solution is considered a master variable. By this, we mean that the pH of the solution can be the determining factor in a variety of parameters, especially chemical speciation, which we will spend considerable time discussing in Sections 2.5.1 and 3.2.2. Therefore, pH is one of the most common parameters measured for a water sample.

Although we commonly relate pH to hydrogen concentration (Eq. (2.2)), the pH of a water is almost always measured with an electrode, and electrodes measure activity instead of concentration. Activity is discussed in Section 2.4.2.

Most natural waters have a pH between 5.5 and 9, but extreme pH values have been observed in natural settings such as geothermal water and eutrophic (organic-rich) systems. When hazardous waste enters natural environments, any pH value is possible, given the vast amount of acidic and caustic wastes that the chemical industry produces. Other topics related to pH that you should review from general chemistry are buffer solutions and the Henderson–Hasselbalch equation, which we will look at closely in Section 2.5.1.

2.4.2 Activity

You will find, as you take more chemistry, that what you learned in general chemistry was a simplification of reality. This is true for units of concentration. For example, molar units are only appropriate in what are called "ideal solutions," in which the molar concentration of an ion or compound is equal to its activity. Activity is a measure of the effective concentration of an ion, accounting for its interactions with other ions that can mask it. Therefore, an ideal solution is one containing very little dissolved salt. Activity (A), the concentration that the solution "sees," is expressed by

$$A = \gamma\,[C] \tag{2.3}$$

where γ is the activity coefficient (almost always equal to or less than 1.00; for extreme conditions it can be greater than 1.00), and $[C]$ is the molar concentration of the chemical or pollutant. The activity coefficient (γ) is a direct function of ionic strength or the amount of other salts in the solution, which helps explain why we must be concerned with activity instead of concentration. If you picture an ion in solution, it is not present as a free cation or anion but is surrounded by water molecules (waters of hydration) and by ions of opposite charge, which serve to balance out the charge of the ion of interest. A representation of such a configuration is shown in Figure 2.5 for ions of $CaSO_4$.

This balancing of charges tends to make the ion less active in solution than expected based on its concentration (by changing the ion mobility), and activity accounts for this canceling out of concentration. As noted above, the activity of a pollutant is equal to the activity coefficient times the molar concentration, and the activity coefficient is related to the salt content of

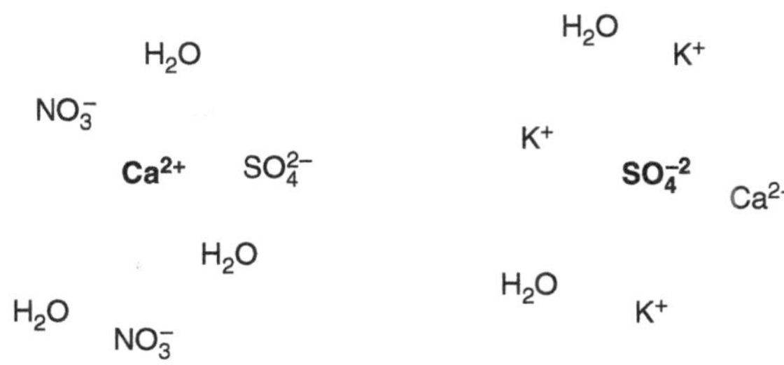

Figure 2.5 Illustration of ions of Ca^{2+} and SO_4^{2-} in aqueous solution.

the water. An activity coefficient of near 1.00 for dilute solutions yields an activity equal to the concentration. As the salt concentration increases, the activity coefficient decreases from 1.00, thus lowering the activity.

So, what is the practical reason for using activity rather than concentration? Remember that our end goal in pollutant fate and transport modeling is to be able to predict risk, and risk is based on toxicity. It has been almost universally found that water containing higher concentrations of nontoxic salts has lower toxicity for the same pollutant concentration. Toxicity is thus governed by activity instead of concentration.

So how can we quantify the effect of activity? As noted earlier, concentration is related to activity by the activity coefficient, which is related to the total ionic strength (μ). We can calculate the ionic strength of any water, if we know the anion/cation composition of the water, by

$$\mu = 0.500\,(C_1Z_1^2 + C_2Z_2^2 + C_3Z_3^2 + \cdots)$$
$$= 0.500\,\Sigma\,C_iZ_i^2 \tag{2.4}$$

where C is molar concentration and Z is charge.

Here is an example calculation: What is the total ionic strength of (a) 0.100 M $NaNO_3$ and (b) 0.100 M Na_2SO_4.

(a) $\mu = 0.500\,[(0.100 * 1^2) + (0.100 * 1^2)] = 0.100$ M
(b) $\mu = 0.500\,[(0.100 * 2 * 1^2) + (0.100 * 2^2)] = 0.300$ M

From this trend we can develop the general rules shown in Table 2.3.

Table 2.3 Estimation of ion strength based on salt composition.

Electrolyte	Molarity	μ
1 : 1	M	1 * M
2 : 1	M	3 * M
3 : 1	M	6 * M
2 : 2	M	2 * M

Table 2.4 Activity coefficients and hydrated radii for aqueous solutions at 25 °C.

Ion	Hydrated ion radius (pm)	Ionic strength of solution				
		0.001	0.005	0.01	0.05	0.1
		Activity coefficients				
Charge ± 1						
H^+	900	0.967	0.933	0.914	0.86	0.83
$(C_6H_5)_2CHCO_2^-$, $(C_3H_7)_4N^+$	800	0.966	0.931	0.912	0.85	0.82
$(O_2N)_3C_6H_2O^-$, $(C_3H_7)_3NH^+$, $CH_3OC_6H_4CO_2^-$	700	0.965	0.930	0.909	0.845	0.81
Li^+, $C_6H_5CO_2^-$, $HOC_6H_4CO_2^-$, $ClC_6H_4CO_2^-$, $C_6H_5CH_2CO_2^-$, $CH_2{=}CHCH_2CO_2^-$, $(CH_3)_2CHCH_2CO_2^-$, $(CH_3CH_2)_4N^+$, $(C_3H_7)_2NH_2^+$	600	0.965	0.929	0.907	0.835	0.80
$Cl_2CHCO_2^-$, $Cl_3CCO_2^-$, $(CH_3CH_2)_3NH^+$, $(C_3H_7)NH_3^+$	500	0.964	0.928	0.904	0.83	0.79
Na^+, $CdCl^+$, ClO_2^-, IO_3^-, HCO_3^-, $H_2PO_4^-$, HSO_3^-, $H_2AsO_4^-$						
$Co(NH_3)_4(NO_2)_2^+$, $CH_3CO_2^-$, $ClCH_2CO_2^-$, $(CH_3)_4N^+$, $(CH_3CH_2)_2NH_2^+$, $H_2NCH_2CO_2^-$	450	0.964	0.928	0.902	0.82	0.775
$^+H_3NCH_2CO_2H$, $(CH_3)_3NH^+$, $CH_3CH_2NH_3^+$	400	0.964	0.927	0.901	0.815	0.77
OH^-, F^-, SCN^-, OCN^-, HS^-, ClO_3^-, ClO_4^-, BrO_3^-, IO_4^-, MnO_4^-, HCO_2^-, H_2 citrate$^-$, $CH_3NH_3^+$, $(CH_3)_2NH_2^+$	350	0.964	0.926	0.900	0.81	0.76
K^+, Cl^-, Br^-, I^-, CN^-, NO_2^-, NO_3^-	300	0.964	0.925	0.899	0.805	0.755
Rb^+, Cs^+, NO_4^+, Tl^+, Ag^+	250	0.964	0.924	0.898	0.80	0.75
Charge ± 2						
Mg^{2+}, Be^{2+}	800	0.872	0.755	0.69	0.52	0.45
$CH_2(CH_2CH_2CO_2^-)_2$, $(CH_2CH_2CH_2CO_2^-)_2$	700	0.872	0.755	0.685	0.50	0.425
Ca^{2+}, Cu^{2+}, Zn^{2+}, Sn^{2+}, Mn^{2+}, Fe^{2+}, Ni^{2+}, Co^{2+}, $C_6H_4(CO_2^-)_2$, $H_2C(CH_2CO_2^-)_2$, $(CH_2CH_2CO_2^-)_2$	600	0.870	0.749	0.675	0.485	0.405
Sr^{2+}, Ba^{2+}, Cd^{2+}, Hg^{2+}, S^{2-}, $S_2O_4^{2-}$, WO_4^{2-}, $H_2C(CO_2^-)_2$, $(CH_2CO_2^-)_2$, $(CHOHCO_2^-)_2$	500	0.868	0.744	0.67	0.465	0.38
Pb^{2+}, CO_3^{2-}, SO_3^{2-}, MoO_4^{2-}, $Co(NH_3)_5Cl^{2+}$, $Fe(CN)_5NO^{2-}$, $C_2O_4^{2-}$, H citrate^{2-}	450	0.867	0.742	0.665	0.455	0.37
Hg_2^{2+}, SO_4^{2-}, $S_2O_3^{2-}$, $S_2O_6^{2-}$, $S_2O_8^{2-}$, SeO_4^{2-}, CrO_4^{2-}, HPO_4^{2-}	400	0.867	0.740	0.660	0.445	0.355
Charge ± 3						
Al^{3+}, Fe^{3+}, Cr^{3+}, Sc^{3+}, Y^{3+}, In^{3+}, lanthanides	900	0.738	0.54	0.445	0.245	0.18
Citrate^{3-}	500	0.728	0.51	0.405	0.18	0.115
PO_4^{3-}, $Fe(CN)_6^{3-}$, $Cr(NH_3)_6^{3+}$, $Co(NH_3)_6^{3+}$, $Co(NH_3)5H_2O^{3+}$	400	0.725	0.505	0.395	0.16	0.095
Charge ± 4						
Th^{4+}, Zr^{4+}, Ce^{4+}, Sn^{4+}	1100	0.588	0.35	0.255	0.10	0.065
$Fe(CN)_6^{4-}$	500	0.57	0.31	0.20	0.048	0.021

Source: Data from Kielland (1937).

For mixtures of ionic salts, the empirically derived extended Debye–Hückel equation can be used to estimate γ:

$$\log \gamma = -\frac{0.512\, Z^2 \sqrt{\mu}}{1 + \dfrac{\alpha \sqrt{\mu}}{305}} \tag{2.5}$$

where

Z is the ionic charge you are calculating γ for,
μ is the ionic strength, and
α is the effective hydrated radius of the ion you are calculating γ (found in Table 2.4).

Note the trend implicit in this equation – small, highly charged ions bind solvent molecules more tightly and have smaller hydrated radii than do larger or less highly charged ions. Also note some generalizations: (i) as ionic strength (μ) increases, the activity coefficient (γ) decreases, (ii) as the ionic charge (Z) increases, the activity coefficient (γ) decreases, and (iii) as the effective hydrated radius (α) decreases, the activity coefficient (γ) decreases.

Example: Calculate the activity coefficient and activity of Ca^{2+} in 0.0200 M $CaCl_2$.

$$\mu = 0.500\,[(0.0200 \times\ 2^2) + (0.0200 \times 2 \times 1^2)]$$
$$= 0.0600\ \mathrm{M}$$
$$\log \gamma = -\frac{0.512(\,2^2)\ \sqrt{0.0600}}{1 + \frac{600\ \sqrt{0.0600}}{305}} = 0.339$$
$$\gamma = 0.459$$
$$A = C\gamma = (0.0200)\,(0.459) = 0.009\,17$$

Note the substantial difference between an activity of 0.009 17 and a concentration of 0.0200 M. As noted, differences become more pronounced as the salt content increases. Waters where activity calculations are important include some groundwaters, inland salt lakes, estuaries, and oceans.

As we continue in this textbook, we will almost always refer to concentration for simplicity, but remember that when you are dealing with a water with high salt content, it is better to work in activities to achieve greater accuracy.

2.4.3 Solubility

Solubility is defined as the maximum concentration of a chemical species that can be present in a solution at equilibrium. Since we will be dealing exclusively with water as our solvent, we are concerned with the aqueous solubility, given in moles per liter or milligram per liter. We will divide our discussions into organic and inorganic pollutants. Although water is considered a universal solvent, it does not necessarily dissolve large quantities of all chemicals. Some organic pollutants, such as short-chained alcohols, acetone, acetonitrile, and a few other organic solvents are completely miscible with water; that is, water and the solvent will completely mix in any proportion. Many other pollutants are incorrectly listed as insoluble in water, which really means that the solubility is very low. For example, the famous pollutants such as 1,1′-(2,2,2-trichloroethylidene)bis[4-chloro-benzene] (DDT) and polychlorinated biphenyls (PCBs) are listed as insoluble in many chemistry handbooks but are actually soluble in the ppb to ppm range. Their low solubilities contribute to these chemicals' toxicity to wildlife. In order to understand their toxicity, we must understand bioconcentration or biomagnification.

Pollutants such as DDT and PCBs are hydrophobic; they do not like being dissolved in water. As noted above, their solubilities are in the ppb to ppm level in pure water. In the environment, they actively partition from (move out of) the water onto surfaces or into biological organisms, even at concentrations below their aqueous solubility. Microorganisms such as algae have a large surface area, and their hydrophobic cell surfaces will readily attract and sorb hydrophobic pollutants. Thus, these pollutants are concentrated into the algae or microbe cell mass. Organisms that feed on algae automatically concentrate more DDT and PCBs since they eat large quantities of algae over a longer lifespan. This bioconcentration of the pollutants continues up the food chain to the top predators, including birds of prey and humans. An example of bioconcentration in an ecosystem is shown in Table 2.5. As you see, the pollutant concentration in the water is only 0.000 05 ppm, but it is bioconcentrated to ~25 ppm in biological species at the top of the food chain. This almost million-fold increase illustrates the dangers of these extremely low levels of pollutants in the environment.

The experimental determination of the exact solubility of a hydrophobic or any low solubility pollutant is very difficult and is wrought with analytical errors. The lower the solubility, the more error in the results and disagreement between laboratory determinations. Some solubility values for the same pollutant disagree by a factor of 10–100; however, these procedures do allow more precise estimates of the relative solubilities of hydrophobic pollutants. As a result of the disagreement in literature values, a modeling method to determine absolute and relative solubilities of compounds is

Table 2.5 Bioconcentration of DDT in Long Island Food Web (USA).

Organism	DDT residues (ppm)
Water	0.000 05
Plankton	0.04
Silverside minnows	0.23
Sheepshead minnows	0.94
Pickerel (predatory)	1.33
Needlefish (predatory)	2.07
Heron (feeds on small aquatic animals)	3.57
Herring gull (scavenger)	6.00
Osprey egg	13.8
Merganser (fish-eating duck)	22.8
Cormorant (feeds on large fish)	26.4

Source: From Woodwell et al. (1967).

becoming more accepted today; solubilities are estimated from highly sophisticated calculations using the chemical parameter program SPARC, developed by the U.S. EPA (https://www3.epa.gov/ceampubl/learn2model/part-two/onsite/sparcproperties.html). This program can relatively accurately predict the aqueous solubility, vapor pressure, and Henry's law constant for any chemical with a known structure and melting point. Table 2.6 lists these parameters for several common pollutants that will be used in the modeling section

Table 2.6 Summary of aqueous solubility, vapor pressure, and Henry's law constants for common environmental pollutants (estimated by SPARC; https://www3.epa.gov/ceampubl/learn2model/part-two/onsite/sparcproperties.html).

Classes of pollutants	CAS #	Melting point (°C)	Molecular weight	Solubility (mg/l)	Vapor pressure (atm)	Henry's law constant (atm l/mol)
C1–C2 Halocarbons						
Trichlorofluoromethane	75-69-4	−111	137.4	2 413	1.228	0.056 9
Dichlorodifluoromethane	75-71-8	−158	120.9	523.3	5.957	0.231
Chlorodifluoromethane	75-45-6	−157.4	86.4	4 005	6.223	0.021 6
1,2-Dichloromethane	75-71-8	−158	120.9	523.3	5.955	0.231 1
Trichloroethene	79-01-6	−84.7	131.388	913.9	0.177 3	0.025 49
Tetrachloroethene	127-18-4	−22.3	165.83	127.3	0.051 1	0.066 57
1,1,1-Trichloroethane	71-55-6	−30	133.4	1 940	0.203 2	0.013 97
Methyl bromide	74-83-9	−93.7	94.94	23 970	1.831	0.003 960
1,2-Dichloroethane	107-06-2	−35.5	98.96	7 903	0.124 4	0.001 558
1,1,2,-Tetrachloroethane	79-72-1	−44	167.9	2 171	0.006 077	0.004 699
Hexachloroethane	67-72-1	187	236.74	12.95	9.931E−5	0.001 816
Polychlorinated biphenyls						
2-Chlorobiphenyl	2051-60-7	34	188.66	5.769	9.179E−6	0.000 300 1
4-Chlorobiphenyl	2051-62-9	78.8	188.66	1.915	2.245E−6	0.000 221 2
2,2′-DCB	13029-08-8	58	223.10	0.752 8	1.304E−6	0.000 386 5
4,4′-DCB	2050-68-2	149.3	223.10	0.077 93	5.732E−8	0.000 164 1
2,2′,3,3′-TCB	38444-93-8	123	291.99	0.008 024	3.641E−9	0.000 131 2
2,2′,5,5′-TCB	35693-99-3	84	291.99	0.037 05	2.755E−8	0.000 217 1
2,3,4,5-TCB	33284-53-6	93	291.99	0.014 78	5.235E−9	0.000 103 4
2,2′,4,5,5′-PCB	37680-73-2	26	326.43	0.029 45	1.266E−8	0.000 140 3
2,2′,4,4′,6,6′-HCB	33979-03-2	110	360.88	0.000 909 0	2.563E−10	0.000 101 8
Carboxyate esters (pyrethrins)						
Pyrethrin I	121-21-1	Liquid at 25 °C	328.45	3.303	8.517E−10	1.198E−7
Pyrethrins	8003-34-7	Liquid at 25 °C	328.45	3.974	9.099E−10	1.141E−7
Pyrethrin II	121-29-9	Liquid at 25 °C	372.46	12.37	4.053E−11	1.221E−9
Organophosphates (www.chemfinder.com)						
Parathion	56-38-2	6.1	291.26	3.064	3.498E−8	3.324E−6
Trimethyl phosphate	512-56-1	−46	140.08	7 750 000	0.026 38	1.279E−7
Triphenyl phosphate	115-86-6	50.5	326.29	38.74	2.666E−11	2.246E−10
Paraoxon	311-45-5	N/A (assume liquid)	275.20	24 070	9.462E−8	1.082E−9
Methyl parathion	298-00-0	36	263.210 5	58.51	3.033E−7	1.365E−6
Disulfoton	298-04-4	−25	274.41	0.498 9	1.959E−7	0.000 107 8
Diazinon	333-41-5	~120	304.35	1.658	3.351E−8	6.152E−6
Dichlorvos	62-73-7	−60	220.98	761 700	8.389E−4	2.434E−7
Acephate	30560-19-1	88	183.17	5 397 000	2.006E−7	6.134E−13
Carbamates						
Carbofuran	1563-66-2	151	221.26	44.88	2.203E−9	1.086E−8
Aldicarb	116-06-3	100	190.37	981.6	4.500E−7	8.723E−8

Table 2.6 (Continued)

Classes of pollutants	CAS #	Melting point (°C)	Molecular weight	Solubility (mg/l)	Vapor pressure (atm)	Henry's law constant (atm l/mol)
Carbaryl	63-25-2	145	201.22	17.59	9.532E−10	1.090E−8
Methomyl	16752-77-5	78	162.21	24 160	1.114E−6	7.479E−9
Propoxur	114-26-1	87	209.24	262.5	6.175E−8	4.922E−8
Mexacarbate	315-18-4	85	222.29	218.4	8.370E−10	8.519E−10
Chlorinated pesticides and by-products						
p,p-DDT	50-29-3	108.5	354.49	0.003 758	2.185E−10	2.061E−5
p,p′-DDE	72-55-9	89	318.03	0.003 216	5.141E−10	5.074E−5
o,p′-DDD	53-19-0	76	320.04	0.034 73	1.010E−9	9.305E−6
p,p′-DDD	72-54-8	109.5	320.04	0.012 93	3.133E−10	7.755E−6
Lindane (HCB)	58-89-9	113	290.83	10.14	4.059E−7	1.164E−5
PCP	77-10-1	46.5	243.39	114.3	4.934E−8	1.051E−7
Kepone	143-50-0	350 (decomp)	490.64	0.210 8	1.290E−10	3.002E−7
Mirex	2385-85-5	485 (decomp)	545.54	2.516E−6	1.640E−10	0.035 56
cis- or α-Chlordane	5103-71-9	106	409.78	0.064 42	1.031E−8	6.556E−5
Heptachlor	76-44-8	95.5	373.32	0.167 7	1.208E−7	0.000 268 8
Aldrin	309-00-2	104	364.91	0.018 39	1.761E−7	0.003 494
Dieldrin	60-57-1	175	380.91	0.119 3	4.688E−9	1.498E−5
Endosulfan	115-29-7	106	406.93	1.211	1.947E−8	6.544E−6
Alachlor	15972-60-8	40	269.77	946.7	1.251E−8	3.566E−9
Aldicarb	116-06-3	100	190.37	981.6	4.499E−7	8.723E−8
Endrin	72-20-8	200 (decomp)	380.91	0.079 49	3.125E−9	1.497E−5
2,3,7,8-Tetrachlorodibenzo-*p*-dioxin	1746-01-6	295	321.97	5.573E−5	2.426E−12	1.401E−5
Phenoxy acid herbicides						
2,4-DB	106-93-4	9.9	187.86	7 266	0.012 60	0.000 325 8
Dalapon	75-99-0	166 5 (decomp)	142.97	69 300	3.125E−4	5.919E−7
Dicamba	1918-00-9	115	221.04	278.6	7.437E−7	5.901E−7
Dichlorprop	120-36-5	117.5	235.07	325.4	9.587E−8	6.925E−8
Dinoseb	88-85-7	40	240.22	2 120	1.305E−7	1.479E−8
MCPA	94-74-6	120	200.62	377.5	2.179E−8	1.158E−8
MCPP	93-65-2	93	214.65	494.8	1.559E−7	6.766E−8
2,4,5-T	93-76-5	153	255.48	63.47	1.714E−9	6.901E−9
2,4-D	94-75-7	140.5	221.04	257.7	1.315E−8	1.128E−8
Substituted phenols						
4-Methylphenol	106-44-5	35.5	108.14	19 210	1.478E−4	8.317E−7
4-Ethylphenol	123-07-9	45.0	122.17	6 240	5.193E−5	1.017E−6
3-Methylphenol	108-39-4	11.8	108.14	25 090	1.969E−4	8.488E−7
2-Chlorophenol	95-57-8	9.8	235.07	325.4	9.587E−8	6.926E−8
Phenol	108-95-2	40.9	94.11	58 840	4.148E−4	6.633E−7
4-Chlorophenol	106-48-9	42.7	128.56	27 890	9.998E−5	4.609E−7
4-Hydroxybenzoic acid	99-96-7	214.5	138.12	3 681	4.056E−10	1.257E−11
3-Chlorophenol	108-43-0	32.6	128.56	38 860	1.307E−4	4.322E−7
4-Hydroxyacetophenone	99-93-4	109.5	136.15	3 201	2.207E−8	9.378E−10
2-Hydroxybenzoic acid	69-72-7	158	138.12	2 190	6.139E−6	3.486E−7
4-Nitrophenol	100-02-7	113.8	139.11	7 514	1.028E−7	1.903E−9

(Continued)

Table 2.6 (Continued)

Classes of pollutants	CAS #	Melting point (°C)	Molecular weight	Solubility (mg/l)	Vapor pressure (atm)	Henry's law constant (atm l/mol)
Triazines						
Atrazine	1912-24-9	173	215.68	37.95	9.535E−9	5.419E−8
Eptam	759-94-4	Liquid at 25 °C	189.32	152.4	3.357E−5	4.172E−5
Sutan	2008-41-5	Liquid at 25 °C	217.37	18.35	1.502E−5	0.000 177 9
Vernam	1929-77-7	Liquid at 25 °C	203.35	47.51	1.190E−5	5.096E−5
Tilam	1114-71-2	Liquid at 25 °C	203.35	50.44	1.084E−5	4.372E−5
Propachlor	1918-16-7	77	211.69	648.7	1.547E−7	5.048E−8
Trifluralin	1582-09-8	49	335.28	2.121	2.858E−8	4.518E−6
Simazine	122-34-9	226	201.66	34.44	2.886E−9	1.690E−8
Propazine	139-40-2	213	229.71	12.24	8.861E−9	1.664E−7
Bromacil	314-40-9	158	261.12	913.7	2.437E−9	6.964E−10
Prowl	40487-42-1	56	281.31	1.044	7.028E−9	1.894E−6
Phthalate esters						
Di or bis(2-ethylhexyl) phthalate esters (DEHPs)	117-81-7	−55	390.56	0.009 507	1.046E−11	4.294E−6
Diethyl phthalate	84-66-2	−40.5	222.24	261.2	1.078E−6	9.169E−7
Dimethyl phthalate	131-11-3	5.5	194.19	2 383	5.391E−6	4.393E−7
Di-*n*-butyl phthalate	84-74-2	−35	278.35	2.191	1.711E−8	1.631E−6
Di-*n*-octyl phthalate	117-84-0	25	390.56	0.003 374	1.739E−12	3.096E−6
Polycyclic aromatic hydrocarbons						
Naphthalene	91-20-3	80.2	128.17	39.06	1.197E−4	0.000 392 6
Phenanthrene	85-01-8	99.2	178.23	0.988 9	1.164E−7	2.098E−5
Anthracene	120-12-7	215.0	178.23	0.085 08	8.126E−9	1.703E−5
Benzanthracene	56-55-3	84	228.29	0.061 21	2.141E−10	7.987E−7
Pyrene	129-00-0	151.2	202.26	0.105 5	6.148E−9	1.178E−5
Fluoranthene	206-44-0	107.8	202.26	0.214 9	2.173E−8	1.280E−5
Fluorine	86-73-7	114.8	166.22	1.963	7.973E−7	6.750E−5
Chrysene	218-01-9	258.2	228.29	0.001 461	6.278E−12	9.808E−7
Benzo(*k*)fluoranthene	207-08-9	217	252.31	0.003 810	1.879E−12	7.258E−7
Benzo(*b*)fluoranthene	205-99-2	168	252.31	0.012 37	6.849E−12	9.185E−7
Benzo(*a*)pyrene	50-32-8	176.5	252.31	0.003 501	5.871E−12	9.356E−7
Benzo(*e*)pyrene	192-97-2	177.5	252.31	0.003 605	6.871E−12	1.149E−6
Aliphatic hydrocarbons						
C_5 (*n*-pentane)	109-66-0	−129.7	72.15	51.81	0.739 3	1.029
C_6 (*n*-hexane)	110-54-3	−95.35	86.18	14.53	0.224 5	1.331
C_7 (*n*-heptane)	142-82-5	−90.6	100.20	3.725	0.069 42	1.867
C_8 (*n*-octane)	111-65-9	−56.82	114.23	0.931 8	0.021 28	2.608
C_9 (*n*-nonane)	111-84-2	−53.5	128.26	0.234 9	0.006 343	3.463
C_{10} (*n*-decane)	124-18-5	−29.7	142.28	0.056 77	0.001 925	4.824
C_{11} (undecane)	1120-21-4	−25.60	156.31	0.022 52	5.839E−4	6.749
C_{12} (dodecane)	112-40-3	−9.66	170.34	0.014 96	1.455E−4	9.555
$C1_3$ (tridecane)	629-50-5	−5.3	184.36	0.010 30	4.376E−5	13.35
C_{14} (tetradecane)	629-59-4	5.82	198.39	0.007 339	3.256E−5	18.76
C_{15} (pentadecane)	629-62-9	9.9	212.42	0.005 402	4.345E−6	26.35
C_{16} (hexadecane)	544-76-3	18.12	226.12	0.004 100	1.443E−6	37.12
C_{17} (heptadecane)	629-78-7	22	240.47	0.003 184	4.814E−7	49.08
C_{18} (octadecane)	593-45-3	28.2	254.50	0.002 490	1.603E−7	61.35

Table 2.6 (Continued)

Classes of pollutants	CAS #	Melting point (°C)	Molecular weight	Solubility (mg/l)	Vapor pressure (atm)	Henry's law constant (atm l/mol)
C_{19} (nonadecane)	629-92-5	32.1	268.52	0.001 998	5.748E−8	82.28
C_{20} (icosane)	112-95-8	36.8	282.55	0.001 631	2.139E−8	110.3
Substituted nitrobenzenes						
2-Methylnitrobenzene	88-72-2	−10	137.14	764.6	2.299E−4	4.125E−5
3-Methylnitrobenzene	99-08-1	15.5	137.14	1394	1.556E−4	1.529E−5
4-Methylnitrobenzene	99-99-0	51.6	137.14	869.1	8.833E−5	1.389E−5
2-Chloronitrobenzene	88-73-3	32.5	157.56	603.3	9.349E−5	2.442E−5
3-Chloronitrobenzene	121-73-3	44.4	157.56	900.3	1.139E−4	1.994E−5
4-Chloronitrobenzene	100-00-5	83.5	157.56	476.4	5.994E−5	1.982E−5
2-Acetylnitrobenzene	577-59-3	28.5	165.15	4 715	2.996E−6	1.050E−7
3-Acetylnitrobenzene	121-89-1	81	165.15	2 953	4.792E−7	2.681e−8
4-Acetylnitrobenzene	100-19-6	81.8	165.15	3 008	4.279E−7	2.349E−8
2,4,6-Trinitrotoluene	118-96-7	80.1	227.13	144.3	1.759E−9	2.769E−9
2,4-Dinitrotoluene	121-14-2	71	182.14	376.5	5.431E−7	2.627E−7
2,6-Dinitrotoluene	606-20-2	66	182.14	186.5	5.453E−7	5.327E−7
Nitrobenzene	98-95-3	5.7	123.11	3 720	4.066E−4	1.346E−5
HMX	2691-41-0	281	296.16	4.519	4.235E−17	2.774E−15
RDX	121-82-4	205.5	222.12	69.45	1.055E−12	4.249E−12
Common solvents						
Pentane	109-66-0	−129.7	72.15	51.81	0.738 3	1.029
Pentadecane	629-62-9	9.9	212.42	0.005 402	4.345E−6	26.35
Benzene	71-43-2	5.49	78.11	2 481	0.135 5	0.004 266
Toluene	108-88-3	−94.95	92.14	775.8	0.041 33	0.004 909
Styrene	100-42-5	−31	104.15	425.5	0.008 913	0.002 181
Pyridine	110-86-1	−40.70	79.10	178 600	0.030 64	1.357E−5
Methylene chloride	75-09-2	−94.94	84.93	25 240	0.537 0	0.001 807
Chloroform	67-66-3	−63.41	119.38	11 930	0.208 8	0.002 090
Carbon tetrachloride	56-23-5	−22.73	153.82	2 348	0.167 1	0.010 95
Trichloroethylene	79-01-6	−84.7	131.39	914.0	0.177 3	0.025 49
p-Dichlorobenzene	106-46-7	53.09	147.00	122.1	0.001 934	0.002 329
Acetone	67-64-1	−94.8	58.08	246 800	0.305 1	7.180E−5
Methyl ethyl ketone	78-93-3	−86.65	72.11	129 200	0.132 3	7.382E−5
Acetonitrile	75-05-8	−43.82	41.05	265 800	0.099 67	1.539E−5
Methanol	67-56-1	−97.54	32.04	1 144 500	0.198 2	5.549E−6
Ethanol	64-17-5	−114.1	46.07	406 200	0.086 56	9.817E−6
Aniline	62-53-3	−6.02	93.13	47 480	7.075E−4	1.388E−6
Nitrobenzene	98-95-3	5.7	123.11	3 720	4.066E−4	1.346E−5
Methyl *tert*-butyl ether	1634-04-4	−108.6	88.15	15 290	0.563 4	0.003 247
Gasoline components						
Benzene	71-43-2	5.49	78.11	2 481	0.135 5	0.004 266
Ethylbenzene	100-41-4	−94.9	106.17	270.4	0.015 69	0.006 160
Methyl *tert*-butyl ether	1634-04-4	−108.6	88.15	15 300	0.563 4	0.003 247
Heptane	142-82-5	−90.6	100.20	3.725	0.069 42	1.867
2-Methylpentane	107-83-5	−153.7	86.18	16.69	0.324 3	1.674
Toluene	108-88-3	−94.95	92.14	775.8	0.041 33	0.004 909
m-Xylene	108-38-3	−47.8	106.17	268.3	0.012 41	0.004 909
o-Xylene	95-47-6	−25.2	106.17	320.0	NA	NA

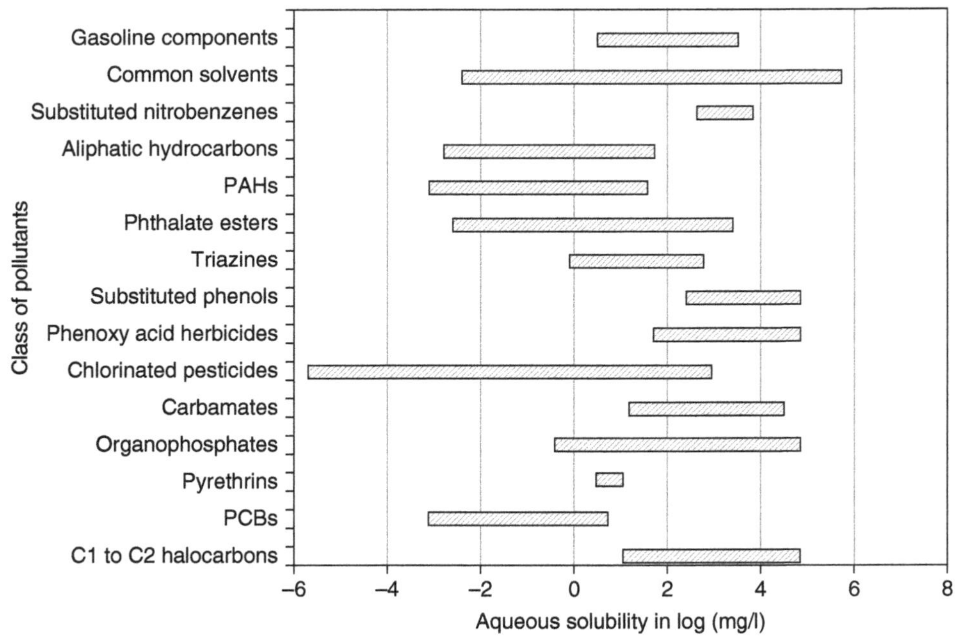

Figure 2.6 Summary and comparison of aqueous solubilities of selected pollutants. Source: Data Source: www3.epa.gov/ceampubl/learn2model/part-two/onsite/sparcproperties.html.

of this textbook. This table and the SPARC model can be used in the modeling chapters to determine the source masses of pollutants in step and pulse models. The aqueous solubilities of different classes of pollutants are also compared in Figure 2.6. Note the wide range of pollutant solubilities.

While we will generally study organic pollutants in the aqueous phase, many sources of organic pollutants are pure liquids known as nonaqueous phase liquids (NAPLs). We discussed one example, from the Idaho National Engineering and Environmental Laboratory, in Chapter 1. The NAPL may be more dense than water and known as a dense nonaqueous phase liquid (DNAPL) or may be less dense than water and known as a light nonaqueous phase liquid (LNAPL). Neither of these liquids mix with water, but as water flows past the NAPL, compounds in the NAPLs slowly dissolve into the water phase. Although there are sophisticated models to predict the release rate of these "pools" of pollutants, one simple way to model them is to use the maximum aqueous solubility predicted from SPARC and use this as the input mass in a step transport model. We will discuss this in the groundwater modeling chapter.

The solubilities of inorganic pollutants (salts), at least as they are calculated in general chemistry, are much easier to determine than those of organics. You should remember the solubility product constant or K_{sp}. Values of K_{sp} for several salts are given in Table 2.7. We will use an applied example from general chemistry to illustrate how solubilities are calculated from these constants. Note that in the following calculations we are concerned with the concentration of the dissolved metal ion in equilibrium with some solid phase.

Say that you have an industrial process that produces copper(I) bromide as a waste product. What will be the maximum Cu^+ solubility if this solid waste is contacted by rainwater? The K_{sp} for CuBr is 5.3×10^{-9}.

Solution:

$$\text{CuBr (aq)} \Rightarrow \text{Cu}^+ \text{ (aq)} + \text{Br}^- \text{ (aq)}$$

$$K_{sp} = [\text{Cu}^+]\,[\text{Br}^-]$$

For every mole of CuBr that dissolves, 1 mol of Cu^+ and 1 mol of Br^- dissolve, so let $[Cu^+] = [Br^-] = x$. Substituting into the K_{sp} equation yields

$K_{sp} = 5.3 \times 10^{-9} = x^2$

$x = (K_{sp})^{1/2}$ or

$x = [Cu^+] = 7.3 \times 10^{-5}$ M Cu^+. Finally, to get solubility in ppm, $(7.3 \times 10^{-5}$ M $Cu^+)$ (63.6 g/mol) $= 4.6 \times 10^{-3}$ g/l or 4.6 mg/l Cu^+ or 4.6 ppm

Now, let us make the problem a bit more complicated. Say that you have a hazardous waste sludge sample containing stoichiometrically equal amount of $PbCO_3$, $PbCl_2$, $PbCrO_4$, PbF_2, $PbSO_4$, and PbS. Which form of lead waste will determine the maximum concentration of lead in any leachate that may come from the waste?

Table 2.7 Selected solubility products for inorganic compounds at 25 °C (Weast and Astle 1980).

Substance	Solubility product at temperature noted (°C)	Substance	Solubility product at temperature noted (°C)
Aluminum hydroxide	4×10^{-13} (15)	Lead iodide	7.47×10^{-9} (15)
Aluminum hydroxide	1.1×10^{-15} (18)	Lead iodide	1.39×10^{-8} (25)
Aluminum hydroxide	3.7×10^{-15}(25)	Lead oxalate	2.74×10^{-11} (18)
Barium carbonate	7×10^{-9} (16)	Lead sulfate	$1.06\times\times10^{-8}$ (18)
Barium carbonate	8.1×10^{-9} (25)	Lead sulfide	3.4×10^{-28} (18)
Barium chromate	1.6×10^{-10} (18)	Lithium carbonate	1.7×10^{-3} (25)
Barium chromate	2.4×10^{-10} (28)	Magnesium ammonium phosphate	2.5×10^{-13} (25)
Barium fluoride	1.6×10^{-6} (9.5)	Magnesium carbonate	2.6×10^{-5} (12)
Barium fluoride	1.7×10^{-6} (18)	Magnesium fluoride	7.1×10^{-9} (18)
Barium fluoride	1.73×10^{-6} (25.8)	Magnesium fluoride	6.4×10^{-9} (27)
Barium iodate, $Ba(IO_3)\cdot2H_2O$	8.4×10^{-11} (10)	Magnesium hydroxide	1.2×10^{-11} (18)
Barium iodate, $Ba(IO_3)\cdot2H_2O$	6.5×10^{-10} (25)	Magnesium oxalate	8.57×10^{-5} (18)
Barium oxalate, $BaC_2O_3\cdot3.5H_2O$	1.62×10^{-7} (18)	Manganese hydroxide	4×10^{-14} (18)
Barium oxalate, $BaC_2O_4\cdot2H_2O$	1.2×10^{-7} (18)	Manganese sulfide	1.4×10^{-15} (18)
Barium oxalate, $BaC_2O_4\cdot0.5H_2O$	2.18×10^{-7} (18)	Mercuric sulfide	4×10^{-53} to 2×10^{-49} (18)
Barium sulfate	0.87×10^{-10} (18)		
Barium sulfate	1.08×10^{-10} (25)	Mercurous bromide	1.3×10^{-21} (25)
Barium sulfate	1.98×10^{-10} (50)	Mercurous chloride	2×10^{-18} (25)
Cadmium oxalate $CdC_2O_4\cdot3H_2O$	1.53×10^{-8} (18)	Mercurous iodide	1.2×10^{-28} (25)
Cadmium sulfide	3.6×10^{-29} (18)	Nickel sulfide	1.4×10^{-24} (18)
Calcium carbonate (calcite)	0.99×10^{-8} (15)	Potassium acid tartrate $[K^+]\ [HC_4H_4O^{6-}]$	3.8×10^{-4} (18)
Calcium carbonate (calcite)	0.87×10^{-8} (25)		
Calcium fluoride	3.4×10^{-11} (18)	Silver bromate	3.97×10^{-5} (20)
Calcium fluoride	3.95×10^{-11} (26)	Silver bromate	5.77×10^{-5} (25)
Calcium iodate, $Ca(IO_3)_2\cdot6H_2O$	22.2×10^{-8} (10)	Silver bromide	4.1×10^{-13} (18)
Calcium iodate, $Ca(IO_3)_2\cdot6H_2O$	64.4×10^{-8} (18)	Silver bromide	7.7×10^{-13} (25)
Calcium oxalate, $CaC_2O_4\cdot H_2O$	1.78×10^{-9} (18)	Silver carbonate	6.15×10^{-12} (25)
Calcium oxalate, $CaC_2O_4\cdot H_2O$	2.57×10^{-9} (25)	Silver chloride	0.21×10^{-10} (4.7)
Calcium sulfate	2.45×10^{-5} (25)	Silver chloride	0.37×10^{-10} (9.7)
Calcium tartrate, $CaC_4H_4O_6\cdot2H_2O$	0.77×10^{-6} (18)	Silver chloride	1.56×10^{-10} (25)
Cobalt sulfide	3×10^{-26} (18)	Silver chloride	13.2×10^{-10} (50)
Cupric iodate	1.4×10^{-7} (25)	Silver chloride	215×10^{-10} (100)
Cupric oxalate	2.87×10^{-8} (25)	Silver chromate	1.2×10^{-12} (14.8)
Cupric sulfide	8.5×10^{-45} (18)	Silver chromate	9×10^{-12} (25)
Cuprous bromide	4.15×10^{-8} (18–20)	Silver cyanide $[Ag^+][Ag(CN)_2^-]$	2.2×10^{-12} (20)
Cuprous chloride	1.02×10^{-6} (18–20)	Silver dichromate	2×10^{-7} (25)
Cuprous iodide	5.06×10^{-12} (18–20)	Silver hydroxide	1.52×10^{-8} (20)
Cuprous sulfide	2×10^{-47} (16–18)	Silver iodate	0.92×10^{-8} (9.4)
Cuprous thiocyanate	1.6×10^{-11} (18)	Silver iodide	0.32×10^{-16} (13)

(Continued)

Table 2.7 (Continued)

Substance	Solubility product at temperature noted (°C)	Substance	Solubility product at temperature noted (°C)
Ferric hydroxide	1.1×10^{-36} (18)	Silver iodide	1.5×10^{-16} (25)
Ferrous hydroxide	1.64×10^{-14} (18)	Silver sulfide	1.6×10^{-49} (18)
Ferrous oxalate	2.1×10^{-7} (25)	Silver thiocyanate	0.49×10^{-12} (18)
Ferrous sulfide	3.7×10^{-19} (18)	Silver thiocyanate	1.16×10^{-12} (25)
Lead carbonate	3.3×10^{-14} (18)	Strontium carbonate	1.6×10^{-9} (25)
Lead chromate	1.77×10^{-14} (18)	Strontium fluoride	2.8×10^{-9} (18)
Lead fluoride	2.7×10^{-8} (9)	Strontium oxalate	5.61×10^{-8} (18)
Lead fluoride	3.2×10^{-8} (18)	Strontium sulfate	2.77×10^{-7} (2.9)
Lead fluoride	3.7×10^{-8} (26.6)	Strontium sulfate	3.81×10^{-7} (17.4)
Lead iodate	5.3×10^{-14} (9.2)	Zinc hydroxide	1.8×10^{-14} (18–20)
Lead iodate	1.2×10^{-13} (18)	Zinc oxalate, $ZnC_2O_4{\cdot}2H_2O$	1.35×10^{-9} (18)
Lead iodate	2.6×10^{-13} (25.8)	Zinc sulfide	1.2×10^{-23} (18)

What could be the maximum concentration of Pb^{2+} in the leachate?

Pollutant	K_{sp}
$PbCO_3$	7.4×10^{-14}
$PbCl_2$	1.7×10^{-5}
$PbCrO_4$	2.8×10^{-13}
PbF_2	3.6×10^{-8}
$PbSO_4$	6.3×10^{-7}
PbS	3×10^{-28}

Solution:

In general, this can be solved by looking at the compound with the highest K_{sp} value, since this will be the one most responsible for the solubility of Pb^{2+} (but for similar K_{sp} values, you must also consider stoichiometry). In this case, $PbCl_2$ will determine the overall solubility of Pb^{2+} from the waste.

$K_{sp} = 1.7\times10^{-5} = [Pb^{2+}][Cl^-]^2$

For each mole of $PbCl_2$ that dissolves, 1 mol of Pb^{2+} ion and 2 mol of Cl^- ions are released. We will let $x = Pb^{2+}$. For every Pb^{2+} there are two Cl^-, so if $x = Pb^{2+}$, $Cl^- = 2x$. Substituting into the K_{sp} equation, yields

$K_{sp} = 1.7\times10^{-5} = [x][2x]^2 = 4x^3$

$x = 0.016\,\text{M}\ Pb^{2+}$

in ppm: $(0.016\,\text{M}\ Pb^{2+})$ (207.2 g/mol) (1000 mg/g) = 3400 mg/l.

There is one complicating variable concerning solubility, both for organic and inorganic pollutants. Numerous field investigations report "dissolved" concentrations of pollutants that exceed their known maximum solubilities. But how can this be? The answer is the presence of a second "dissolved" phase, usually in the form of colloidal inorganic or organic particles. There is no conclusive way to isolate free ion in the dissolved phase from these phases for measurement. Actually, the dissolved phase is commonly defined operationally as the pollutant concentration present in a sample that has been filtered through a 0.20- or 0.45-μm filter. But there are many naturally occurring inorganic and organic particles (which can contain sorbed pollutants) that can pass through these filters. The most common second phase that has been found in field samples is natural organic matter (NOM) present in the dissolved phase (dissolved organic matter, DOM). NOM is the organic matter left over from microbial decay of natural plant and animal material. NOM can vary greatly in molecular weight, and in general no two NOM molecules are the same. Scientists have attempted to characterize NOM using a variety of techniques (IR, NMR, MS) in order to develop a "representative" structure. One such structure is shown in Figure 2.7. The purpose of these structures is not to have a working molecular model of NOM but to illustrate the presence of important functional groups and hydrophobic centers in the molecule. Every pollutant studied has some affinity for sorbing or binding to NOM molecules. We will discuss NOM in greater detail in Section 2.7.2.

The variety of forms in which a pollutant may be present gives rise to the term speciation. For example, Cd^{2+} may be present as the free hydrated cation (an ion surrounded by water of hydration) or as the bound

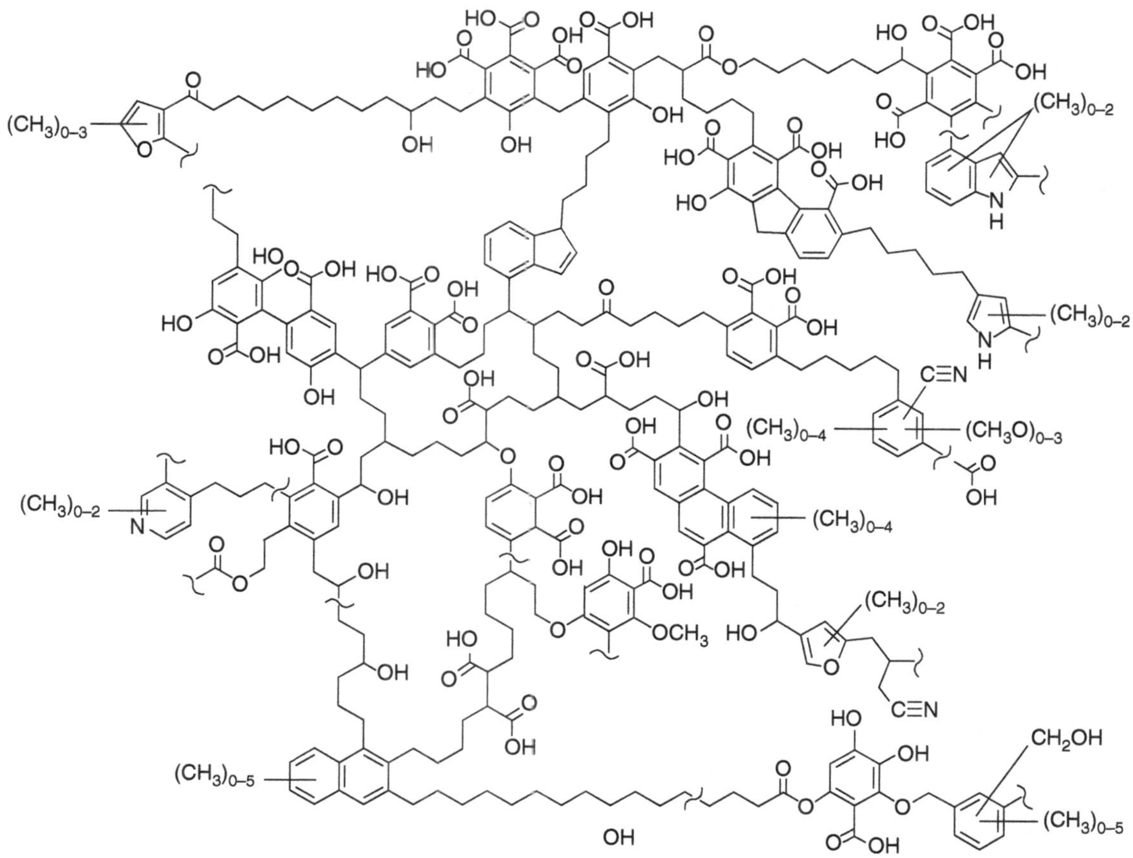

Figure 2.7 Representative structures of natural organic matter. Source: Schulten and Schnitzer (1993). Reprinted with permission of Springer Publishing.

NOM-Cd^{2+} complex. But note that given the variety of ionic binding sites in the NOM molecule, a variety of NOM-Cd^{2+} complexes can be present, which greatly complicates the transport modeling process. We will discuss this further in Chapter 3 and in the groundwater modeling chapter.

2.4.3.1 Vapor Pressure

Vapor pressure is the pressure of a compound's vapor phase at equilibrium with the compound's pure phase, which can be a liquid or solid depending on the temperature. As a reference point we use conventional standard conditions (25 °C and 1.00 atm of pressure) to describe standard vapor pressures. As with aqueous solubilities, and especially for the broad range of pollutants of interest to this textbook, vapor pressures for many compounds are difficult to measure experimentally. So, again we will rely on the SPARC program to generate a range of values for selected compounds. These are also shown in Table 2.6 and are summarized by type of pollutant in Figure 2.8. Note that the range extends from 0.77 atm to $10^{-11.7}$ atm. Vapor pressure is important in the fate and transport modeling of atmospheric pollutants, since the vapor pressure determines both whether a chemical will volatilize and, in general, how fast it volatilizes in an open system. Pollutants with high vapor pressures volatilize faster than those with low vapor pressure. This parameter will be important in Chapter 9, where we will study the fate and transport of pollutants in the atmosphere.

2.4.4 Henry's Law Constant

A Henry's law constant (K_H) is the ratio of the equilibrium vapor pressure to aqueous solubility of a substance (at a given temperature), expressed by

$$X\,(\text{aq}) \leftrightarrow X\,(\text{g})$$

$$K_H = \frac{\text{Pressure in atms } (P_x)}{\text{M Solubility (mol/l)}} \tag{2.6}$$

K_Hs can be calculated by taking the ratio of experimentally determined or predicted vapor pressures and aqueous solubilities or can be measured using the dynamic purge technique described by Mackay et al. (1979). (Note: A variety of units are used in Henry's law constants.) Since the K_H is a ratio of two experimentally

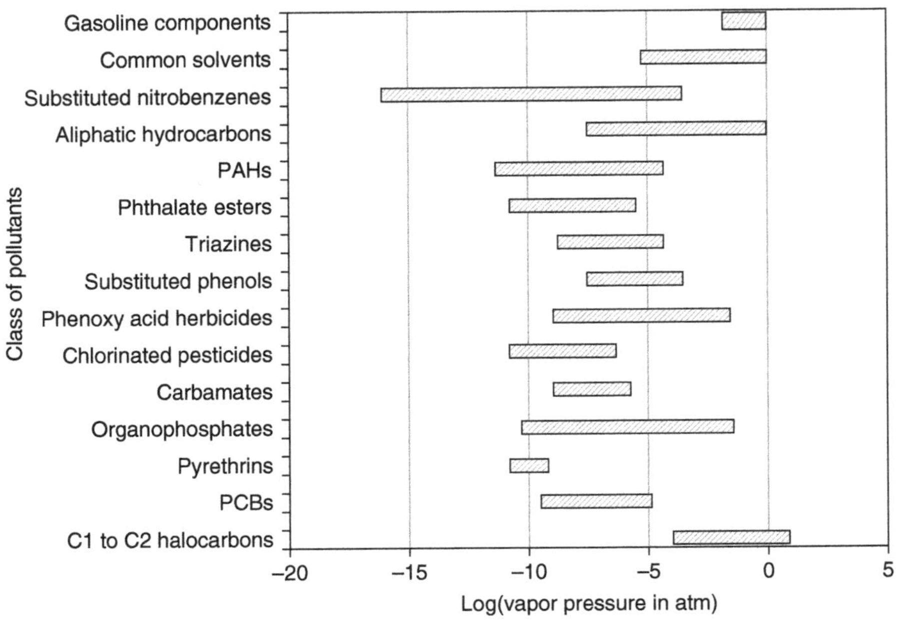

Figure 2.8 Summary and comparison of vapor pressures of selected pollutants. Date Source: www3.epa.gov/ceampubl/learn2model/part-two/onsite/sparcproperties.html.

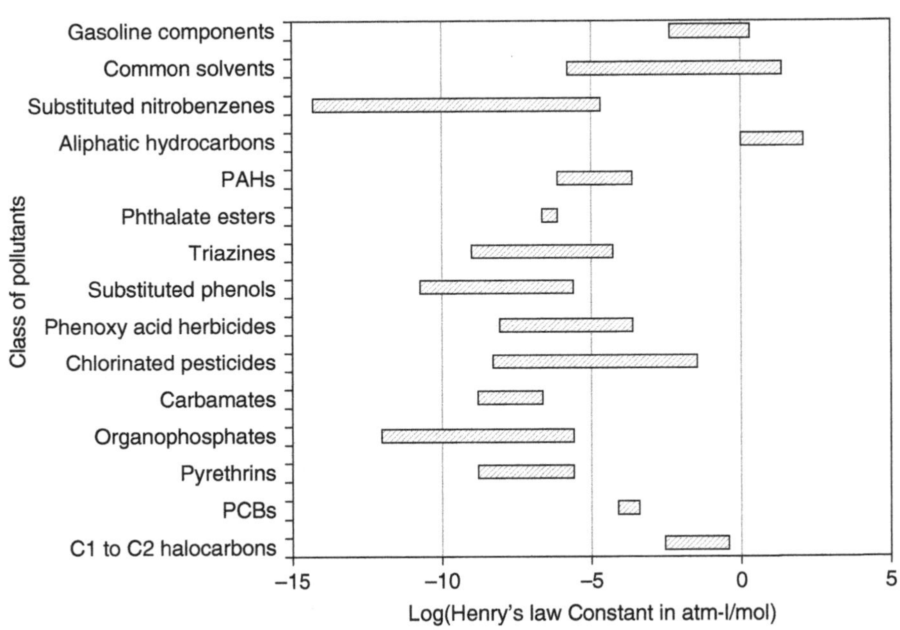

Figure 2.9 Summary and comparison of Henry's law constants of selected pollutants. Data Source: www3.epa.gov/ceampubl/learn2model/part-two/onsite/sparcproperties.html.

difficult measurements, its experimental determination is subject to the same errors as its principle measurements. Thus, again, we will turn to the SPARC estimation program for K_H values. These are summarized in Table 2.6 and in Figure 2.9 by type of pollutant. Note the large range in values reflecting the large range in vapor pressures and aqueous solubilities.

K_Hs are important in environmental chemistry, perhaps more important than vapor pressure, for estimating the volatilization of pollutants from water. Although many hydrophobic pollutants have low vapor pressure, they have high K_H values; when they are placed in water they readily escape to the overlying gas phase. This equilibrium preference for the gaseous phase gives many pollutants a higher than expected volatilization rate. Classic examples are DDT and PCBs. This will be expressed in our river and lake fate and transport models as a first-order removal rate.

2.5 Reactions and Equilibrium

The three basic types of chemical reactions are precipitation, acid–base, and oxidation–reduction. We covered precipitation to the extent necessary in Section 2.4.2 (and we will cover it again in Section 3.2.2). We will spend

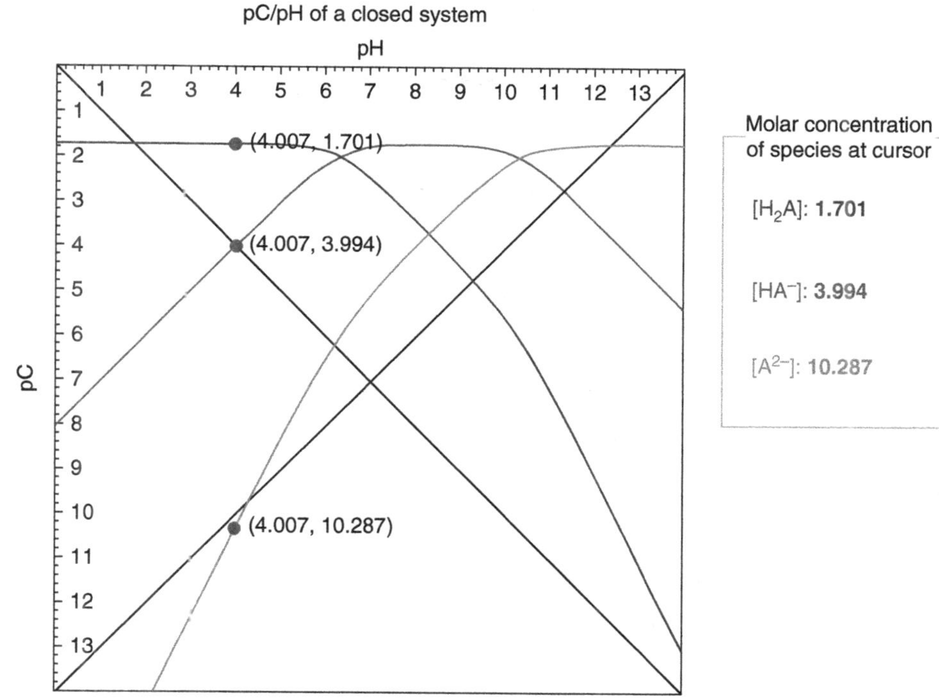

Figure 2.10 A pC–pH diagram for a closed 0.0200 M carbonate system. Source: Anders, diagram from Fate®.

a little more time covering acid–base and oxidation–reduction reactions in this section. First, we will extend your knowledge of acid–base chemistry through the concept of buffers and pC–pH diagrams.

2.5.1 Acid–Base Chemistry

By far the most important buffer system in nature is the carbonate system, and we will use it here to introduce you to a new way of looking at pH and acid–base equilibrium, and to introduce the concept of chemical speciation. The diagrams we will be constructing are referred to as pC–pH diagrams (also referred to as distribution or α plots), where the pC represents the −log of concentration of any chemical species in solution. Note the *C* in this case does not represent the chemical element carbon, but the concentration of any compound.

The concentration of a weak acid or base in a solution (for example H_2CO_3, HCO_3^-, or CO_3^{2-}) can be calculated using simple equilibrium expressions at any given pH value. In some cases, it is useful to look at the equilibrium distribution of each of the protonated and nonprotonated species in solution at the same time. A pC–pH diagram (Figure 2.10) is an excellent tool for viewing these concentrations simultaneously. As the name implies, this diagram shows the concentrations of all chemical species (expressed as the negative log of concentration), with pC values on the vertical axis and pH on the horizontal. To construct a pC–pH diagram, the total concentration of the acid or base is needed along with the corresponding equilibrium equations and acid dissociation constants (K_a).

We will first construct a pC–pH diagram for a system that is closed to the atmosphere, such that no additional atmospheric CO_2 can enter the water. For calculation purposes, we assume that carbonate in the water comes solely from the dissolution of carbonate-containing minerals. Refer to Figure 2.10 during the following discussion.

All pC–pH diagrams have two lines in common: the line describing the concentration of hydroxide (OH^-) as a function of pH and the line describing the concentration of hydronium ion (H_3O^+) as a function of pH. These are based on the equilibrium relationship

$$H_2O \Longleftrightarrow H^+ + OH^-$$

where

$$K_w = [H^+][OH^-] = 1.00 \times 10^{-14} \text{ (at 25°C)}$$

By rearranging and taking the negative log of each side,

$$-\log K_w = -\log[H^+] - \log[OH^-]$$
$$14 = pH + pOH$$
$$pOH = 14.0 - pH$$

Thus, when pH equals 0, pOH equals 14. This results in a line from (pH = 0.0, pC = 14.0) to (pH = 14.0, pC = 0.0). The slope of this diagonal line is

$$\frac{\Delta(-\log[OH])}{\Delta\, pH} = +1$$

Similarly, a line can be drawn representing the hydronium ion concentration as a function of pH. By definition, $-\log[H^+] = pH$, therefore

$$\frac{\Delta(-\log[H^+])}{\Delta pH} = -1$$

When the pH equals 0, the $-\log[H^+]$ equals 0. This results in a line from (pH = 0.0, pC = 0.0) to (pH = 14.0, pC = 14.0).

The next line (or set of lines) normally drawn on a pC–pH diagram is the one representing the total concentration of acid or base, C_T. We will use a total carbonate concentration of 0.0200 M and assume it is constant, since we are dealing with a closed system. When pC–pH diagrams are drawn by hand, C_T is drawn as a straight horizontal line at pC_T on the y-axis, as a guide for filling in the curves for the different acid/base species. This line is actually broken into regions of two or more lines for the dominant chemical species, depending on the number of protons present in the acid. Vertical (dotted) lines representing the pK_a values (negative logs of the K_as) are then added. These vertical lines become the guides for adding the curves representing the different carbonate species (CO_3^{2-}, HCO_3^-, H_2CO_3). At the lowest pK_a, along the C_T line, for example falls the intersection of the lines representing the most acidic and second most acidic species (HCO_3^- and H_2CO_3 in the case of the carbonate system), one with a negative whole number slope and one with a positive whole number slope. We will limit our discussion of the mathematical derivation of these lines to a diprotic system (the carbonate system), but after reviewing the equations you should be able to derive the governing equations for triprotic and monoprotic systems.

For the diprotic system, the equilibrium equations for H_2A (H_2CO_3), HA^- (HCO_3^-), and A^{2-} (CO_3^{2-}) (where A refers to the anionic species, CO_3^{2-} for the carbonate example) are:

$$H_2A \leftrightarrow HA^- + H^+$$

where

$$K_{a1} = \frac{[HA^-][H^+]}{[H_2A]}$$

$$HA^- \leftrightarrow A^{-2} + H^+$$

where

$$K_{a2} = \frac{[A^{-2}][H^+]}{[HA^-]}$$

The mass balance equation (a description of all carbonate forms) is

$$C_T = [H_2A] + [HA^-] + [A^{2-}].$$

When the equilibrium equations are solved for H_2A, HA^-, and A^{2-} (in terms of C_T, $[H^+]$, and the equilibrium constants), and combined with the mass balance equation, three equations are obtained

$$[H_2A] = C_T\left(\frac{1}{1 + \frac{K_1}{[H^+]} + \frac{K_1K_2}{[H^+]^2}}\right)$$

$$[HA^-] = C_T\left(\frac{1}{\frac{[H^+]}{K_1} + 1 + \frac{K_2}{[H^+]}}\right)$$

$$[A^{2-}] = C_T\left(\frac{1}{\frac{[H^+]^2}{K_1K_2} + \frac{[H^+]}{K_2} + 1}\right).$$

If a pH-dependent constant, α_H, is defined as

$$\alpha_H = \frac{[H^+]^2}{K_1K_2} + \frac{[H^+]}{K_2} + 1 \quad (2.7)$$

substituting Eq. (2.7) into the previous equations for the diprotic system, the three major HA species equations can be simplified to:

$$[H_2A] = \frac{C_T\,[H^+]^2}{K_1\,K_2\,\alpha_H} \quad (2.8)$$

$$[HA^-] = \frac{C_T\,[H^+]}{K_2\,\alpha_H} \quad (2.9)$$

$$A^{2-} = \frac{C_T}{\alpha_H} \quad (2.10)$$

By taking the log transform of Eqs. (2.7)–(2.9), we now have equations describing every line in the pC–pH diagram. The utility of a pC–pH diagram is that all of the ion concentrations can be estimated at the same time for any given pH value. The computer program (the pC–pH Simulator®) used to create this diagram, included with your textbook, allows the user to select an acid system, enter the pK_a values, and draw the pC–pH diagram. After the diagram is drawn, the user can point the cursor at a given pH and the concentration of each species will be given. One last point, in a fixed total carbonate concentration system (closed system), there is only one important point that determines the equilibrium pH, that is the intersection of the $[HCO_3^-]$ line and the $[H^+]$ line. In Figure 2.10, for a total carbonate concentration of 0.02 M, the pH is 4.00. The intersection of these lines in a closed system will also be the equilibrium pH.

Now we will develop a pC–pH diagram for an open system, one that is open to a gas that can dissolve and contribute chemical species (dissolved gases and ions) to the solution. The pC–pH diagrams for open systems are similar to those described for the closed systems. The primary difference is that in an open system, a component of the system exists as a gas and the system is open to inputs from the atmosphere. In other words, the system can exchange matter with the atmosphere. The most important environmental examples of such systems are lakes, rivers, and oceans, which exchange carbon dioxide with the atmosphere and therefore contain dissolved carbon dioxide (CO_2 (aq)), carbonic acid (H_2CO_3), bicarbonate ion (HCO_3^-), and carbonate ion (CO_3^{2-}).

The reactions occurring in this system are

1. $CO_2\ (aq) \leftrightarrow CO_2\ (g)$
2. $CO_2\ (g) + H_2O \leftrightarrow H_2CO_3$,
3. $H_2CO_3 \leftrightarrow HCO_3^- + H^+$,
4. $HCO_3^- \leftrightarrow CO_3^{2-} + H^+$,
5. $H_2O \leftrightarrow H^+ + OH^-$.

The equilibrium relationships, corresponding to the line number, for this system are, respectively:

$$K_2 = \frac{[H_2CO_3]}{P_{CO_2}} = 10^{-1.47}$$

$$K_3 = \frac{[H^+][HCO_3^-]}{[H_2CO_3]} = 10^{-6.35}$$

$$K_4 = \frac{[H^+][CO_3^{2-}]}{[HCO_3^-]} = 10^{-10.33}$$

$$K_5 = [H^+][OH^-] = 10^{-14}$$

where P_{CO_2} is the partial pressure of CO_2 in the atmosphere. For an atmospheric CO_2 concentration of 380 ppm, or a pressure of 380×10^{-6} atm, using the K_2 equation above, yields a $[H_2CO_3]$ of 1.28×10^{-5}. On the pC, negative log scale used in open systems (Figure 2.11), this results in a value of −4.89.

Open-system pC/pH diagrams, such as the one shown in Figure 2.11, contain lines describing the concentration of hydroxide (OH^-) and hydronium ion (H^+) identical to those for closed systems. However, because open systems can exchange matter with the atmosphere, the total inorganic carbon (C_T) concentration is not constant as it is for a closed system, where all carbon present was derived from an initial input of calcium carbonate. Rather, in an open system, C_T varies as a function of pH. Still, the total inorganic carbon concentration is the sum of all inorganic carbon species, as it was for closed systems. In this case:

$$C_T = [H_2CO_3] + [HCO_3^-] + [CO_3^{2-}].$$

The concentration of H_2CO_3, HCO_3^-, and CO_3^{2-}, as a function of pH and partial pressure, P_{CO_2}, can be calculated from the equilibrium relationships given previously. The equations for these lines are:

For

$$[H_2CO_3] = (K_2)(P_{CO_2}) = (P_{CO_2})10^{-1.47}$$

and taking the negative log yields

$$-\log[H_2CO_3] = -\log(P_{CO_2}) + 1.47$$

For

$$[HCO_3^-] = \frac{K_3\,[H_2CO_3]}{[H^+]} = \frac{K_3\,(P_{CO_2})\,10^{-1.47}}{[H^+]} = \frac{10^{-6.35}\,(P_{CO_2})\,10^{-1.47}}{[H^+]}$$

and taking the negative log yields

$$-\log[HCO_3^-] = -\log(P_{CO_2}) + 7.82 - pH$$

$$\text{For } CO_3^- = \frac{K_4(10^{-6.35})\,P_{CO_2}\,(10^{-1.47})}{[H^+]^2} = \frac{10^{-10.33}\,(10^{-6.35})\,P_{CO_2}\,(10^{-1.47})}{[H^+]^2}$$

and taking the negative log yields

$$-\log[CO_3^{2-}] = -\log(P_{CO_2}) + 18.15 - 2pH$$

As mentioned previously and demonstrated by the above equations, the concentrations of H_2CO_3, HCO_3^-, and CO_3^{2-} vary as a function of both pH and P_{CO_2}. This means that as P_{CO_2}, the global carbon dioxide concentration in the atmosphere, has varied naturally over the years, often associated with climate changes, the concentration of H_2CO_3, HCO_3^-, and CO_3^{2-} in surface waters has changed. It also means that P_{CO_2} changes caused by global warming will alter the surface water concentrations of these species.

A pC–pH diagram for the open carbonate system (P_{CO_2} = 380 ppm CO_2 or 3.80×10^{-4} atm) in the atmosphere is shown in Figure 2.11. The intercept of the $[H^+]$ line and the $[HCO_3^-]$ line represents the pH of the system, which decreases with increasing atmospheric CO_2 concentration.

As you can see, the pH of the system determines the chemical speciation of protonated and deprotonated species, and pC–pH diagrams can be used to determine the dominant form of protonated–deprotonated compound at a specific pH of interest. So how does this relate to the buffering of pH? Recall from your general chemistry course, the concept of buffers and locate the two important buffer points in Figures 2.10 and 2.11. These are located where the concentrations of H_2CO_3–HCO_3^- and HCO_3^-–CO_3^{2-} are approximately close to each other, within 0.5–1.0 pH units of the two pK_a values (6.33 and 10.33; refer to any titration in a general chemistry text). At or near these pH values, the pH remains relatively stable with small inputs of acid or base, the effects of which are effectively absorbed by the weak acid–base pairs. We will discuss the effect of pH on metal speciation further in Section 3.2.2.

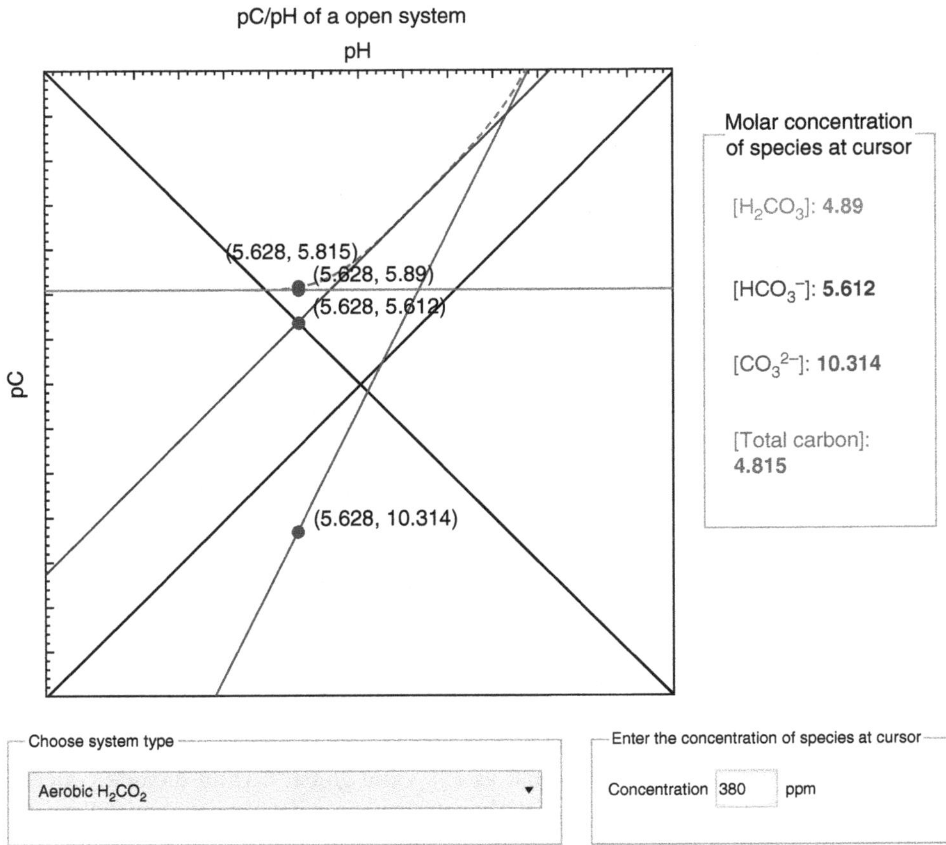

Figure 2.11 A pC–pH diagram for an open carbonate system in contact with an atmosphere of 380 ppm CO_2. Source: Anders, diagram from Fate®.

2.5.2 Oxidation–Reduction Chemistry

Oxidation–reduction (redox) is probably the most common, and the most complicated, type of reaction in the environment. In general chemistry, you mostly studied redox reactions from the standpoint of batteries and electrochemical cells, but there is much more to this complicated area of chemistry and biochemistry. Earlier in this chapter, we described pH as a master variable. Another important chemical parameter of an aqueous solution, which determines, and in many cases drives, other chemical reactions, is the apparent electrode potential of the aqueous system. We refer to this potential of the aqueous system as the E_H, measured in volts. But where does this term come from? We start by looking at the activity of electrons in solution (their availability to participate in reduction reactions), designated by $\{e^-\}$, and use the same notation as for pH

$$p\varepsilon = -\log\{e^-\} \tag{2.11}$$

Thus, $p\varepsilon$ is a measure of the electron activity. Under typical conditions in surface waters (pH of 7.0, in the presence of dissolved oxygen), the $p\varepsilon$ is approximately 14. But it decreases to approximately 4 in the presence of reduced iron and drops further to approximately −4 when sulfide and methane are present. $p\varepsilon$ is related to E_H by

$$E_H = \frac{2.3\,R\,T}{n\,F}\,p\varepsilon \tag{2.12}$$

where R is the ideal gas law constant, T is the absolute temperature (in K), and F is the Faraday constant (96 485 C/mol of electrons). At 25°C and for a 1 mol transfer of electrons ($n = 1$), the equation reduces to

$$E_H = 0.059\,p\varepsilon \tag{2.13}$$

The oxidation state of most natural waters is controlled by microbial activity; thus, the reactions controlling the chemistry of these waters are said to be biologically mediated. In order to understand these reactions and their implications, we must first understand the concept of terminal electron acceptors (TEAs).

Among biological energy-producing oxidation reactions, we are most familiar with the use of oxygen to oxidize food (as occurs in our bodies); yet there are many other ways of utilizing carbon-based compounds as food. Food, in the general sense, consists of reduced carbon in the form $(CH_2O)_n$. When we eat (oxidize) this food, we remove electrons from the reduced carbon, split off water, and oxidize the carbon to CO_2, which we exhale. But what happens to the electrons we have taken off the carbon? There must be an acceptor for these electrons, since there is no free current running through our bodies – we do not light up bulbs if we hold them in our hands (or mouth). This is where molecular oxygen comes into the equation in human bodies; it takes the electrons freed up from the carbon and is reduced to water (O_2 goes from an oxidation state of 0 to an oxidation state of −2 in water). These paired redox reactions are illustrated for glucose in the equations below, where glucose is our reduced carbon source (food) and oxygen is the TEA (where the released electrons go). The standard free energy change for each half reaction ($\Delta G°$) under the conditions in natural water is shown to the right (Schwarzenbach et al. 1993).

$$C_6H_{12}O_6\text{ (glucose)} + 6H_2O\text{ (l)}$$
$$\Rightarrow 6CO_2\text{ (g)} + 24H^+\text{ (aq)} + 24e^-$$
$$-984.0\text{ kJ/mol glucose}$$

$$O_2(g) + 4H^+\text{ (aq)} + 4e^-$$
$$\Rightarrow 2H_2O\text{ (l)} \quad -313.2\text{ kJ/mol }O_2$$

You should recall from general chemistry that you can add half reactions to obtain an overall energy value for the complete process. In order to add these half reactions, you must have an equal number of electrons on opposite sides of the two equations and cancel out the electrons when you add the reactions. To do this, we must multiply the oxygen equation by 6, and when we do this you must also multiply the $\Delta G°$ value by 6. This yields:

$$C_6H_{12}O_6\text{(glucose)} + 6H_2O$$
$$\Rightarrow 6CO_2 + 24H^+ + 24e^-$$
$$-984.0\text{ kJ/mol glucose}$$

$$6O_2 + 24H^+ + 24e^-$$
$$\Rightarrow 12H_2O \quad -1879.2\text{ kJ/6 mol }O_2.$$

Adding the two equations together and reducing the sum to its simplest form yields

$$C_6H_{12}O_6\text{(glucose)} + 6O_2$$
$$\Rightarrow 6H_2O + 6CO_2 \quad -2863.0\text{ kJ/mol glucose}$$

Note that this reaction yields 2863 kJ for each mole of glucose oxidized. Again, in this reaction electrons are taken from the glucose and added to gaseous oxygen, producing water and releasing energy from the chemical bonds. Neither reaction is possible without coupling to the other half reaction. As mentioned, this is the way people think of food becoming oxidized, but in reality there are many more ways the overall reaction is completed by other organisms, especially in aqueous environments not in contact with our highly oxygen-rich atmosphere.

In the reaction discussed above, oxygen (O_2) was the TEA, but many other TEAs also exist. Other common TEAs that are present in the deep water of lakes and rivers, in anoxic lake and river sediments, and in anoxic groundwater systems include

$$\mathbf{2NO_3^-} + 12H^+ + 10e^-$$
$$\Rightarrow N_2 + 6H_2O \quad -714\text{ kJ/mol}$$
$$\mathbf{MnO_2(s)} + HCO_3^- + 3H^+ + 2e^-$$
$$\Rightarrow MnCO_3(s) + 2H_2O \quad -100.4\text{ kJ/mol}$$
$$\mathbf{FeOOH(s)} + HCO_3^- + 2H^+ + e^-$$
$$\Rightarrow FeCO_3(s) + 2H_2O \quad 4.6\text{ kJ/mol}$$
$$\mathbf{SO_4^{2-}} + 9H^+ + 8e^-$$
$$\Rightarrow HS^- + 4H_2O \quad 170.4\text{ kJ/mol}$$
$$\mathbf{CO_2} + 8H^+ + 8e^-$$
$$\Rightarrow CH_4 + 2H_2O \quad 188.0\text{ kJ/mol}$$

The chemicals in bold print represent the TEAs of interest. Each of these half reactions can be manipulated and added to the glucose half reaction above to obtain the energy produced by the microbe per mole of glucose. The magnitude of these $\Delta G°$ values determines which TEA will be used first (i.e. the combination yielding the most energy will be favored and used first, since it gives the microbes that utilize it an evolutionary advantage, then the next most energetic, and so on).

The reason for discussing these reactions is to note the oxidation state (E_H) where each reaction occurs in nature and polluted systems. Not all environments contain oxygen, and the chemistry (and fate and transport) of metals and organic pollutants can significantly change depending on the oxidation potential of the surrounding system. Figure 2.12 shows the oxidation potential where each of the reducing reactions can occur. Note the E_H of the systems. Highly oxygenated systems have an E_H value of approximately +0.81 V, and this value decreases as you move through the other TEAs (oxidized and reduced species). Extremely anaerobic water can have E_H values of −0.40 V, indicating a highly reducing environment. In such environments, transformation reactions of the pollutants can occur. We will discuss these TEA reactions further in the lake and groundwater chapters (Chapters 5 and 8, respectively). In addition, specific transformation reactions occurring in anaerobic groundwater will be discussed in Section 2.8.

The redox potential of a water is obviously important in the fate and transport of metal pollutants, since the oxidation state of transition metals is strongly influenced by E_H. But what about organic pollutants? As we will see in Section 2.8, organic pollutants are subject to a variety of transformation reactions. Now, we will use one of these reactions, the reduction of nitrobenzene pollutants, to illustrate the effect of the pH and E_H on the concentration of reactants and the resultant speed (rate) of the transformation reaction. This example is taken from experiments conducted using anaerobic groundwaters

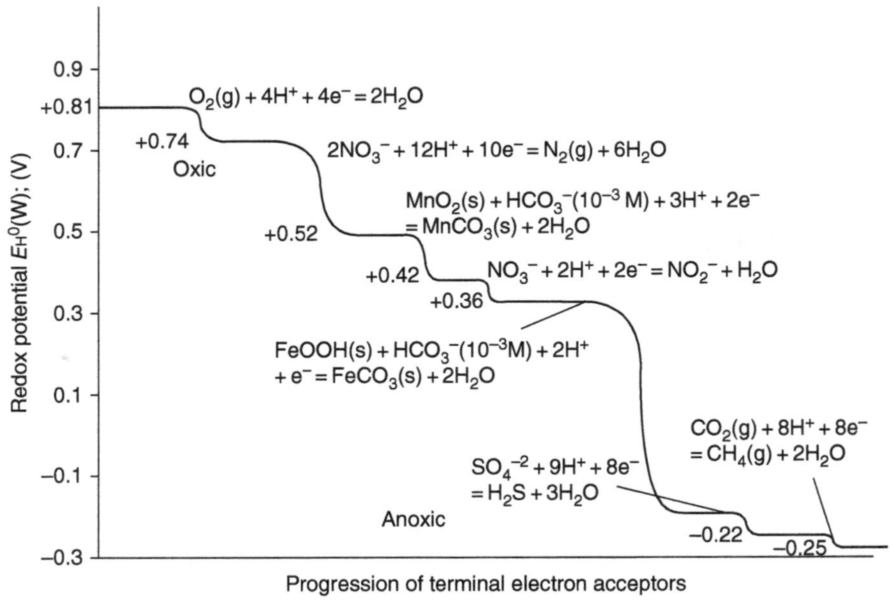

Figure 2.12 The electrochemical potential of water as a function of the terminal electron acceptor being used. Source: Data from Schwarzenbach et al. 1993.

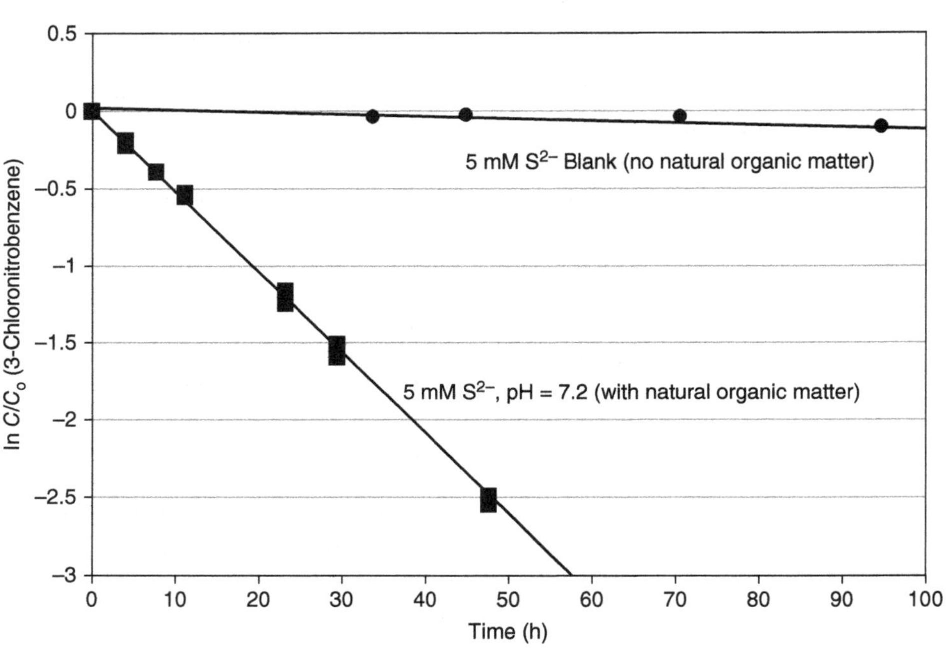

Figure 2.13 Reduction of 3-chloronitrobenzene at a pH of 7.2 and E_H of −0.207 V. Source: Dunnivant et al. (1992). Copyright 1992. Reprinted with permission of American Chemical Society.

by Dunnivant et al. (1992). When organic matter is present in groundwater, the water may become anaerobic from high microbial activity, which will lower the E_H of the water to reducing conditions. Many organic chemicals can be transformed (degraded) to their reduced counterparts by biotic (involving microorganisms) and abiotic (not directly involving microorganisms) reactions under these conditions. Figure 2.13 shows such a reduction reaction for 3-chloronitrobenzene at a pH of 7.2 and E_H of −0.207 V. Note the exponential shape of the plot (concentrations are given in natural log units), which follows first-order kinetics (discussed in Chapter 3). Reduction products 3-chlorophenylhydroxylamine (3-ClPhA) and 3-chloroaniline (3-ClAn) are formed and increase in concentration as 3-chloronitrobenzene is removed from the system. This reduction follows the expected reaction path shown in Figure 2.14. 3-Chloronitrosobenzene, the second degradation product in Figure 2.14, is not shown in Figure 2.15 since it is so reactive that no concentration builds up during the experiment (nitrosobenzene is immediately converted to 3-chlorophenylhydroxylamine). Thus, not only does this redox reaction change the chemical form of the pollutant (a nitrobenzene), but it creates a more toxic form of the pollutant (an aniline).

But what are the effects of changing the pH and E_H of the system? These effects are shown in Figures 2.16 and 2.17, respectively. Increasing the pH and decreasing the

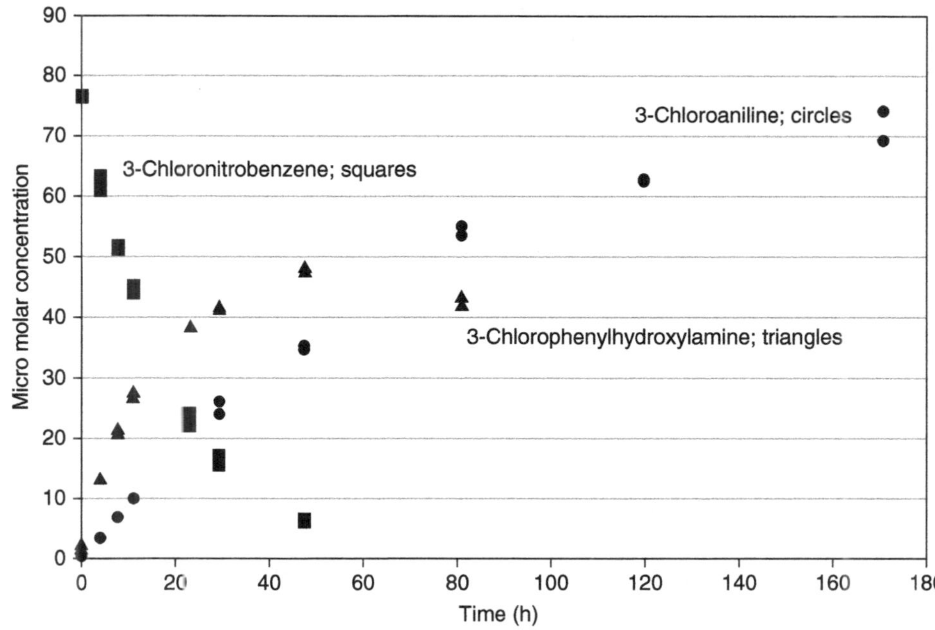

Figure 2.14 Reduction products of 3-chloronitrobenzene. Source: Dunnivant et al. (1992). Copyright 1992. Reprinted with permission of American Chemical Society.

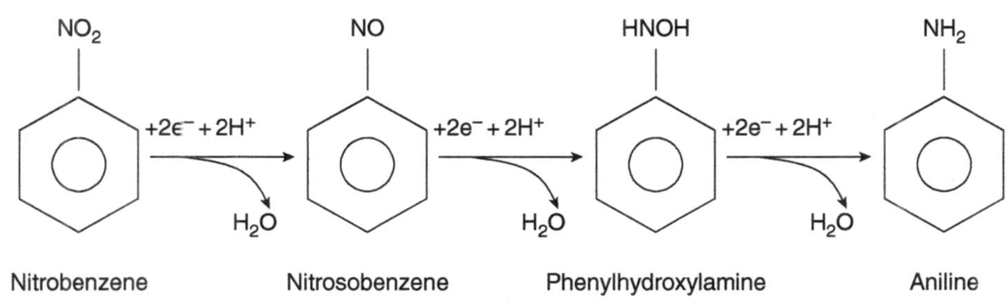

Figure 2.15 Reduction pathway of 3-chloronitrobenzene. Source: Dunnivant et al. (1992). Copyright 1992. Reprinted with permission of American Chemical Society.

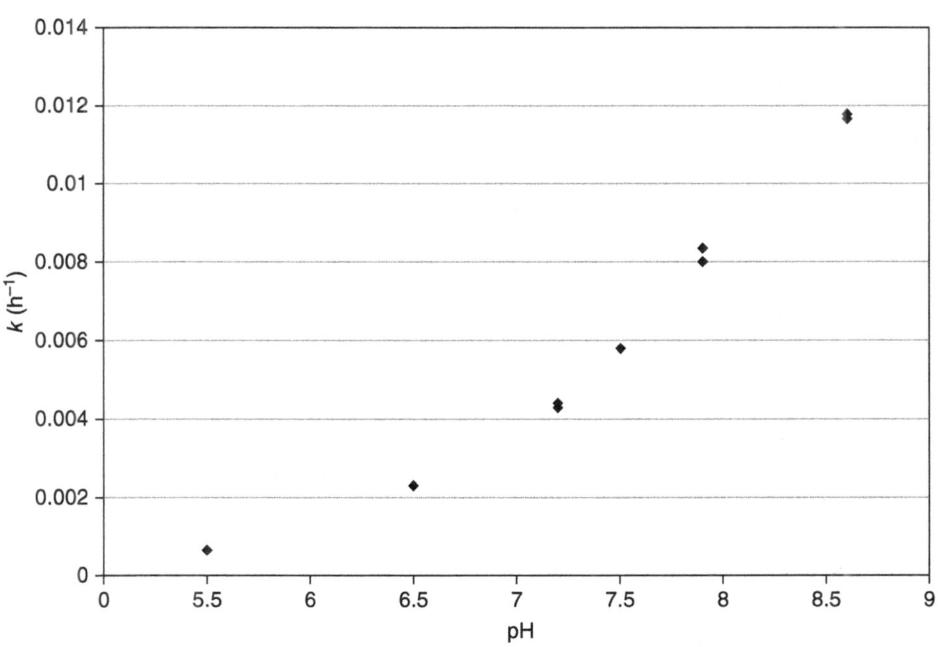

Figure 2.16 The reduction of 3-chloronitrobenzene as a function of pH. Source: Dunnivant et al. (1992). Copyright 1992. Reprinted with permission of American Chemical Society.

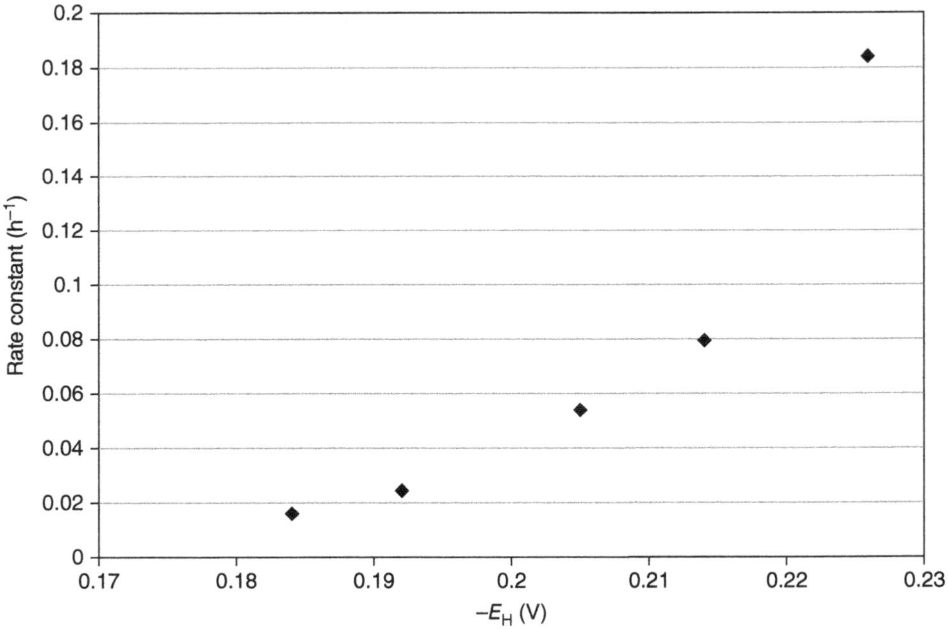

Figure 2.17 The reduction of 3-chloronitrobenzene as a function of E_H. Source: Dunnivant et al. (1992). Copyright 1992. Reprinted with permission of American Chemical Society.

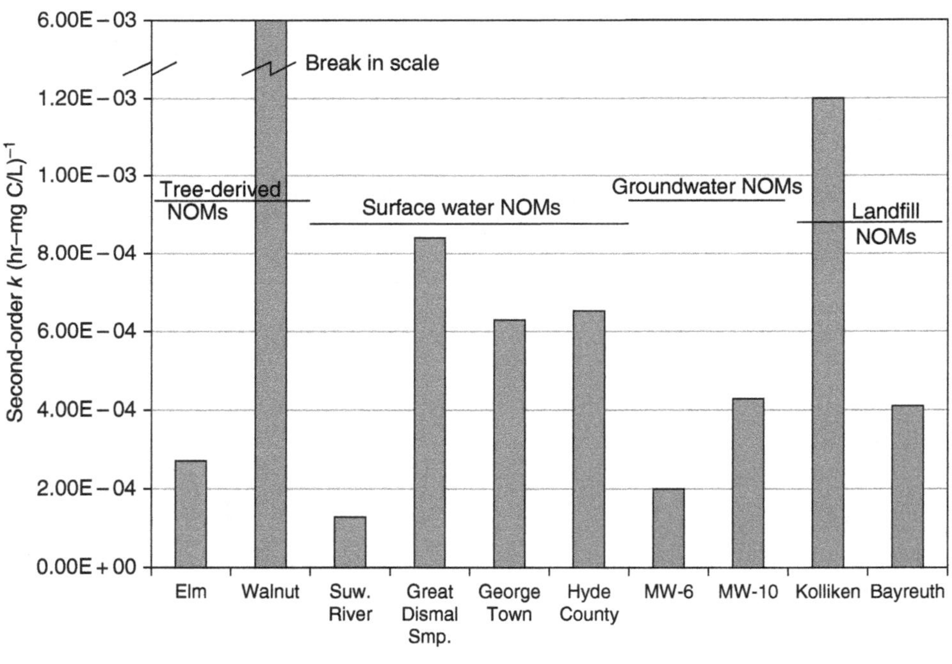

Figure 2.18 Comparison of second-order rate constants for a variety of NOM samples. Source: Dunnivant et al. (1992). Copyright 1992. Reprinted with permission of American Chemical Society.

E_H of the system increases the rate of reaction (k_{NOM}) in an exponential manner. An increasing negative E_H represents increasing reducing conditions (potential). In many of our transport models discussed in Chapters 4–10, we will include a degradation term; therefore, it is important to carefully characterize your system with respect to all chemical parameters that will affect the reaction rate. Another important parameter to characterize is the concentration of the chemical agent involved in the reduction reaction. In the experiments discussed

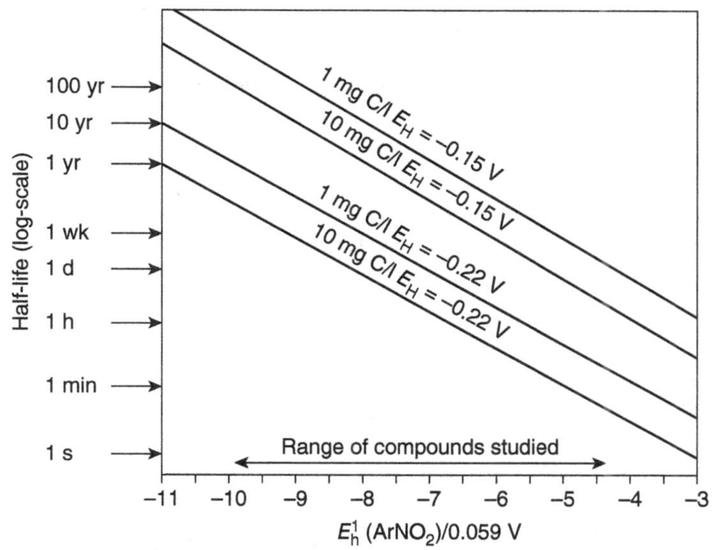

Figure 2.19 A summary of the effects of NOM concentration and E_H on the reduction of 3-chloronitrobenzene. Source: Dunnivant et al. (1992). Copyright 1992. Reprinted with permission of American Chemical Society.

in Figures 2.13–2.17, the active reducing agent is NOM. But how do the reactivities of different types of organic matters differ? This is illustrated in Figure 2.18. Surprisingly, the reactivities of the 10 NOMs shown, as judged by second-order reaction rate, are very similar and vary only over a factor of approximately 15. This indicates that most organic matter will reduce nitrobenzenes under anaerobic conditions, with the reaction rate highly dependent on E_H (Figure 2.17). Figure 2.19 summarizes the reduction of several substituted nitrobenzenes as a function of NOM concentration (in milligram per liter) and E_H of the system. The E_H value on the x-axis is the E_H associated with the transfer of the first electron to the substituted nitrobenzene molecule (there are six electrons required in the complete reduction reaction). This study shows the importance of carefully evaluating each chemical parameter in determining the fate and transport of a reactive pollutant, although rarely can such a complete characterization of a system be conducted.

2.6 Complexation

Chemical speciation, or the way in which chemicals distribute themselves between different forms and phases, is central to studies of fate and transport. In order to understand the behavior of a chemical in the environment, we need to know its physical and chemical properties, as determined by the form in which it exists. In order to be held in solution, molecules of a dissolved species must be subject to interactions with other molecules of solvent or other solutes. Most free metal ions in aqueous solution exist as hydrated or aquated cations (e.g. Ca^{2+}–$6H_2O$). The exact number of hydration waters varies for different metals, but four to six waters of hydration are the most common.

A complex, or coordination compound, is a dissolved species formed from two or more simpler species (for example Pb^{2+} reacting with Cl^- to form $PbCl^+$). Complexes consist of the central metal cation and a ligand or anion. Depending on the combination of cations and ligands, the complex can have an overall positive, neutral, or negative charge. The number of ligands (or water molecules) that surround a metal cation is referred to as the coordination number. Ligands come in a variety of forms. Basically, a ligand is any chemical species that can complex the metal cation. A few common examples relevant to environmental chemistry include Cl^-, OH^-, CO_3^{2-}, HCO_3^-, HPO_4^{2-}, and $H_2PO_4^-$, as well as naturally occurring organic compounds such as amino acids, humic and fulvic acids, and synthetic organic compounds such as ethylenediaminetetraacetic acid (EDTA) and nitrilotriacetic acid (NTA). The charge and shape of the ligand determines how it will complex the metal cation. For example, the simple anions only have one binding site and are referred to as unidentate. Other ligands, such as EDTA and NTA, have multiple charges and molecular shapes that allow more than one site on the ligand to bind with the metals. These are referred to as multidentate or chelating agents.

Complexation can occur in two ways, as inner-sphere or outer-sphere complexes. An outer-sphere complex consists of a hydrated metal cation associated with (bonded to) a ligand but retaining its waters of hydration. An inner-sphere complex results from replacement of hydration waters with the ligand. The bonding forces of inner-sphere complexes are more polar-covalent in nature and are stronger than the bonds present in outer-sphere complexes. In general, cations with greater charge and lesser radius form more inner-sphere-type complexes with a given ligand. Another way of phrasing this is to say "the higher the surface charge density, the more inner-sphere the complex." Outer-sphere complexes are frequently formed by metals, such as Na^+, K^+, Ca^{2+}, Mg^{2+}, and Sr^{2+} with ligands such as Cl^-, HCO_3^-, SO_4^{2-}, and CO_3^{2-}. Inner-sphere complexes are frequently formed by metals such as Ag^+, Cd^{2+}, Zn^{2+}, and Hg^{2+} with ligands such as S^{2-} and SH^-. We will return to complexation in Section 3.2.2, where we will work equilibrium problems associated with complex solutions.

2.7 Equilibrium Sorption Phenomena

We have already discussed several equilibrium processes, and these are summarized in Table 2.8 along with some new ones for this section. Most chemists, especially those dealing with pollutant fate and transport phenomena, prefer to work with a system at equilibrium. This makes the mathematical expressions much simpler and allows us to ignore many poorly understood kinetic processes. In this section, we will look at pollutant sorption (attraction) phenomena between the aqueous phase and other phases present in natural water. We will use several terms to describe these processes, but in general they all mean the same thing. For example, when metals associate with particles in water, it is usually through an ion exchange mechanism on the surface of the particle and it is technically correct to use the term "adsorption" to describe the process. However, when hydrophobic pollutants associate with particles, it is more of a solvation process, since the interaction is not site- or charge-specific, and we use the term "partitioning." Some researchers use the term "sorption" to include both processes, since some organic compounds are slightly polar in nature and the associated process can be a mixture of partitioning and adsorption. You should try to keep these terms straight, since some researchers become very agitated when the terms are used incorrectly.

2.7.1 Sorption Surfaces

2.7.1.1 Clays

There are a variety of surfaces in natural waters, including inorganic and organic colloids (mineral phases and humin – one form of NOM). Colloidal particles are defined as very small particles that do not settle out of the solution during the time scale of interest, which can range from hours to decades in length. Colloidal particles can consist of very small inorganic minerals, NOM (generally classified as dissolved), or a combination of both. In addition, the mineral phases can be coated with precipitants such as iron or manganese oxides and hydroxides. Due to small particles' high ratio of surface area to volume, they can account for a large amount of adsorbed and/or partitioned mass of the pollutant in solution. NOM was mentioned at the end of Section 2.4.3, where representative structures were

Table 2.8 A summary of important equilibrium constants used in environmental chemistry.

K	Name	Mathematical expression	Description
K_{eq}	Typical equilibrium constant	$K_a = \frac{[A^-][H^+]}{[HA]}$	A common expression used to represent the equilibrium concentration in a chemical reaction
K_H	Henry's law constant	$\frac{\text{Partial pressure of } x}{\text{Water concentration of } x \text{ (M)}}$	Used to describe the equilibrium partitioning of a pollutant between the gas and liquid phase (units of atm l/mol)
K_{DOM}	Organic matter–water partition coefficient	$\frac{\text{Concentration of } x \text{ in DOM}}{\text{Concentration of } x \text{ in water}}$	Used to describe the partition or binding of pollutants between dissolved organic matter and water (l/kg)
K_{ow}	Octanol–water coefficient	$\frac{\text{Concentration of } x \text{ in octanol}}{\text{Concentration of } x \text{ in water}}$	Used to model the partitioning of pollutants in biota (bioconcentration) (unitless)
K_d	Sediment–water distribution coefficient	$\frac{\text{Concentration of } x \text{ on sediment (mg/kg)}}{\text{Concentration of } x \text{ in water (mg/l)}}$	Used to show how pollutants adsorb to a solid phase (units of l/kg)
K_p	Sediment–water partition coefficient	$\frac{\text{Concentration of } x \text{ on sediment (mg/kg)}}{\text{Concentration of } x \text{ in water (mg/l)}}$	Used to show how pollutants partition onto or into another phase (i.e. organic matter on a sediment particle) (units of l/kg)
K_{oc}	Organic carbon–water partition coefficient corrected for the presence of organic matter	$\frac{K_d \text{ or } K_p}{\text{Fraction of organic matter}}$	Used to correct the K_d or K_p for the presence of organic matter coatings (units are the same as K_d or K_p)

given. NOM comes in a variety of sizes and forms, and there seem to be an endless number of ways to characterize and describe it (discussed later). For the moment, we will classify NOM as either dissolved, sorbed to particles (colloidal and large particles), or humin (insoluble chunks of NOM). There is no officially defined size of colloids, but environmental chemists usually filter natural water samples through a 0.20- to 0.45-μm filter and call every thing that passes through the filter "dissolved," including colloidal. The reader should be aware that there are other definitions of dissolved, but we will use this most common definition in our discussions.

If you have had a course in geology, you know that many minerals exist in nature. Soil scientists interested in sorption phenomena study many forms of particulate matter (those that do not pass a 0.45-μm filter), especially aluminum oxides, iron oxides and hydroxides, manganese oxides and hydroxides, and clay minerals. Clays and NOM coatings on inorganic particles are important in the adsorption phenomena of all pollutants, including metals and organics. Iron and manganese minerals and precipitants, meanwhile, are important primarily for sorption of metals and polar or ionic organics.

The term clay has two meanings: a clay particle is any particle smaller than 2 μm in size, regardless of composition, whereas a clay mineral is distinguished by its chemical composition and crystallographic structure. These two definitions tend to overlap, since most particles in the <2 μm fraction of most soils and sediments are some form of clay mineral. Examples of common clay minerals in nature are shown in Table 2.9, along with chemical relevant properties such as surface area and surface charge.

We will discuss two clay mineral phases, kaolinite and montmorillonite, shown in Figures 2.20 and 2.21, respectively. We will begin with a summary and then elaborate on the terms used. Clay minerals are phyllosilicates, hydrous aluminum silicate sheet structures. Clays minerals are composed of alternating sheets of silicon in tetrahedral coordination with oxygen, and aluminum in octahedral coordination with oxygen. Kaolinite is composed of one-to-one (1,1) layers, each composed of one of each kind of sheet, with a chemical structure of $Al_2O_3{-}2SiO_2{-}2H_2O$. Montmorillonite is a 2 : 1 layered clay (each octahedral sheet is bounded by two tetrahedral sheets, to form a layer), with a chemical structure of $Na_2O{-}7Al_2O_3{-}22SiO_2{-}nH_2O$ or $CaO{-}7Al_2O_3{-}22SiO_2{-}nH_2O$.

In order to understand how and why adsorption of metals to clays occurs, we must further expand on clays' chemical structure and three-dimensional shape. Clays have a characteristic structure of layers composed of two alternating types of sheets. One sheet consists of Al^{3+}, O^{2-}, and OH^- ions, where the negative ions form an octahedral structure around the Al^{3+}. The relative numbers of Al^{3+}, O^{-2}, and OH^- must satisfy the valences of the entire continuous structure in two dimensions.

Table 2.9 Properties of representative clay minerals from each main group.

Common mineral name	Chemical formula	Surface area (m^2/g)	Structural charge (cmol/kg)
1 : 1 Dioctahedral classification			
Kaolinite	$Al_2Si_2O_5(OH)_4$	7–30	2–15
1 : 1 Trioctahedral classification			
Serpentine	$(Mg,Fe)_3Si_2O_5(OH)_4$	560	<1
2 : 1 Dioctahedral classification			
Nontronite	$(CaO_{0.5}, Na)_{0.3}Fe_2^{3+}(Si, Al)_4O_{10}(OH)_2 \cdot nH_2O$	477	No data
Montmorillonite	$(Na,Ca)_{0.33}(Al,Mg)_2(Si_4O_{10})(OH)_2{\cdot}nH_2O$	611–749	80–150
Muscovite	$KAl_2(AlSi_3O_{10})(F,OH)_2$	60–80	10–40
Pyrophyllite	$Al_2Si_4O_{10}(OH)_2$	65–80	No data
2 : 1 Trioctahedral classification			
Talc	$Mg_3Si_4O_{10}(OH)_2$	1–80	<1
Vermiculite	$(Mg,Fe^{+2},Fe^{+3})_3[(Al,Si)_4O_{10}](OH)_2{\cdot}4H_2O$	550–700	10–180
2 : 1 : 1 Tetrahedral–octahedral–tetrahedral classification		70–100	15–40
Chlorite group	$(Mg,Fe)_3(Si,Al)_4O_{10}(OH)_2{\cdot}(Mg,Fe)_3(OH)_6$	25–150	10–40

Source: Data from https://www.srs.fs.usda.gov/pubs/ja/ja_barton002.pdf; https://www.srs.fs.usda.gov/pubs/ja/ja_barton002.pdf; https://theses.lib.vt.edu/theses/available/etd-11182004-120456/unrestricted/chapter1.pdf; Kampf et al. (2000) and Sparks (2003).

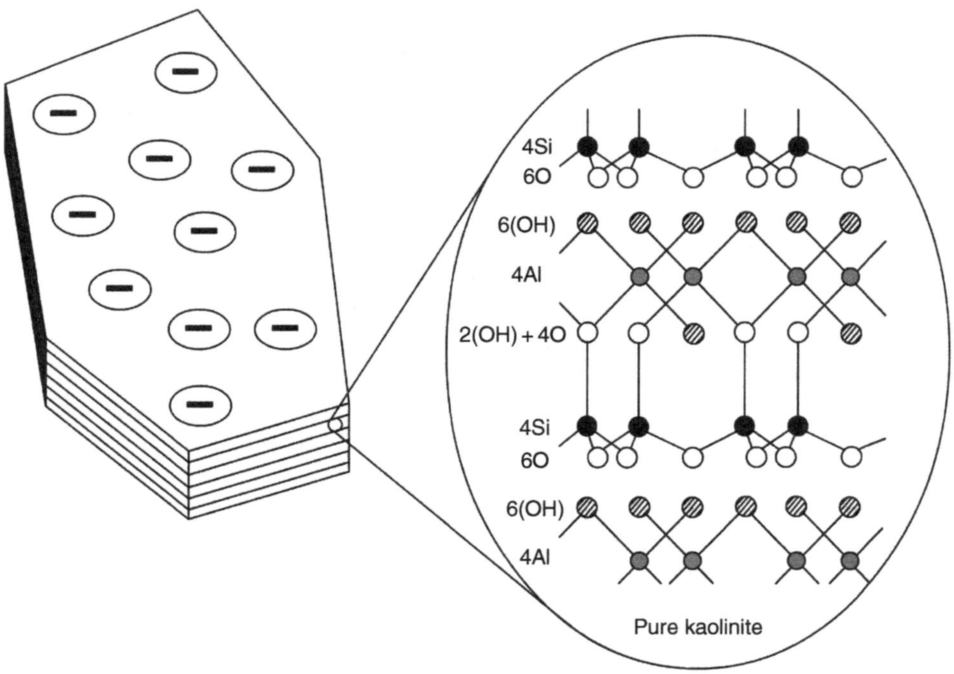

Figure 2.20 Structure of kaolinite.

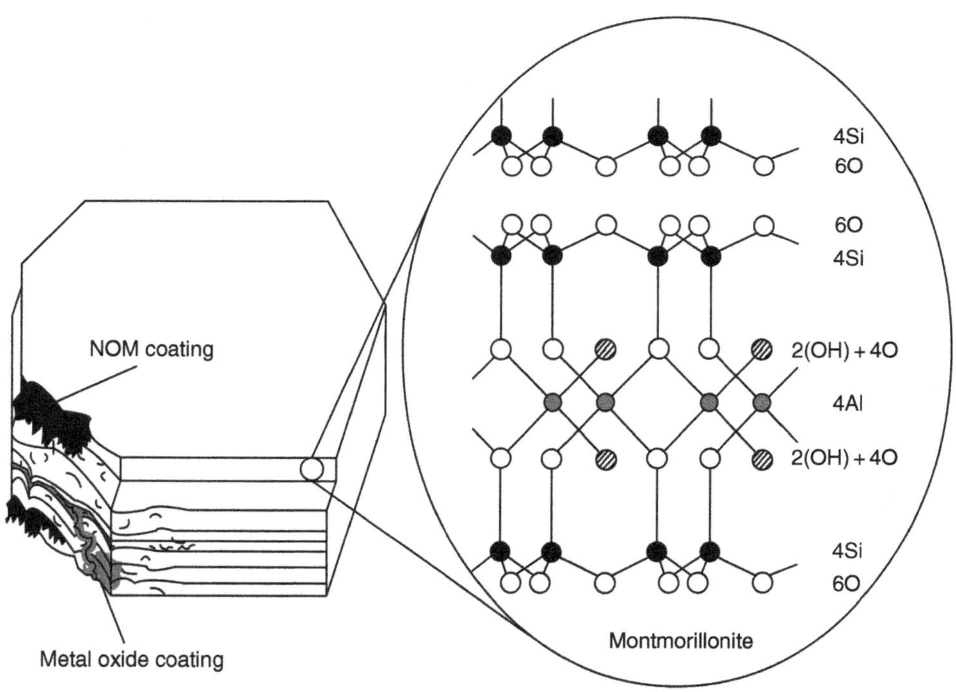

Figure 2.21 Structure of montmorillonite.

This sheet is commonly referred to as the gibbsite sheet or octahedral sheet, since it has the same general chemical formula [$Al_2(OH)_6$] as the mineral gibbsite. The second type of sheet is composed of Si^{4+}, O^{2-}, and OH^- ions. The Si^{4+} ion forms the center of a tetrahedron of oxygen atoms, while the bases of the tetrahedrons form hexagonal rings. This sheet is referred to as the silica sheet or tetrahedral sheet of the clay structure.

Clay structures consist of layers composed of various combinations of octahedral and tetrahedral sheets. The simplest combination, one of each sheet, forms a kaolinite clay. Each octahedral sheet is linked to one tetrahedral sheet through the sharing of oxygens by the Si and Al atoms. This results in the structures shown in Figure 2.20 for kaolinite. Clays with this structure are referred to as 1 : 1 type clays, since for every octahedral sheet there is one tetrahedral sheet. Note how combinations of O^{2-} and OH^- are used to satisfy the valence charge and result in a neutral structure. The resulting clay crystal is built up with a succession of the 1 : 1 (gibbsite–silicate) layers, one on top of another. Successive layers of kaolinite are relatively difficult to separate because of the hydrogen bonding between gibbsite and silicate sheets in adjacent 1 : 1 layers. The hydrogen in the OH^- from the gibbsite sheet is bound to the O^{2-} in the silicate sheet of the adjacent layer. This rigid structure will become important in Section 2.7.2, when we discuss isomorphic substitution and surface charge.

Another common form of clay is the 2 : 1 structure, where an octahedral sheet is sandwiched between two tetrahedral sheets. An idealized clay of this type is montmorillonite, illustrated in Figure 2.21. These 2 : 1 clays have very interesting properties with respect to absorption phenomena. For example, each 2 : 1 layer is rigidly held together, but adjacent 2 : 1 layers may be loosely held together depending on the chemicals or ions that are present in the interstitial area (the space between adjacent 2 : 1 layers). In the absence of interstitial ions, dry montmorillonite layers are held together by a combination of electrostatic forces (resulting from isomorphic substitution discussed later) and van der Walls dispersion forces (between adjacent O^{2-} groups from each layer) (Sposito 1984). The interstitial spaces between adjacent layers in 2 : 1 clays can be expanded in the presence of water (or other solvents), as water molecules commonly occupy these interstitial areas. Water acts to hold the layers loosely together through hydrogen bonding or by the presence of hydrated ions that may be present in the interstitial space. In contrast, adjacent kaolinite layers are held so tightly together that essentially no ions can migrate between the layers. The importance of the expandable nature of 2 : 1 clays is that it provides more surface area for diffusion and sorption of metal ions or organic pollutants.

There are many other possible configurations of gibbsite and silicate sheets, including the brittle platy minerals called micas. Muscovite is basically a montmorillonite layering structure with potassium between the two 2 : 1 layers. Potassium acts to collapse the 2 : 1 layers and holds the layers tightly together. Chlorite, another common mineral, is composed of a Mg–Al gibbsite-type layer sandwiched between two 2 : 1 layers.

2.7.1.2 Isomorphic Substitution

The chemical structure of clays lends itself to imperfections, and these imperfections are referred to as isomorphic substitutions. Essentially any cation with a coordination number of 4 or 6 can be substituted for Si^{4+} and Al^{3+} in individual sheets. Montmorillonite rarely, if ever, is found in the pure form. However, this substitution does not occur in kaolinite, which always has a chemical structure of $Al_4Si_4O_{10}(OH)_8$. For montmorillonite, the most common substitution is in the tetrahedral sheets, where Al^{3+} replaces Si^{4+}. Greater varieties of substitution occur in the octahedral sheet. Most commonly, the substitution of Al^{3+} for Si^{4+} and Mg^{2+} for Al^{3+} leaves a deficiency of positive charges in the montmorillonite layers. These substitutions occur in the crystal lattice when the clay forms, resulting in a permanent charge. The charge deficiency may be compensated for in a variety of ways: (i) by replacement of O^{2-} by OH^-, (ii) by introduction of excess cations into the octahedral sheet, which may have some of its cation sites unfilled, and (iii) by adsorption of cations onto the surface of individual layers. Although all of these may happen, the last will be our main focus, since it occurs after the clay is formed and can account for removal of metals from solution. Overall, isomorphic substitution results in permanent, nonspecific, diffuse charges that are spread across the clay surface.

2.7.1.3 The Inorganic Hydroxyl Group

In our previous discussions, we have presented the clay sheet as a continuous two-dimensional surface. However, clay particles are <2 μm in diameter; therefore, there will also be considerable clay edges present in a soil or sediment sample. The most common and reactive functional group in clays is the hydroxyl group that is exposed on the outer periphery of the clay on the truncated ends of the sheets. Two types of edges occur, with silanol groups originating from the silicon tetrahedral edge and aluminol groups originating from the aluminum octahedral edge. These are similar in reactivity and only differ in the fact that silanol groups do not form inner sphere complexes (discussed earlier). The charge of silanol and aluminol groups is highly pH dependent, so that these groups will in general be protonated at low pH values and deprotonated (anionic) at high pH values. This is important later when we discuss the adsorption of metal ions.

2.7.1.4 Measurement of Surface Charge

The actual charge of a soil/sediment suspension can be determined by a variety of experiments. The most common measure of charge is the point of zero charge (PZC), which is the pH value of a soil suspension at the point when the total net particle charge vanishes (Sposito 1984). This can be determined by titrating a sample and measuring the mobility of the particles under an applied voltage. Another measure of charge is the

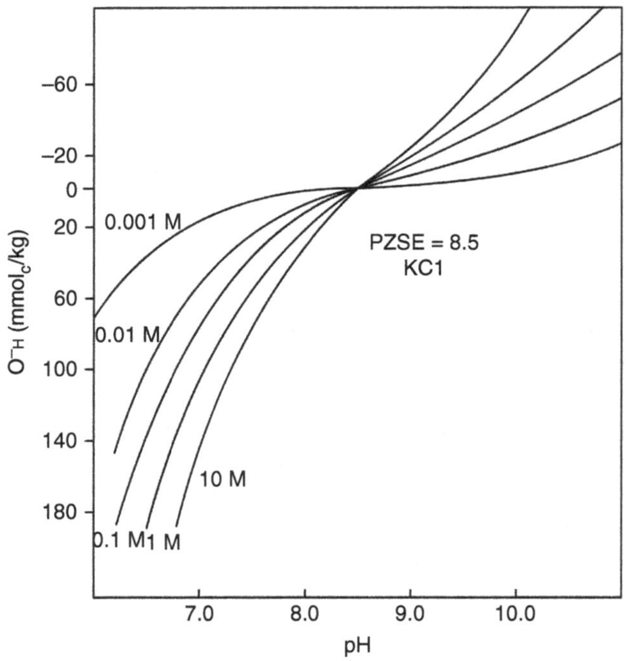

Figure 2.22 Hypothetical PZSE determination by a simulated titration.

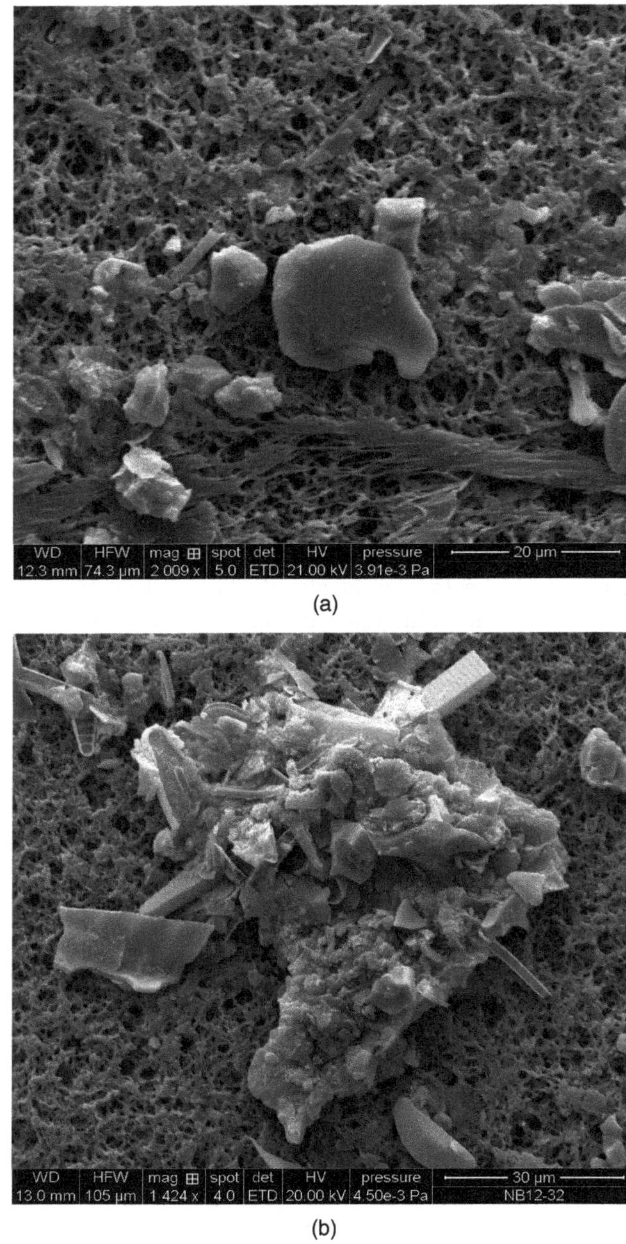

Figure 2.23 Scanning electron microscope image of an individual (a) and aggregated aluminosilicate (b) particle from a river sediment sample.

point of zero salt effect (PZSE). The PZSE is determined by locating the common point of intersection for several graphs of surface charge (a general plot of σ_H is shown in Figure 2.22) versus pH, each determined at a fixed ionic strength of the background electrolyte.

Factors Affecting Metal Sorption: Clearly both of these variables (pH and ionic strength) will affect adsorption of metal pollutants. For example, as the surface charge of a particle changes, by a change in either ionic strength or pH, the affinity of the surface for a metal pollutant will change. As the pH is increased, the particle surface becomes more anionic on average and absorbs more and more of the metal from solution. As we will see in Section 2.7.4, as more metal is adsorbed onto the solid, the observed K_d (the ratio of pollutant concentration on the clay to the water phase) will increase.

Other factors that affect the adsorption of metals include the oxidation state of the metal, composition of salts contributing to ionic strength, other surface-complexed cations, and the concentration of suspended solids. Summaries of these effects can be found in Ames and Rai (1978), Bell and Bates (1988), Looney et al. (1987), and Tichnor (1993).

2.7.1.5 Clay Particles in Nature

Clays in natural environments are rarely free of coatings; they are not "clean." They are normally coated with inorganic precipitates or organic molecules resulting from the degradation of plant and animal material (illustrated in Figure 2.21). An example of an individual clay particle and clay aggregate are shown in Figure 2.23. Both of these types of coatings will affect surface charge, sometimes imparting a charge of their own. This brings us to the next topic, a discussion of coating on mineral surfaces.

2.7.2 Organic Matter

One of the most important factors influencing sorption phenomena is the presence of organic matter in a sample. Virtually all samples have some organic matter present, but the type and concentration can vary dramatically. In principle, the sources of organic matter are obvious – any plant, animal, or excrement of these can be incorporated into water or soil sample. As you can imagine, the chemical variability of the resulting compounds is unlimited. However, on introduction into a natural system, they undergo complex microbial and abiotic transformations that produce a set of compounds generally referred to as fulvic, humic, and humin materials. For simplicity and consistency, the term "natural organic matter" will be used in this textbook to refer to any organic compound present in the sample. Compounds entering a natural system include proteins (polypeptides and nucleotides), lipids (fats, waxes, oils, and hydrocarbons), carbohydrates (cellulose, starch, hemicellulose, lignin), and porphyrins and plant pigments (chlorophyll, hemin, carotenes and xanthophylls) (Stumm and Morgan 1996). The products of microbial digestion and degradation of these compounds make up NOM. A typical NOM sample contains a mixture of "fresh" organic matter additions as well as "aged" organic matter. Thus, a single NOM sample will contain thousands to tens of thousands of chemically different structures. Generally, 20–30% of the compounds in a NOM sample can be identified by conventional means as protein-like materials, polysaccharides, fatty acids, and alkanes (Schnitzer 1986). The remaining 70–80% of the NOM consists of complex, altered residues of plants and animals. Molecular weights of NOM found "dissolved" in water range from 500 to 5000 amu (Thurman 1985).

There have been intensive efforts to characterize the structure of NOM. These efforts are summarized in Thurman (1985), Hayes et al. (1989), and Suffet and MacCarthy (1989). However, no one expects to establish a single structural formula to describe NOM. We have simply attempted to identify important functional groups, molecular sizes, and chemical properties for these compounds. One generic proposed structure for NOM was shown earlier in Figure 2.7 from Schulten and Schnitzer (1993). The key point of this figure is that there are numerous ionic sorption sites and hydrophobic centers in the large NOM molecules. These will be important when we discuss sorption of pollutants by NOM.

Other characterization attempts have concentrated on the chemical functionality of NOM. Researchers have devised chromatographic techniques to separate or fractionate NOM based on chemical properties, such as hydrophilicity and hydrophobicity (Leenheer 1980; Leenheer and Huffman 1976). These provide the environmental chemist with a means of characterizing different NOM molecules based on their reactivity, but you must realize that, as with most classifications, these may be rather arbitrary. A simple version of this classification identifies the functional groups observed in different NOM molecules. These groups are summarized in Table 2.10 (partially based on Killops and Killops 1993).

Another important way of characterizing NOM is by changing the pH of a water sample and observing the behavior of various NOM components (Hayes et al. 1989). Fulvic acids are the fraction of NOM that is soluble under all pH conditions. Humic acids are defined as the organic matter that is precipitated from an aqueous solution when the pH is decreased below 2. Given that the pH of most natural waters is between 5.5 and 9, both humic and fulvic acids will be present in most natural water samples. In contrast, humin is the fraction of NOM that is not soluble in water at any pH value. Thus, humin will be present in or associated with soil or sediment.

The most important characterization of NOM in water defines whether it is present in the dissolved form or sorbed to a solid. For this characterization, we use another operational definition. DOM is defined as the organic matter that will pass through a 0.45-μm filter (Gelman type A/E glass fiber filters are usually used for this distinction). The organic matter retained by the filter is considered to be in the particulate form, usually sorbed to inorganic particles. NOM can be attached to inorganic particles through a variety of binding mechanisms, including hydrogen bonding, van der Waals dispersion forces, cationic bridging, and hydrophobic effects. But where is the NOM spacially located on inorganic particles, in clumps or even coating all over the particle's surface area. The answer to this question is both as illustrated in Figure 2.24. The location of organic carbon is shown by white pixels, dispersed across the particles and in one large clump. This will become very important in Chapter 3 when desorption kinetics is discussed.

2.7.3 Sorbates

First, we will consider the adsorption of ionic pollutants, specifically metals. Metals, being cations in aqueous solutions, will be adsorbed to anionic sites (negative charges on a clay or deprotonated functional groups on the NOM). Solution conditions that favor the formation of negative sites will favor increased adsorption of metals. A good exercise at this point would be to return to Table 2.10 and Figure 2.7 and identify functional groups that may be important in attracting cationic pollutants.

Now consider the binding of organic pollutants, specifically nonpolar, nonionizable pollutants such as PCBs. These are commonly referred to as hydrophobic pollutants. These types of pollutants are not attracted to ionized functional groups – in fact, they are repelled

Table 2.10 Functional groups observed in different NOM molecules.

Symbol	Name	Resulting compound
ROH	Hydroxyl	Alcohol Phenol
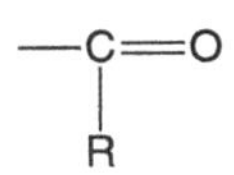	Carbonyl	Aldehyde Ketone Quinone
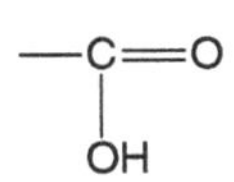	Carboxyl	Carboxylic acid
—O—	Oxo	Ether
$—NH_2$	Amino	Amine
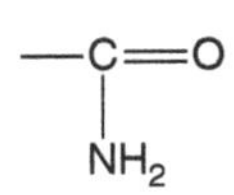	Amido	Amide
—SH	Thio	Thiol
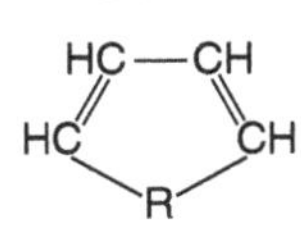	$R=CH_2$ indenyl R=O furanyl R=NH pyrryl R=S thiophenyl	Indene Furan Pyrrole Thiophene
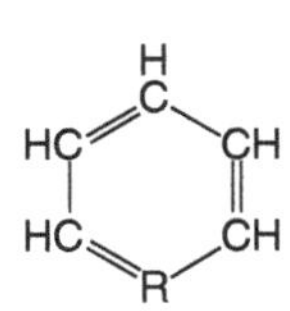	R=CH phenyl R=N pyridinyl	Benzene Pyridine
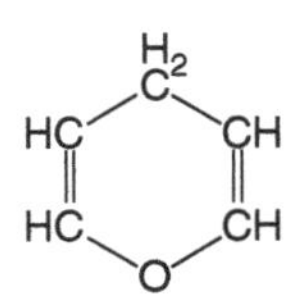	Pyranyl	Pyran
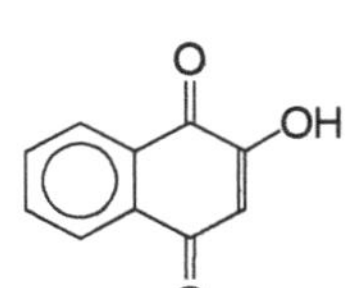	LAW	H_2LAW
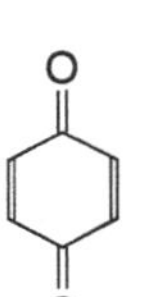	Quinone	Hydroquinone
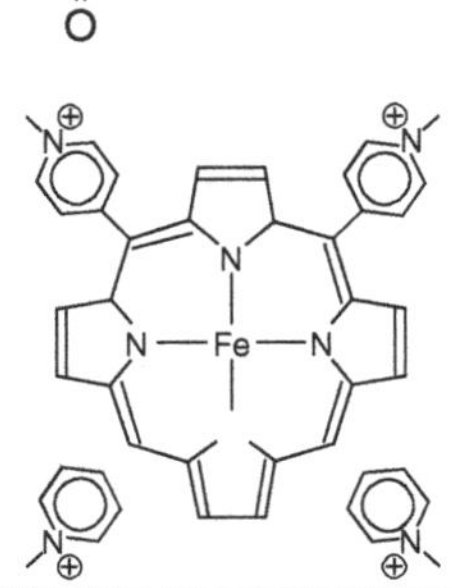	Iron(II) porphyrin	Iron(III) porphyrin

Figure 2.24 The location of NOM on sediment particles. The white pixels represent the presence of organic carbon; note individual and clumped carbon.

by these groups. Hydrophobic pollutants are attracted to hydrophobic centers in the NOM molecule and hydrophobic mineral surfaces. The intermolecular forces responsible for these attractions are van der Waals dispersion forces, which simply follow the old saying "like dissolves like" (e.g. hydrophobic liquid coatings dissolve hydrophobic pollutants). In this regard, the attraction is not really adsorption but is more like a solution or dissolving phenomenon. As noted earlier, environmental chemists regard this type of attraction as partitioning or sorption, where the pollutant dissolves or partitions into the hydrophobic center of the NOM. The difference between adsorption and sorption is the basis for environmental chemists using distribution coefficients (K_d) to describe the adsorption of metals and partition coefficients (K_p) to describe the partitioning of hydrophobic pollutants to environmental particles. As with the adsorption of metals, favored by conditions that create negative sites, sorption of hydrophobic pollutants is favored by conditions that promote the formation of hydrophobic centers in the NOM or coiling of the NOM.

2.7.4 Distribution and Partition Coefficients, K_d and K_p

One of the most important parameters that can determine the fate of a pollutant in an aqueous system, especially in rivers, lakes, and groundwater, is its distribution coefficient (K_d) or partition coefficient (K_p) between different media. These coefficients are a measure of how a pollutant distributes itself between the water phase and the particulate (or solid) phase. Pollutants on the solid phase are considerably less bioavailable and therefore less toxic. These sorbed pollutants can also settle out of solution in lakes or become immobile in groundwater and be effectively removed, at least temporally, from the system. An example of the buildup of pollutants in sediments is given below. We will present ways of calculating these coefficients in Chapter 3.

Distribution coefficients are concerned with adsorption, defined as the net accumulation of matter (pollutants) at the interface between a solid and a liquid. The matter (pollutant) that accumulates at the surface is referred to as the adsorbate. The solid surface on which the pollutant accumulates is the adsorbent. Partition coefficients, as we discussed earlier, are concerned with the partitioning and induced-dipole interactions between two nonpolar compounds (i.e. PCB and hydrophobic regions of NOM). Even though adsorption is not occurring in this process, the effects on the system are similar, and the terms adsorbate and adsorbent are still used.

As soil and sediments wash into lakes and streams, the particles aggregate and form larger particles that will settle in calm (quiescent) waters. When these particles have accumulated pollutants, the settling of these particles to the sediments can act as a removal mechanism (referred to as a sink) for pollutants. Over time, and when clean water and sediments return to the water body, the pollution will be buried by clean material and removed from interaction with the ecosystem. Such an example is shown in Figure 2.25 for PCBs in Lake Hartwell in South Carolina (USA). Note that as you move down into the sediment from the water–sediment interface, you are moving back in time. Figure 2.25a shows an area of the lake that is subject to considerable mixing and input of pollutants. This is indicated by the variable but high presence of PCBs in each section of the sediment column. In Figure 2.25b, the PCB-contaminated sediments start to be buried by cleaner, more recent deposited sediment. Figure 2.25c shows an even further burial of contaminated sediments. Thus, the accumulation of pollutants on soil–sediment particles is an important factor in fate and transport processes. We will return to distribution coefficients and partition coefficients in the lake, stream, and groundwater modeling chapters.

2.7.5 Ion Exchange Phenomena for Ionic Pollutants

Another way of representing the adsorption process is as an ion exchange reaction. Here, we visualize the negative surface or edge sites of a clay or environmental particle as being saturated, or nearly saturated, by native cations such as H^+, Na^+, and K^+. Cationic pollutants, such as heavy metals, generally have a higher charge (or charge density) and therefore a higher affinity for these sites and displace the native, readily exchangeable ions. Thus, we can model the process as ion exchange. There will be a finite number of sites for cations to adsorb to the surface, and the relative abundance of these sites on the particles of a soil or other material is expressed as the ion exchange capacity. A more formal definition is the number of moles of adsorbed ions that can be desorbed from a unit mass of solid under a given set of conditions (i.e. temperature, pressure, solution composition, solid–solution ratio, etc.). Soil scientists refer to this measure as the cation exchange capacity (CEC), which is defined as the concentration of sorbed cations that can be readily exchanged for other cations. The CEC is usually reported in units of meq/100 g soil, as an exchangeable charge per mass, and is an indication of a soil's ability to store nutrients or absorb metals.

As a general rule, the affinity of a soil or sediment for a metal cation will increase with the tendency of the cation to form inner-sphere complexes. For a series of uniform valence metals, this tendency is directly related to ionic radius, R, for two reasons:

(1) the ionic potential (z/R; charge/radius) decreases with increasing R, and
(2) a larger radius implies a greater tendency for the metal to polarize (distort) in response to an electric field (the surface charge of the soil particle).

Using these two guidelines, selectivity sequences for metal ions can be established based on ionic radii

$$Cs^+ > Rb^+ > K^+ > Na^+ > Li^+$$
$$Ba^{2+} > Sr^{2+} > Ca^{2+} > Mg^{2+}$$
$$Hg^{2+} > Cd^{2+} > Zn^{2+}$$

where adsorption of the ion increases from left to right. Unfortunately, ionic radius alone is not sufficient to predict selectivity for transition metals. Extensive experimentation has established the following order of selectivity (the Irving–Williams order) (Stumm and Morgan 1981)

$$Mn^{2+} < Fe^{2+} < Co^{2+} < Ni^{2+} < Cu^{2+}$$

Outer-sphere complexes, meanwhile, are responsible for the effect of pH on metal adsorption. As pH increases above the pH_{ZPC} (the pH where there is no surface charge), the net surface charge of the particle increases, thus increasing the electrostatic attraction of a mineral surface for the metal. Usually NOM is present to some degree and competes with the surface for the metal. If the NOM is present in the dissolved phase, the adsorption to the surface will decrease if the NOM has a greater affinity for the metal than does the surface. If the affinity of the NOM for the metal is less than that of the

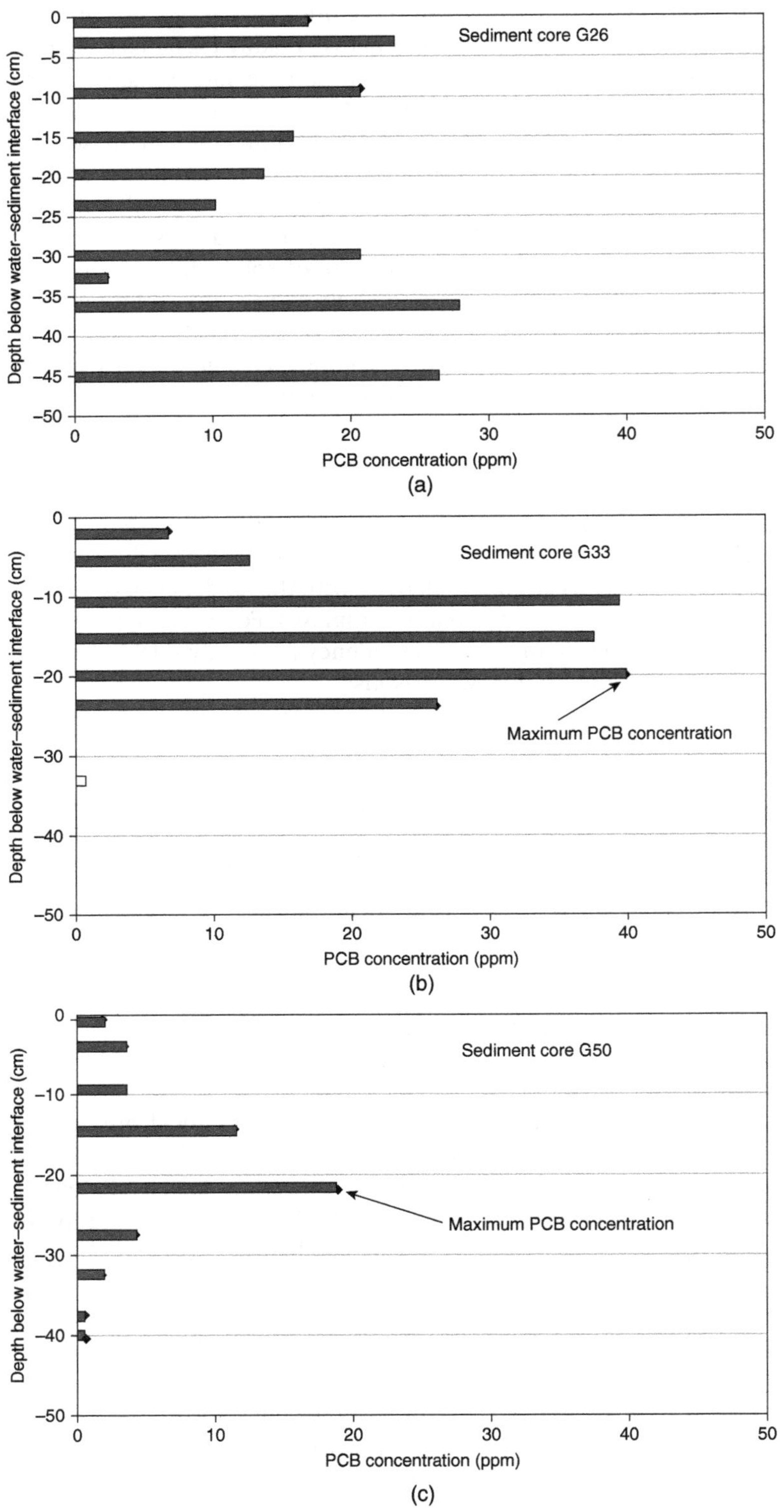

Figure 2.25 (a–c). PCB concentration as a function of sediment depth in Lake Hartwell sediments. Source: Data from Germann (1988).

surface, little or no change in the surface adsorption will occur due to the presence of NOM. This scenario will be complicated further if the NOM subsequently adsorbs to the mineral surface, a common phenomenon in nature.

2.8 Transformation/Degradation Reactions

In this section, we will discuss various ways that pollutants are removed from the system through transformation and degradation reactions. In general, these terms can be used interchangeably, but it should be emphasized that just because the original pollutant is removed from a system, the degradation process may not be beneficial to or safe for the environment. For example, many pollutants can be transformed or degraded to other or more toxic pollutants (e.g. the reduction of nitrobenzenes to anilines and the conversion of carbon tetrachloride to vinyl chloride). Transformations can be divided into abiotic (chemical reactions without the aid of microorganisms), photochemical, nuclear, and biological (microbial). Abiotic, photochemical, and biological transformations are generally more important for organic pollutants. A notable exception is the methylation of inorganic mercury.

2.8.1 Abiotic Chemical Transformations/Degradations

Abiotic chemical degradations refer to the removal of a pollutant exclusively through chemical reactions without the direct aid of "active" microorganisms. In other words, live, viable microorganisms do not use their internal enzymes in the process. Abiotic reactions can, however, include indirect effects of microbial activity, which may regulate the pH or E_H of the environment and thus allow or control the presence of abiotic oxidizing or reducing agents. The abiotic nature of these oxidizing and reducing agents is complicated because once a microbial cell, animal material, or plant material dies, intracellular chemicals (such as enzymes) are released and incorporated into NOM. As you can see, this is a gray area and subject to interpretation. There is a general consensus among chemists that if a degradation process can occur in a system that has been sterilized (by filtration, irradiation, autoclaving, etc.), it is abiotic.

The large variety of abiotic reactions, summarized in Table 2.11, include nucleophilic substitution (dehalogenation), elimination, dealkylation, sulfoxide and sulfone reduction, heterocyclic ring cleavage, hydrolysis, oxidation, nitro reduction, and azo reduction.

Many of these reactions, such as nucleophilic substitution, elimination, hydrolysis, and dealkylation, are detailed in organic chemistry textbooks and occur in aqueous environments. The general nature of abiotic reduction reactions is similar to that of reactions performed by microorganisms and is illustrated in Figure 2.26. Here, a bulk electron donor (such as sulfide) is supplied by the environment (abiotic or biotic) and reduces an abiotic electron mediator, which in turn reacts with the pollutant. In most cases, the bulk electron donor is present in excess and reacts slowly or not at all with the pollutant. But what are the abiotic electron mediators? A number have been found to exist, including iron and manganese minerals and a variety of functional groups and centers in NOM, including bound transition metals, quinines, and porphyrin structures.

In general, abiotic chemical transformations/degradations are easier to study than biochemical reactions involving microorganisms. This is because chemists can eliminate microbiology in their experiments through sterilization, and then may easily control system-specific pH, E_H, and chemical concentrations in the laboratory. In addition, chemical laboratory studies are more transferable between the laboratory and the environment, since we can measure the chemical conditions of a system and reproduce them in the laboratory. Biologists, on the other hand, have difficulty selectively growing microorganisms in the field and relating rates of biochemical reactions from the laboratory to the field. Abiotic reactions are also thought to be especially important for the degradation of some pollutants by nucleophilic substitution, elimination, and hydrolysis, and reduction reactions can be important in certain environments (in anaerobic lakes, sediments, and aquifer systems).

2.8.2 Photochemical Transformation/Degradation Reactions

No discussion of environmental chemistry or fate and transport would be complete without specifically considering photochemical reactions, a special form of abiotic reactions that break apart organic molecules. These reactions occur both in the atmosphere and in surface waters. In aqueous systems, these enter into fate and transport reactions as first-order or

Table 2.11 Abiotic reactions.

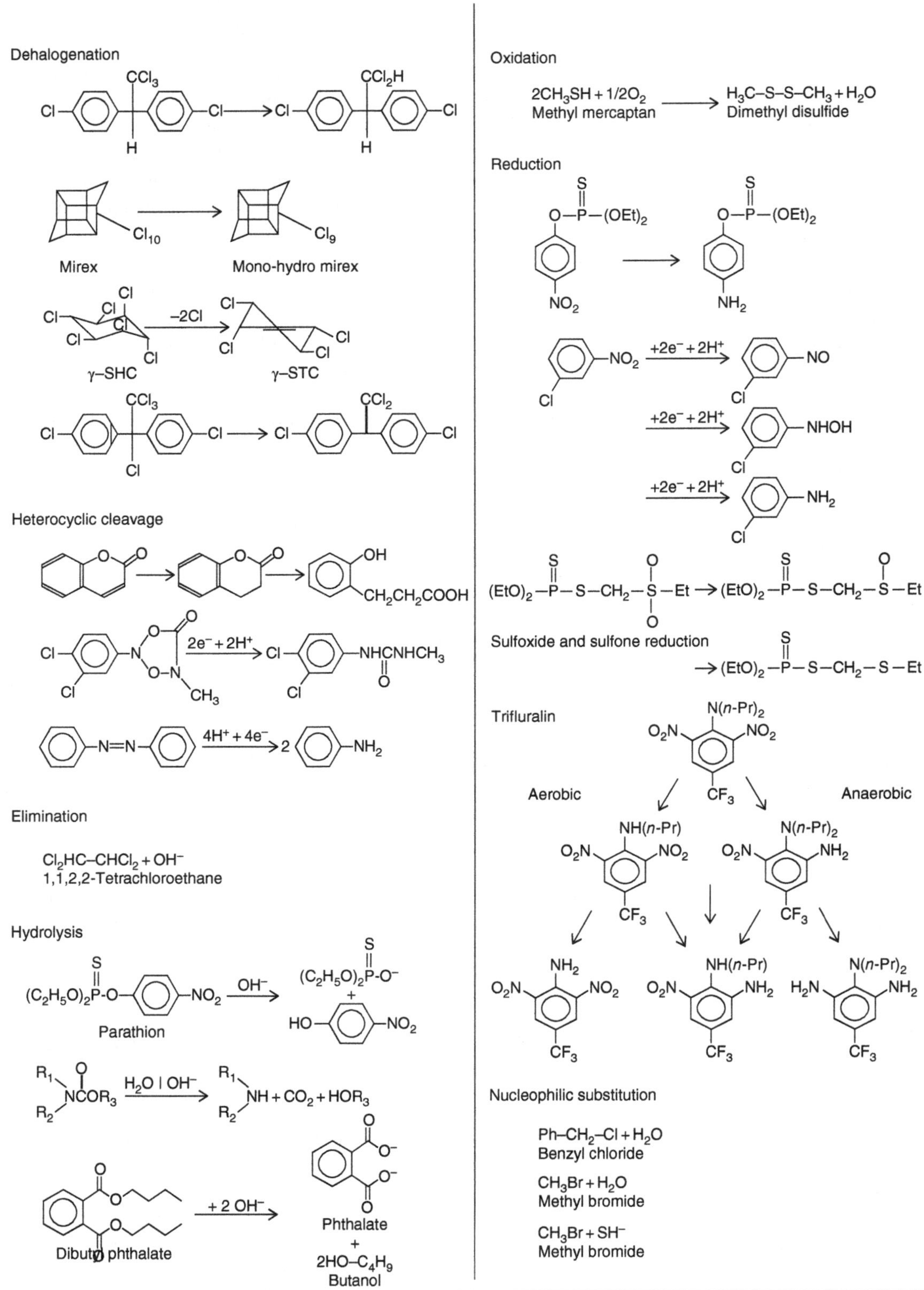

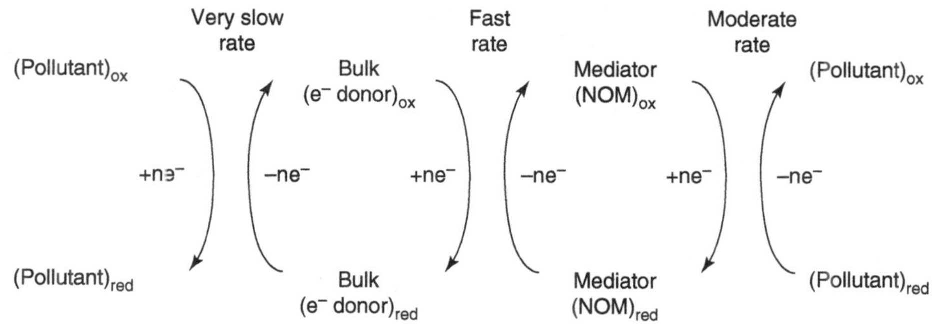

Figure 2.26 Generalized reaction scheme for the reduction of a pollutant by abiotic reactions. Source: Dunnivant et al. (1992). Copyright 1992. Reprinted with permission of American Chemical Society.

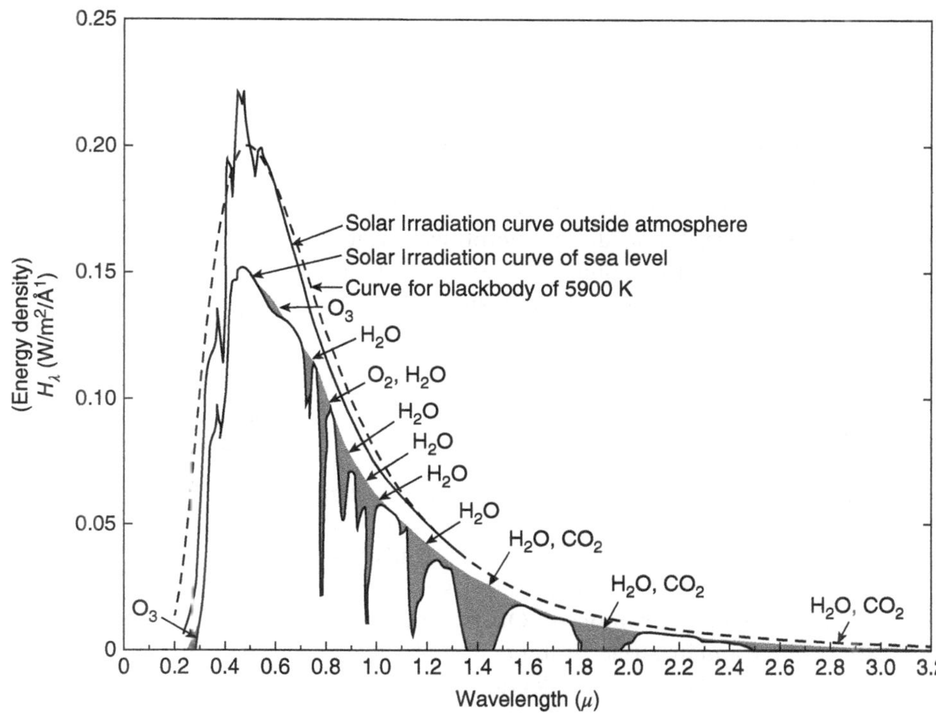

Figure 2.27 The distribution of electromagnetic radiation (EMR) at the Earth's surface. Source: From Jursa (1985).

pseudo-first-order reaction rates, which are dependent on the concentration of one reactant or one reaction parameter (reaction kinetics are discussed further in Chapter 3). In atmospheric systems, second-order reaction kinetics are used. In order to understand photochemical reactions, we must first review the availability of wavelengths present in the atmosphere at the Earth's surface and the energy associated with these wavelengths.

Figure 2.27 shows the distribution of the spectrum of electromagnetic radiation at the Earth's surface. The top solid line shows the distribution of wavelengths entering the Earth's atmosphere, while the attenuated (lower) solid line shows the distribution at the Earth's land surface. This lower line is more important to our discussions. Radiation from ultraviolet (200 nm, 0.2 μm in Figure 2.25) to visible (750 nm or 0.75 μm) wavelengths can be involved in photochemical reactions. An easy calculation from general chemistry relates wavelength of a photon to energy of a mole of these photons. This can be achieved using the equations:

$$E = h\nu = h\,c/\lambda$$
$$= 6.022 \times 10^{23} h\,c/\lambda \text{ for a mole of photons}$$

where h is Planck's constant (6.63×10^{-34} J s), ν is the frequency of light (s^{-1}), c is the speed of light in a vacuum (2.99×10^8 m/s), and λ is the wavelength (m).

Table 2.12 relates energy of some single bonds to the corresponding wavelengths of light (i.e. the wavelength needed to break the bond). As seen from these calculations, UV and visible wavelengths possess sufficient energy to break covalent bonds in many molecules, but

Table 2.12 Relationship between bond energy and minimum wavelength of light required to break the bond.

Bond	Bond energy (kJ/mol)	Wavelength λ (nm)
O—H	465	257
C—H	415	288
N—H	390	307
C—O	360	332
C—C	348	344
C—Cl	339	353

infrared radiation possesses only enough energy to cause an increase in bond vibration and does not break chemical bonds. A photochemical reaction in which photons are absorbed by a pollutant molecule and the molecule is broken apart is referred to as direct photolysis. But just because a molecule absorbs a photon of sufficient energy does not mean that it will be destroyed, since there are a number of ways the molecule can dissipate the energy. Thus, the quantum yield (the number of photons actually producing a photochemical reaction) of the absorbed photons must also be considered.

Other indirect types of photolysis can also occur. For example, an atmospheric molecule such as O_2 can absorb energy from a photon, become excited, and transfer the energy to a pollutant molecule by collisions. If the transfer of energy is sufficient to break a bond in the pollutant, destruction can occur. Other compounds in the atmosphere and in surface water can be photosensitized (made more likely to photodegrade a pollutant), which may lead to the formation of free radicals that can subsequently react with pollutant molecules. An important radical formation in the atmosphere is the reaction of an excited oxygen atom with water to form the hydroxyl radical (sometimes called the vacuum cleaner of the atmosphere):

$$O_3 \Rightarrow O_2 + O^*$$
$$O^* + H_2O \Rightarrow 2OH^*$$

Another, less common, reaction also produces these radicals:

$$H_2O + h\nu \Rightarrow HO^* + H$$

The OH^* radical is very reactive, both in the formation of smog and in the destruction of atmospheric pollutants. An important creator of free radicals in surface water is NOM, which can form singlet oxygen, a strong oxidizer of pollutant molecules.

While photochemical reactions are very important in the formation of smog, and in some cases are important removal mechanisms of atmospheric pollutants, the atmospheric fate and transport models we will use in the textbook do not allow for the inclusion of degradation terms. This is because the atmospheric fate and transport models are only valid on short time and distance scales, before most photochemical degradation reactions have time to become important. Photochemical reactions can, and will, be included as first-order degradation terms in lake and river transport models.

2.8.3 Nuclear Reactions

Nuclear reactions are often one of the least studied types of reactions by undergraduates, but are perhaps the easiest to understand. The three most common and important types of nuclear decay involve the release of alpha (α), beta (β), or gamma (γ) forms of radiation. These are summarized in Table 2.13. α particles consist of ejected helium nuclei (two protons and two neutrons). These particles have the greatest mass of radioactive emissions, but cannot easily penetrate substances. β radiation consists of high-speed electrons emitted from an unstable nucleus and have intermediate penetrating power. γ radiation consists of high-energy photons and has the greatest penetrating power.

All known radioactive decay reactions follow simple first-order kinetics. Half-lives range from fractions of a second to billions of years. An extensive list of the important environmental radionuclides is in Fate®.

Sources of anthropogenic radionuclides released into the environment include medical research and treatment, nuclear power production, nuclear weapons production, and nuclear weapons testing, with the latter three accounting for most of the waste generation and environmental contamination.

2.8.4 Biological Reactions

Microorganisms offer an amazing diversity of biochemical reactions to degrade organic pollutants. In fact, virtually every organic compound that has been tested

Table 2.13 Properties of common forms of radiation.

Property	Type of radiation		
	α	β	γ
Charge	2+	1−	0
Mass	6.64×10^{-24} g	9.11×10^{-28} g	0
Relative penetrating power	1	~100	~10 000
Composition	He nucleus	Electron	High-energy photon

can be degraded by microorganisms under laboratory settings and at sufficiently faster rates to be potentially significant. Microorganisms use enzymes (natural catalysts) to degrade pollutants relatively quickly. You will recall from general chemistry that catalysts do not affect the direction of the reaction (ΔG or the equilibrium), but they speed up the reaction (approach to equilibrium) by lowering the activation energy. Yet despite the potential for these reactions, most toxic organic chemicals (pollutants) are relatively stable in the environment; that is they do not rapidly degrade. Of course, in the absence of microbial processes, there are abiotic and photochemical degradation reactions, as discussed above, but most of these require special conditions. If pollutants did rapidly degrade when released into the environment, after all, pollutants would not be a problem.

There are as many ways that microorganisms can degrade pollutants as there are microbial enzymes. We like to think of microorganisms mineralizing pollutants to stable inorganic forms of carbon, hydrogen, phosphorous, sulfur, and so on, but as in chemical degradations, not all biological degradations are this much complete. Complete mineralization usually requires a diversity of microorganisms to complete the process.

When an established (acclimated) colony of microorganisms is present, the rate-limiting step is usually uptake through the cell wall, and the kinetics of the reaction can be simplified to pseudo-first-order rate laws. Of course, the difficult part of this assumption is determining if and when an established community of microorganisms is present in a lake or river, and it is especially difficult to test these conditions in groundwater aquifers. Another highly complicating factor in isolating and documenting biochemical reactions is that abiotic degradations often occur at the same time and under the same conditions. Chemists can eliminate microorganisms from their experiments, but suffice it to say biologists cannot eliminate chemistry from their experiments.

We mentioned in the first paragraph of this section that almost every organic compound can be degraded by some type of microorganism. The point of testing for biodegradability is twofold: to determine whether degradation can occur and to determine the rate of the reaction. The determination of whether a degradation will happen is a bit complicated but easy compared to extrapolating rates. The laboratory rate must be extrapolated to an estimated field rate, since the rate is largely determined by the number and distribution of microbial cells in the system. Documenting this in sediments or soils is virtually impossible. For example, let us look at the PCB contamination in lake sediments. Researchers for PCB manufacturers and users have documented fairly well that microbes can degrade PCBs under laboratory conditions and have proposed that PCB contamination in lake and river sediments should not undergo costly remediation since microbes will take care of the problem given sufficient time. But what is sufficient time? Laboratory removal rates of PCBs from sediment slurries are relatively fast (weeks to months), but river systems such as the Hudson River have been contaminated for decades with little reduction in PCB contamination. Obviously, transferring results from laboratory experiments to the natural environment is difficult.

When considering whether an organic pollutant can be degraded by microorganisms, it is important to mimic the environmental setting as close as possible. This includes pH, E_H, nutrient concentrations, pollutant concentrations, and temperature, just to name a few parameters. Students should note that there are many types of microorganisms in the environment that have a variety of biochemical enzymes to perform these reactions, and a variety of TEAs can be used (refer to the Section 2.5.2). Biologically mediated oxidation reactions of organics in the environment can be especially important, due to their potentially rapid rates of removal of pollutants.

At the end of Chapter 3, more quantitative aspect of fate and transport chemistry, we will discuss more detailed chemical speciation and the concept of fugacity or the escaping tendency of pollutants from one environmental compartment to another.

2.9 Fugacity Concepts and Modeling

Concepts like equilibrium, solubility, vapor pressure, sorption, and chemical activity can be integrated into a modeling tool often used to assess chemical fate–fugacity modeling. This approach contrasts with the Exposure Analysis Modeling System (EXAMS) approach ostensibly because EXAMS has greater regulatory utility, nevertheless useful for assessing chemical fate and interpreting the media of most interest based on chemical properties. Both will be explored in greater detail in Chapter 14 of this book. Fugacity approaches are most useful in multimedia environments and gained acceptance by European Nations in dealing with environmental fate in the EU REACH (2017) regulation governing consumer products imported into member countries and will be discussed in Chapter 10 in greater detail, as it does affect all imports into the European market.

EXAMS, on the other hand, is an interactive (command line) software application for formulating aquatic ecosystem models and rapidly evaluating the fate, transport, and exposure concentrations of synthetic organic chemicals including pesticides, industrial materials, and leachates from disposal sites (Burns et al. 1982). A central feature of EXAMS is that it contains an integrated Database Management System (DBMS) specifically designed for storage and management of project databases required by the software. Built-in user interaction is provided by a context-sensitive help menus, an online data dictionary, and plotting capabilities for review of output data.

2.10 Summary

In this chapter, we have illustrated that many chemical factors are important in the fate and transport of pollutants in aquatic and atmospheric systems. In fact, the chemistry of a system can dictate the type of fate and transport model that we use. We have covered a lot of information and concepts in this chapter, and a good review of these would be to make an outline of the chapter. As you list the concepts, predict how these will affect inorganic versus organic pollutants. This organization will be very important when we make our summary at the end of Chapter 3.

Concepts

2.1 List the global volumes of freshwater in groundwater, lakes, rivers, and glaciers in increasing order. Which is the most available for human use?

2.2 List the average residence times of water in increasing order in the following compartments: lakes, oceans, rivers, groundwater, and glaciers.

2.3 List two world regions that will probably have shortages of fresh water in the near future.

2.4 List and describe the five most important unique properties of water.

2.5 What intermolecular force is responsible for water's unique properties?

2.6 Explain the term bioconcentration.

2.7 Define a colloid.

2.8 List two important sorption surfaces commonly found in water systems.

2.9 How does isomorphic substitution affect clay surface charges and sorption phenomena?

2.10 What is meant by natural organic matter (NOM)? How does it relate to DOM?

2.11 List five functional groups in NOM.

2.12 Contrast adsorption and partitioning with respect to the surface concentration of a pollutant.

2.13 How do sorption reactions affect pollutant fate and transport in aquatic systems? Specifically, will a pollutant be more or less mobile in the sorbed state?

2.14 List the four major transformation reactions of pollutants and give a two-to-three sentence description of each.

Exercises

For problems requiring the pC–pH Simulator, please refer to http://edusolns.com/pCpH/2016/closed/or http://edusolns.com/pCpH/2016/open/.

2.1 By how much do water's strong intermolecular forces increase the actual temperature range of liquid water over the theoretical range?

2.2 Mathematically show how a ppb is approximately equal to μg/l. Why is this equality only approximate?

2.3 Calculate the water solubility of H_2CO_2 for a water that is in contact with an atmosphere containing 1000 ppm CO_2 (1000×10^{-6} atm). K_H is given in the text of this chapter.

2.4 Define what is meant by the term "pollutant water solubility."

2.5 How can a pollutant's water solubility affect its fate and transport in water systems?

2.6 Calculate the solubility of BaF_2 at 18 °C.

2.7 Calculate the solubility of silver chloride at 25 °C.

2.8 Calculate the pH of a water solution containing 5.89×10^{-5} M H^+.

2.9 Calculate the pH of a water solution containing 4.92×10^{-4} M OH^-.

2.10 Hand-draw a pC–pH diagram for a closed system containing 5.29×10^{-3} M total carbonate. Calculate the concentration of H_2CO_3, HCO_3^-, and CO_3^{2-} at a pH of 5.90. Check your results using the pC–pH Simulator.

2.11 Hand-draw a pC–pH diagram for a closed system containing 8.37×10^{-5} M total sulfide. Check your results using the pC–pH Simulator.

2.12 Hand-draw a pC–pH diagram for an open system of water in contact with an atmosphere containing 50.5 ppm CO_2 (50×10^{-6} atm). Check your results using the pC–pH Simulator.

2.13 The concentration of CO_2 in the atmosphere is predicted to rise to ~700 ppm (partial pressure of 700×10^{-6} atm) by the year 2100. First, use the pC–pH Simulator to estimate the pH of rainwater falling through an atmosphere of 380 ppm CO_2 (the approximate current concentration). Next, use the simulator to predict the pH of rainwater in the year 2100.

2.14 Use the $\Delta G°$ values from Section 2.5.2 for the half reactions of NO_3^-, MnO_2, FeOOH, SO_4^{2-}, and CO_2 to calculate the combined energy generated (in kJ/mol) from the oxidation of a mole (1.00) of glucose with each. Which trichloroethylene (TCE) yields the most energy?

2.15 If you have a water with an E_H of 0.55 V, would you consider it to be an oxidizing or reducing environment?

2.16 If you have a water with an E_H of −0.48 V, would you consider it to be an oxidizing or reducing environment?

References

Ames, L.L. and Rai, D. (1978). *Processes Influencing Radionuclide Mobility and Retention. Element Chemistry and Geochemistry, Conclusions and Evaluation*, vol. 1, EPA 520/6-78-007. Richland, WA: Battelle Pacific Northwest Laboratories.

Bell, J. and Bates, T.H. (1988). Distribution coefficients of radionuclides between soils and groundwater and their dependence on various test parameters. *Sci. Total Environ.* **69**: 297–317.

Burns, L.A., Cline, D.M., and Lassiter, R.P. (1982). *Exposure Analysis Modeling System (EXAMS): User Manual and System Documentation*. EPA-600/3-82-023. U.S. EPA.

Dunnivant, F.M., Macalady, D.L., and Schwarzenbach, R.P. (1992). Reduction of substituted nitrobenzenes in aqueous solutions containing natural organic matter. *Environ. Sci. Technol.* **26**: 2133–2141.

EU REACH (2017). State of Compliance '18. https://www.assentcompliance.com/stateofcompliance/?utm_source=bing&utm_medium=cpc&utm_campaign=sl&msclkid=9a2ab8cb9c4a101c3420e2e3f9c331be (accessed 7 November 2017).

Food and Agriculture Organization of the United Nations (2017). Aquastat Data Base. http://www.fao.org (accessed 27 November 2017); public domain.

Germann, G.G. (1988). The distribution and mass loading of polychlorinated biphenyls in Lake Hartwell sediments. Masters thesis. Environmental Systems Engineering, Clemson University.

Hayes, M.H.B., MacCarthy, P., Malcolm, R.L., and Swift, R.S. (eds.) (1989). *Humic Substances II: In Search of Structure*. New York: Wiley.

Jursa, A.S. (ed.) (ed.) (1985). *Handbook of Geophysics and the Space Environment*. Air Force Geophics Laboratory, U.S. Air Force, Document Accession Number: ADA 167000.

Kampf, N., Scheinost, A.C., and Schulze, D.G. (2000). Oxide minerals. In: *Handbook of Soil Science*, F 125–F 168. Coca Raton, FL: CRC Press.

Kielland, J. (1937). Individual activity coefficients of ions in aqueous solutions. *J. Am. Chem. Soc.* **59**: 1675.

Killops, S.D. and Killops, V.J. (1993). *An Introduction to Organic Geochemistry*. Harlow, Essex/New York: Longman Scientific & Technical/Wiley.

Leenheer, J.A. (1980). Study of sorption of complex organic solute mixture on sediments by dissolved organic carbon fractionation analysis. In: *Contaminants and Sediments*, vol. II (ed. R.A. Baker), 267–278. Ann Arbor, MI: Ann Arbor Science.

Leenheer, J.A. and Huffman, E.W.D. Jr. (1976). Classification of organic solutes in water by using macroreticular resins. *J. Res. US Geol. Surv.* **4**: 737–751.

Looney, B.B., Grant, M.W., and King, C.M. (1987). *Estimation of Geochemical Parameters for Assessing Subsurface Transport at the Savannah River Plant*, DPST-85-904. Aiken, SC: Savannah River Laboratory.

Mackay, D., Shiu, W.Y., and Sutherland, R.P. (1979). Determination of air-water Henry's law constants for hydrophobic pollutants. *Environ. Sci. Technol.* **13**: 333–337.

Schnitzer, M. (1986). Binding of humic substances by soil mineral colloids. In: *Interactions of Soil Minerals with Natural Organic and Microbes* (ed. P.M. Huang and M. Schnitzer). Soil Science Society of America Spec. Pub. No. 17.

Schulten, H.R. and Schnitzer, M. (1993). A state of the art structural concept for humic substances. *Naturwissenschaften* **80**: 29–30.

Schwarzenbach, R.P., Gschwend, P.M., and Imboden, D.M. (1993). *Environmental Organic Chemistry*, 407. New York: Wiley.

Sparks, D.L. (2003). *Environmental Soil Chemistry*, 2e. New York: Academic Press.

Sposito, G. (1984). *The Surface Chemistry of Soils*. New York: Oxford University Press.

Stumm, W. and Morgan, J.J. (1981). *Aquatic Chemistry: An Introduction Emphasizing Chemical Equilibria in Natural Waters*, 2nde, 515. New York: Wiley.

Stumm, W. and Morgan, J.J. (1996). *Aquatic Chemistry: Chemical Equilibria and Rates in Natural Waters*, 3rde. New York: Wiley.

Suffet, I.H. and MacCarthy, P. (1989). *Aquatic Humic Substances: Influence of Fate and Treatment of Pollutants*, Advances in Chemistry Series # 219. Washington, DC: American Chemical Society.

Thurman, E.M. (1985). *Organic Geochemistry of Natural Waters*. Boston, MA: Martinus Nijhoff/Junk.

Tichnor, K.V. (1993). Actinide sorption by fracture-infilling materials. *Radiochim. Acta* **60**: 33–42.

Weast, R.C. and Astle, M.J. (eds.) (1980). *Handbook of Chemistry and Physics*, 61ste, B-242. Boca Raton, FL: CRC Press.

Woodwell, G.M., Wurster, C.F., and Isaacson, P.A. (1967). DDT residues in an east coast estuary: a case of biological concentration of a persistent insecticide. *Science* **156**: 821–824.

3

Quantitative Aspects of Chemistry Toward Modeling

3.1 Introduction

In the previous chapter, we introduced several important chemical concepts and processes that contribute to pollutant fate and transport. In this chapter, we will consider more quantitative mathematical operations that allow us to incorporate these processes into models of environmental processes. Pollutant fate and transport modeling is a two-part process: chemistry and dispersion in each system. The detailed chemistry concepts that are presented here are important but can be overwhelmed by the large degree of uncertainly in dispersion and mixing in real-world systems. We should also note that fate and transport models, even when using sophisticated numerical methods of analysis, cannot incorporate all of the chemistry that we will discuss. Some of the chemistry, such as chemical speciation, must be taken into account in the risk assessment portion of the process. The key to this chapter, and chemistry as related to fate and transport phenomena, is to note how each process will affect mobility of a pollutant in an aqueous and in an atmospheric system and how it will affect toxicity (which is generally associated with the free, unbound form of the pollutant). In the closing section of this chapter, we will attempt to "bring it all together" as we relate each chemical process to pollutant fate and transport.

3.2 Calculation of the Free Metal Ion Concentration in Natural Waters

3.2.1 Calculating Chemical Equilibria

Students begin their study of chemical equilibrium in general chemistry with basic equilibrium expressions and Le Chatelier's principle. However, these courses rely on many simplifications regarding equilibria. These simplifications were fine when you were working in simple solutions – solutions containing only one or two salts at low concentration – but are rarely acceptable in environmental chemistry (as in groundwater, estuary, and ocean water). Courses in quantitative analysis, a sophomore-level chemistry course, go on to consider more complex equilibrium that utilize activity coefficients. Now we must further develop your concepts of chemical equilibrium in order to study metal speciation in natural and industrial aqueous solutions.

First, we will review the basic principles that you learned (and should remember) from general chemistry. Then we will make the chemical system a little more complicated, to a point where we can still do calculations but only with considerable effort. Finally, we will make the problems even more complicated (and reflective of reality) and find that we cannot solve them by hand. This is where we can use chemical speciation modeling programs such as GEOCHEM, MINTECH, and MINEQL+, discussed at the end of this chapter, that contain numerical solving routines to work these problems.

First, let us discuss general chemistry. In your introduction to chemistry, we kept systems and reactions simple, maybe a little too simple, considering that students coming out of this course usually have some common misconceptions concerning ions in solution. For example, we teach simple reactions such as the following precipitation reaction:

$$Pb^{2+} + 2Cl^- \leftrightarrow PbCl_2(s)$$

However, as we will see in this section, lead is present in solution with chloride not only as Pb^{2+} but also as $PbCl^+$, $PbCl_2(aq)$, $PbCl_3^-$, and $PbCl_4^{2-}$. You may wonder why this chemical speciation matters? There are several answers to this question. The most relevant to environmental chemistry are, first, that this provides us with a much deeper understanding of equilibrium processes; second, that speciation determines how a metal will be transported through aquatic systems; and third, that different forms of a specific metal ion will exhibit different toxicities or bioavailabilities. In general, the most toxic form of each metal ion is the hydrated free metal form (e.g. Pb^{2+} or Cd^{2+} instead of $PbCl^+$ or $CdCl^+$). However, there are important exceptions to this rule, such as methylmercury and dimethylmercury, highly toxic forms of mercury. Thus, we now have two

Pollutant Fate and Transport in Environmental Multimedia, First Edition. Frank M. Dunnivant and Elliot Anders.
© 2019 John Wiley & Sons, Inc. Published 2019 by John Wiley & Sons, Inc.
Companion website: www.wiley.com/go/Dunnivant/pollutant_fate&Transp_EnvironMultimedia

important inputs for risk assessment models concerning bioaccumulation and toxicity: pollutant concentration (e.g. ppm, ppb, etc.) reaching the receptor organism, determined by fate and transport models, and chemical speciation of the pollutant in contact with the receptor organism, expressed in activities.

3.2.1.1 Review of Equilibrium Calculations from Previous Courses

In general chemistry, you were given reactions such as

$$aA + bB \leftrightarrow cC + dD$$

where a, b, c, and d are the respective (stoichiometric) number of atoms or molecules of chemical species A, B, C, and D in the balanced equation. The equilibrium expression for this reaction was expressed as

$$K = \frac{[C]^c\ [D]^d}{[A]^a\ [B]^b} \tag{3.1}$$

where K is the equilibrium constant. You were told (and you probably did an experiment in laboratory to demonstrate) that K was constant for a system, even as the concentrations of the species A, B, C, and D varied. However, this was an idealized experiment. More typically, the equilibrium constant has been found to vary as the concentration changes, as illustrated in Figure 3.1. You should recall Le Chatelier's principle from general chemistry that states if a system is at equilibrium, such as the reaction above, and you add additional A or B, the system will react to readjust the concentrations of A, B, C, and D to re-establish the same K value.

There is a relatively simple way to correct for this observation. When activities rather than concentrations are used in the equilibrium expression, the equilibrium constant **does** remain constant:

$$K = \frac{(A_C)^c\ (A_D)^d}{(A_A)^a\ (A_B)^b} = \frac{[C]^c\,\gamma_c\ [D]^d\,\gamma_D}{[A]^a\,\gamma_A\ [B]^b\,\gamma_B}$$

where A is the activity and γ is the activity coefficient (from Chapter 2) of the species indicated by subscript. Note that, in order for K to remain constant, all activities must remain constant. In order for this to occur, if the concentration of a species increases, its activity coefficient must decrease. The activity coefficient is a function of total concentration of all ions in solution (also known as ionic strength), as seen in Eq. (3.1). Recall that $C = A$. For distilled water, the concentration is equal to the activity, which means that the activity coefficient must be equal to 1.00.

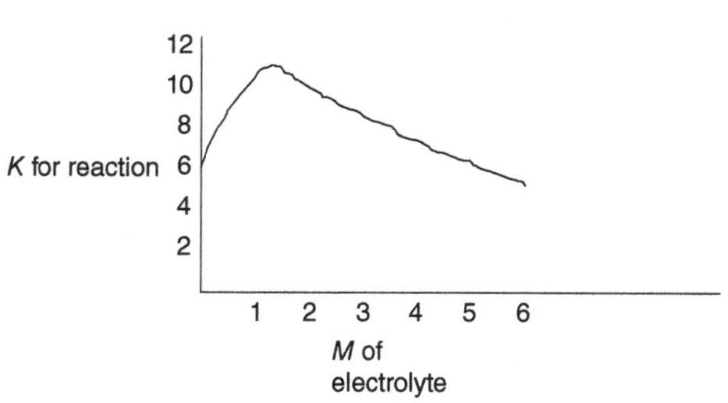

Figure 3.1 Generalization of K as a function of molar concentration of electrolyte.

General rule: If you add an inert salt (a salt that dissolves but does not undergo a chemical transformation) to a solution, generally you will increase the solubility of another salt, when the two salts do not share a common ion. (*Note*: This is not true when the anion of the added salt acts as a ligand to form insoluble complexes with the cation of interest.)

Let us work an example to illustrate this rule. Consider the solubility of $CaSO_4$ in distilled water ($K_{sp} = 2.4 \times 10^{-5}$). In distilled water (in the strict sense, K_{sp}s are only accurate for distilled water; these values are compiled from many μ values and are extrapolated to $\mu = 0$), the concentration is equal to the activity, and

$$K_{sp} = A_{Ca^{2+}}\ A_{SO_4^{2-}} = [Ca^{2+}]\,[SO_4^{2-}]$$

Since the stoichiometry is one-to-one, we can let x equal the Ca^{2+} ion concentration and the SO_4^{2-} ion concentration. Thus, $K_{sp} = x^2$, and $x = [Ca^{2+}] = [SO_4^{2-}] = (2.4 \times 10^{-5})^{1/2} = 4.9 \times 10^{-3}$ M.

Recall that the solubility of $CaSO_4$ is limited by the attraction between the Ca^{2+} and SO_4^{2-} ions. Anything that we do to decrease these interactions will increase the solubility. One way that we can decrease these interactions is to add another electrolyte. In the distilled water, only the attractions to polar water molecules decreased these attractions. Now, let us look at the solubility of $CaSO_4$ in a 0.0200 M solution of KNO_3. As the Ca^{2+} and SO_4^{2-} ions go into solution, the calcium ion will be surrounded by the − (negative dipole) end of H_2O, as well as the negative ions SO_4^{2-} and NO_3^-, while sulfate will be surrounded by the positive end of H_2O and the positive ions K^+ and Ca^{2+}.

Returning to the problem, we also assume that the ionic strength from the $CaSO_4$ is insignificant compared to the ionic strength of KNO_3 (based on their relative concentrations). Thus, KNO_3 determines the ionic strength. By using this ionic strength to calculate the activity coefficients, we can plug these coefficients into the solubility expression to calculate the concentrations of the two ions in solution. This was covered in Section 2.4.2 and will be repeated here.

$$\mu = 0.500\,(C_1 Z_1^2 + C_2 Z_2^2)$$
$$= 0.500\,([K^+](1^2) + [NO_3^-](1^2))$$
$$\mu = 0.500\,(0.0200 + 0.0200) = 0.0200\ \text{M}$$

$$\log \gamma = -\frac{0.512\, Z^2 \sqrt{\mu}}{1 + \frac{\alpha \sqrt{\mu}}{305}} \quad \text{at } 25^\circ\text{C}$$

$$\log \gamma_{Ca} = -\frac{0.512\, 2^2 \sqrt{0.0200}}{1 + \frac{600 \sqrt{0.0200}}{305}} = -0.230$$

$$\gamma_{Ca} = 0.590$$

$$\log \gamma_{SO_4^{2-}} = -\frac{0.512\, 2^2 \sqrt{0.0200}}{1 + \frac{450 \sqrt{0.0200}}{305}} = -0.240$$

$$\gamma_{SO_4^{2-}} = 0.580$$

$$K_{sp} = A_{Ca^{2+}}\, A_{SO_4^{2-}} = [Ca^{2+}]\gamma_{Ca^{2+}}\, [SO_4^{2-}]\gamma_{SO_4^{2-}} = x^2\, (\gamma_{Ca^{2+}})\,(\gamma_{SO_4^{2-}})$$

$$K_{sp} = 2.4 \times 10^{-5} = x^2\, (\gamma_{Ca^{2+}})\,(\gamma_{SO_4^{2-}})$$

$$\text{or } x^2\, (0.590)(0.580) = 2.4 \times 10^{-5}$$

$$x = [8.37 \times 10^{-3}]$$

So, x, the concentration of Ca^{2+} and SO_4^{2-}, equals 8.37×10^{-3} M.

Note that the concentration of $CaSO_4$ in distilled water (calculated above) was 4.9×10^{-3} M, while in the presence of another ionic salt the concentration increased to 8.37×10^{-3} M. Thus, the general rule was followed. Common ion problems, which you also worked in general chemistry and quantitative analysis, are worked in a similar manner but the presence of a common ion, sulfate in the above problem, would *decrease* the calcium ion concentration.

Now, we will develop a more general method for solving equilibrium/speciation problems. In the remainder of this discussion, we will ignore activities, in order to make the calculations a little more manageable. However, if you are concerned with obtaining the most accurate estimate of concentration, you must use equations such as the Debye–Hückel equation to first calculate the ionic strength of the solution (as was done in the example above).

The approach we will use below uses the algebraic observation that if you have as many equations to describe a system (here, an aqueous solution) as you have unknowns (here, concentrations), then you can mathematically solve the problem. To solve these types of problems, we will use the charge balance equation, the mass balance equation, and the basic equilibrium expressions.

The charge balance is an algebraic statement of the sum of the molar concentrations of cations and anions and can be represented by

Positive charges on the ions
= negative charges on the ions

For example, say that you have K_3PO_4 in aqueous solution. The charge balance equation would be

$$[H^+] + [K^+] = [OH^-] + [H_2PO_4^-] + 2[H_2PO_4^{2-}] + 3[PO_4^{3-}]$$

Note that the concentrations for each of the above are in molar units, and we multiply each concentration by the charge of the ion to make this a charge balance. For example, say that PO_4^{3-} is present at 0.300 M. We are concerned with the "charge" concentration, not the molar concentration of the ions. Since there are three charges per ion, the value used in the charge balance is $3 * [PO_4^{3-}] = 3 * 0.300\ \text{M} = 0.900\ \text{M}$.

The mass balance states that the quantity of a particular element in all species put into a solution must equal the amount of that element delivered to the solution. Thus, although a chemical input into a solution may dissociate into or form a number of different species, the sum of the concentrations of a particular element in the system must be equal to its concentration from the input chemical(s). The following examples illustrate this concept.

For the reaction $CH_3COOH \leftrightarrow CH_3COO^- + H^+$, if the total acetic acid added is equal to 0.0500 M, then the mass balance would be:

$$\text{Total mass of acid and associated species} = 0.0500\ \text{M} = [CH_3COOH] + [CH_3COO^-]$$

For a 0.105 M solution of Na_2S, the reaction would be $Na_2S \leftrightarrow 2Na^+ + S^{2-}$, but S^{2-} reacts with water to form HS^- and H_2S. The mass balance for all sulfur species would be

$$\text{Total S} = 0.105\ \text{M} = [S^{2-}] + [HS^-] + [H_2S]$$

The general steps for this approach to equilibrium are

Step 1: Write the pertinent reactions.
Step 2: Write the charge balance.
Step 3: Write the mass balance.
Step 4: Write the equilibrium expressions with appropriate constants.
Step 5: Count the equations and unknowns.
Step 6: If the number of equations is equal to the number of unknowns, then solve.

In the following example, we will look at a salt that does not react with water (other than hydration reactions characteristic of all metal ions). The problem statement is this: calculate the concentration of Hg_2^{2+} in a saturated solution of Hg_2Cl_2. Again, in this example we will ignore activities.

Step 1: Write the pertinent reactions

$$Hg_2Cl_2 \leftrightarrow Hg_2^{2+} + 2Cl^- \quad K_{sp} = 1.2 \times 10^{-18}$$

Also, you should always include the dissociation of water, since it will affect the charge balance in aqueous solutions:

$$H_2O \leftrightarrow H^+ + OH^- \quad K_w = 1.00 \times 10^{-14}$$

Step 2: Write the charge balance

$$[H^+] + 2[Hg_2^{2+}] = [Cl^-] + [OH^-]$$

Step 3: Write the mass balance

Since neither Hg_2^{2-} nor Cl^- reacts with water, the $[H^+]$ and $[OH^-]$ remain constant. Recall that in pure distilled water (not open to the atmosphere):

$$[H^+] = [OH^-]$$

$$[Cl^-] = 2[Hg_2^{2+}]$$

Step 4: Write the equilibrium expressions and constants

$$K_{sp} = [Hg_2^{2-}]\,[Cl^-]^2 = 1.2 \times 10^{-18}$$

$$K_w = [H^+]\,[OH^-] = 1.00 \times 10^{-14}$$

Step 5: Count the equations and unknowns

Step 6: Solve

For pure water:

$$[H^+] = [OH^-]$$

$$K_w = [H^+]\,[OH^-] = [H^+]^2 = 1.00 \times 10^{-14}$$

$$[H^+] = [OH^-] = 1.00 \times 10^{-7}$$

For Hg_2Cl_2:

$$[Cl^-] = 2[Hg_2^{2+}]\ \text{(from mass balance)}$$

$$K_{sp} = [Hg_2^{2+}]\,[Cl^-]^2$$

$$= [Hg_2^{2+}]\,(2[Hg_2^{2+}])^2 = 1.2 \times 10^{-18}$$

or

$$[Hg_2^{2+}] = (K_{sp}/4)^{1/3} = 6.7 \times 10^{-7}\ \text{M}$$

This is the same answer that you would obtain if you solved the problem as you did in general chemistry. But what if we were considering a soluble species that reacts with water? In this case, the general chemistry approach would not work, and you would have to use the five-step process illustrated above. For example, consider the solubility of HgS, which can be the predominant source of Hg ion in the sediments and hypolimnion of lakes.

$$HgS(s) \leftrightarrow Hg^{2+} + S^{2-}\ K_{sp} =\sim 5 \times 10^{-54}$$

$$S^{2-} + H_2O \leftrightarrow HS^- + OH^- K_{\beta1} = 0.80$$

$$HS^- + H_2O \leftrightarrow \quad H_2S + OH^- \qquad K_{\beta2} = 1.1 \times 10^{-7}$$

and, as always,

$$H_2O \leftrightarrow H^+ + OH^-\ K_w = 1.00 \times 10^{-14}$$

Recall that S^{2-} is a strong base, so $[H^+]$ will not be equal to $[OH^-]$ as in the earlier example.

Step 1: See above equations for HgS(s)

Step 2:

$$2[Hg^{2+}] + [H^+] = 2[S^{2-}] + [HS^-] + [OH^-]$$

Step 3: Since the composition of HgS is one-to-one for the Hg and S and the Hg and S dissociate completely, the total molar concentration of Hg is equal to the total concentration of S species

$$[Hg^{2+}] = [S^{2-}] + [HS^-] + [H_2S]$$

Step 4:

$$K_{sp} = [Hg^{2+}]\,[S^{2-}] = 5 \times 10^{-54}$$

$$K_{\beta1} = \frac{[HS^-]\,[OH^-]}{[S^{2-}]} = 0.80$$

$$K_{\beta2} = \frac{[H_2S]\,[OH^-]}{[HS^-]} = 1.1 \times 10^{-7}$$

$$K_w = [H^+]\,[OH^-] = 1.00 \times 10^{-14}$$

Steps 5 and 6: You have six equations and six unknowns, so you can solve this by substitution. But we can also make assumptions that make the calculation much easier. For example, most of the time you know what the pH of the system is or you know the pH at which you want to evaluate the metal speciation. Let us evaluate the speciation at a pH value of 8.00. This assumption lowers the number of unknowns to five since we know the value for $[H^+]$.

$$\text{pH} = 8.00, \qquad [H^+] = 10^{-8}$$

$$[OH^-] = \frac{10^{-14}}{10^{-8.00}} = 1.00 \times 10^{-6}$$

Rearrangement of $K_{\beta1}$ equation (above) yields

$$[HS^-] = \frac{K_{\beta1}\,[S^{2-}]}{[OH^-]}$$

Rearrangement of the $K_{\beta2}$ equation (above), yields

$$H_2S = \frac{K_{\beta2}\,[HS^-]}{[OH^-]}$$

and with substitution of the $[HS^-]$ expression yields

$$[H_2S] = \frac{(K_{\beta1})\,(K_{\beta2})\,[S^{2-}]}{[OH^-]^2}$$

Now, we have the values of $[HS^-]$ and $[H_2S]$ expressed in terms of $K_{\beta1}$, $K_{\beta2}$, $[OH^-]$, and $[S^{2-}]$. All we need in order to solve the problem is an expression for $[S^{2-}]$ expressed in terms of the K values and $[OH^-]$.

From the mass balance equation, we have

$$[Hg^{2+}] = [S^{2-}] + [HS^-] + [H_2S]$$

Substitution of the $[HS^-]$ and $[H_2S]$ expressions from the previous set of equations yields

$$[Hg^{2+}] = [S^{2-}] + \frac{K_{\beta_1}[S^{2-}]}{[OH^-]} + \frac{K_{\beta_1}K_{\beta_2}[S^{2-}]}{[OH^-]^2}$$

On rearrangement, this yields

$$[Hg^{2+}] = [S^{2-}]\left[1 + \frac{K_{\beta 1}}{[OH^-]} + \frac{K_{\beta 1}K_{\beta 2}}{[OH^-]^2}\right]$$

or

$$S^{2-} = \frac{[Hg^{2+}]}{1 + \frac{K_{\beta 1}}{[OH^-]} + \frac{K_{\beta 1}K_{\beta 2}}{[OH^-]^2}}$$

Substitution of the $[S^{2-}]$ expression into the K_{sp} expression yields

$$K_{sp} = [Hg^{2+}][S^{2-}]$$

$$K_{sp} = [Hg^{2+}]\left[\frac{[Hg^{2+}]}{1 + \frac{K_{\beta 1}}{[OH^-]} + \frac{K_{\beta 1}K_{\beta 2}}{[OH^-]^2}}\right]$$

Rearrangement yields an expression that allows you to calculate the free metal ion concentration.

$$[Hg^{2+}] = \sqrt{K_{sp}\left(1 + \frac{K_{\beta 1}}{[OH^-]} + \frac{K_{\beta 1}K_{\beta 2}}{[OH^-]^2}\right)}$$

As we stated in the original problem, we made the derivation easier by selecting a single pH value for which to calculate the Hg^{2+} concentration. At a pH value of 8.00, $[OH^-]$ is equal to 1.00×10^{-6} M, and, using the equation above, the $[Hg^{2+}]$ is equal to 2.10×10^{-24} M. This equation can be used at other pH values or put into a spreadsheet to generate a plot of $[Hg^{2+}]$ versus pH value.

3.2.2 Equilibrium Applied to More Complex Speciation Problems

Although the problems and derivations above have been a bit tedious, they have been quite manageable. Now, let us make the problem a little more difficult (and a little more true to reality). Three cases are relevant. Case I (Section 3.2.2.1) illustrates the speciation of a metal in the presence of excess metal salt. This situation could occur in mining operations or in industrial treatment processes and plating operations, as well as in laboratory experiments. Case II (Section 3.2.2.2) is more illustrative of situations occurring in the natural environment. Here, we know the total aqueous concentration of the metal, and it is always below the maximum solubility limit. We also know the predominant ligand in solution. From this data (and the equilibrium constants), we can calculate the species distribution for the solution. Case III (Section 3.2.2.3) looks at the combined effects of pH and E_H.

3.2.2.1 Case I: Complexation in the Presence of Excess Solid

Problem

You are interested in studying the speciation of Pb^{2+} in the presence of iodide ion, and you want to determine the predominant chemical form (species) of Pb^{2+}–I^- complexes in solution as a function of I^- concentration. Excess PbI_2 salt is placed in a beaker containing distilled water. Thus, the concentration of hydrated Pb^{2+} ion in solution is controlled by the solubility of the PbI_2 salt ($K_{sp} = 7.9 \times 10^{-9}$, from Table 2.7). We will treat iodide as independent of the solubility of this salt, with other sources available. Lead can form four complexes with iodide, according to the expressions shown below. Complexation data are from Table 3.1.

$Pb^{2+} + I^- \Longleftrightarrow PbI^+$ $\quad K_1 = [PbI^+]/([Pb^{2+}][I^-]) = 1.00 \times 10^2$

$Pb^{2+} + 2I^- \Longleftrightarrow PbI_2(aq)$ $\quad \beta_2 = [PbI_2(aq)]/([Pb^{2+}][I^-]^2) = 1.40 \times 10^3$

$Pb^{2+} + 3I^- \Longleftrightarrow PbI_3^-$ $\quad \beta_3 = [PbI_3^{2-}]/([Pb^{2+}][I^-]^3) = 8.30 \times 10^3$

$Pb^{2+} + 4I^- \Longleftrightarrow PbI_4^{2-}$ $\quad \beta_4 = [PbI_4^{2-}]/([Pb^{2+}][I^-]^4) = 3.00 \times 10^4$

To determine the major species, draw a complexation diagram (pPb–pI), much like the pC–pH diagram described above and illustrated in the pC–pH diagrams in Chapter 2, for the system at equilibrium. Use I^- concentrations ranging from 0.0001 to 10 M (pI = −2 to 4). Using your diagram, estimate the concentrations of Pb^{2+}, PbI^+, $PbI_2(aq)$, PbI_3^-, and PbI_4^{2-} at an I^- concentration of 0.3162 M (arbitrarily chosen). Then, using the equations shown above and your understanding of equilibrium, calculate the exact concentration of each species at an I^- concentration of 0.3162 M.

Solution

In the presence of excess PbI_2, the concentration of Pb^{2+} in solution will be controlled by the K_{sp} for PbI_2. Thus, since we know the K_{sp}, we can calculate the log of the concentrations of Pb^{2+} at I^- concentrations of 0.0001 and 10 M:

$$K_{sp} = [Pb^{2+}][I^-]^2 = 7.9 \times 10^{-9}$$

$$[Pb^{2+}] = \frac{7.9 \times 10^{-9}}{[I^-]^2}$$

Table 3.1 Stability constants (log *K* or *β*) for the formation of complexes and solids from metals and ligands.

	OH^-		CO_3^{2-}		SO_4^{2-}		Cl^-		Br^-		F^-		NH_3		$B(OH)_4$
Sr^{2+}			SrLs	9.0	SrL	2.6					SrL$_2$s	8.5			
					SrLs	6.5									
Ba^{2+}			BaL	2.8	BaL	2.7					BaL$_2$s	5.8			
			BaLs	8.3	BaLs	10.0									
Cr^{3+}	CrL	10.0			CrL	3.0	CrL	0.23			CrL	5.2			
	CrL$_2$	18.3									CrL$_2$	9.2			
	CrL$_3$	24.0									CrL$_3$	12.0			
	CrL$_4$	28.6													
	Cr$_3$L$_4$	47.8													
	CrL$_3$s	30.0													
Al^{3+}	AlL	9.0													
	AlL$_2$	18.7									AlL	7.0			
	AlL$_3$	27.0									AlL$_2$	12.6			
	AlL$_4$	33.0									AlL$_3$	16.7			
	Al$_3$L$_4$	42.1									AlL$_4$	19.1			
	AlL$_3$s	33.5													
Fe^{3+}	FeL	11.8			FeL	4.0	FeL	1.5	FeL	0.6	FeL	6.0			
	FeL$_2$	22.3			FeL$_2$	5.4	FeL$_2$	2.1			FeL$_2$	10.6			
	FeL$_4$	34.4									FeL$_3$	13.7			
	Fe$_2$L$_2$	25.0													
	FeL$_3$s	42.7													
	FeL$_3$s	38.8													
Fe^{2+}	FeL	4.5	FeLs	10.7	FeL	2.2					FeL	1.4			
	FeL$_2$	7.4													
	FeL$_3$	11.0													
	FeL$_{2s}$	15.1													
Co^{2+}	CoL	4.3	CoLs	10.0	CoL	2.4	CoL	0.5			CoL	1.0	CoL	2.0	
	CoL$_2$	9.2											CoL$_2$	3.5	
	CoL$_3$	10.5											CoL$_3$	4.4	
	CoL$_4$	15.7											CoL$_4$	5.0	
Ni^{2+}	NiL	4.1	NiLs	6.9	NiL	2.3	NiL	0.6			NiL	1.1	NiL	2.7	
	NiL$_2$	9.0									NiL$_2$	4.9			
	NiL$_3$	12.0											NiL$_3$	6.6	
	NiL$_2$s	17.2											NiL$_4$	7.7	
													NiL$_5$	8.3	
Cu^{2+}	CuL	6.3	CuL	6.7	CuL	2.4	CuL	0.5			CuL	1.5	CuL	4.0	
	CuL$_2$	11.8	CuL$_2$	10.2	Cu$_4$(OH)$_6$Ls	68.6							CuL$_2$	7.5	
	CuL$_4$	16.4	CuLs	9.6									CuL$_3$	10.3	
	Cu$_2$L$_2$	17.7	Cu$_2$(OH)$_2$Ls	33.8									CuL$_4$	11.8	
	CuL$_2$s	19.3	Cu$_3$(OH)$_2$L$_2$s	46.0											
	CuL$_2$s	20.4													
Zn^{2+}	ZnL	5.0	ZnLs	10.0	ZnL	2.1	ZnL	0.4			ZnL	1.2	ZnL	2.2	
	ZnL$_2$	11.1			ZnL$_2$	3.1	ZnL$_2$	0.2					ZnL$_2$	4.5	
	ZnL$_3$	13.6					ZnL$_3$	0.5					ZnL$_3$	6.9	
	ZnL$_4$	14.8					Zn$_2$(OH)$_3$Ls	26.8					ZnL$_4$	8.9	
	ZnL$_2$s	15.5													
	ZnL$_2$s	16.8													

Table 3.1 (Continued)

	OH^-		CO_3^{2-}		SO_4^{2-}		Cl^-		Br^-		F^-		NH_3		$B(OH)_4$	
Pb^{2+}	PbL	6.3			PbL	2.8	PbL	1.6	PbL	1.8	PbL	2.0				
	PbL_2	10.9	PbLs	13.1	PbLs	7.8	PbL_2	1.8	PbL_2	2.6	PbL_2	3.4				
	PbL_3	13.9					PbL_3	1.7	PbL_3	3.0	PbL_2s	7.4				
	PbL_2s	15.3					PbL_4	1.4	PbL_2s	5.7						
							PbL_2s	4.8								
Hg^{2+}	HgL	10.6	HgLs	16.1	HgL	2.5	HgL	7.2	HgL	9.6	HgL	1.6	HgL	8.8		
	HgL_2	21.8			HgL_2	3.6	HgL_2	14.0	HgL_2	18.0			HgL_2	17.4		
	HgL_3	20.9					HgL_3	15.1	HgL_3	20.3			HgL3	18.4		
	HgL_2s	25.4					HgL_4	15.4	HgL_4	21.6			HgL_4	19.1		
							HgOHL	18.1	HgL_2s	19.8						
Cd^{2+}	CdL	3.9	CdLs	13.7	CdL	2.3	CdL	2.0	CdL	2.1	CdL	1.0	CdL	2.6		
	CdL_2	7.6			CdL_2	3.2	CdL_2	2.6	CdL_2	3.0	CdL_2	1.4	CdL_2	4.6		
	CdL_2s	14.3			CdL_3	2.7	CdL_3	2.4					CdL_3	5.9		
							CdL_4	1.7					CdL_4	6.7		
Ag^+	AgL	2.0	Ag_2Ls	11.1	AgL	1.3	AgL	3.3	AgL	4.7	AgL	0.4	AgL	3.3	AgL	0.6
	AgL_2	4.0			Ag_2Ls	4.8	AgL_2	5.3	AgL_2	6.9			AgL_2	7.2	$AgHL_2s$	22.9
	AgLs	7.7					AgL_3	6.4	AgL_3	8.7						
							AgLs	9.7	AgL_4	9.0						
									AgLs	12.3						
	SiO_3^{2-}	S^{2-}	$S_2O_3^{2-}$		PO_4^{3-}		$P_2O_7^{4-}$		$P_3O_{10}^{5-}$		CN^-					
Sr^{2+}			SrL	2.0	SrL	5.5	SrL	5.4	SrL	7.2						
					SrHL	14.5	SrOHL	7.7	SrHL	13.6						
					SrH_2L	20.3	Sr_2Ls	12.9	SrOHL	9.3						
					SrHLs	19.3										
Ba^{2+}			BaL	2.3	BaHLs	19.8			BaL	6.3						
			BaLs	4.8					BaHL	12.9						
									Ba_2Ls	16.1						
Fe^{3+}	FeHL	22.7	FeL	3.3	FeHL	22.5					FeL_6	43.6				
					FeH_2L	23.9										
					FeLs	26.4										
Fe^{2+}			FeLs	18.1	FeHL	16.0					FeL_6	35.4				
					FeH_2L	22.3										
					Fe_3L_2s	36.0										
Co^{2+}	CoLs	21.3	CoL	2.1	CoHL	15.5	CoL	7.9	CoL	9.7						
	CoLs	25.6					CoHL	14.1	CoHL	14.8						
Ni^{2+}	NiLs	19.4	NiL	2.1	NiHL	15.4	NiL	7.7	NiL	9.5	NiL	7.3				
	NiLs	24.9					NiHL	14.4	NiHL	14.7	NiL_4	30.2				
	NiLs	26.6									NiH_2L_4	40.8				
											$NiHL_4$	36.1				

(Continued)

Table 3.1 (Continued)

	OH^-		CO_3^{2-}		SO_4^{2-}		Cl^-		Br^-		F^-		NH_3		$B(OH)_4$	
Cu^{2+}	CuLs	36.1			CuHL	16.5	CuL	9.8	CuL	11.1	CuL_2	16.3				
					CuH_2L	21.3	CuHL	15.5	CuHL	15.5	CuL_3	21.6				
							CuL_2	12.5			CuL_4	23.1				
							CuH_2L	19.2								
Zn^{2+}	ZnL	16.6	ZnL	2.4	ZnHL	15.7	ZnL	8.7	ZnL	10.3	ZnL	5.7				
	ZnLs	24.7	ZnL_2	2.5	ZnH_2L	21.2	ZnL_2	11.0	ZnHL	14.9	ZnL_2	11.1				
			ZnL_3	3.3	Zn_3L_2	35.3	ZnOHL	13.1	ZnOHL	13.6	ZnL_3	16.1				
			Zn_2L_2	7.0							ZnL_4	19.6				
											ZnL_2s	15.9				
Pb^{2+}	PbLs	27.5	PbL	3.0	PbHL	15.5	PbL	9.5								
			PbL_2	5.5	PbH_2L	21.1	PbL_2	10.2								
			PbL_3	6.2	Pb_3L_2s	43.5										
			PbL_4	7.3	PbHLs	23.8										
Hg^{2+}	HgL	7.9	HgL_2	29.2			HgOHL	18.6			HgL	17.0				
	HgL_2	14.3	HgL_3	30.6							HgL_2	32.8				
	HgOHL	18.5									HgL_3	36.3				
	HgLs	52.7									HgL_4	39.0				
	HgLs	53.3									HgOHL	29.6				
Cd^{2+}	CdL	19.5	CdL	3.9			CdL	8.7	CdL	9.8	CdL	6.0				
	CdHL	22.1	CdL_2	6.3			CdOHL	11.8	CdHL	14.6	CdL_2	11.1				
	CdH_2L_2	43.2	CdL_3	6.4					CdOHL	12.6	CdL_3	15.7				
	CdH_3L_3	59	CdL_4	8.2							CdL_4	17.9				
	CdH_4L_4	75.1	Cd_2L_2	12.3												
	CdLs	27.0														
Ag^+	AgL	19.2	Al	8.8	Ag_3Ls	17.6					AgL_2	20.5				
	AgHL	27.7	AgL_2	13.7							AgL_3	21.4				
	$AgHL_2$	35.8	AgL_3	14.2							AgOHL	13.2				
	AgH_2L_2	45.7	Ag_2L_4	26.3							AgLs	15.7				
	AgH_2Ls	50.1	Ag_3L_5	39.8												
			Ag_6L_8	78.6												
	Ethylene-diamine		NTA		EDTA		CDTA		IDA		Picolinate		Cysteine		Desferri-ferrioxamine B	
Sr^{2+}			SrL	6.3	SrL	10.5	SrL	12.4	SrL	3.1	SrL	1.8			SrL	3.1
					SrHL	14.9									SrL_2	3.0
Ba^{2+}			BaL	5.9	BaL	9.6	BaL	10.5	BaL	2.5	BaL	1.6				
					BaHL	14.6	BaHL	17.8								
Cr^{3+}					CrL	26.0			CrL	12.2						
					CrHL	28.2			CrL	23.2						
					CrOHL	32.2										
Al^{3+}			AlL	13.4	AlL	18.9	AlL	22.1	AlL	9.9						
			AlOHL	22.1	AlHL	21.6	AlHL	24.3	AlL_2	17.5						
					AlOHL	26.6	AlOHL	28.1								
			FeL		$Al(OH)_2L$	30.0										
Fe^{3+}			FeL_2		FeL	27.7	FeL	32.6	FeL	12.5	FeL_2	13.9			FeL	31.9
					FeHL	29.2	FeOHL	36.5			$FeOHL_2$	24.9			FeHL	32.6
					FeOHL	33.8										
					$Fe(OH)_2L$	37.7										

Table 3.1 (Continued)

	OH^-		CO_3^{2-}		SO_4^{2-}		Cl^-		Br^-		F^-		NH_3		$B(OH)_4$	
Fe^{2+}			FeL	4.3	FeL	9.6	FeL	16.1	FeL	20.8	FeL	6.7	FeL	5.3	FeHL	18.7
			FeL_2	7.7	FeL_2	13.6	FeHL	19.3	FeHL	23.9	FeL_2	11.0	FeL_2	9.7	FeH_2L	21.0
			FeL_3	9.7	FeOHL	12.6	FeOHL	20.4					FeL_3	13.0		
							$Fe(OH)_2L$	13.7								
Co^{2+}	CoL	6.0	CoL	11.7	CoL	18.1	CoL	21.4	CoL	7.9	CoL	6.4			CoL	11.2
	CoL_2	10.8	CoL_2	15.0	CoHL	21.5	CoHL	24.7	CoL_2	13.2	CoL_2	11.3			CoHL	18.0
	CoL_3	14.1	CoOHL	14.5							CoL_3	14.8			CoHL	23.6
Ni^{2+}	NiL	7.4	NiL	12.8	NiL	20.4	NiL	22.1	NiL	9.1	NiL	7.2	NiL	10.7	NiL	11.8
	NiL_2	13.6	NiL_2	17.0	NiHL	24.0	NiHL	25.4	NiL_2	15.7	NiL_2	12.5	NiL_2	20.9	NiHL	18.3
	NiL_3	17.9	NiOHL	15.5	NiOHL	21.8					NiL_3	17.9			NiH_2L	23.8
Cu^{2+}	CuL	10.5	CuL	14.2	CuL	20.5	CuL	23.7	CuL	11.5	CuL	8.4		Cu(II)–Cu(I)	CuL	15.0
	CuL_2	19.6	CuL_2	18.1	CuHL	23.9	CuHL	27.3	CuL_2	17.6	CuL_2	15.6			CuHL	24.1
	CuOHL	11.8	CuOHL	18.6	CuOHL	22.6									CuH_2L	27.0
Zn^{2+}	ZnL	5.7	ZnL	12.0	ZnL	18.3	ZnL	21.1	ZnL	8.2	ZnL	5.7	ZnL	10.1	ZnL	11.0
	ZnL_2	10.6	ZnL_2	14.9	ZnHL	21.7	ZnHL	24.4	ZnL_2	13.5	ZnL_2	10.3	ZnL	19.1	ZnHL	17.5
	ZnL_3	13.9	ZnOHL	15.5	ZnOHL	19.9					ZnL_3	13.6	ZnHL	16.4	ZnH_2L	22.9
Pb^{2+}	PbL	7.0	PbL	12.6	PbL	19.8	PbL	22.1	PbL	8.3	PbL	5.0	PbL	12.5		
	PbL_2	8.5			PbHL	23.0	PbHL	25.3			PbL_2	8.6				
Hg^{2+}	HgL	14.3	HgL	15.9	HgL	23.5	HgL	26.8	HgL	11.7	HgL	8.1	HgL	15.3		
	HgL_2	23.2			HgHL	27.0	HgHL	30.3			HgL_2	16.2				
	HgOHL	24.2			HgOHL	27.7	HgOHL	29.7								
	$HgHL_2$	28.0														
Cd^{2+}	CdL	5.4	CdL	11.1	CdL	18.2	CdL	21.7	CdL	6.6	CdL	5.0			CdL	8.8
	CdL_2	9.9	CdL_2	15.1	CdHL	21.5	CdHL	25.1	CdL_2	11.1	CdL_2	8.3			CdHL	16.2
	CdL_3	11.7	CdOHL	13.4							CdL_3	11.4			CdH_2L	22.7
Ag^+	AgL	4.7	AgL	5.8	AgL	8.2	AgL	9.9			AgL	3.6				
	AgL_2	7.7			AgHL	14.9					AgL_2	6.1				
	AgHL	11.9														

Source: Selected data from Morel and Hering (1993). Orginal data from Smith and Martell (1975 and 1976), Martell and Smith (1974 and 1977), Whitifield (1974), Baes and Mesmer (1976), Sanda and Hanson (1979), Sainte Marie et al. (1964), and Morgan (1967).

For $[I^-] = 0.0001$ M, $[Pb^{2+}] = 0.790$ M

$$\log(0.790) = -0.102$$

For $[I^-] = 10.0$ M, $[Pb^{2+}] = 7.9 \times 10^{-11}$

$$\log(7.9 \times 10^{-11}) = -10.102$$

Refer to Figure 3.2 for the following discussion of the construction of a complexation diagram. The Pb^{2+} line on the pPb–pI diagram should be drawn from (−4, −0.102) to (1, −10.102) (where the first number is the log molar I^- concentration and the second is the log molar Pb^{2+} concentration.) This line is the basis for the remaining calculations.

Note the equilibrium constants shown in the problem statement. The first expression uses K to represent the equilibrium between Pb^{2+} and one I^- ion. The letter "K" is used to represent stepwise formation constants. Notice that the remaining constants are represented by beta, β, which represents the cumulative formation constants (the addition of more than one ligand treated as a single reaction). For example, β_2 represents the formation of PbI_2 from the reaction of one Pb^{2+} and two I^- ions. Since the $[Pb^{2+}]$ concentration is determined by the I^- concentration, the concentration of PbI^+, $PbI_2(aq)$, PbI_3^-, and PbI_4^{2-} can be calculated for each Pb^{2+} concentration. Thus, we can add lines to the diagram representing these species.

For $[PbI^+]$:

An I^- concentration of 0.0001 results in a Pb^{2+} concentration of 0.790 M.

Substitution into

$$K_1 = \frac{[PbI^+]}{[Pb^{2+}][I^-]} = 100$$

yields a $[PbI^+] = K_1\ [Pb^{2+}]\ [I^-] = 7.90 \times 10^{-3}$.
An I^- concentration of 10 M results in a Pb^{2+} concentration of 7.9×10^{-11}.
This results in a $[PbI^+]$ concentration of 7.9×10^{-8}.
Thus, a line should be drawn for the log $[I^-]$, log $[PbI^+]$ pairs from (−4, −2.10) to (1, −7.10).

For $[PbI_2(aq)]$:

Again, an I^- concentration of 0.0001 results in a Pb^{2+} concentration of 0.790 M.
Substitution into

$$\beta_2 = \frac{[PbI_2(aq)]}{[Pb^{2+}]\ [I^-]^2} = 1.4 \times 10^3$$

yields $[PbI_2(aq)] = \beta_2\ [Pb^{2+}]\ [I^-]^2 = 1.11 \times 10^{-5}$.An I^- concentration of 10 M results in a Pb^{2+} concentration of 7.9×10^{-11}.This results in a $[PbI_2(aq)]$ concentration of 1.11×10^{-5}.|Thus, a line should be drawn for the log I^-, log $PbI_2(aq)$ pairs or from (−4, −4.956) to (1, −4.956).

For $[PbI_3^-]$:

An I^- concentration of 0.0001 results in a Pb^{2+} concentration of 0.79 M.
Substitution into

$$\beta_3 = \frac{[PbI_3^-]}{[Pb^{2+}]\ [I^-]^3} = 8.3 \times 10^3$$

yields $[PbI_3^-] = \beta_3\ [Pb^{2+}][I^-]^3 = 6.56 \times 10^{-9}$.An I^- concentration of 10 M results in a Pb^{2+} concentration of 7.90×10^{-11}.This results in a $[PbI_3^-]$ concentration of 6.56×10^{-4}.Thus, a line should be drawn for the log I^-, log PbI_3^- pairs or from (−4, −8.183) to (1, −3.183).

For $[PbI_4^{2+}]$:

An I^- concentration of 0.0001 results in a Pb^{2+} concentration of 0.790 M.
Substitution into

$$\beta_4 = \frac{[PbI_4^{2+}]}{[Pb^{2+}]\ [I^-]^4} = 3.00 \times 10^4$$

yields $[PbI_4^{2-}] = \beta_4\ [Pb^{2+}][I^-]^4 = 2.37 \times 10^{-12}$.An I^- concentration of 10 M results in a Pb^{2+} concentration of 7.90×10^{-11}.This results in a $[Pb_4^{2-}]$ concentration of 2.37×10^{-2}.Thus, a line should be drawn for the log I^-, log PbI_4^- pairs or from (−4, −11.63) to (1, −1.63).

Using this approach, the pPb–pI diagram in Figure 3.2 can be drawn. Note that the total [Pb] in solution is the sum of all the Pb species. Now you can use this figure and the equations given above to solve the original problem.

Figure 3.2 also illustrates another very important point concerning the solubility of lead and complex ion formation. In general chemistry, you learned that the solubility of Pb^{2+} was governed by the K_{sp} and, in this case, the concentration of I^- ion. Because of the common ion effect (another concept from general chemistry; Le Chatelier's principle), as $[I^-]$ increased you calculated that $[Pb^{2+}]$ would decrease. If you based your calculation

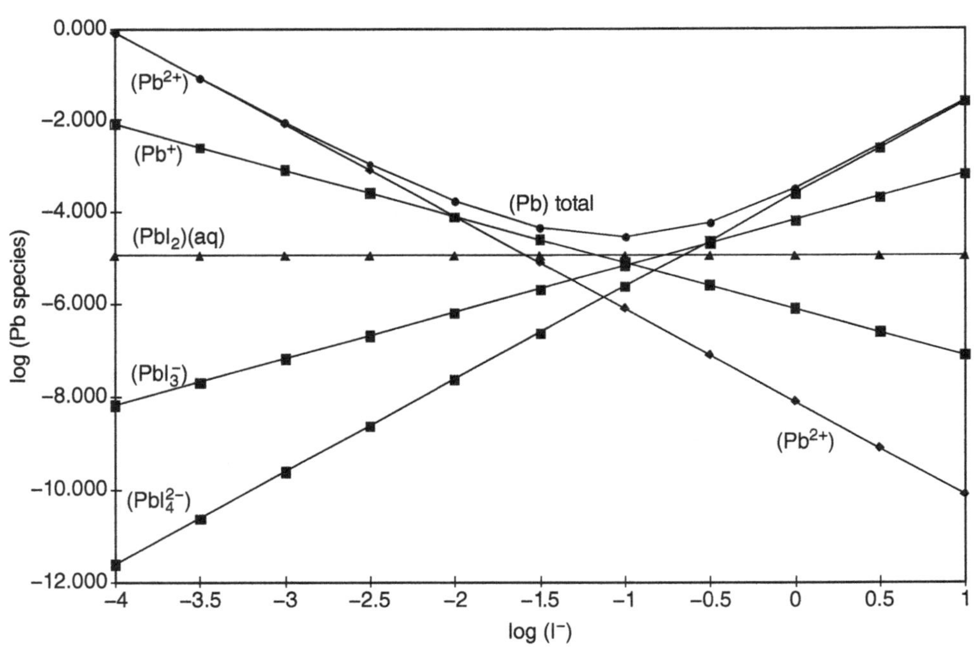

Figure 3.2 Complexation diagram for Pb^{2+} and I^- species.

solely on this assumption, the solubility would follow the $[Pb^{2+}]$ line in Figure 3.2. However, note the total concentration of Pb in solution. We see that the total Pb decreases to a point but after the formation of complex ions becomes important, at approximately 0.1 M, the total Pb in solution actually increases. Most metals that form complex ions follow this trend, which is important in toxicity and risk assessment.

3.2.2.2 Case II: Complexation in the Absence of a Solid

Problem

Nitrilotetraacetic acid (NTA) acid was a common component in detergents and was a chemical of concern (COC) in sewage effluent in the past. It is known for its high metal complexing power; thus it keeps metals in solution, increasing their mobility, and it may reduce their toxicity. Draw a pCd–pNTA diagram for a cadmium system (total cadmium concentration = 0.015 M) using total NTA concentrations ranging from 1.00×10^{-1} to 1.00×10^{-14} M. Use the following data:

$$Cd^{2+} + NTA^{3-} \leftrightarrow CdNTA^-$$
$$K_1 = [CdNTA^-]/([Cd^{2+}][NTA^{3-}]) = 6.31 \times 10^9$$
$$Cd^{2+} + 2NTA^{3-} \leftrightarrow Cd(NTA)_2(aq)$$
$$\beta_2 = [CdNTA_2^{4-}(aq)]/([Cd^{2+}][NTA^{4-}]^2) = 1.58 \times 10^{15}$$

Solution

First, note that this is a simplified case and we are ignoring the cadmium hydroxide species and NTA protonation. This problem is easier than in the previous section since we only have one metal concentration to be concerned with, $[Cd^{2+}] = 0.015$ M. The total Cd concentration (C_T) is the result of three chemical species:

$$C_T = [Cd^{2+}] + [CdNTA^-] + CdNTA_2^{4-}$$

First, we will develop an expression to calculate the free ion $[Cd^{2+}]$. The equilibrium expressions K_1 and β_2 are next rearranged to solve for $[CdNTA^-]$ and $[CdNTA_2^{3-}]$. These expressions are substituted into the C_T equation, and the equation is rearranged to solve for $[Cd^{2+}]$.

$$C_T = [Cd^{2+}] + [CdNTA^-] + [CdNTA_2^{4-}]$$
$$C_T = [Cd^{2+}] + K_1[Cd^{2+}][NTA^-] + \beta_2[Cd^{2+}][NTA^{3-}]^2$$
$$C_T = [Cd^{2+}](1 + K_1[NTA^{3-}] + \beta_2[NTA^{4-}]^2)$$
$$[Cd^{2+}] = \frac{C_T}{1 + K_1[NTA^{3-}] + \beta_2[NTA^{4-}]^2}$$

Similar approaches can be used to derive expressions for $[CdNTA^-]$ and $[CdNTA_2^{4-}]$, as shown below. In deriving these equations, you must first express the mass balance as the sum of all metal species. Next, you will rearrange the K expressions in terms of each metal species and substitute the resulting expressions into the mass balance (using the K expressions containing each complexed metal species to solve the equation for that species). Finally, solve the mass balance equation for each complexed metal species as shown below.

$$[CdNTA^-] = \frac{C_T}{\frac{1}{K_1[NTA^{3-}]} + 1 + \frac{\beta_2[NTA^{3-}]}{K_1}}$$
$$[CdNTA_2^{4-}] = \frac{C_T}{\frac{1}{\beta_2[NTA^{3-}]^2} + \frac{K_1}{\beta_2[NTA^{3-}]} + 1}$$

Lines for each chemical species can be drawn by varying the concentration of NTA and calculating the concentration of Cd^{2+}, $CdNTA^-$, and $CdNTA_2^{4-}$. Such an approach results in the diagram shown in Figure 3.3.

NTA, suggested by the diagram, is an excellent surrogate to illustrate the binding of metal pollutants by natural organic matter (NOM) and can also be used to explain how metal and organic pollutants' aqueous concentrations can exceed those predicted by solubilities calculated from pure water calculations or experiments. NOM can bind to pollutants, pulling more into solution than would be predicted. Still, NOM-bonded pollutants are less bioavailable than free aqueous pollutants. This will be discussed in the next section.

3.2.2.3 Case III: The Combined Effects of pH and E_H

Finally, we will look at the combined effects of pH and E_H on speciation. We have discussed that pH and E_H are important chemical parameters in aqueous solutions, and they directly affect the chemical state of metal and organic pollutants, as well as the reactive state of surfaces (sorption and reactive surfaces such as metal oxide coatings and NOM functional groups) in the solution.

Figure 3.4 is referred to as a stability diagram and shows the equilibrium form of Fe as a function of pH and E_H. As you see, these two chemical parameters strongly influence the chemical nature of Fe in solution and as a solid phase. Diagrams such as this are important in determining which form of Fe (Fe^{2+}, Fe^{3+}, or solids) will be present under the pH and E_H of a natural water. In addition, if Fe^{2+}, present in anaerobic systems, is mobilized and transported to an aerobic region of lake or groundwater and is subsequently oxidized, it will precipitate and form a reactive sorption surface. During the oxidation process, Fe may reduce a pollutant through an abiotic transformation reaction, or after precipitation

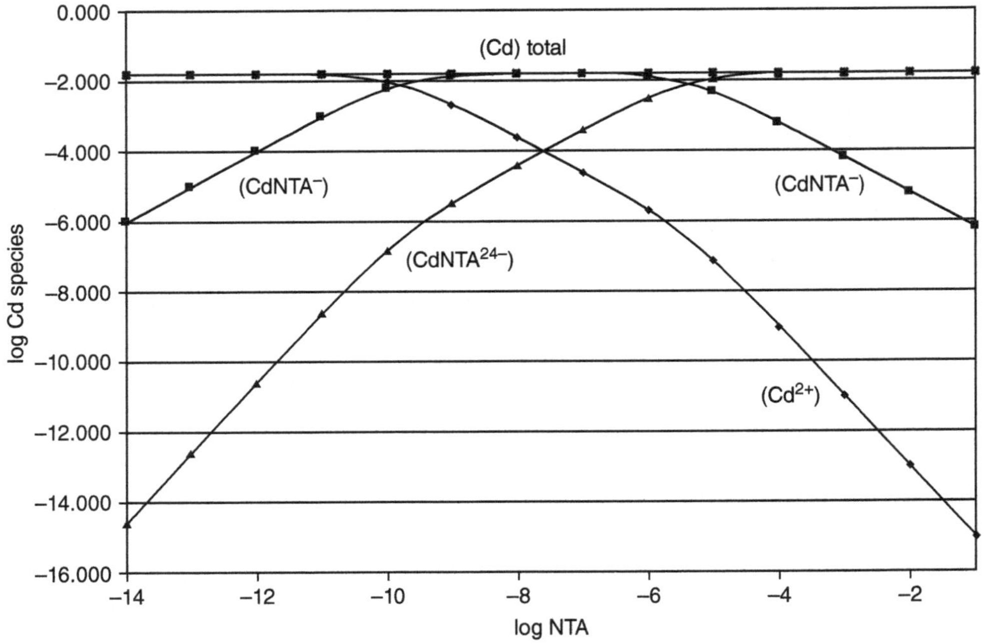

Figure 3.3 Speciation of Cd^{2+} in solution with NTA.

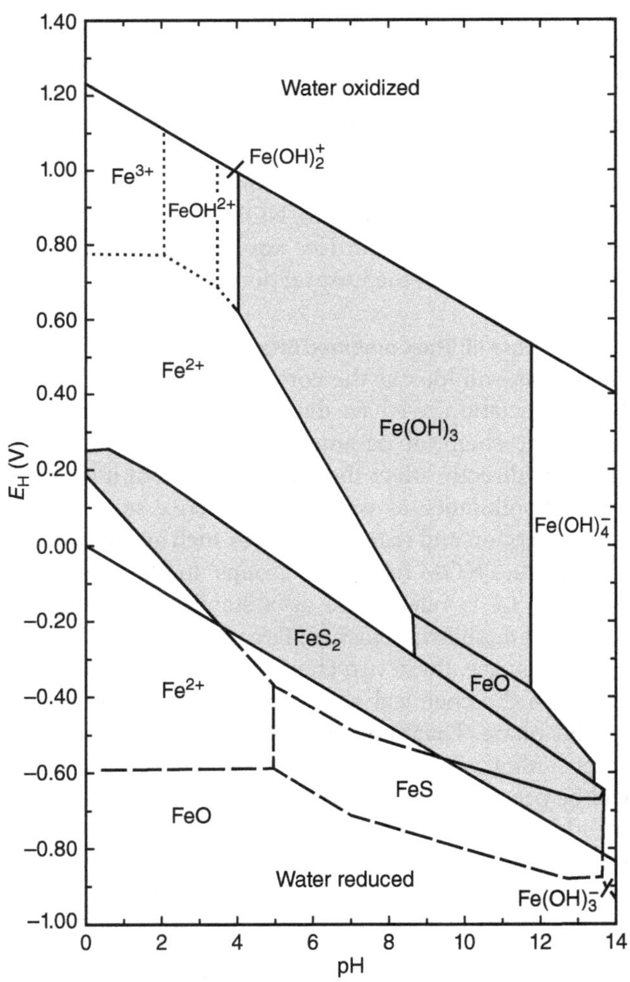

Figure 3.4 E_H–pH diagram for dissolved iron. Source: Hem (1985).

the Fe^{3+} surface may adsorb a metal pollutant from the water. Thus, the fate of another metal can be greatly affected by changing pH and E_H conditions of a water containing Fe. Stability diagrams can be constructed for all redox-active metal pollutants. Numerous pH–E_H diagrams, also referred to as Pourbaix diagrams, can be found in Brookins (1988) and Lindsay (1979).

The three cases presented above are instructional, but are still rather simplistic in design. What if you have a common ion or other salts present that also complex with the cation of interest? Or say that you have chloride, nitrate, sulfate, and carbonate present at similar concentrations, and you also have Ca^{2+} and Mg^{2+} present. Also imagine what would happen if you were concerned with the speciation as a function of solution pH and E_H. In these cases, you can throw the calculator away. There are simply too many variables (unknowns) to solve the problem using normal, exact analytical solutions; however, this is exactly what we encounter in aquatic systems. Problems such as this one led to the development of numerical solution models in the late 1960s and early 1970s. There are several of these programs, some designed for fresh water systems and others for higher ionic strength systems such as these found in marine water or groundwater. One of the most common, and user-friendly, computer programs to solve equilibrium problems such as the ones described above is MINEQL+. This program allows the user to estimate the equilibrium speciation of all cations and anions as a function of pH and E_H. Your instructor may choose to have you use this program or may use a demonstration of how it is used to solve more complicated equilibrium problems. In any case, you should be aware that there is a way to solve these complicated problems.

3.3 Methods for Determining K_d and K_p

As discussed in Chapter 2, distribution coefficients are central to our modeling of equilibrium sorption processes in environmental media. First, we will mathematically define several terms (refer to Table 2.6 for a description of their use).

For partitioning between the aqueous phase and dissolved organic matter (DOM):

$$\text{Pollutant}_{\text{aqueous}} \leftrightarrow \text{Pollutant}_{\text{DOM}}$$

$$K_{\text{DOM}} = \frac{\text{Concentration of pollutant in} \times \text{DOM (mg/kg)}}{\text{Concentration of pollutant} \times \text{in water (mg/l)}} \quad (3.2)$$

For the distribution/partition between the aqueous phase and particle in solution:

$$\text{Pollutant}_{\text{aqueous}} = \text{Pollutant}_{\text{particle}}$$

$$K_{\text{d}} = \frac{\text{Concentration of pollutant} \times \text{on particle (mg/kg)}}{\text{Concentration of pollutant} \times \text{in water (mg/l)}} \quad (3.3)$$

or

$$K_{\text{p}} = \frac{\text{Concentration of pollutant on} \times \text{organic particle (mg/kg)}}{\text{Concentration of pollutant} \times \text{in water (mg/l)}} \quad (3.4)$$

To describe partitioning as it relates to the concentration of organic matter:

$$K_{\text{organic carbon}} = \frac{K_{\text{p}} \text{ or } K_{\text{d}}}{\text{Fraction of organic matter} \times \text{is sample}} \quad (3.5)$$

As you can see, to calculate the K value from any of these expressions, you need the concentration of pollutant in each of two phases, each with its own set of concentration units. Generally, in order to collect this data in a laboratory, a solution of known pollutant mass and water, with soil, sediment, or DOM is mixed together for three days and the phases (solid and aqueous) are then separated. Determination of K_{DOM} and concentrations of pollutants in DOM phases is complicated, and the approach of all three K expressions is basically the same, so we will concentrate on K_d and K_p as an example here. After mixing, separation of the soil–water or sediment–water suspension is completed by filtration of the sample through a 0.45-µm filter. The concentration of pollutant in the aqueous phase (or all phases) is then determined. The determination of aqueous concentrations is relatively simple, but solid phase concentrations require highly involved extraction procedures. In our example, since we know the total mass of pollutant added to each system (each sample vial), we can measure the concentration (mass) of pollutant in the dissolved phase, subtract this mass from the total mass, and thus calculate the concentration of pollutant in the solid phase. An example of this calculation is shown in Table 3.2 and in the two "A Closer Look" examples at the end of this chapter.

Equilibrium coefficients are by definition constants, but these are considerably more complicated than the typical K expressions for equilibrium reactions in general chemistry. As we have seen, K_d for a metal in a sediment/water suspension is dependent, at a minimum, on pH, the type of other cations present, ionic strength, surface charge, and solids-to-water ratio. Thus, many

Table 3.2 Experimental data for the determination of a distribution coefficient for Cd^{2+} on EPA reference sediment B-2.

Total mass (mg) of pollutant added to flask	0.00725
Mass of pollutant recovered in blank (mg)	0.00720
Mass of pollutant measured in water phase (mg)	0.00542
Volume of water (l)	0.0300
Concentration of pollutant in water phase (mg/l)	0.181
Mass of pollutant on solid phase (mg)	0.00178
Mass of solid phase (kg)	3.58×10^{-5}
Concentration of pollutant on solid phase (mg/kg)	49.7
K_d	275

Sediment B2 is an EPA, highly characterized sediment with respect to chemical and physical composition.

variables must be controlled in an experiment or must be known to be constant in the environmental medium of interest. Given the ever-changing conditions in the environment, K_d measured in the laboratory can only be an approximation. Furthermore, we are assuming reversible sorption, which is not always the case, especially for metals. Laboratory K_d measurements for metals tend to increase for decreasing total metal concentrations in the aqueous phase, probably due to adsorption competition or saturation of adsorption sites. This is referred to as the Donnan effect.

The usefulness of K_p values, however, stems from the fact that they are usually constant over a wide range of pollutant concentrations in the water phase. This is illustrated in Figure 3.5, illustrating equilibrium partitioning of aqueous methoxychlor (a pesticide) with clay (Karickhoff et al. 1979). The slope of the linear plot is K_p and is independent of water phase concentration. It is also important to note that Figure 3.5 uses the term K_p, not K_d. This is because methoxychlor is a hydrophobic pollutant and does not undergo site-specific adsorption. Methoxychlor partitions to surfaces, particularly to NOM sorbed onto the clay. Although it is independent of pollutant water-phase concentration, the magnitude of the sorption of hydrophobic pollutants is highly dependent on the amount of organic matter present; the higher the organic matter content, the more pollutant sorbed. This is illustrated in Figure 3.6 (Karickhoff et al. 1979).

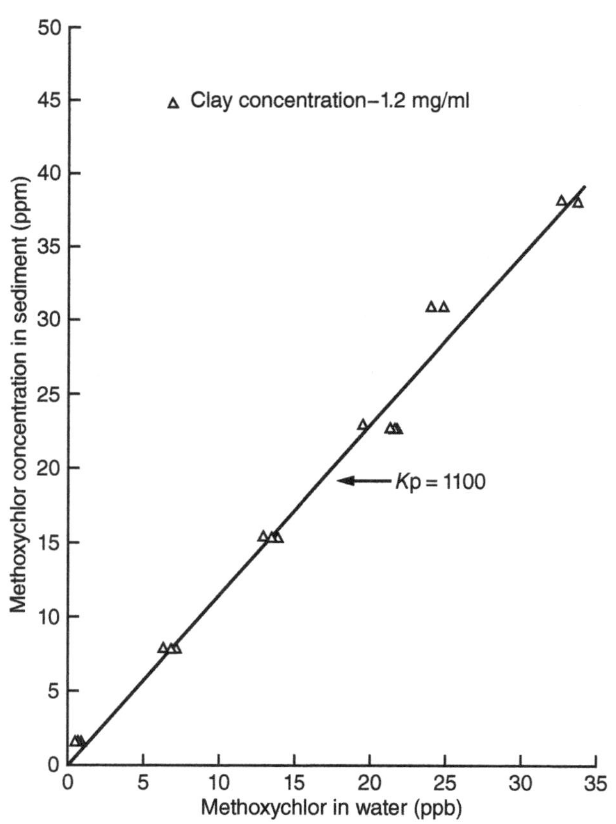

Figure 3.5 Adsorption isotherm for methoxychlor on clay. Source: Taken from Karickhoff et al. (1979), figure 2, p. 244.

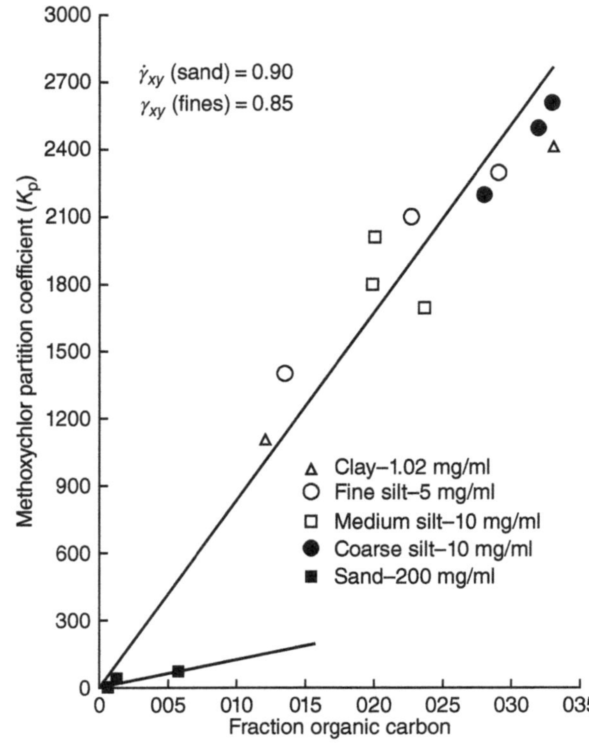

Figure 3.6 Methoxychlor K_p as a function of sediment organic carbon. Source: Taken from Karickhoff et al. (1979), figure 4, p. 245.

Also note the slope of the two lines in Figure 3.6, which are defined by Eq. (3.5) given earlier ($K_{oc} = K_p/f_{oc}$). When the partition coefficient is divided by the fraction of organic carbon present in a sample, we refer to the resulting equilibrium constant as K_{oc}. Thus, the partitioning of a pollutant can be directly related to the organic content of a sediment and, in general, if you know K_p, K_{oc}, and the organic content of a sediment, you can calculate the equilibrium water concentration of a pollutant. You should mathematically work through this approach. Unfortunately, measuring K_p or K_{oc} for every pollutant would require a very labor-intensive effort. Therefore, we estimate K_{oc} and therefore K_p based on an observation made by Karickhoff et al. (1979). They observed that K_{oc} values are directly correlated to octanol–water equilibrium coefficients (K_{ow}) for the same pollutants, which can be easily and quickly measured in the laboratory by placing a known mass of pollutant in a vial containing a known volume of octanol and water, allowing it to equilibrate, and measuring the pollutant concentration in each liquid phase. Thus,

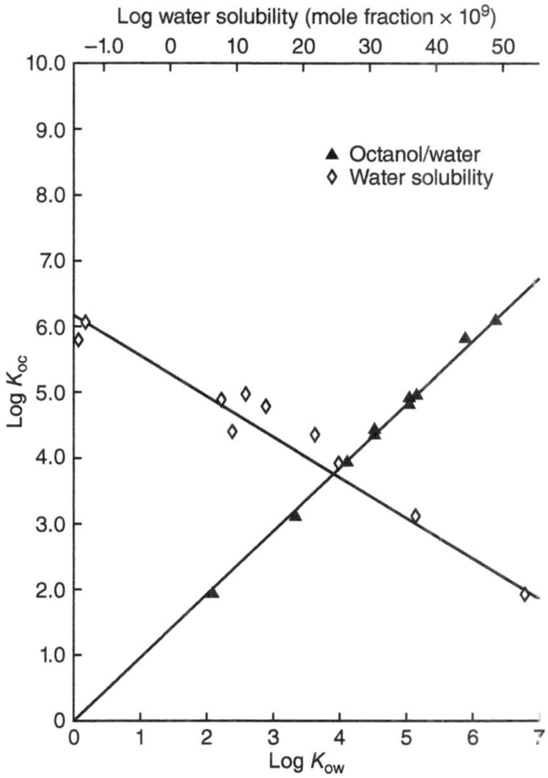

Figure 3.7 K_{oc} as a function of compound water solubility and octanol/water distribution coefficient. Source: Taken from Karickhoff et al. (1979), figure 5, p. 247.

by measuring K_{ow} for a compound in the laboratory and knowing the general relationship between K_{oc} and K_{ow}, you can calculate its corresponding K_{oc} and K_p. An example of such a K_{ow}–K_{oc} relationship observed for one particular set of hydrophobic compounds is $\log K_{oc} = 1.00\ K_{ow} - 0.21$. This equation is illustrated in Figure 3.7 (Karickhoff et al. 1979).

One final note concerning K_d and K_p should be added. In theory, equilibrium coefficients should be independent of suspended solids concentrations. However, numerous investigations for metals and organic pollutants have indicated a strong dependence with respect to solids concentration. In almost every case, K and the concentration of pollutant on the sediment decreased with increasing solid concentration. A summary of these findings can be found in O'Connor and Connolly (1980).

3.4 Kinetics of the Sorption Process

Another factor that must be considered in determining equilibrium coefficients is the time scale required to reach equilibrium. Most studies of inorganic and organic pollutants find that a three-day equilibrium period is sufficient to obtain a constant K_d or K_p. However, this does not mean that true equilibrium concentrations within particle aggregates have been reached. For example, thus far we have limited our discussion to individual particles, but many particles in nature are large aggregates of many smaller inorganic and organic particles. Thus, pollutants must diffuse to the center of the aggregate until a constant concentration is reached (Wu and Gschwend 1986). Time scales for these diffusion processes range from days to months (Karickhoff 1984; Karickhoff and Morris 1985; Coates 1984). Thus, most environmental particles are in an ever-changing state of sorption equilibrium.

Furthermore, while the sorption process, in terms of measured K_p values, occurs over a three-day period, the desorption process takes much longer. An example is the desorption of a polychlorinated biphenyl (PCB) from a sediment suspension shown in Figure 3.8. Three sediment suspensions are shown, and statistical analysis using two kinetic terms, rather than only one, clearly shows that the desorption process of pollutants can be divided into two steps with distinct rates, one rapid and adjacent to the y-axis in Figure 3.8 and one slow that spreads across the entire figure. These distinct steps have been extensively studied and have been observed for a variety of pollutants. Karickhoff and Morris (1985) describe this two-step release in terms of a labile (rapid) and a non-labile (slow) component. At present, the mechanisms for these two steps are unclear. The labile component may be due to the release of the pollutant from the surface of an aggregate or NOM, while the non-labile component could be accounted for by the slow diffusion of pollutant from the interior of aggregates. This explanation is partially supported by Wu and Gschwend (1986), who proposed a radial diffusion model for the desorption from a sediment aggregate. Also note that the labile desorption rates shown in Figure 3.8 are a function of the suspended solid concentration while the non-labile rates are not.

A variety of models have been proposed the "pure" desorption process – experiments designed to only observe the desorption process, with no sorption occurring. The most prominent are shown in Table 3.3. These models were compared in light of the observations discussed concerning the scanning electron micrograph and carbon imaging of Figure 2.24. Dunnivant (unpublished results) proposed a new model to describe the labile and non-labile desorption components as a disperse NOM component and a diffusion-based aggregate NOM component (see Figure 3.9).

$$M_{\text{total}} = A - Ae^{-k_1 t} + B - Bk_D t^{\frac{1}{2}} \tag{3.6}$$

where M_{total} is the total mass desorbed, A is the total mass in the labile phase, B is the total mass in the nonlabile phase, t is the time, k_D is the diffusion coefficient for the

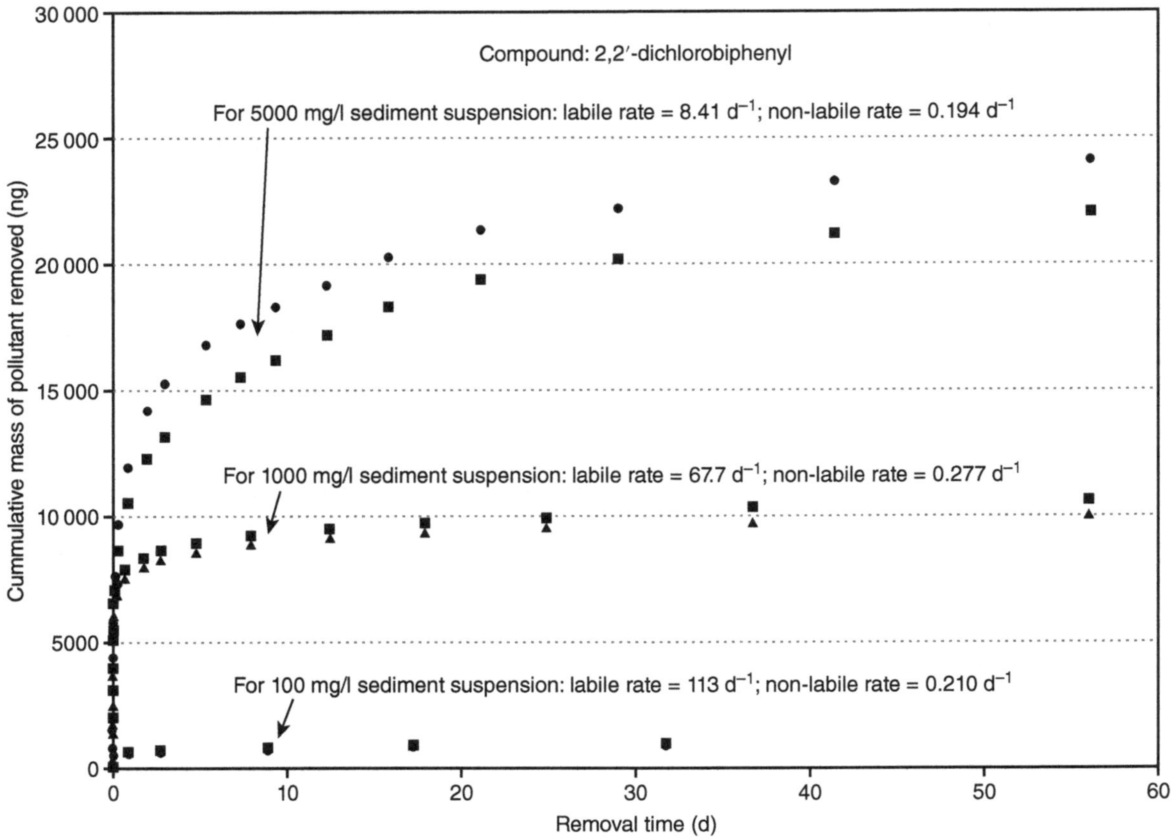

Figure 3.8 Release of a PCB from sediment suspensions. Source: Data from Dunnivant (1988).

Table 3.3 A summary of desorption modeling equations and justification for sediment suspension.[a]

Equation	Model and basis of justification
$f = f_O \exp - (k_1 + k_2)t$	1980 Two-box: it fits experimental data indicating two kinetic release components (Karickhoff 1979)
$f_s(t) = k_s t^{1/2}$	1985 Desorption kinetic model that includes a diffusive release constant (Karickhoff and Morris 1985)
$K^{obsd}_{oc-p} = (1 + K^{true}{}_{oc-NSP}DOC)^{-1}$	1985 Porous diffusion: it is based on persistent organic pollutant (POP) diffusion out of a sediment aggregate and originates from pollutant diffusion in granulated activated carbon (Gschwend and Wu 1985)
$T(ng) = A - A^{-\lambda_1 t} + B - B^{-\lambda_2 t}$	1988 Recent, more common two-box[b]: it fits experimental data indicating two kinetic release components (Dunnivant et al. 2005)

a) A two-site gamma model for two kinetic NOM release components was developed in 1999 but does not apply to this research because of its focus on soil cores, rather than on sediment suspensions (Ahn et al. 1999).

b) The model was first used in Clemson University's 1988 Dissertation by Frank M. Dunnivant but was not published until 2005 (Dunnivant 1988; Dunnivant et al. 2005).

nonlabile phase, and k_1 is the desorption kinetic rate constant.

The kinetic release of pollutants from sediment suspension can be important sources of pollutants to a system when a sediment is suspended into clear water. This will be important in the lake and stream modeling chapters. But what about the combined desorption–sorption processes occurring during the dredging of contaminated sediment in a shipping channel or Superfund site. These combined, and competitive, processes were evaluated by Handlin et al. (2014) and found that when a mix of contaminated and non-contaminated sediment are studied, sorption processes dominate the competitive process, resulting

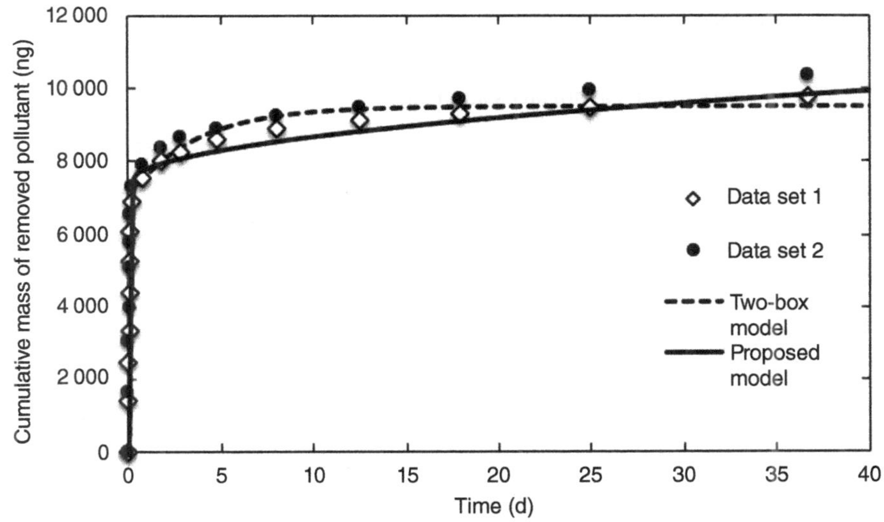

Figure 3.9 Comparison of the standard two-box, two first-order desorption model to the new disperse-aggregate NOM desorption model.

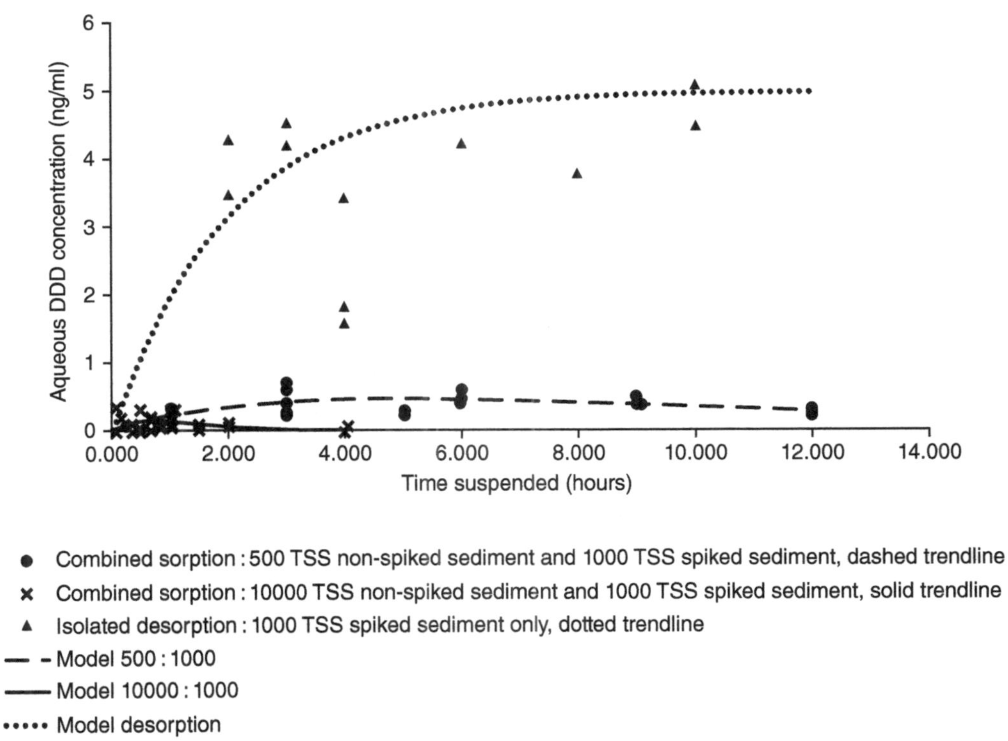

Figure 3.10 The net result of the pollutant desorption–sorption process with contaminated and non-contaminated sediments are mixed. Source: Handlin et al. (2014). Copyright 2014. Reprinted with permission from Environmental Toxicology and Chemistry.

in only slight desorption of hydrophobic pollutants to the dissolved (water) phase (see Figure 3.10). The main point is that most desorption investigations have focused on the pure desorption process and have overestimated the mass of pollutant desorbed in actual environmental setting. In dredging events, considerable masses and volumes of non-contaminated are mixed in the water column with contaminated sediment, yet little net pollutant desorption to the water column may result.

3.5 Sorption Isotherms

No discussion on sorption phenomena would be complete without the derivation of isotherm equations (sorption studies conducted at constant temperature). So, now that we understand the nature of environmental particles and what a distribution/partition coefficient is, we will look a little closer at the mathematical theory of sorption phenomena at surfaces. The processes that promote adsorption/sorption can be divided into three types:

(1) Chemical reactions at surfaces such as
surface hydrolysis
surface complexation
surface-ligand exchange, and
hydrogen bond formation
(2) Electrostatic interactions, and
(3) Hydrophobic expulsion.

Adsorption/sorption isotherms are determined by equilibrating a system with varying pollutant concentrations (or solid concentrations) and determining the equilibrium solid and liquid phase concentrations for each, as described above. As discussed earlier, K_d and K_p can be calculated from laboratory experiments; however, a measurement of K at a single dissolved phase concentration is not sufficient to determine whether the K_d or K_p applies at other concentrations. The effect of the dissolved phase concentration on K values is crucial, because pollutant concentrations constantly vary in nature. Therefore, we need to determine adsorption/sorption isotherms in order to establish the general shape of the sorption relationship and to determine the sorbed-to-dissolved pollutant relationship.

Four types of isotherms have been observed and are shown in Figure 3.11. Each of these corresponds to a certain type of adsorption process or set of conditions. Note that three of these are not linear over the entire range of pollutant concentrations tested, indicating that K is not constant with varying pollutant concentration. The most common type of isotherm encountered for soil and sediment suspensions is the L-type (upper right-hand plot), and an equation of this curve can be easily derived.

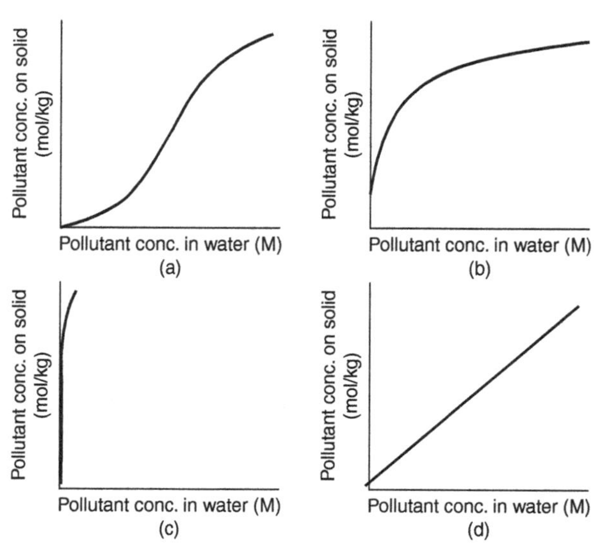

Figure 3.11 General classes of adsorption isotherms. (a) S-curve, (b) L-curve, (c) H-curve, and (d) C-curve.

The L-type isotherm is referred to as the Langmuir isotherm. Derivation of the equation for this curve requires an assumption that unoccupied adsorption sites (S) on the surface of the adsorbent become occupied by adsorbate (A) in solution, such that

$$S + A \Longleftrightarrow SA$$

where SA is the adsorbate on the surface sites. At equilibrium,

$$K_d \text{ or } K_p = \frac{\text{Products}}{\text{Reactants}} = \frac{[SA]}{[S][A]}$$

Therefore, at equilibrium, the maximum concentration of surface sites, S_T, is $[S_T] = [S] + [SA]$.

$$[S] = \frac{[SA]}{K\,[A]}$$

Solving the equation above for [S] yields, and substitution into the previous S_T equation yields

$$[S_T] = [SA] + \frac{[SA]}{K\,[A]} = [SA]\left(1 + \frac{1}{K\,[A]}\right)$$

Substituting $[A]K/[A]K$ for 1 and solving for [SA] yields

$$[SA] = [S_T]\,\frac{K\,[A]}{1 + K[A]}$$

Now, we need to express the equation in terms of the mass of adsorbent. If we normalize [SA] and $[S_T]$ to the mass of adsorbent where

$$\Gamma_A = \frac{[SA]}{\text{mass adsorbent}}$$

$$\Gamma_{max} = \frac{[S_T]}{\text{mass adsorbent}}$$

the above equation becomes

$$\Gamma_A = \Gamma_{max}\,\frac{K[A]}{1 + K[A]}$$

This equation is presented graphically in Figure 3.10. The first figure shows the general shape of the isotherm (L-type) while the second plots $1/\Gamma$ versus $1/[A]$ where A is the notation for the pollutant. Note the slope of the line and Γ_{max} located at the y-intercept.

There are several assumptions in the Langmuir model, including

- thermodynamic equilibrium exists until monolayer coverage is achieved,
- adsorption energy is independent of the degree of site coverage (i.e. that all surface sites have equal energies),
- if more than one adsorbate (pollutant) is present, they do not interact on the surface (as in the case of hydrophobic pollutant partitioning), and
- if more than one adsorbate (pollutant) is present and they do interact, they both sorb by the same mechanism.

Thus, the following equations allow the Langmuir equation to be applied to systems having multiple adsorbates and adsorbents.

For two adsorbates, A and B, the equation would be

$$\Gamma_A = \frac{\Gamma_{max} K_A [A]}{1 + K_A[A] + K_B[B]}$$

For two adsorbents, 1 and 2, the equation would be

$$\Gamma_A = \frac{\Gamma_1 K_1 [A]}{1 + K_1[A]} + \frac{\Gamma_2 K_2 [A]}{1 + K_2[A]}$$

Another common isotherm, which applies to solids with heterogeneous surface properties (i.e. surfaces where all sites do not have equal energies) is the Freundlich isotherm. The Freundlich equation can be represented by

$$\Gamma_A = K[A]^n$$

where Γ_A = [SA]/mass adsorbent, [A] = concentration of adsorbate in solution, K = the equilibrium constant, and n is a constant. The Freundlich isotherm is represented graphically in Figure 3.12. Unlike the Langmuir isotherm, the Freundlich does not assume monolayer coverage and thus does not limit the maximum amount of pollutant adsorbed or sorbed. This type of isotherm is important in describing partitioning reactions of hydrophobic compounds with NOM. In general, determination of surface coating (sorption or adsorption in moles of pollutant per square centimeter of particle surface) is difficult to measure, so concentration terms (moles of pollutant per mass of sediment) are used in practice. Thus, the experimental results would have the solid-phase concentration on the *y*-axis and the liquid phase concentration on the *x*-axis (as seen in Figure 3.13).

While the theoretical approach to Freundlich and Langmuir is attractive, in practice, scientists use experimentally measured K_ds and K_ps and assume a linear isotherm such as the ones represented by the linear portion of the *L* isotherm and the complete plot of the C-type Freundlich isotherm. This is usually appropriate,

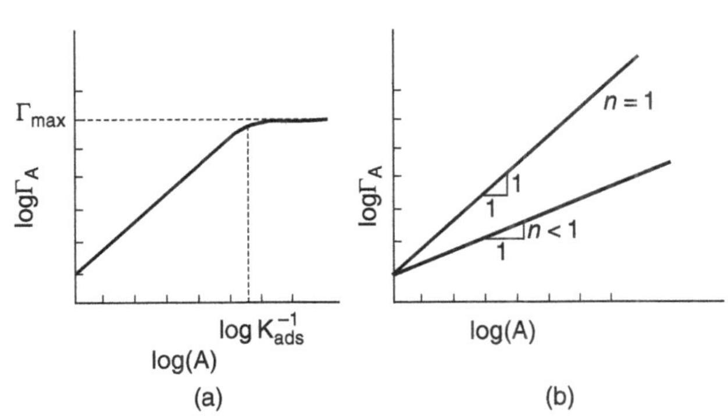

Figure 3.12 (a) Langmuir and (b) Freundlich adsorption isotherm.

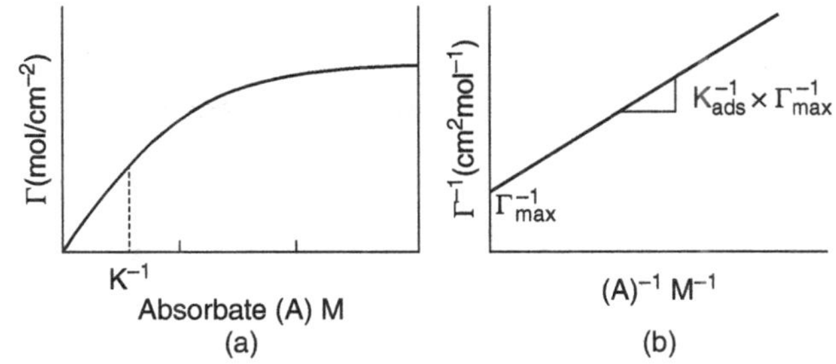

Figure 3.13 (a) Solid-phase concentration versus (b) liquid phase concentration.

since linear isotherms hold for the relatively low concentrations of pollutants found in the environment.

3.6 Kinetics of Transformation Reactions

In Chapters 2 and 3, we have looked at various chemical processes that are important in understanding how chemistry affects fate and transport. These have included acid–base, precipitation, and oxidation–reduction processes and sorption equilibrium. These processes affect the transport and in some cases can account for any removal of pollutants from the system under study (see previous Figures 2.16 and 2.17 on the degradation of substituted nitrobenzenes in anaerobic systems). As noted earlier, most sorption equilibrium reactions are reversible; if they were not, we would not be concerned with these reactions and would treat them as removal reactions. This section deals with transformation and degradation reactions that are responsible for the removal of the original pollutant from the system, and the rates of these reactions.

To begin our discussion, it is important to know when kinetics or equilibrium dominate in an environmental system. For the most part, acid–base and metal adsorption–desorption reactions are of little concern since these reactions are rapid and considered to be equilibrium processes. Precipitation, oxidation–reduction, and many organic sorption–desorption reactions are dominated by moderate to slow kinetic rates.

Fortunately for the modelers of environmental systems, most of these transformation/degradation reactions follow first-order kinetics, and when they do not, we can usually make reasonable assumptions that allow us to use pseudo-first-order kinetics. First-order kinetics follows the rate expression

$$\text{Rate} = -\frac{\Delta C}{\Delta t} = -kC \tag{3.7}$$

where $\Delta C/\Delta t$ is the change in concentration with change in time, k is the first-order rate constant with units of reciprocal time, and C is concentration (molar, ppm, or any consistent unit).

On integration, this equation yields

$$\ln \frac{C_t}{C_o} = -kt \tag{3.8}$$

where ln is the natural log, C_t is the pollutant concentration at any time (t), and C_o is the initial concentration of pollutant. One of the useful features of first-order reactions is the concept of half-life. The half-life ($t_{1/2}$) is defined as the time when one-half of the initial pollutant concentration has been degraded or removed and is related to the rate constant k by

$$\ln(0.50) = -kt_{\frac{1}{2}}$$

or

$$\frac{0.693}{k} = t_{\frac{1}{2}}$$

After one half-life, one-half of the pollutant will remain; after two half-lives, one-fourth of the pollutant will remain; after three half-lives, one-eighth of the pollutant will remain, and so on. If you do not fully understand or remember first-order reactions, you should review your general chemistry textbook.

As we mentioned, most environmental transformation/degradation reactions are first order or pseudo-first order. Let us look a little more closely at pseudo-first-order examples. Many chemical reactions are of the form:

$$\text{Rate} = k\,[\text{pollutant}]\,[\text{some oxidizing or reducing reactant}]$$

The oxidizing or reducing reactant may be NOM, the surface of a mineral phase, a photon of light or a microbial cell. In general, the concentrations of these oxidizing/reducing reactants are high and relatively constant compared to the low concentrations of pollutant. Thus, we can assume that the concentration of this reactant does not significantly change with time and we can reduce the rate expression to

$$\text{Rate} = k'\,[\text{pollutant}], \text{where } k' = k\,[\text{some oxidizing or reducing reactant}]$$

This approach greatly simplifies the kinetics of the reactions and easily allows the inclusion of kinetics in the fate and transport equations, as shown in the subsequent chapters.

It should be noted that some fate and transport equations rely on other forms of rate constants, including zero- and second-order reactions, but these require different derivations from those given for equations used in this textbook. However, generally there are several transformations/degradations happening at one time, and unless all of the rate expressions describing these degradations are of the same order (zero, first, etc.) they cannot be added together into one rate expression in the fate and transport equation (as discussed in Section 2.8). This gives the use of first-order rate expressions an added advantage, given that most reactions can be adequately described by first-order kinetics.

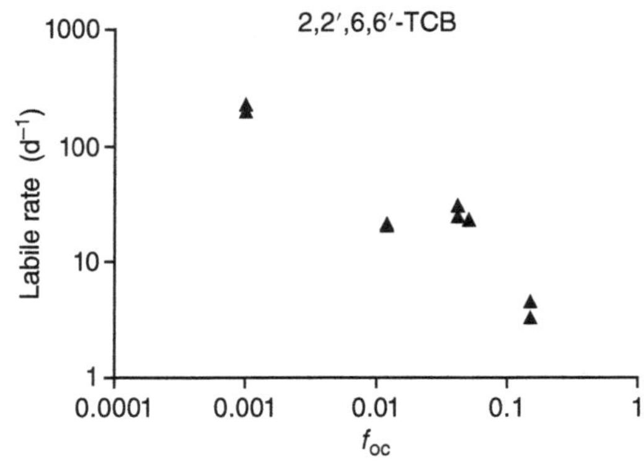

Figure 3.14 Plot of 2,2′,6,6′-tetrachlorobiphenyl data illustrating the linear relationship between labile desorption rate constant and fraction of organic matter in a series of sediment samples. Source: Ginsbach et al. (2010). Copyright 2010. Reprinted with permission from Environmental Toxicology and Chemistry.

For example, the importance of NOM in the sorption–desorption process has been noted several times in Chapters 2 and 3. Of the desorption process, the rapid, labile desorption process is certainly the most important with respect to the time scales of a pollution event. Ginsbach et al. (2010) found this process to also be first-order as illustrated in Figure 3.14.

3.7 Numerical Chemical Speciation Models

This chapter introduced more complex ion–pollutant interactions that quickly became too complicated to solve with algebraic approaches. Several numerical models have been developed to solve chemical, mostly inorganic, speciation. The basic principle behind each of these models is a mass balance, meaning that the chemical concentration of pollutant, anions, cation, and NOM are known, along with all equilibrium binding constants (Ks). Given this input, a numerical method iteratively solves the many possible ion–pollutant associations until an "agreeable" solution is found, with agreeable meaning that all mass balances are observed. Several such models have been developed including MINEQL and its later version (http://www.mineql.com/), GEOCHEM and its later versions (http://www.plantmineralnutrition.net/pubs/Shaff-et-al_2010.pdf), PHREEQC (https://wwwbrr.cr.usgs.gov/projects/GWC_coupled/phreeqc/

Table 3.4 A summary of chemical factors affecting fate and transport for each category of pollutants.

Chemical factor (by section)	Metals	Radionuclides (a class of metals)	Ionizable organics	Hydrophobic organics
For organics				
Section 2.4				
pH	Very important	Very important	Potentially important	Usually not important
Solubility	Very important	Very important	Very important	Very important
Vapor pressure	Not important	Not important	Can be important	Can be important
HLC	Not important	Not important	Can be important for uncharged form	Usually important
For inorganics				
Acid–base	Very important	Very important		
Redox	Very important	Very important		
Precipitation	Very important	Very important		
Section 2.5				
Sorption	Very important	Very important	Very important	Very important
Section 2.6 (Organic redox transformations)				
Abiotic			Can be important	Can be important
Photochemical			Can be important	Can be important
Biological			Can be important	Can be important

fs/FactSheetFS-031-02.html), WATEQ and its versions (https://wwwbrr.cr.usgs.gov/projects/GWC_chemtherm/software.htm), SOLMINEQ.88 (https://pubs.usgs.gov/wri/1988/4227/report.pdf), and others. These models allow detailed "microscopic" examinations of chemical speciation that allow more accurate pollutant risk assessments.

One final point should be made at the closing of this chapter that pollutant fate and transport modeling has three main components: modeling, chemistry, and risk assessment. The majority of error of these components is the mixing and dispersion of large physical systems (thus in fate and transport models), with smaller error associated with chemical speciation calculations. It follows that risk assessment offers simple approaches to considerable more complicated real-world events, especially when two or more pollutants are present.

3.8 Putting It All Together: Where Chemistry Enters Into the Modeling Effort

We have covered many important and complicated chemical concepts in Chapters 2 and 3. But where do they all fit into the pollutant fate and transport modeling approach, and when is each important? It depends on the pollutant and the environmental system under study, but we will attempt to summarize the role of chemistry in the modeling effort. We will divide our discussions into a metal pollutant and a hydrophobic (organic) pollutant, and important chemical processes are summarized in Table 3.4. Note that ionizable organic pollutants, such as phenols, will fall in between these two extremes.

3.8.1.1 Case I: A Metal Pollutant

Add Case I (Section 3.8.1.1) and Case II (Section 3.8.1.2) impact sheets from lecture.

In our models for aqueous systems, the source of a metal pollutant is largely controlled by K_{sp}, since the dissolved (mobile) phase of the metal cannot exceed its thermodynamically determined solubility (if it does, the metal will precipitate). Vapor pressure and Henry's law constants are of little consequence for metals, which are usually not volatile.

Recall that the purpose of fate and transport modeling is to determine the concentration of pollutant reaching a receptor (human or animal), which in turn serves as the input to risk assessment models (Chapter 11). Although chemical speciation, discussed in Section 3.2.2, is not included in even the most sophisticated fate and transport models, it is considered in risk assessment. Many of the parameters discussed in Chapters 2 and 3 influence the speciation of metals. For example, pH can determine a metal's solubility, speciation, and sorption

Table 3.5 Factors affecting adsorption/desorption of a metal pollutant.

For a metal pollutant				
Parameter	**Degradation/transformation**	**Mobility**	**Bioaccumulation**	**Sorption**
pH				
E_H				
Solubility				
Complexation				
Vapor pressure				
HLC				
Acid/base				
Abiotic/niotic				
Photic				
DOM				
NOM				
Colloids				
Sorption				
Particle size				
Kinetics				

to mineral surfaces. Metals are more readily available in their hydrated free metal form at low pH values. The adsorption of all metals to particles increases with pH of the solution due to the surface charge availability (thus more dissolved metal is present at low pH values). E_H can greatly affect oxidation states of transition, lanthanide, and actinide metals and therefore change their solubility, speciation, and degree of sorption. Thus, E_H is another important parameter affecting mobility in aquatic systems. As we discussed, increasing ionic strength (salt content) can also have profound effects on metal speciation. *A good generalization is that high ionic strength waters exhibit lower toxicity than low ionic strength waters.* This is due to the complexation of the toxic-free metal ion, generally the most toxic form of the metal, by anions in solution.

Sorption phenomena can greatly influence transport in aquatic systems by impeding further distribution by advection or diffusion mechanisms. Metals adsorbed to mineral surfaces or NOM are less bioavailable than dissolved metal species. Also, adsorbed metals generally share the fate of the particle. In lakes and streams, most particles and the metal ions sorbed to them settle to the bottom of the system and are incorporated into the sediments. Metals adsorbed to dissolved NOM usually stay in the moving water and are transported throughout and exit from the system under study. Transformation reactions (biological, chemical, and photochemical) are of little consequence to metals, with the exception of radioactive decay for radionuclides or ingested from food sources. One other rare exception is the methylation of mercury (a biological reaction) that creates a very toxic form of the pollutant.

As a summary of concepts covered in this chapter, see Figure 3.4. As an exercise to test your understanding, refer to and fill out each blank in Table 3.5 with the keywords: important, maybe, and not important. Explain your answers on a separate sheet.

3.8.1.2 Case II: Hydrophobic Pollutants

Hydrophobic pollutants represent the other extreme of types of pollutants since these pollutants do not "like" being dissolved in a polar fluid (water). Here, the concentration of pollutants in an aquatic system is controlled by the aqueous solubility and/or Henry's law constant. For atmospheric systems, vapor pressure determines the mass input.

The pH and ionic strength of the system have little to no effect on the fate of hydrophobic pollutants. (Of course, pH does have a large effect on ionizable organics, and the presence of a salt would serve to decrease the solubility of these compounds.) E_H, however, can greatly influence biotic and abiotic degradation reactions. Some pollutants are easily degraded in aerobic environments by microbes, while other pollutants are more easily degraded by biotic and abiotic processes under anaerobic or reducing conditions. Thus, biological and abiotic processes can be very important removal mechanisms for organic pollutants.

In the atmosphere and in surface waters, photochemical degradations can also be important.

Sorption phenomena are very important for hydrophobic pollutants, since these pollutants would rather be on any surface than dissolved in a polar solvent like water. Thus, sorption to NOM and mineral surfaces is important. In general, K_p values for hydrophobic pollutants are orders of magnitude greater than K_d values for metal pollutants. As with metal pollutants, sorption phenomena will be important in lake, stream (especially estuaries), and groundwater systems.

So, how do we put chemistry into the fate and transport modeling approach? To understand this, we must introduce the concepts of box models and mass balance. In environmental modeling, it is important to define your system, and we do this by using boxes (or compartments by others). For example, if we are studying the transport of a pollutant in the atmosphere or in a groundwater system, we define the section of the system we are interested in with a box of physical dimensions equal to that of the system under study. Next, we account for all the pollutant mass entering, reacting in, being retained in, or exiting the system (a mass balance). Of course, this means we use a lot of mathematics, and the second section of this text will deal with the development and use of models to describe the fate and transport of pollutants in lakes, rivers, groundwater, and atmospheric systems. Similar to the last section, test your knowledge of this chapter for hydrophobic pollutants by filling out Table 3.6.

3.9 Basic Approach to Fate and Transport Modeling

The basic approach for our mass balance in each of the following chapters will be represented by

Change of mass in system with time	=	Sum of all inputs	+	Sum of internal sources	−	Sum of all outputs	−	Sum of all internal sinks
Δmass/Δt		Mass of pollutant input		Any source or generation of the pollutant from within the system		Mass of pollutant exiting the system		Removal from the system by sorption or degradation reactions

We have already mentioned most of these terms in this book. For example, the "mass of pollutant input" can be controlled by point and non-point sources; pulse and step inputs; and solubility, vapor pressure, and Henry's law constants. The term "any source or generation of the pollutant from within the system" can be illustrated by desorption from the sediments (K_d or K_p) or by the generation of an atmospheric pollutant by photochemical reactions. The "mass of pollutant exiting the system" can be represented by the outflow from a lake, river, or the specific section of an aquifer or the atmosphere. We will show detailed examples of each of these in the fate and transport chapters.

Finally, let us concentrate on the last term in the mass balance equation, the sum of internal sinks. This is where kinetic transformation/degradation reactions come into the equation. Recall that many of our reactions were found to be or were simplified to be first-order reactions. This makes life much easier for the modelers, since if all of the reactions are of the same rate order (first, in our case), we can add the individual rate constants together and have one overall first-order rate constant. Thus, say we have a pollutant that is biodegraded with a rate constant of 0.05 $days^{-1}$ and is photochemically degraded such that k equals 0.005 $days^{-1}$. Instead of deriving a much more complicated equation with two kinetic variables, we can simply add the two k values together to obtain an overall rate constant of 0.055 $days^{-1}$ and have one kinetic term in the equation. Any number of first-order rate constants can be added together, but may compound result interpretation. This will become clearer in the fate and transport chapters when we give each general transport equation.

3.9.1.1 A Closer Look: Calculation of a Partition Coefficient from Experimental Data

A K_{oc} is to be determined for the sorption of 2,2′4,4′,6,6′-hexachlorobiphenyl (HCB, a PCB congener) on a sediment sample. The organic content of the sediment is 2.05%. A 2.05 mg/l solution of HCB is prepared, and ~40 ml of the solution is placed in a vial containing 0.102 g of dry sediment. The final solution volume is 40.0 ml. The sample is mixed for 3 days, the aqueous and sediment phases are separated with a 1.0 μm glass fiber filter, and the aqueous phase is measured for HCB. A concentration of 0.506 mg/l is measured in the aqueous phase. Analysis of a blank vial (containing water and HCB but no sediment) measures a concentration of 1.88 mg/l of HCB in the dissolved phase. What are the K_p and K_{oc} for the sample?

Table 3.6 Factors affecting sorption/desorption of a hydrophobic pollutant.

For an organic pollutant				
Parameter	**Degradation/transformation**	**Mobility**	**Bioaccumulation**	**Sorption**
pH				
E_H				
Solubility				
Complexation				
Vapor pressure				
HLC				
Acid/base				
Abiotic/biotic				
Photic				
DOM				
NOM				
Colloids				
Sorption				
Particle size				
Kinetics				

1.	Total mass (mg) of HCB added each flask	0.0820	
2.	Mass recovered in blank (mg)	0.0752	
3.	Mass of HCB in water phase (mg) of mixture	0.0202	
4.	Volume of water (l)	0.0400	
5.	Concentration of HCB in water phase (mg/l)		0.506
6.	Mass of HCB on solid phase (mg) (box 2–3)	0.0550	
7.	Mass of solid phase (kg)	1.02×10^{-4}	
8.	Concentration of HCB on solid phase (mg/kg)		539
9.	K_p		1065
10.	K_{oc} (K_p/f_{oc}) (l/kg)		51 951

Row 1: First we should determine the mass of HCB in each flask. This is done by multiplying the HCB concentration by the volume of solution: (2.05 mg/)*(0.0400 l) = 0.0820 mg.

Row 2: Note that not all of the HCB added to each flask was recovered during the analysis of the blank (1.88 mg/l * 0.0400 = 0.0752 mg). This observation is common in laboratory experiments. Thus, we must assume that all the flasks only have 0.0752 mg of recoverable HCB in them. Sources of loss of HCB could include volatilization from the aqueous phase during solution preparation or sorption to the vial walls or vial top.

Row 3: Next the concentration of HCB in the vial containing sediment is calculated: (0.506 mg/l * 0.040 l = 0.0202 mg). We will use the concentration in the water to calculate K_p.

Row 6: Next the mass of HCB associated with the sediment is determined by subtracting the mass measured in the dissolved phase from the mass in the blank: 0.0752–0.0202 = 0.0550 mg.

Row 8: Next the concentration of HCB on the sediment is determined: 0.0550 mg/1.02×10^{-4} kg = 539 mg/kg.

Row 9: The K_p is calculated from the ratio of sediment-phase concentration to dissolved-phase concentration: 539 mg/kg/0.506 mg/l = 1068 l/kg.

Row 10: K_{oc} is calculated by multiplying the K_p by the fraction of organic carbon present in the sample: 1068 l/kg * 0.205 = 219 l/kg.

3.9.1.2 A Closer Look: Determination of an Average K_d Based on a Set of K_d Measurements

A more accurate determination of K_d or K_p can be made when the parameter is measured over a range of pollutant concentrations. For example, a set of K_d experiments were conducted to investigate the adsorption of Pb on a soil. It is important to note that all of the vials contained the same mass of sediment. This is important because K_d can be a function of suspended solid concentration (sediment). The experiments were conducted in a manner similar to that described in our other examples. The following results were then compiled:

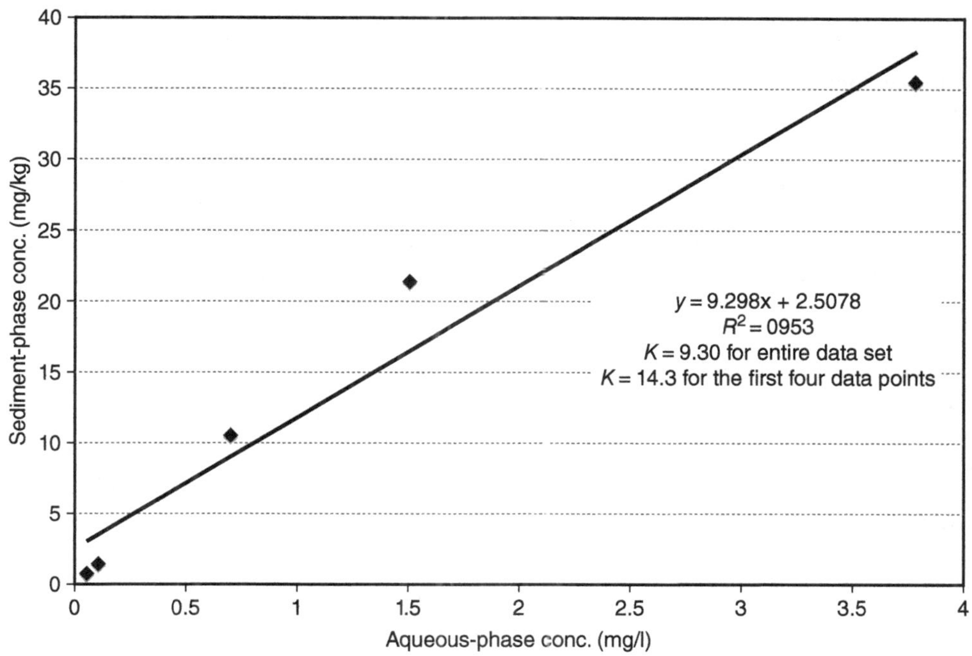

Figure 3.15 Example of equilibrium sorption results (*K*) for a hydrophobic pollutant as a function of pollutant concentration (mass of pollutant added to the system).

Aqueous-phase concentration (mg/l)	Sediment-phase concentration (mg/kg)
0.0500	0.725
0.103	1.42
0.698	10.5
1.50	21.4
3.78	35.5

To solve this problem, we must first plot the data and determine the slope of the line (which equals the K_d). The results are shown in Figure 3.15. Note that if all the data are used, the plot is a straight line, until it levels off at the highest concentration. This is not uncommon in laboratory experiments when excessively high dissolved metal phase concentrations are used. This phenomenon results primarily from complete coverage of the sorption sites on the sediment and results in an excess concentration in the dissolved phase. When this occurs, it is more accurate to estimate K_d from the lower data points. A linear regression of the first four data points in the table yields a slope, and K_d, of 14.3 l/kg.

Exercises

3.1 Which is a more accurate representation of how toxic a pollutant is, activity or concentration?

3.2 What is the numeric range of the activity coefficient?

3.3 Calculate the ionic strength for the following:
a) A mixture of 0.050 M $CaCl_2$, 0.025 M NaCl, and 0.045 M KNO_3
b) A mixture of 0.097 M $CaCl_2$, 0.015 M KCl, and 0.405 M $NaNO_3$

3.4 Using your results from Problem 3.9.3 and the extended Debye–Hückel equation, calculate the activity of each cation and anion in the mixtures.

3.5 Acid rain is known to be harmful to the environment for multiple reasons, but few people realize that free Al^{3+} ions are one of its most harmful products. Ionic aluminum is toxic to fish at levels as low as 6.2 ppm Aluminum forms five ligand complexes with OH^- (complexation constants are given in Table 3.1), but the formation of these is understandably minimized at lower pHs. Draw a speciation plot using a spreadsheet showing the concentrations of each species at pHs ranging from 0 to 14. Assume that the solution is in equilibrium with the solid phase, $Al(OH)_3$. K_{sp} values are given in Table 2.7.

3.6 Mining operations have often been major contributors to groundwater and surface pollution. They introduce toxic metal concentrations as well as increasing the total ion concentration of an aquatic system. An especially troublesome phenomena encountered in mining operations is acid mine drainage (AMD). This occurs when groundwater runs over old mining sites or through old mining tunnels. At these sites, many minerals and metals have been exposed from the surrounding rock and are free to react with the water and the atmosphere. Pyrite (FeS_2) is a common mineral found at mine sites that can cause damage to aquatic ecosystems. When pyrite is exposed to air and water, it reacts to form sulfuric acid (H_2SO_4) and iron hydroxide $Fe(OH)_3$. This not only raises the pH of the stream, but introduces a solid metal hydroxide which can be toxic to fish and aquatic life. However, the pyrite reaction does not always result in solid iron hydroxide. Consider the following complexation reactions, and produce a speciation diagram (using a spreadsheet) for the various iron hydroxide complexes. Determine what the dominant species will be at a pH of 4.5, a common pH found in streams that suffer from AMD.

$Fe^{3+} + OH^- \Rightarrow FeOH^{2+}$ $K_1 = 6.31 \times 10^{11}$
$Fe^{3+} + OH^- \Rightarrow FeOH^{2+}$ $K_1 = 6.31 \times 10^{11}$
$Fe^{3+} + 2(OH^-) \Rightarrow Fe(OH)_2^+$ $\beta_2 = 2.00 \times 10^{22}$
$Fe(OH)_3(s) \Rightarrow Fe^{3+} + 3(OH^-)$ $K_{sp} = 1.10 \times 10^{-36}$
$Fe^{3+} + 4(OH^-) \Rightarrow Fe(OH)_4^-$ $\beta_4 = 2.51 \times 10^{34}$

3.7 You are curious about the speciation of Ni^{2+} in a solution with OH^- ions. Excess $Ni(OH)_2$ salt is added to a beaker with distilled water. Your goal is to determine the predominant species of Ni^{2+} and OH^- complexes based on the OH^- concentration. Since there is excess salt in the solution, $Ni(OH)_2$ controls the main species present and its concentration by the K_{sp} of the salt (5.48×10^{-16}). Draw a speciation plot using OH^- concentrations from 1.00×10^{-18} to 1.00×10^{-8} M and the following log K values for binding between the Ni^{2+} and OH^- ions:
for $NiOH^-$: 4.1
for $Ni(OH)_2$: 9.0
for $Ni(OH)_3^-$: 12.0
for $Ni(OH)_4^{2-}$: 17.2

3.8 Silver bromide is used in photography as a developing emulsion. Given the K_{sp}s (Table 2.7) and stability constants (Table 2.6) for Br^- ligands, determine how the concentration of each Ag species in solution varies as the Br^- concentration changes. Construct a graph of the log[Br^-] from −1.0 to 6.0 versus log[Ag species] from 0.0 to −35. Include a line for total [Ag] as [Br^-] varies.

3.9 Cobalt is usually emitted during the production of steel and other alloys, and specifically in the production of airline engines and gas turbines. Ionic cobalt can be carcinogenic, but the amount of consumption must be very high in order for cancer to occur. Create a speciation diagram using a spreadsheet for a cobalt hydroxide system without any solid phase present (Section 3.2.2.2). The total concentration of Co^{2+} species is 20 ppm. The hydroxide concentrations should range from 1×10^{-1} to 1×10^{-14}.
Helpful equations

$$Co^{2+} + OH \Rightarrow CoOH \quad K_1 = 5.01\ 187 \times 10^{-5}$$

$$Co^{2+} + 2OH \Rightarrow Co(OH)_2 \quad \beta_2 = 6.30\,957 \times 10^{-10}$$

$$Co^{2+} + 3OH \Rightarrow Co(OH)_3 \quad \beta_3 = 3.16\,228 \times 10^{-11}$$

3.10 Mercury sulfate was tested in a renegade agricultural operation as a mildew inhibitor. A few acres were dusted with $HgSO_4$, and then irrigation water and rainfall washed the compound into a nearby holding basin. The EPA caught wind of the illegal activity and sent you in to investigate. Since there is no solid phase present, you will be given the K and β values for Hg and SO_4^{2-} ligands. Your assignment is to show how the concentration of each Hg species changes as the SO_4^{2-} concentration is varied. Create a labeled graph of the log[Hg] from 0.0 to −30.0 versus log [SO_4^{2-}] from −10.0 to 10.0. Be sure to include how the total [Hg] changes overall.

3.11 An aspiring restaurateur purchases a cheap plot of land to open a restaurant, complete with a less than pristine pond. In an effort to make the restaurant more scenic, she valiantly undertakes the task of cleaning up garbage in the pond. In doing so, she discovers a 55-gal drum labeled "Barium Waste." She then hires you, a contractor, to determine what has been chemically disposed of in the pond. You find that the concentration of dissolved barium in the pond is 10.5 ppm, with no solid phase present in the sediments. Using the equations in this chapter, determine the concentration of Ba^{2+} in the water as a function of CO_3^{2-} concentration. Use CO_3^{2-} concentrations of 0.000 100, 0.001, 0.100, 1.00, and 5.00 M.

3.12 Ethylenediaminetetraacetic acid (EDTA) is a widely used chemical in industry and is very good at complexing metal ions. We can use EDTA as a complexing agent for mercury, which lowers the mercury's toxicity but at the same time increases its mobility by keeping it in solution. Draw a speciation diagram for a closed system (Section 3.2.2.2) of 10 ppm Hg as a function of EDTA concentration. Use a log[EDTA] concentration range from 5 to −35.

$$Hg^{2+} + EDTA^{2-} \rightarrow HgEDTA \; K_1 = 3.16 \times 10^{23}$$

$$Hg^{2+} + HEDTA^{-} \rightarrow HgHEDTA^{+} \; \beta_1 = 1.00 \times 10^{27}$$

3.13 Cadmium metal ions form three complexes with SO_4^{2-} in the absence of a solid. Draw a speciation plot of Cd^{2+} in the presence of SO_4^{2-} ions using a Cd^{2+} concentration of 40.5 ppm and a concentration of SO_4^{2-} ranging from 1.00×10^{-6} to 1.0 M. The following binding log Ks will be useful:
for $CdSO_4$: 2.3
for $Cd(SO_4)_2^{2-}$: 3.2
for $Cd(SO_4)_3^{4-}$: 2.7

3.14 The leachate from a local landfill has been suspected of containing high cadmium concentrations, and the city downstream from the watershed has begun monitoring the streams for cadmium. In order to better predict transport, they need to determine the distribution coefficient for cadmium between the stream water and the local soil. Two liters of sample water were collected, filtered, and dried. The dried filtered particulate matter weighed 10.0 g and was used to determine the total suspended solids (TSSs) (5000 mg/l). Sediment samples of the local soil (250 mg) were prepared in ~50 ml sample bottles to create the same TSS as the local streams. Cadmium (0.375 mg) was added to each of the sample vials, including the blanks, and water was added for final solution volumes of 50.0 ml. The solutions were mixed for three days and then filtered with 0.20 μm filters and analyzed by flame atomic absorption spectrometry. The equilibrium cadmium concentration in the aqueous phase was 5.00 mg/l. Calculate the K_d for Cd^{2+} on the soil.

3.15 Aldrin is a chlorinated pesticide that was used to regulate termite populations until the 1970s. After application, this non-biodegradable pesticide found its way into freshwater systems, poisoning organisms. A sample of lake water sediment was taken to determine how well Aldrin adsorbs onto solid sediment. For this experiment, 0.1147 mg of Aldrin were added to a flask containing 100.0 ml of water and 4.26×10^{-4} kg of lake sediment. Through gas chromatographic-electron capture detection (GC-ECD), the equilibrium concentration of Aldrin in the aqueous phase was determined to be 0.0180 ppm Determine the mass of Aldrin in the aqueous and solid phases and calculate the K_p value for the pollutant Aldrin.

Total mass (mg) of pollutant added to flask	0.1147
Mass of pollutant in aqueous phase (mg)	
Volume of water (ml)	100.0
Concentration of pollutant measured in aqueous phase (mg/l)	0.0180
Mass of pollutant on solid phase (mg)	
Mass of solid phase (kg)	4.26×10^{-4}
Concentration of pollutant on solid phase (mg/kg)	
K_p	

3.16 Kepone (chlordecone), a carcinogenic, tan to white crystalline solid or powder that is insoluble in water, was used as an insecticide, fungicide, and larvacide on bananas, tobacco, and other domestic plants. The US EPA banned the use of Kepone in 1975, but the chemical is still used in some countries. Before the termination of its production in the United States, large amounts of Kepone were dumped into the upper James River. This Kepone poses a threat to the fish and other marine animals, as well as to the groundwater supply. Most of the Kepone in the James River has settled into the sand and sediments at the bottom. A K_p needs to be determined for the sorption of Kepone on a sediment sample. A solution of 0.02 453 mg/l Kepone is prepared. One hundred milliliter (100.0 ml) of the solution is placed in a vial with 0.000 100 kg of dry sediment. The sample is mixed for three days, and the aqueous and solid phases are separated using a 0.20 μm glass fiber filter. The mass of Kepone in the aqueous phase is measured with gas chromatography. From this, the concentration of Kepone in the aqueous phase is determined to be 0.02 108 mg/l. Analysis of the blank shows no loss of kepone from absorption onto the vial wall. Find the K_p for the sample using the following chart.

1	Total mass (mg) of kepone added to each flask
2	Mass of kepone in water phase (mg) of mixture
3	Volume (l) of water
4	Concentration of kepone in water phase (mg/l)
5	Mass of kepone on solid phase (mg)
6	Mass of sediment in vial (kg)
7	Concentration of kepone on solid phase (mg/kg)
8	K_p

3.17 The EPA just received a report of a small-scale dumping site of lead acid batteries in a city under your jurisdiction of investigation. The site contains various car and tractor batteries located on a downhill slope near a stream. It is your job to determine the extent of the contamination of lead pollution in the stream, accounting for the concentration in the aqueous phase as well as the sorption to the solid particles in the stream – in other words, you need to determine the K_d (distribution coefficient) of lead. To determine the K_d, you add 0.100 g clay to 0.1000 l water and then add 0.04 976 mg Pb^{2+} to the sample. After allowing the samples to equilibrate for a minimum of three days, you analyze them on a flame atomic absorption spectroscopy system and find that there is 0.500 mg/l Pb in the aqueous phase of each sample. You also run a blank and find that the amount of Pb lost to sorption in the test tubes is 0.01 000 mg/l. Use these values to complete the table and determine the K_d of Pb in the samples.

1. Total Pb in sample (mg)	0.0 510
2. Volume water in sample (l)	0.1 000
3. [Pb] aqueous phase of sample (mg/l)	0.0 500
4. [Pb] lost in blank (mg/l)	0.01 000
5. Mass Pb in aqueous phase (mg)	
6. Solid phase mass (mass of clay) (g)	0.1 000
7. Mass of Pb in solid phase (mg)	
8. [Pb] solid phase (mg/kg)	
9. K_d	

3.18 We have discussed equilibrium in terms of K_d and K_p, and kinetics in terms of first-order sorption and desorption rates. Explain how each can be important in determining the fate and transport of pollutants in natural water systems.

3.19 Select a metal pollutant. Using the information from Chapters 2 and 3, explain all chemical processes that can influence its fate and transport in an aqueous system. Which processes will increase the transport? Which processes will decrease the transport? Limit your discussion to three typed pages. A good place to start is to outline the important processes in each chapter.

3.20 Select a hydrophobic pollutant. Using the information from Chapters 2 and 3, explain all chemical processes that can influence its fate and transport in an aqueous system. Which processes will increase the transport? Which processes will decrease the transport? Limit your discussion to three typed pages. A good place to start is to outline the important processes in each chapter.

3.21 Calculations for the comparison of sea, river, and groundwater composition. Below are ionic compositions for three types of water: river water, groundwater, and sea water.

Ion	Average river water[a] (mg/l)	Average river water, M (mol/l)	Average river water (mM)
HCO_3^-			96
Ca^{2+}			38
Mg^{2+}			34
Na^+			27
Cl^-			2.2
SO_4^{2-}			1.2
K^+			0.59
F^-			0.053

[a] Baird (1995).

Ion	Average ground water[a] (mg/l)	Average ground water, M (mol/l)	Average ground water (mM)
HCO_3^-	93.6		
Ca^{2+}	54.1		
Mg^{2+}	13.1		
Na^+	37.0		
Cl^-	16.5		
SO_4^{2-}	70		
K^+	1.55		

[a] https://web.viu.ca/earle/geol304/304g.pdf (accessed 22 January 2018).

Ion	Average sea water[a] (mg/l)	Average sea water, M (mol/l)	Average sea water (mmol/l)
HCO_3^-	140		
Ca^{2+}	400		
Mg^{2+}	1 260		
Sr^{2+}	13		
Na^+	10 600		
Cl^-	19 000		
SO_4^{2-}	2 650		
K^+	380		
Br^-	65		
BO_3^{3-}	26		

[a] https://www.lenntech.com/composition-seawater.htm (accessed 22 January 2018).

For each set of data, calculate in the missing data, perform an anion–cation balance, and calculate the ion strength and activity coefficient.

Bibliography

Ahn, I.S., Lion, L.W., and Shuler, M.L. (1999). Validation of a hybrid "two-site gamma" model for naphthalene desorption studies. *Environ. Sci. Technol.* **33**: 3241–3248.

Baes, C.F. Jr. and Mesmer, R.E. (1976). *The Hydrolysis of Cations*. New York: Wiley.

Baird, C. (1995). *Environmental Chemistry*, 336. New York: W.H. Freeman and Company.

Bell, J. and Bates, T.H. (1988). Distribution Coefficients of Radionuclides Between Soils and Groundwater and their Dependence on Various Test Parameters. *Sci. Total Environ.* **69**: 297–317.

Brookins, D.G. (1988). *Eh-pH Diagrams for Geochemistry*. New York: Springer-Verlag.

Coates, J.T. (1984). Sorption equilibria and kinetics for selected polychlorinated biphenyls on river sediments. PhD dissertation. Environmental Systems Engineering, Clemson University, Clemson, SC.

Coates, J.T. and Elzerman, A.W. (1986). Desorption kinetics for selected PCB congeners from river sediments. *J. Contam. Hydrol.* **1**: 191–210.

DeLaune, R.D., Reddy, C.N., and Patrick, W.H. Jr., (1981). Effects of pH and redox potential on concentration of dissolved nutrients in an estuarine sediment. *J. Environ. Qual.* **10** (3): 276–279.

Dunnivant, F.M. (1988). Congener-specific PCB chemical and physical parameters for evaluation of environmental weathering of aroclors. PhD dissertation. Environmental Systems Engineering, Clemson University, Clemson, SC.

Dunnivant, F.M., Coates, J.T., and Elzerman, A.W. (2005). Labile and non-labile desorption rate constants for 33 PCB congeners from lake sediment suspensions. *Chemosphere* **61**: 332–340.

Elzerman, A.W. and Coates, J.T. (1987). Hydrophobic organic compounds on sediments: equilibria and kinetics of sorption. In: *Sources and Fates of Aquatic Pollutants*, ACS Advances in Chemistry Series #216 (ed. R.A. Hites and S.J. Eisenreich). Washington, DC: American Chemical Society.

Ginsbach, J.W., Taylor, K.L., Olsen, R.M. et al. (2010). The fraction of organic carbon predicts labile desorption rates of chlorinated organic pollutants in laboratory-spiked geosorbents. *Environ. Toxicol. Chem.* **29** (5): 1049–1055.

Gschwend, P.M. and Wu, S.C. (1985). On the constancy of sediment water partition coefficients of hydrophobic organic pollutants. *Environ. Sci. Technol.* **19**: 90–96.

Weast, R.C. and Astle, M.J. (eds.) (1980). *Handbook of Chemistry and Physics*, 61e, B-242.

Handlin, M., Molina, A., James, N.M. et al. (2014). Release rates of persistent hydrophobic pollutant in suspensions containing both contaminated and uncontaminated sediment. *J. Environ. Toxicol. Chem.* **33** (2): 375–381.

Harris, D.C. (1999). *Quantitative Chemical Analysis*, 5e, 180. W.H. Freeman and Company, Table 8-1, selected data (original source of data: J. Keilland, J. Am. Chem. Soc. 1939, 59:1675).

Hem, J.D. (1985). Study and interpretation of the chemical characteristics of natural waters. U.S. Geological Survey Water Supply Paper 2254.

Karickhoff, S.W. (1984). Organic pollutant sorption in aquatic systems. *J. Hydraul. Eng.* **110** (6): 707–735.

Karickhoff, S.W. and Morris, K.W. (1985). Sorption dynamics of hydrophobic pollutants in sediment suspensions. *Environ. Toxicol. Chem.* **4**: 467–479.

Karickhoff, S.W., Brown, D.S., and Scott, T.A. (1979). Sorption of hydrophobic pollutants on natural sediments. *Water Res.* **13**: 241–248.

Lindsay, W.L. (1979). *Chemical Equilibria in Soils*. New York: Wiley.

Martell, A.R. and Smith, R.M. (1974). *Critical Stability Constants*, vol. 1, Amino Acids. New York: Plenum.

Martell, A.E. and Smith, R.M. (1975). *Critical Stability Constants*, vol. 2, Amines. New York: Plenum.

A.E. Martell and R.M. Smith, Vol. 4, Inorganic Ligands, New York: Plenum, 1976.

Martell, A.E. and Smith, R.M. (1977). *Critical Stability Constants*, vol. 3, Other Organic Ligands. New York: Plenum.

Morel, F.M.M. and Hering, J.G. (1993). *Principles and Applications of Aquatic Chemistry*, 332–341. New York, selected data from: Wiley.

Morgan, J.J. (1967). *Principles and Application in Water Chemistry* (ed. S.D. Faust and J.V. Hunter). New York: Wiley.

O'Connor, D.J. and Connolly, J.P. (1980). The effect of concentration of absorbing solids on the partition coefficient. *Water Res.* **14**: 1517–1523.

Sainte Marie, J., Torma, A.E., and Guebeli, A.O. (1964). The stability of thiocompleses and solubility products of metal sulphides: I. Cadmium sulphide. *Can. J. Chem.* **42**: 662–668.

Sanda, W.G. and Hanson, P.J. (1979). *Chemical Modeling in Aqueous Systems*, ACS Symposium Series 93 (ed. E.A. Jenne). Washington, DC: American Chemical Society.

Sposito, G. (1984). *The Surface Chemistry of Soils*. New York: Oxford University Press.

Stumm, W. and Morgan, J.J. (1981). *Aquatic Chemistry: An Introduction Emphasizing Chemical Equilibria in Natural Waters*, 2e. New York: Wiley.

Stumm, W. and Morgan, J.J. (1996). *Aquatic Chemistry: Chemical Equilibria and Rates in Natural Waters*, 3e. New York: Wiley.

Whitfield, M. (1974). The ion-association model and the buffer capacity of the carbon dioxide system in seawater at 25°C and 1 atm. total pressure. *Limnol. Oceanogr.* **19**: 235–248.

Wu, S. and Gschwend, P.M. (1986). Sorption kinetics of hydrophobic organic compounds to natural sediments and soils. *Environ. Sci. Technol.* **20**: 1213–1217.

Part III

Modeling

[Mathematics] The Handmaiden of the Sciences

Eric Temple Bell

For every problem, there is one solution which is simple, neat and wrong

Henry Louis Mencken

Pollutant Fate and Transport in Environmental Multimedia, First Edition. Frank M. Dunnivant and Elliot Anders.
© 2019 John Wiley & Sons, Inc. Published 2019 by John Wiley & Sons, Inc.
Companion website: www.wiley.com/go/Dunnivant/pollutant_fate&Transp_EnvironMultimedia

4

An Overview of Pollutant Fate and Transport Modeling

Models are often used to represent more complicated or larger systems. Perhaps the most common model we use is a road map. Scientists and engineers use models to help understand how something in the past happened or to predict what will happen in the future. These models can be very simple or very complicated, depending on the system being imitated or the accuracy desired in the calculations. In this chapter, we will develop some simple mathematical models to introduce the modeling approaches. Then we will attempt to explain how fate and transport models are developed, starting with the relatively simple models presented in later chapters and ending with the modeling approaches used by professionals to study very complicated environmental systems. Finally, we will look at how good our modeling approaches are and what we do with the final numbers produced by these models.

4.1 Modeling Approaches

4.1.1 Algebraic Solutions

A linear model is the simplest and easiest to understand, since we tend to think in a linear manner. Figure 4.1 shows a common plot used in chemistry to calibrate an instrument. Instruments that measure pollutant concentrations need daily calibration, accomplished by analyzing solutions with known chemical concentrations. The instrument responds to these different known concentration levels with a proportional signal. These outputs are created by the instrument's interaction with the sample, and expressed in units such as millivolts and transmission (a function of absorbance) or calculated values such as peak height and peak area. We use this data to make a calibration plot such as the one shown in Figure 4.1. Thus, we derive a linear relationship ($y = 10.0x + 0.05$, or the type: $y = mx + b$) between instrument response and concentration, and we can then use the instrument to measure a solution containing a chemical or pollutant. If the instrument produces a response of 65 units (as shown on the y-axis) for a sample, we can trace the response over to the calibration line, and then down to the concentration line and determine the concentration of pollutant in the sample (6.49 mg/l in this example). This is typically how instrumental measurements are made.

Similarly, using a road map model to go from point A to C, we might first chose Highway 1 from Boston (point B), then Interstate 95 to Miami (point C). Clearly, the road is linearly connected: point B follows A, and C follows point B. Another useful interpretation is that events are related by causes and effects in that one follows the other in a stepwise linear manner.

This concept can be extended to environmental systems, where we look at flow rate and travel time in a system. If the water or wind flow rate is 2.00 m/s, then we can calculate that it will take 10 seconds to travel 20.0 m. Similarly, using a road map linear relationships are easy to understand, but unfortunately many environmental processes are not linear. As we discussed in Section 3.4, many processes are first order with respect to reaction rates (a nonlinear relationship). Figure 4.2a,b shows two common exponential plots that follow the first-order kinetic model. When the concentration values are transformed using a natural log function, a linear plot of $\ln(C)$ versus time is obtained. However, there is no simple way to take the log transformation in our relatively more complicated modeling equations and linearize our results. We must learn to identify and interpret log functions and plots; these functions operate much like the linear model from Figure 4.1 by providing an equation that can be used to predict concentrations for the nonlinear models. As a review exercise, you should transform the data from Figure 4.2a into a linear $\ln(C)$ versus time plot.

4.1.2 Modeling Using Differential Equations

The models used to predict the concentrations of pollutants in environmental media mostly fall into a class of equations referred to as differential equations, which fall into a special class of calculus that we will discuss in a moment. First, it is important that we show the complete

Pollutant Fate and Transport in Environmental Multimedia, First Edition. Frank M. Dunnivant and Elliot Anders.
© 2019 John Wiley & Sons, Inc. Published 2019 by John Wiley & Sons, Inc.
Companion website: www.wiley.com/go/Dunnivant/pollutant_fate&Transp_EnvironMultimedia

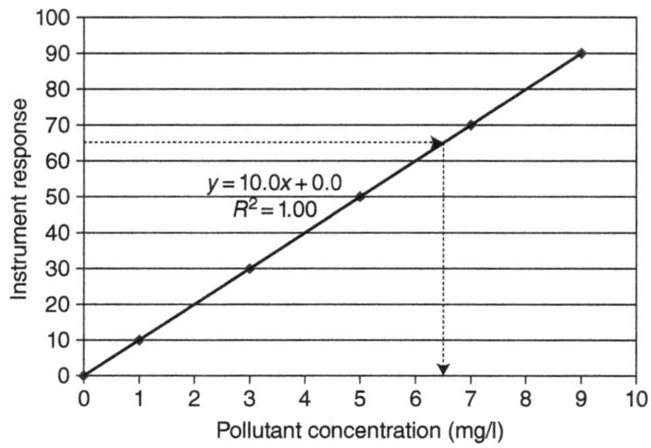

Figure 4.1 A linear calibration model.

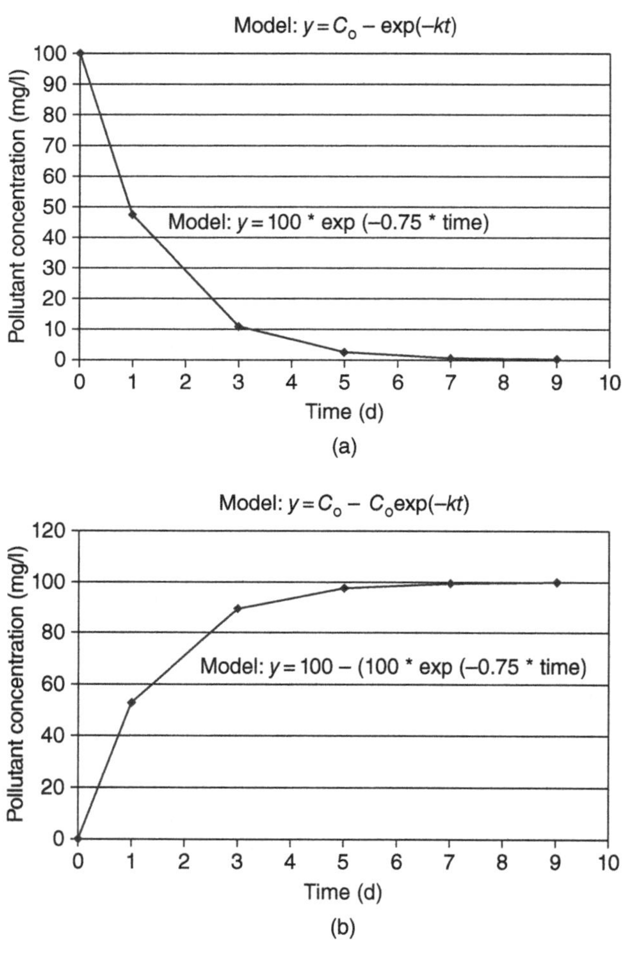

Figure 4.2 Two common exponential (first-order) models. (a) Decreasing exponential and (b) increasing exponential.

development of one fate and transport equation, so that you understand the entire process. To do this, we will use the lake pulse and step scenarios, which are the simplest of the systems that we will cover in this textbook. It is not critical that you understand every step, especially if you have not taken calculus. Our goal is to show the overall process in the development of a governing fate and transport equation (like the equations used in Fate®).

The first step in developing the governing equations for the fate of a pollutant in a lake system is to set up a mass balance for the system. We intuitively should know that mass cannot be created or destroyed only changed in form even if converted to heat energy by combustion. Even in this case, mass is conserved. First, quantify all of the mass inputs of pollutant to the system (Eq. (4.1)). This can be expressed as

$$W = Q_{\mathrm{w}}\,C_{\mathrm{w}} + Q_{\mathrm{i}}\,C_{\mathrm{i}} + Q_{\mathrm{trib}}\,C_{\mathrm{trib}} + P\,A_{\mathrm{s}}\,C_{\mathrm{p}} + V\,C_{\mathrm{s}} \tag{4.1}$$

where

W = the mass input of pollutant to the lake per unit time, kg/time

Q_{w} = the inflow rate of the wastewater, m^3/time

C_{w} = the pollutant concentration in the wastewater, kg/m^3

Q_{i} = the inflow rate of the main river, m^3/time

C_{i} = the pollutant concentration in the main inlet river, kg/m^3

Q_{trib} = the net inflow rate from all other tributaries, m^3/time

C_{trib} = the net pollutant concentration of the tributaries, kg/m^3

P = annual precipitation, m/time

A_{s} = mean lake surface area, m^2

C_{p} = the net pollutant concentration in precipitation, kg/m^3

V = the average lake or system volume, m^3

C_{s} = the average pollutant release from suspended lake sediments, kg/m^3-time

A useful exercise is to check the units of each term (mass/unit time) or rate is each part of the mass balance. We will see later in this text that rate for a given plane is known as flux and has units of mass/time (area) or mg/m^2 day.

In most situations, the mass inputs from the smaller tributaries and precipitation are minor compared to the major input source, and the terms C_{trib} and C_{p} are ignored. We will further simplify the mass input expression here by assuming that the contribution from contaminated sediments is negligible, although this may not always be the case in certain lakes where benthic or sediment processes are critical. These assumptions simplify the input expression to Eq. (4.2).

$$W = Q_w C_w + Q_i C_i \tag{4.2}$$

Next, in order to set up a mass balance for the pollutant across the entire system, we need to incorporate outflow into the expression. The equation for the mass outflow from the system is analogous to the input equation, but with only one outlet usually considered. We also assume that there are no additional sources of pollutant and that chemical degradation yields a constant first-order chemical removal rate for the pollutant. Thus, the mass balance becomes

$$\text{Change in mass} = \text{inflow} - \text{outflow} + \text{sources} - \text{sinks}$$

$$V\,dC = (Q_w C_w\,dt + Q_i C_i\,dt) - Q_e C\,dt + 0 - V C k\,dt, \quad \text{or}$$

$$V\,dC = W\,dt - Q_e C\,dt - V C k\,dt \tag{4.3}$$

where

dC or ΔC = the change in pollutant concentration in the lake

dt or Δt = the incremental change in time dC/dt or $\Delta C/\Delta t$ is the fundamental theorem of differential calculus meaning a "rate of change"

Q_e = the outlet or effluent flow from the lake (m^3/time)

Q_w = the waste flow to the lake (m^3/time)

Q_i = the inlet or influent flow from the lake (m^3/time)

C = the average lake concentration, kg/m^3

k = the first-order removal rate for the pollutant, 1/time

Note that since we are looking at changes in concentration and time, we must use an operator to indicate this. The operator is the d or Δ symbol.

The last equation, on substitution and rearrangement, yields

$$[Q_e C - W(t)]\,dt + V dC = -V C k\,dt \tag{4.4}$$

On rearrangement, this reduces to

$$V\frac{dC}{dt} + (Q_e + k V)\; C = W(t) \tag{4.5}$$

where Q_e, k, and V of the lake, and thus the quantity $(Q_e + kV)$, are assumed to be constant (usually, they are reasonably constant).

The average detention time (t_0) of water (and thus the pollutant) in the lake, from the time of input to outlet, is defined as

$$t_0 = \frac{V}{Q_e} \tag{4.6}$$

Substitution and further rearrangement into Eq. (4.5) yields

$$V\frac{dC}{dt} + C\,V\left(\frac{1}{t_0} + k\right) = W \tag{4.7}$$

Equation (4.7) is the unintegrated, first-order differential equation (indicated by the dC/dt, called the derivative of concentration C with respect to time t, and representing the rate of change of concentration with respect to time). It expresses how concentration of pollutant in the lake changes with time in response to the lake volume and the rates of flow and degradation of the pollutant. The technique used for solving this equation, for the function $C_{(t)}$, depends on the nature of the input source (instantaneous or pulse versus continuous or step).

4.1.2.1 Integration for the Instantaneous Pollutant Input Model

When the mass input with time from all sources, $W(t)$, is zero, we approach what is referred to as an instantaneous input. In this case, an instantaneous input is characterized as a one-time, finite (known) addition of pollutant mass to the lake. Thus, there is no pollutant added over time, and $W(t)$ is zero. For example, the release of a pollutant by a marine shipping accident would be an instantaneous input, as would a short release from an industry located on the lake. In order to solve this equation for $C_{(t)}$, we must integrate Eq. (4.7) (shown in the appendix in Chapter 5, lake systems). Under these conditions, integration, using basic calculus techniques with $W = 0$, yields

$$C_{(t)} = C_0\,e^{-\left(\frac{Q_e}{V}+k\right)t} \quad \text{or} \quad C_{(t)} = C_0\,e^{-\left(\frac{1}{t_0}+k\right)t} \tag{4.8}$$

where C_0 is the initial pollutant concentration.

This equation, specifically in its second form above, would be used to simulate the pollutant concentration versus time in a lake where an instantaneous release occurred.

But what is the purpose of deriving this equation? When you take a mathematical equation involving a non-negative function and integrate it, you are often developing a way of calculating the area under the curve or the function that the equation represents. A plot of the integrated equation (4.8) is shown in Figure 4.3. The trace of the curve is an exact solution to the governing equation and represents the pollutant concentration at a given time. The area under the curve is the mass balance of the system and represents the total mass of pollutant leaving the lake system as a function of time.

This is a form of mathematical integration, allowing us to take Eq. (4.8) and find a source function for it (an equation that defines the curve or line for a plot of

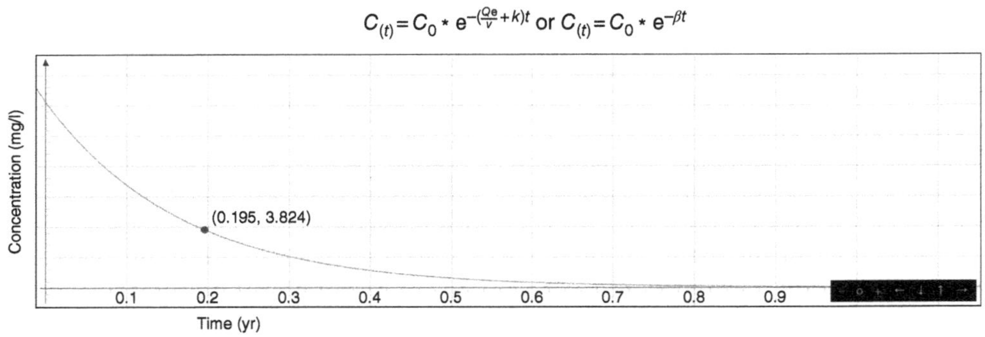

Figure 4.3 Output from Fate for an instantaneous input of pollution to a lake system. Source: Fate®.

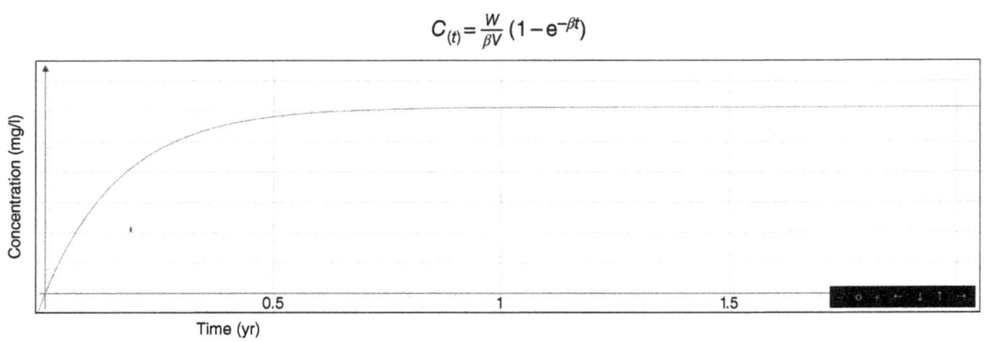

Figure 4.4 Output from Fate for a step input of pollution to a lake system. Source: Fate®.

concentration as a function of time). Many differential equations given in subsequent fate and transport modeling chapters are too complicated to solve using normal integration techniques learned in calculus. These more advanced integrations use Laplace transformations, the original equation is first simplified by algebraic substitutions to make it simpler and hence the term transformation. Next, the resulting simplified equation is integrated using normal calculus techniques. Finally, a reverse transformation is taken on the equation, in which the simplification is reversed and a more complicated form of the equation results, but the entire equation has now been integrated. This technique will be used in later chapters.

4.1.2.2 Integration for the Continuous (Step) Pollutant Input Model

We now will return to Eq. (4.7) to derive an equation describing the constant release of a pollutant into a lake (the derivation is shown in its entirety in the appendix following this chapter). This type of release is known as a step input, and an example would be the constant release from an industrial source (constant mass per time). Under these conditions, $W(t)$ is not zero (as assumed in the previous derivation) and normally there is some background concentration of pollutant in the lake system (such that C_o in the lake cannot be considered to be zero). Here, the net pollutant concentration in the lake (and the water leaving the lake in the effluent river) is the result of two opposing forces: (i) the concentration decreases caused by "flushing" of the lake via the effluent river and by first-order pollutant decay and (ii) the pollutant concentration increases due to the constant input from the source. If the waste load is constant (which we will assume), integration of Eq. (4.7) yields

$$C_{(t)} = \frac{W}{\beta V}(1 - e^{-\beta t}) + C_0\, e^{-\beta t} \tag{4.9}$$

where $\beta = 1/t_0 + k$ and C_0 is the background concentration of pollutant in the lake. If the background concentration in the lake is negligible, the equation reduces to

$$C_{(t)} = \frac{W}{\beta V}(1 - e^{-\beta t}) \tag{4.10}$$

Equations (4.9) and (4.10) can be used to estimate the concentration of a pollutant in a lake that receives a constant input of pollutant. A plot of this last equation is shown in Figure 4.4, where the line represents the pollutant concentration at a given time in the lake and the area under the plot represents the total mass of pollutant leaving in the outflow.

Equation (4.10) is distinguished from Eqs. (4.8) and (4.9) by a distinct boundary condition, which we place on the system to better define it. In the first example, the pulse input, the boundary condition was that the input pollutant concentration from the inlet river and the overall input of pollution with time, was zero. This simplifies the equation as noted earlier. In the second example, pollution entered the lake at a constant input.

4.1.3 General Approach for the Models Used in This Text

Differential equation techniques were used to derive all the governing equations given in the subsequent modeling chapters, which will express changing pollutant concentration with time in rivers, lakes, groundwater, and atmospheric systems. The derivation of these equations will be shown in detail, with the exception of the three-dimensional atmospheric equation. Our goal is to show the use of these equations and learn which model parameters affect pollutant concentrations.

But how realistic is our exact differential solution approach to modeling? For simple systems where you are only looking at a basic understanding, these approaches are fine. However, industry, the government, and the public demand a much more involved (and sometimes more accurate) representation of the environmental system being modeled. Thus, more complicated methods have been developed, and these are the ones used by professionals today. However, you should note that these are based on the same chemical and physical processes and related equations that we used in the examples given above. The difference is that while we will hold parameters such as water or air velocity, partition or distribution coefficients, and kinetic degradation rates constant in the governing equations used in the following chapters, professionals use modeling efforts that can allow these parameters to vary with location and time in the system (as shown in the closing of each fate and transport chapter). The technique used by professionals, numerical methods of analysis, is the subject of the coming section.

4.1.4 Numerical Methods of Analysis

Unlike in the differential equation solution approach, where we found a solution to Eq. (4.7), in numerical methods no governing equation is sought. Here we obtain our solution to the problem by simply calculating concentrations across the system subject to our same boundary conditions and/or based on a few known concentrations at defined points in the system. Two common numerical methods of analysis are finite element and finite difference approaches. We will discuss each separately.

First we will discuss the finite difference approach. We begin by placing a grid over the system under study, such as the one shown in Figure 4.5. Each corner point of the squares in the grid, called a nodal point, is represented by an x and y coordinate, and can have a unique equation for calculation of pollutant concentration. The surface area represented by the square is defined to have uniform properties such as water or air flow, water or air velocity, mixing, pollutant distribution coefficient, etc., but different nodal points can have different values for each of these parameters. This allows a more realistic mathematical representation of real-world environmental systems. The goal of the numerical analysis is to calculate pollutant concentrations for each node, located at the center of four nodal points (shown in the upper right-hand nodal cell). It should be noted that the representation in Figure 4.5 is a gross simplification and that the modeling of an actual environmental system could

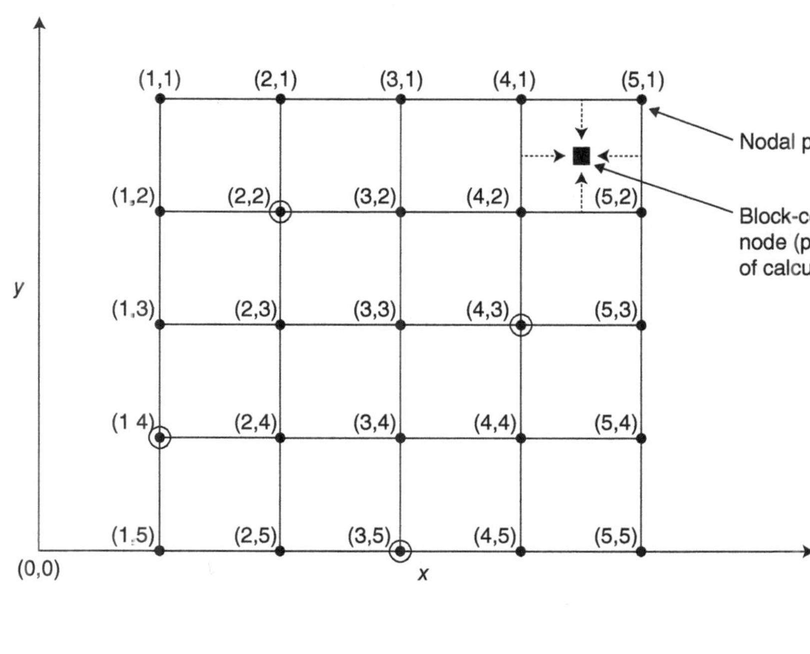

Figure 4.5 Illustration of a finite difference approach for estimating pollutant concentrations.

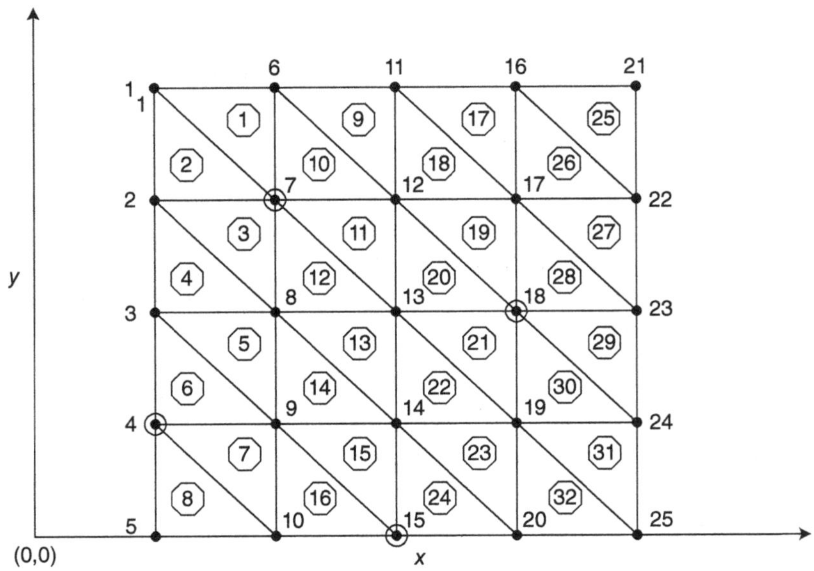

Figure 4.6 Illustration of a finite element approach for estimating pollutant concentrations.

have hundreds to thousands of nodal points composing the system.

We will return to our lake system example for the following discussion, but now we do not assume that mixing is constant across the lake, as was necessary in the basic calculus solution. Water, and associated pollutants, can enter the lake at point (1,1) and leave at point (5,5). The grid is overlaid onto a map of the lake, and the appropriate boundary conditions are applied. For simplification purposes, we will assume that our lake is a square corresponding to our nodal points.

Say we want to predict the pollutant concentration at the circled points, nodal points (1,4), (2,2), (3,5), and (4,3). The approach used in differential equations, in which we integrated and obtained a general solution equation, will not work here, since mixing in each note is different. To solve the equation using finite differences, we use Eq. (4.7) directly. Since each nodal cell can have a unique version of Eq. (4.7) to account for a different magnitude of mixing, we will calculate the concentration of each node (middle of each nodal cell). This is an iterative process governed by one or more boundary conditions. In this case, our boundary condition is the known total mass of pollutant that entered the system. After the computer running the finite difference method completes one set of calculations of each node, it can add up the mass in the system and see if it matches the known total input mass. If it does not, the method makes adjustments to the equations (in our case mixing) and recalculates the entire grid. It repeats this process until an acceptable mass balance is achieved. Completion of this process will yield pollutant concentrations for each node in the system.

The finite element approach differs from finite difference in that, in finite element, interpolation functions are used to define the concentration throughout the domain of each triangle instead of at one point, as in the finite difference method. Figure 4.6 represents a finite element grid, similar to the finite difference grid used in Figure 4.5. A triangular grid is used in Figure 4.6, but square or rectangular grids are also possible. Again, Eq. (4.7) is iterated over the grid until an acceptable mass balance is obtained. Such finite element techniques are more commonly used than the finite difference approach in pollutant fate and transport modeling.

It should be noted that all modeling techniques should be calibrated and validated with field-measured pollutant concentrations. A conceptual comparison of the differential equations and numerical method approaches is shown in Figure 4.7. As noted, if the model results are inconsistent with the field measurements, adjustments to the underlying model may be necessary to successfully model a system.

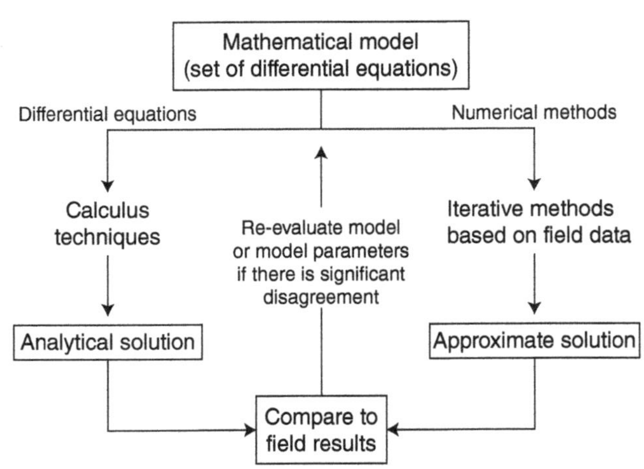

Figure 4.7 A conceptual comparison of the differential equations and numerical methods approaches.

4.2 Quality of Modeling Results

When one talks about a model, it is important to talk about the validity of the model process (i.e. do we understand how nature is behaving or is there uncertainty in this understanding? Is it an approximation of reality?). If our model algorithm includes constants or rate parameters, are they precisely defined or measured, or an approximation as well? Strictly speaking, the mathematical expression is not a model until all its constants and parameters are determined. Prior to that step, the expression is just a code as in computer code or just an equation purporting to describe environmental behavior in a cause and effect environment.

So, how good are typical modeling results? As usual, this depends on a number of factors, such as how realistically your model mimics the system under study, how far past your calibration you are attempting to predict pollutant concentrations, and how long into the future you are extrapolating. Most importantly, it depends on who is asking and answering the question. For example, some modelers will defend their predictions to the death, since they can defend the physics behind the mathematical models. However, it is important to distinguish reality (the environmental system) from the model (a mathematical equation). It is not uncommon to hear a modeler state that the field results are wrong because "they do not match my model results!" One should be careful when dealing with avid modelers.

Some modelers prefer to use statistics that evaluate simulated output to its observed value and compute correlation coefficient using goodness-of-fit statistics. Others prefer Monte Carlo techniques that ascribe distributions to the parameters and rate coefficients.

One of the authors of this textbook (Dunnivant) once gave a presentation of a major field-monitoring project to a group of modelers. The system under study was an uncharacterized, highly fractured, unsaturated groundwater system where we were attempting to gather data to estimate site-specific dispersion (mixing) estimates for a large hazardous waste site located nearby. During the question and answer session that followed the presentation, one modeler asked why we wasted so much money conducting an experiment and measuring pollutant concentrations and mixing since an accomplished modeler could have easily written a model to predict the results. The answer, my colleague, is model validation/calibration. No model is of any value unless based on or verified with experimental data. Many modeling results have been challenged in the courts with a decision made on the basis of misuse of what is simply called best available practice (BMP) or good engineering judgment. Both are nebulous terms with no set established guidelines, although EPA has published guidelines or workshops offered by experienced modelers with stellar reputations.

Occam's razor, a commonly noted idea, is a principle that states that the most simple answer is usually the best. Unfortunately, this rarely applies to equations that model environmental systems. By their very nature, environmental systems are highly complicated, and the accuracy of our experimental and modeling efforts to characterize dispersion in these systems is weak at best. Thus, models need to be as complicated as the system being modeled. Model predictions (explanatory modeling from Chapter 1) should always be compared to experimental data, and the overlay of these two data sets should show consistency. For predictive modeling, we unfortunately have no confident way to judge accuracy, except to wait and let time evaluate our efforts. In each of the closing sections of the following modeling chapters, we will discuss the limitations of our modeling equations. This will hopefully prepare the reader to know where to ask questions of the modeler and how to judge their modeling results.

A corollary to Occam's razor is that the best model is often the simplest and that a model can be quickly developed, inexpensive to develop, or accurate in its forecast, but not all three. Accuracy implies slower and more expensive to develop. This lesson was encountered during the previous several decades as states struggled to complete a huge backlog of total maximum daily load (TMDL) assessments that usually required environmental modeling of some type. The EPABASINS was created by its Office of Research Development for this express purpose (US EPA 1998).

4.3 What Do You Do with Your Modeling Results?

As noted in the Chapter 1, the goal of pollutant identification, laboratory analysis of field samples, and fate and transport modeling is to provide data to be used in risk assessment and exposure analysis. Otherwise why would we do all of this work? Risk assessment (and economics) drive virtually every hazardous waste remediation (Superfund) effort today. The predictions used in these efforts rely on models to apply observed chemical and physical principles to the sites under evaluation, in order to predict future concentrations of pollutants at various sites in the system. Chapter 11 of this textbook will focus on risk assessment calculations that depend on predicted or measured downgradient pollutant concentrations.

Bibliography

Burden, R.L. and Faires, J.D. (1993). *Numerical Analysis*, 5e. Boston: PWS-Kent Publishing Company.

Center for Regulatory Environmental Modeling EPA(CEAM) (2018). Center exposure assessment modeling ceam. https://www.epa.gov/ceam/about-epa-center-exposure-assessment-modeling-ceam (accessed 7 August 2018).

EPA (1998). Better Assessment Science Integrating Point and Nonpoint Sources. BASINS2.0, 823-B-98-005, Office of Science and Technology, Washington, D.C.

EPA (2009). 100/K-09/003, March. Guidance on the Development, Evaluation, and Application of Environmental Models.

EPA (2017). EPABASINS. https://www.epa.gov/exposure-assessment-models/basins (accessed 4 October 2017).

Ross, S.L. (1980). *Introduction to Ordinary Differential Equations*, 3e. New York: Wiley.

Considine, D.M. (ed.) (1976). *Van Nostrand's Scientific Encyclopedia*, 5e. New York: Van Nostrand Reinhold.

Wang, H.F. and Anderson, M.P. (1982). *Introduction to Groundwater Modeling: Finite Difference and Finite Element Methods*. San Francisco, CA: W.H. Freeman and Company.

(1975). Water Quality Modeling Workshop. Manhattan College Summer Institute, Riverdale, NY.

5

Fate and Transport Concepts for Lake Systems

Case Study 1: Lake Onondaga

Lake Onondaga is located north of the city of Syracuse in New York State. The lake has a surface area of 11.9 km^2 (4.6 mi^2) and receives water from a drainage basin of 648 km^2 (248 mi^2). In the late 1880s and early 1900s, Lake Onondaga was a prized recreation resort for the citizens of Syracuse, but as the city grew and became more industrialized, the lake became more and more polluted. Currently, it abuts industrial areas and highways.

From the time Syracuse was founded, the city directly dumped raw sewage into the lake. In 1884, the Solvay Process Company began the production of soda ash and released high concentrations of salt (chloride, sodium, and calcium) directly into the lake. Swimming was banned in the lake in 1940, primarily due to public health issues relating to the untreated sewage. In 1946, Allied-Signal Corp. began chlorine production using the mercury cell process and directly discharged the mercuric waste into the lake. In 1970, fishing was banned due to mercury pollution. As a result of these uncontrolled pollution events, steps were taken to slowly improve the quality of water in the lake. The city of Syracuse slowly upgraded its sewage treatment facilities by installing primary treatment in 1925, constructing the METRO wastewater facility in 1960, and upgrading the METRO to secondary and tertiary treatment in 1979 with Federal and State subsidies. In 1977, Allied-Signal closed a chlorinated benzene plant and one chlorine production facility. In 1986, the soda ash manufacturing operation was closed. In 1995, Lake Onondaga was added to the Federal Superfund National Priority List because of persistent chemical pollution. In more recent years, remediation efforts have been planned but little direct action has been taken, other than limiting current pollutant inputs to the lake, conducting studies, and planning for the future. For example, the 1990 Onondaga Lake Management Conference initiated research and planned remediation projects. In 1992 the Corps of Engineers completed the Lake Onondaga Water Technical Report, outlining possible lake remediation alternatives, and was one of the early total maximum daily load (TMDL) efforts. In 1994, aquatic habitat restoration projects began. In 1996, Allied-Signal started a remedial investigation and feasibility study (RI-FS) for the lake. Even with all of these remediation-planning efforts, Lake Onondaga is still considered by most environmentalists the most contaminated lake in the United States.

What are the results of the previous pollution emission, and what can be done to remediate the lake? The primary pollutants include (i) sewage waste consisting of nutrients such as phosphorus, ammonia, nitrite, and many harmful bacteria, (ii) turbidity in the lake resulting from excess microbial growth due to the abundant nutrients present in the lake, (iii) high salinity from the soda ash production, and (iv) mercury from the chlorine production facilities.

Given adequate time and sufficient flow through the lake, sewage discharges can be removed from a lake system and some biochemical oxygen demand (BOD) will be assimilated by the processes of biochemical oxidation and surface reaeration. In low flow-through systems such as Lake Onondaga, nutrients are mostly recycled in the lake during summer stratification and fall overturn (something we will learn about in this chapter). Some nutrients are removed very slowly by burial in the lake sediments. However, more soluble nutrients such as ammonia, nitrite, and nitrate are difficult to remove by any process associated with nitrification and the inherent nitrogen cycle in nature.

The largest health hazard comes from the mercury released into the lake. It has been estimated that 165 000 lb of mercury were released to the lake between 1946 and 1970. Since metals do not degrade, all of the mercury is still present in the lake and will remain in the lake unless direct and expensive remedial actions are taken (such as dredging or natural burial by sedimentation, which will be discussed in this chapter). You should recall our discussions on bioconcentration from Chapter 2. This is a common process in lake systems, and mercury that was buried in the lake sediments of Lake Onondaga has undergone methylation by microbes and has been bioconcentrated in many fish species and

Pollutant Fate and Transport in Environmental Multimedia, First Edition. Frank M. Dunnivant and Elliot Anders.
© 2019 John Wiley & Sons, Inc. Published 2019 by John Wiley & Sons, Inc.
Companion website: www.wiley.com/go/Dunnivant/pollutant_fate&Transp_EnvironMultimedia

their macroinvertebrate food source. This led to a ban on fishing in and eating fish from the lake. It is especially problematic for high risk populations like pregnant women and the elderly.

A variety of cleanup efforts are underway for Lake Onondaga, and water quality has significantly improved since the 1970s. Catch and release fishing was reinstated in 1986. A consortium of agencies and public interest groups have agreed on eight goals for Lake Onondaga. These include

(1) development of a eutrophication model for the Seneca River,
(2) development of a lake productivity model,
(3) development of a hydrodynamic model for the lake outlet,
(4) funding of studies on the release of nutrients and toxic substances from lake sediments under changing dissolved oxygen (DO) levels,
(5) establishment of a long-term baseline water quality program,
(6) drafting of an urban/suburban nonpoint source pollution plan,
(7) drafting of a fish and wildlife management plan, and
(8) development of a demonstration project of manipulated littoral zone habitat structures; the project indicated that fencing and wave breaks could significantly increase plant survival, growth, and diversity, and that these habitats also increase survival of young fish.

Information on Lake Onondaga was obtained from the Onondaga Lake Partnership website (www.onlakepartners.org). Such public stakeholder input became mandatory in federally mandated TMDL developments.

Case Study 2: Lake Erie, A More Positive Example

Lake Erie is one of the fourth largest Great Lake bordering the United States and Canada. Because of the noted pre-environmental excesses causing eutrophication, scientists declared Lake Erie dead in the 1960s, and was not expected recover for more than 50 years. Swift environmental remediation actions in the 1970s showed marked improvement and today Lake Erie environmental actions are noted as a "near" success stories. While small dead zones, devoid of dissolved oxygen, are likely to have been present prior to industrial and farming pollution, the dead zone spread throughout the lake in the 1960s, much like the current one in the Gulf of Mexico at the mouth of the Mississippi River below New Orleans arising in the 1990s. Dead zones have considerable reduced and today the lake is well on its way to recovery. In fact, EPA established a special program office to deal with the Great Lakes and its many international issues (https://www.epa.gov/greatlakes). These issues include invasive species, sediment contamination, and deposition from air pollution.

Chapter Overview

In this chapter we will learn physical and chemical processes that explain the fate and transport of pollutants in lake systems such as Lake Onondaga, the Great Lakes, and many small lakes. First, we will look at the formation, geological history, and seasonal history of lakes. Then we will focus on water sampling techniques and finally pollutant transport, by conceptually and mathematically describing mixing processes and chemical reactions specific to lakes. We will look at two basic models for predicting the transport of pollutants in lakes, based on pulse and step inputs. We will next look at some of the ways to remediate a lake after a contamination event has occurred, and most importantly we will learn the limitations of what we can actually do to return the lake to pristine conditions. Finally we will extend our simple pulse and step models to numerical models used by professionals to model large lake systems.

5.1 Introduction

Inland surface waters cover slightly less than 2% of the Earth's surface and only account for far less than 1% of the terrestrial fresh water in the world. While this may seem a small percentage, lakes are common features in temperate and subarctic regions of the Northern Hemisphere. Lakes are a highly used source of recreation and food, and the quality of water in these systems is extremely important to local populations and economies. As we saw in the bioconcentration example in Table 2.5, an extremely small concentration of pollutant in water can result in significant and health-threatening pollutant concentrations in the biota living in, on, or near a polluted water body.

In this chapter we will look at types of lakes and how they are formed, pollutant input sources specific to surface water bodies, the chemical nature of lakes as a function of seasons and how this affects pollutant transformation reactions, the pulse and step fate and transport models for lakes, the limitations of our relatively simple models, and ways to remediate contaminated lakes.

5.2 Types of Lakes and Lake-forming Events

Lakes come in a large variety of shapes, sizes, and depths. Of course the most prominent ones are those that appear as large blue shapes on national and global maps, and although these lakes contain the largest percentage of fresh surface water on Earth, they account for only a small percentage of total number of lakes. For example, Lake Baikal in Siberia contains approximately 20% of the Earth's fresh surface water, while another 12% is contained in Lake Superior in North America. An illustration and comparison of the surface area of the largest lakes of the world is shown in Figure 5.1. A summary of the major lakes of the world, with respect to location, surface area, length, and depth, is given in Table 5.1. Table 5.2 contains volume data of the Great Lakes of North America.

While the large lakes in Figure 5.1 are impressive, they are not representative of lakes in general and due to their size require special fate and transport modeling approaches (numerical methods of analysis in Chapter 4). We are more concerned with smaller lakes, which are more common and more widely distributed, and more evenly mixed. To put the size of lakes into perspective, refer to Figure 5.2a that shows the global distribution of lakes versus surface areas. Lake Nicaragua in North America (one of the smaller lakes located on the lower left-hand side of Figure 5.1) has a surface area of $8030\,km^2$. This area in the lower right-hand corner of Figure 5.2a is represented by fewer than 50 lakes worldwide. In contrast, look at more common lake sizes, for example, $0.1–1.0\,km^2$. Figure 5.2a indicates that there are between one million and 10 million lakes in this size range. Similar observations can be made using Figure 5.2b for the depth of lakes. Students may find value in exploring a web site of the United States Geological Survey (USGS) to explore the distribution of water resources on Earth, as delineated by surface, subsurface, polar, permafrost, and glaciers at https://water.usgs.gov/edu/earthwherewater.html. Freshwater sources pale in volume comparison to the saline waters of the oceans and seas.

Lakes are formed through a variety of geologic events millions of years ago, but mostly through glacial, volcanic, and tectonic activity. Other lake-forming events include landslides, dissolution of limestone bedrock, and

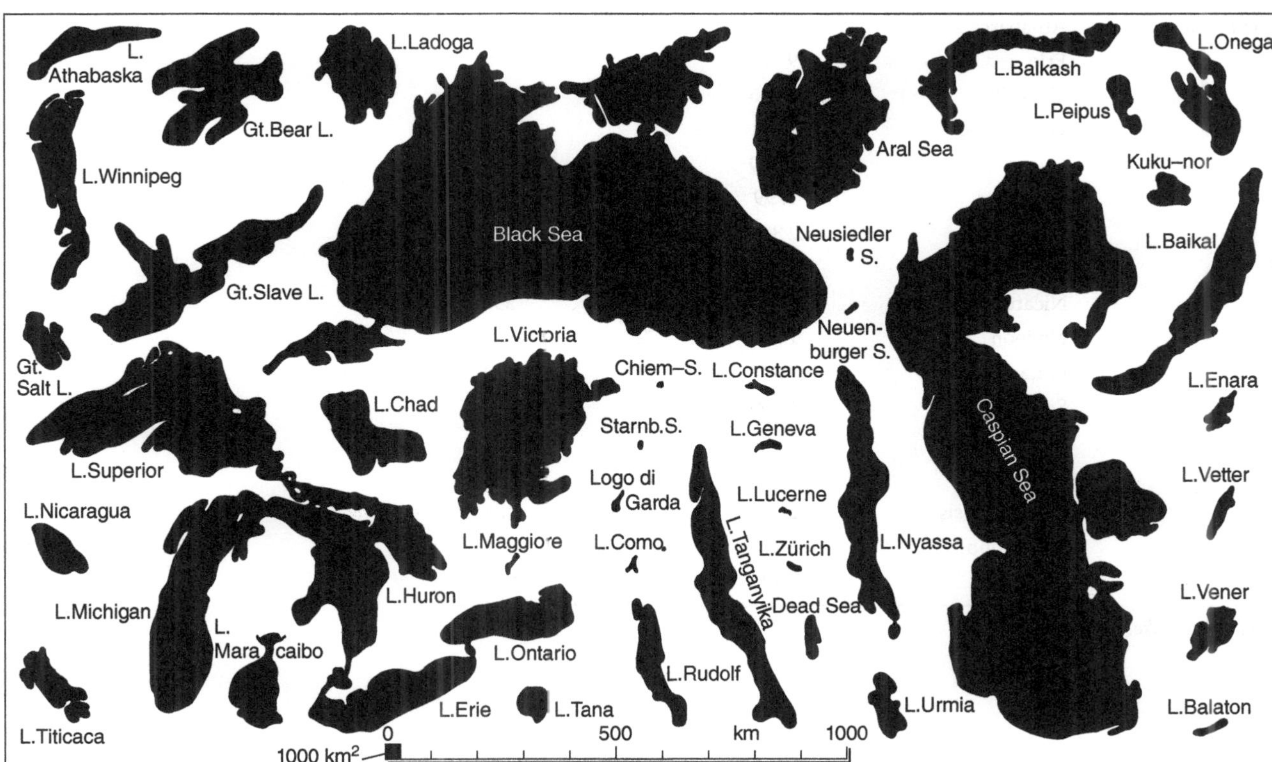

Figure 5.1 Approximate comparison of the surface area of the world's largest lakes. (*Note*: The Black Sea is basically an ocean body, since it is directly connected to the ocean.) Source: Ruttner 1963. Reprinted with permission of the University of Toronto Press, Inc.

Table 5.1 Condensed summary of major natural lakes of the World (The World Almanac, 2004).

Continent	Lake name	Area (mi^2)	Length (mi)	Maximum depth (ft)
Africa	Albert	2075	100	168
	Chad	839	175	24
	Nyasa (Malawi)	11150	360	2280
	Tanganyika	12700	420	4823
	Turkana (Rudolf)	2473	154	240
	Victoria	26828	250	270
Asia	Aral Sea[a)]	13000	260	220
	Baikal	12162	395	5315
	Balkhash	7115	376	85
	Issyk Kul[a)]	2355	115	2303
	Tonle Sap	2500	—	45
	Urmia	1815	90	49
Asia-Europe	Caspian Sea[a)]	143244	760	3363
Australia	Eyre[a)]	3600	90	4
	Gairdner[a)]	1840	90	—
	Torrens[a)]	2230	130	—
Europe	Ladoga	6835	124	738
	Onega	3710	145	328
	Vanern	2156	91	328
North America	Athabasca	3064	208	407
	Erie	9910	241	210
	Great Bear	12096	192	1463
	Great Slave	11031	298	2015
	Huron	23000	206	750
	Manitoba	1799	140	12
	Michigan	22300	307	923
	Nettilling	2140	67	—
	Nicaragua	3100	102	230
	Nipigon	1872	72	540
	Ontario	7340	193	802
	Superior	31700	350	1330
	Reindeer	2568	143	720
	Winnipeg	9417	266	60
	Winnipegosis	2075	141	38
South America	Maracaibo	5217	133	115
	Titicaca	3200	122	922

a) Salt water lake.

Table 5.2 Water volumes in The Great Lakes of North America (The World Almanac 2004).

Lake	Volume (km³)
Erie	484
Huron	3 543
Michigan	4 918
Ontario	1 638
Superior	12 234

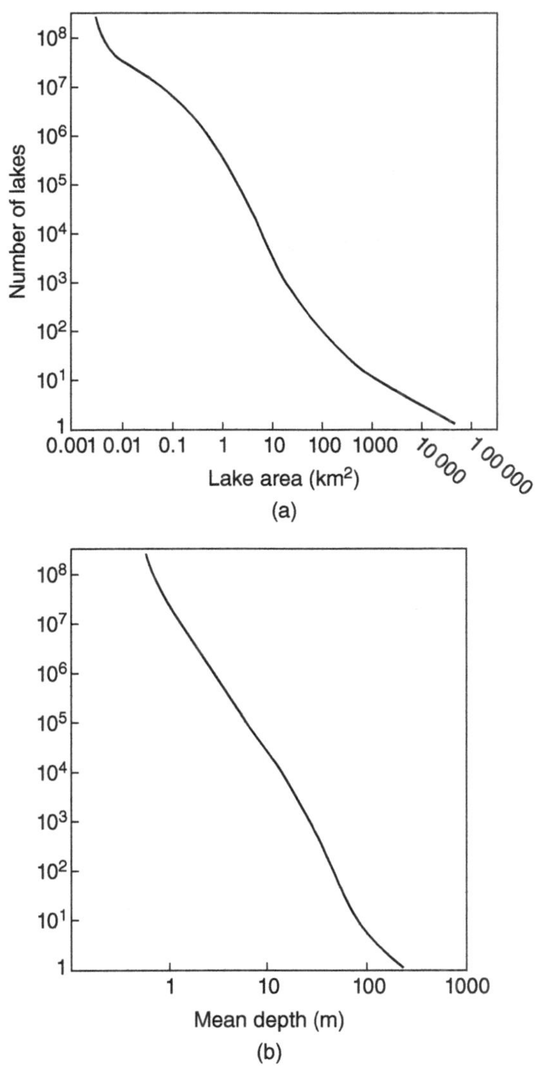

Figure 5.2 The number of lakes correlated to (a) surface area and (b) depth. Source: Wetzel (1990). Reprinted with permission of Taylor & Francis.

human-made reservoirs. A summary and description of these formation events is given in Table 5.3. From a fate and transport modeling standpoint, land features around a lake and physical features of the lake are important. For example, lakes located in flat regions are subject to frequent mixing by the wind. Important physical features include surface area and depth. With the exception of glacial lakes, lakes in flat regions tend to be shallow and are easily mixed. Moderate and deep lakes thermally stratify very easily, which inhibits mixing during summer months.

The last category in Table 5.3, human-made reservoirs, is a point of contention among lake and river scientists. Which category do these reservoirs fit into – lakes or rivers? Human-made reservoirs tend to have a more longitudinal flow direction, like a river, and in fact, many were created by a hydropower or lock-and-dam structures. One of the more prominent ones in the United States is Lake Mead near Las Vegas, NV, and formed by Hoover Dam. Its waters supply the needs of over 20 million people in Arizona, California, and Nevada. Major rivers in the Northwest like the Columbia, Snake, and Fraser Rivers are dammed for hydropower reasons, as are many in the Northeast. However, large surface areas and deeper depths make these systems more similar to lakes. In addition, the increased depths of these lake systems set up conditions that lead to thermal stratification (discussed later in this chapter), which is definitely specific to lakes. Similarly, they can be stratified from a water quality prospective. Modeling of human-made reservoirs requires special considerations, as we will discuss later in this chapter. In the Northeast United States, some hydropower dam structures are fitted with "fish ladders" to facilitate fish passage to upriver spawning areas, and other dams are to be removed entirely to restore the habitat features present in early decades.

One additional note should be made with respect to the size and outflow of a lake. As noted, lakes come in a variety of sizes. In the derivation of our simple models used here, we will assume a constantly and evenly mixed system (or reactor in engineering terms). As lakes increase in size this assumption becomes increasingly incorrect. For example, take mixing in Lake Michigan, one of the Great Lakes located on the United States and Canada border. Figure 5.3 shows the seasonal mixing of surface waters (https://www.glerl.noaa.gov/pubs/fulltext/1999/19990004.pdf). One can easily imagine the difficulty in modeling such a large, non-uniform system. This is where the closing section of this chapter will be useful, numerical modeling of more complex systems. Models like Water Analysis System Program (WASP, discussed later in Chapter 14) are well suited for such modeling tasks.

Table 5.3 Forces that form lakes.

Type of formation	Description
Tectonic basins	These lakes are depressions formed by the movement of deeper portions of the Earth's crust
Fault lake	A process where depressions occur between the bases of a single fault displacement or in void formed by the dropping of one block (in the middle) or by the uplifting of two blocks on either side of a middle block. An example is Lake Tanganyika in Africa
Graben lake	This similar to faulting, but a long trough is downfaulted. An excellent example is Lake Baikal in Asia
Uplifting of marine sea floors	This isolates a portion of the sea but uplifting a portion of the sea floor and creating an area of trapped salt water. This uplifting usually results in a mountain range between the enclosed water body and the oceanic area. Examples include the Caspian Sea and the Sea of Aral
Upwarping	This is similar to uplifting but only minor uplifting of the Earth's crust occurs (no mountain ranges are formed). This gentle uplifting process can create a lake basin in the middle of uplifted areas. Examples include the Lake Okeechobee in Florida (USA) and Lake Victoria in Africa. Upwarping contributed to some of the formation of the Great Lakes in North America with the glacial ice sheets melted and released pressure on the Earth's crust
Volcanic activity	These lakes occur in the craters or calderas of old volcanoes. An example is Crater Lake in Oregon (USA). Also, lava flows can dam rivers or streams and form lakes
Landslides	Lakes can be formed by landslides that dam a stream. However, these lakes tend to be short-lived since the stream will eventually break through the earthen dam and form a new stream channel
Glacial activity	Glaciers are one of the most important lake forming processes. Lake sizes formed by glacier activity range from very small kettle lakes to some of the largest lakes in the world, such as the Great Lakes in North America
Glacial ice scour lakes	These lakes are formed by ice moving over relatively flat rock surfaces that are jointed and contain fractures. In mountains where amphitheater-like formations (cirques) are present, glacier action results in cirque lakes (a lake at the end of the cirque). When the glacier action is along coastal regions, fjord lakes are formed in narrow, deep basins. These are common in areas of Norway and western Canada. Glacial action, in the form of retreating large ice sheets, is most known for the formation of large lakes such the Great Slave Lake and Great Bear Lake in Canada and the Great Lakes of the St. Lawrence drainage in North America
Kettle lakes	Retreating glaciers can also deposit large pieces of ice in the glacial sediments that later melt to form small kettle lakes
Morainal damming	Glaciers deposit large amount of sediment, or moraine, along the sides and at the terminus. These moraines can dam the valleys they are deposited in and dam adjacent valleys
Lakes formed by river activity	The erosive power of some rivers is considerable and can create lakes along the course of the river. These types of lakes include plunge-pool lakes below waterfalls. Floodplain lakes form in low-lying areas adjacent to rivers, levee lakes, and oxbow lakes are formed when a river changes course and cuts off a meander
Solution lakes	These lakes are formed in areas of the world containing extensive deposits of limestone ($CaCO_3$). Water slowly dissolves large pockets at or below the land surface and, over time, a lake is formed. Solution pockets (caves) formed below the surface can result in a collapse of the roof and form an exposed lake. Karst formations commonly have solution lakes. Solution lakes are found in the Adriatic, the Balkan Peninsula, the Alps of Central Europe, and in Michigan, Indiana, Kentucky, Tennessee, and Florida in the United States
Human-made reservoirs	Humans have created lakes through the damming of rivers for thousands of years (beavers have done this much longer). The size of human-made reservoirs range from small ponds to large reservoirs approaching the surface area of the Great Lakes of North America

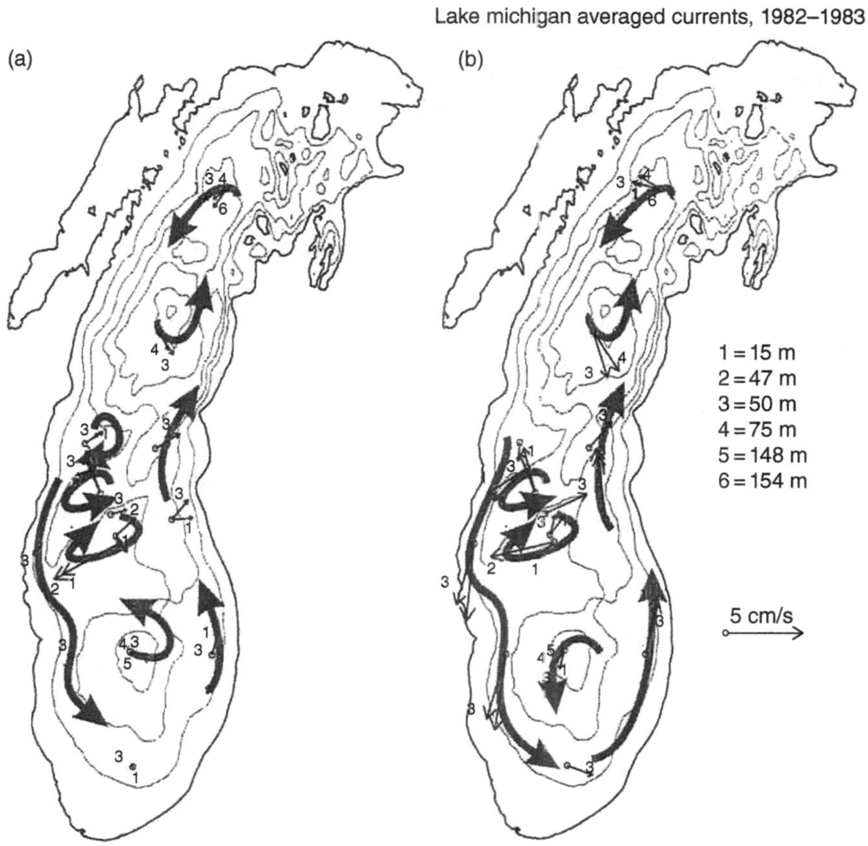

Figure 5.3 Seasonal mixing of the surface waters of Lake Michigan. (a) Summer, (b) winter, and (c) annual. Source: https://www.glerl.noaa.gov/pubs/fulltext/1999/19990004.pdf.

5.3 Input Sources

Input sources of contaminants (pollutants) to lakes are similar to those discussed in Chapter 1. We will be concerned with point and non-point sources and use pulse and step inputs to develop our fate and transport models. Common point inputs include industrial sites and feedlot sources, as well as effluent from domestic sewage plants. Non-point sources can include runoff from farming practices and agricultural and forested settings where soil erosion and runoff are common. Underground non-point sources can include leachate from domestic and hazardous waste landfills or storage tanks. The physical characteristics of lakes can add unique input sources. Most lakes that are used for recreational activities experience pollution by outboard motors, marinas that are both notorious for releasing petroleum-related and paint compounds. Some notorious hazardous chemicals are used in paint formulations: thankfully lead additives have been banned for several decades now but there are other chemicals of concern (pesticides, and additives in commercial products). Although these compounds are relatively insoluble in water, some of these compounds dissolve and are spread (dispersed) throughout a lake. This is a non-point, but continuous, source of pollutants.

Another unique and important pollutant source for lakes, due to their large surface area and relatively long residence times, is aerial inputs as dry or wet deposition. Aerial application of pesticides to adjacent farmland can result in significant inputs of pollutants. These applications are usually treated as non-point seasonal inputs and can be in aerosol form. Long-range transport of atmospheric pollutants can also result in inputs to lakes. In areas of the world where certain pesticides were never used or have been banned, scientists still detect inputs of these compounds to lakes. Such an example can be found for the large lakes in the United States. 1,1′-(2,2,2-Trichloroethylidene)bis[4-chloro-benzene] (DDT) was banned in the early 1970s, but surface lake waters and atmospheric samples above these waters still show measurable concentrations of these pollutants. Studies have found that long-range atmospheric transport of pesticides from Central America can reach the Great Lakes and remote alpine lakes in North America. Further, sources from Central Africa flow East to the

Caribbean affecting many more people. This type of input can be treated as a non-point continuous input.

One last source of pollutants to lakes can be from within the lake itself. As we discussed in Chapter 2, particles in the water attract many inorganic and organic pollutants, and most of these particles aggregate and settle in quiescent regions of a lake. However, if these sediments are resuspended into relatively unpolluted lake water, desorption of the pollutants to the water can occur. Resuspension events of importance in most lakes include bioturbation (the mixing of lake sediments with water from biota in the lake), violent storm events, and dredging of harbors and shipping channels. Bioturbation is a constant process and would be treated as a continuous non-point input, while the latter two events would be treated as pulse non-point inputs.

The low degree of mixing in some lake systems combined with a relatively low input of atmospheric oxygen sets up another unique condition in lakes: eutrophication. In general, the limiting nutrient in freshwater aquatic systems is phosphate, but large inputs of any nutrient may result in the uncontrolled growth of algae in lake systems. The input of carbon substrates, from sources such as farm runoff and domestic sewage, often results in uncontrolled growth. During the daylight hours, large algal blooms are not a direct threat to a lake, but during respiration at night algae consume DO in the lake water. This creates a large oxygen demand on the system, usually one that cannot be met by the limited DO in lake water (maximum of ~12 mg DO/l) or by relatively slow diffusion of oxygen from the atmosphere and low mixing in lake systems. This sets up a condition known as eutrophication, and, as we will see in the next section, this can cause serious problems during months when a lake is thermally and chemically stratified.

5.4 Stratification of Lake Systems

Most lakes undergo some form of thermal stratification during the year. Stratification is a process in which differential heating or cooling occurs and two "climates" are set up in the lake. If a lake undergoes only one stratification event during the yearly cycle, it is referred to as a monomictic lake. The most common stratification scenario is illustrated in Figure 5.4b. Figure 5.4a shows the unstratified lake during the fall, winter, and spring, when the temperature is nearly uniform regardless of depth, due to a high degree of mixing (assuming that the lake surface does not freeze in winter). However, if we take a snapshot in time during the middle of summer, we would commonly find the temperature profile illustrated in Figure 5.4b. Here, a warm body of water, heated by the sun, develops at the surface, and a cooler, denser body of water resides in the bottom of the lake and is cooled by the Earth. The lake is divided into three zones by temperature, and thus density, differences that prevent mixing: the epilimnion (surface water), the thermocline (the area where a rapid change in temperature occurs), and the hypolimnion (the bottom of the lake). A representative

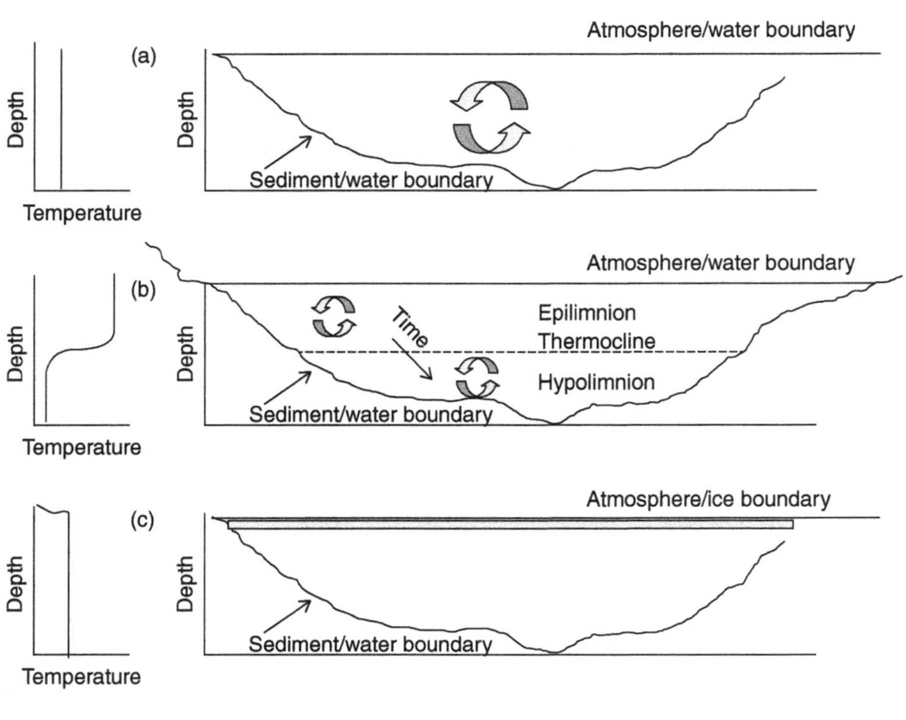

Figure 5.4 Stratification of lakes during annual cycles. The left-hand side of the figure shows the temperature with depth profile of the entire lake. The right-hand side of the figure shows the different types of seasonal stratification in the lake.

temperature profile is shown to the left of each lake diagram.

Some lakes also freeze in winter, which sets up a different temperature profile (refer to Figure 5.4c). Here, the less dense ice floats on the lake surface, and although the water beneath the ice is not actively mixed, it is usually assumed to be consistent with respect to temperature and chemical concentrations with depth. If a lake freezes in winter and stratifies in summers, it has two thermal stratification events and is termed dimictic.

Let's return to the concept of summer stratification, since it can be one of the most important cycles in a lake. How does the stratification occur? During early summer, the sun heats the surface of the lake and a thin warm body of water forms. A strong wind from a weather frontal system can disrupt this initial stratification, but if the weather remains stable for a few days or weeks, the sun continues to heat the water and the depth of the warm water (epilimnion) increases and becomes a stable entity of the lake (refer to Figure 5.5a for a mid-summer temperature versus depth profile). As the summer heating continues, the epilimnion continues to increase in depth in the lake as illustrated in Figure 5.5b. Then, during autumn, the epilimnion, which contains the majority of the water in most lakes, cools to temperatures similar to that of the hypolimnion and the wind associated with a passing weather front is sufficient to cause the two water bodies to mix. This mixing is referred to as the fall overturn.

During summer stratification, if sufficient biodegradable organic matter is present in the hypolimnion, microorganisms will consume all of the dissolved oxygen in these waters. This oxygen cannot be replenished, due to the hypolimnion's isolation from the atmosphere. Thus, a chemocline will be present, associated with the thermocline. In most cases, the system will turn anoxic (devoid of oxygen) due to microbial activity, and a set of interesting biochemical reactions will occur that directly affects the chemistry of the lake water. A chemocline is illustrated in Figure 5.5c. The nitrate-ammonia transition shown in Figure 5.4d,e is an example of the chemistry that can occur in the chemocline. We discussed the sequence of reactions that occur under these conditions in Figure 2.12, and you should review this figure. Note that the terminal electron acceptor (nutrient) must be present in order for the E_H to be buffered (poised) at each level in Figure 2.12. Nitrate, sulfate, and carbon dioxide are common constituents of lake water, so these redox reactions will almost certainly be present, and the E_H associated with each half-reaction is shown in Table 5.4. Recall from our discussions in Chapter 2 concerning Figure 2.12 that a carbon substrate must simultaneously be oxidized when the terminal electron acceptor is reduced. Figure 5.6 shows the final equilibrium profiles

Table 5.4 E_H values for common terminal electron acceptors (TEAs).

Balanced half reaction	$\Delta E°$ (water) (V)
$O_2 + 4H^+ + 4e^- \Rightarrow 2H_2O$	+0.81
$NO_3^- + 10H^+ + 8e^- \Rightarrow NH_4^+ + 3H_2O$	+0.36
$SO_4^{2-} + 9H^+ + 8e^- \Rightarrow HS^- + 4H_2O$	−0.22
$CO_2 + 8H^+ + 8e^- \Rightarrow CH_4 + 2H_2O$	−0.25

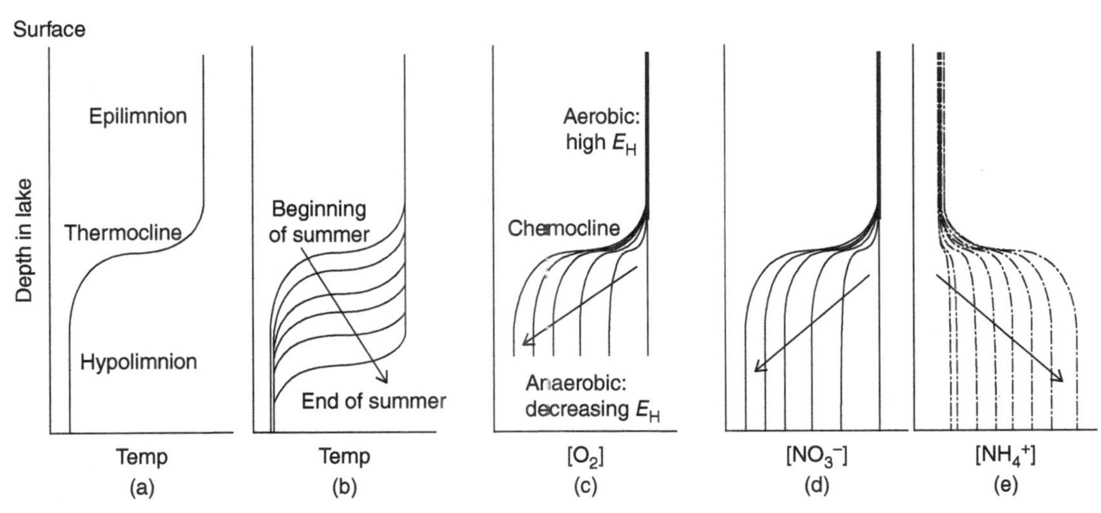

Figure 5.5 Illustration of increase in depth of the thermocline and chemocline as summer heating progresses. The transitioning line in (b) through (e) show the change in analyte concentrations as the summer heating progresses and as the thermocline decreases in lake volume.

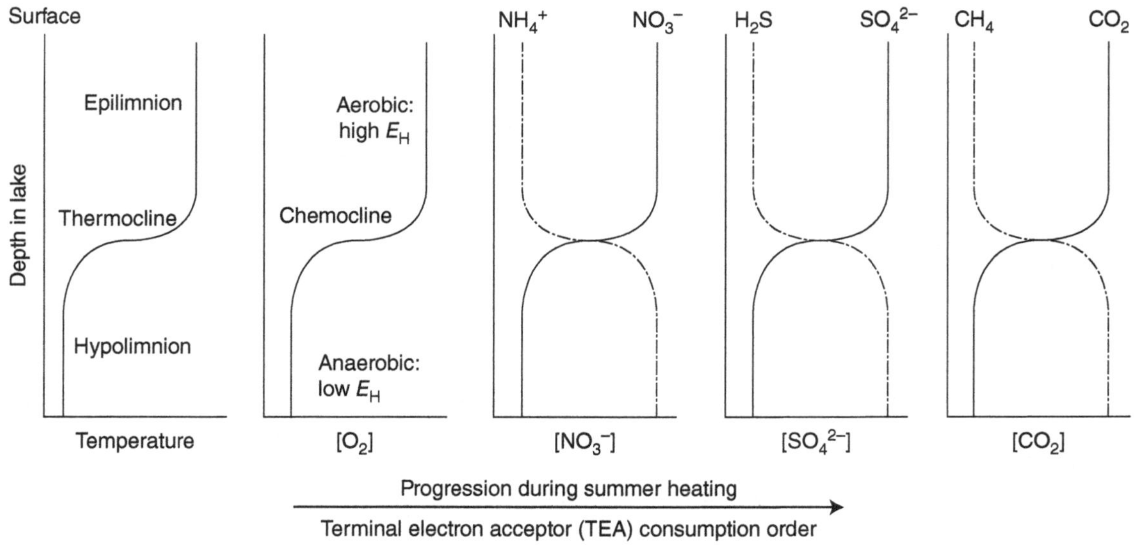

Figure 5.6 Terminal electron acceptor profiles.

of each redox couple after the system has become highly reducing (a low E_H value).

So, what does this have to do with fate and transport modeling in lakes? First, if we are concerned with the release of a pollutant, either a pulse or step input, in a stratified lake, we are no longer modeling the total volume of the lake, since the hypolimnion and epilimnion do not mix during summer. If we did use the total volume of the lake, we most likely would grossly underestimate the pollutant concentration in the lake water, since the pollutant emissions are now present in a smaller volume but at higher concentrations. We can account for this in our models by knowing the epilimnion–hypolimnion volumes. Second, the fate of pollutants trapped in the reduced (anoxic) water of the hypolimnion requires separate consideration. In the surface waters, pollutants can be oxidized by microorganisms and photochemical reactions, but in the hypolimnion many other types of transformation reactions occur for inorganic and organic pollutants under the anoxic conditions. Pollutants may be present in the hypolimnion from a previous release to the water column, or they may be released slowly from contaminated lake sediments. For example, substituted nitroaromatics (organic wastes from chemical and munition industries) can be reduced to their respective anilines by biotic and abiotic pathways under anoxic conditions. Inorganic mercury, already an important toxin, can be biologically methylated to methylmercury, a much more toxic compound, in the hypolimnion. And remember, in the fall, the water from the hypolimnion will be mixed with surface water, re-exposing biota to these toxins. Fall overturn is usually a renewing experience for the lake, since many nutrients are returned to the waters from the sediments; however, depending on the pollutants and transformation reactions that can occur during the summer, it may also be a deadly time.

5.5 Environmental Sampling of Lake Systems

The first step, and in many cases the most important step, in any environmental monitoring plan is sampling. This may seem like an easy part of the process, but if a representative sample of a site is not properly taken, then results obtained from analyzing the sample on a \$100 000 instrument will be worthless. A bad sample can result from taking a sample at an inappropriate location, not taking the sample properly, not properly preserving the sample, or storing the sample too long. Many of these issues will not directly concern you now since most governmental and non-governmental agencies and industries have developed clear sampling and analysis plans (SAPs). These will be clearly stated in your standard operating procedure (SOP) where you work, so it would be pointless to teach you one set of these procedures without knowing where you will be working in the future. EPA is notorious for producing detailed guidance materials and field procedures (including sampling). EPA Region IV's SOPs are found in the following website (nepis.epa.gov/ in references), Therefore, the purpose of this chapter is to introduce you to some of the standard sampling equipment used in environmental sampling. For lakes, we will divide the areas into surface water, subsurface water, and sediment/sludge, although many of these techniques are also relevant to hazardous waste.

It should be noted that most of the sampling equipment can be made of plastic, glass, Teflon, or stainless steel depending on your analyte. For example, plastic or Teflon is usually used when you are analyzing metals, while stainless steel, glass, or Teflon is used when you are analyzing for organic compounds. Many of the sampling tools shown below can be custom-made of specific materials.

Biota: First, we will look at simple biota samplers. Figure 5.7 shows a plankton net that can be held in place in a stream or towed behind a boat. Water and plankton enter the wide mouth of the net and are funneled toward the narrow collection strainer at the right side of the photograph. The mesh size of the netting can be changed to select for different organisms. Figure 5.8 shows a sampling system for macroinvertebrates (mostly insect larva) attached to bottom materials (rocks, leaves, and sticks). This system is used by selecting the area to be sampled, placing the 1-ft × 1-ft brace securely over the stream media and allowing the water to flow over the sampling area but into the net (the net goes downstream of the sampling area), and brushing off the macroinvertebrates into the net. After all of the stream media has been removed, the macroinvertebrates are washed into the end of the net and placed in containers for sorting and identification.

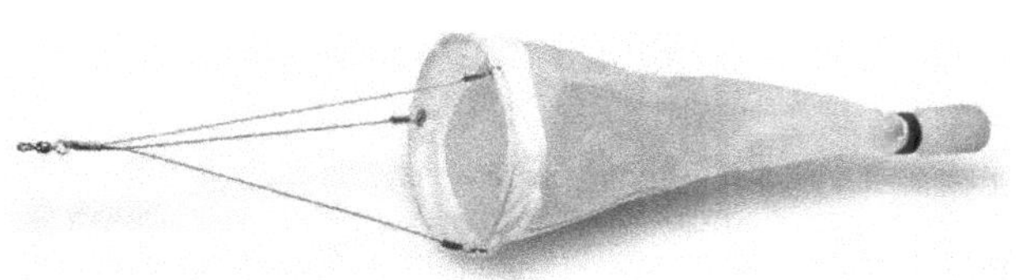

Figure 5.7 Plankton new sampler Source: Courtesy of Forestry Suppliers, http://www.forestry-suppliers.com/product_pages/Products.asp?mi=50501&itemnum=77233&title=Wildco%AE%208%94%20Student%20Plankton%20Net%20Sampler.

Figure 5.8 Surber-type stream bottom sampler. Source: Courtesy of Forestry Suppliers, http://www.forestry-suppliers.com/product_pages/Products.asp?mi=50421&itemnum=77437&title=Wildco%AE%20Surber%20Type%20Stream%20Bottom%20Sampler.

Water: Water (liquid) samplers come in a variety of shapes and sizes suited for depending specific purposes. Grab samples of surface waters can be obtained by simply dipping the appropriate sample container into the water. For hard-to-reach waters or waters/liquids that are potentially hazardous a robotic sample arm can be used (Figure 5.9). Samples can also be taken as a function of depth in a system using automated samplers, such as a van Dorn style sampler

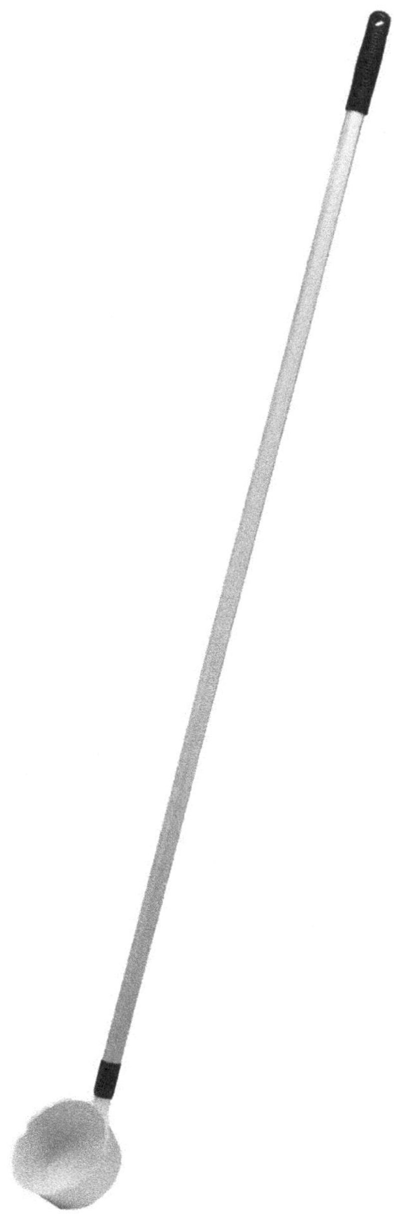

Figure 5.9 Grab sampler for obtaining surface water and hazardous samples. Source: Courtesy of Forestry Suppliers, http://www.forestry-suppliers.com/product_pages/Products.asp?mi=14240&itemnum=53927&title=Non-Collapsible%20Dipper.

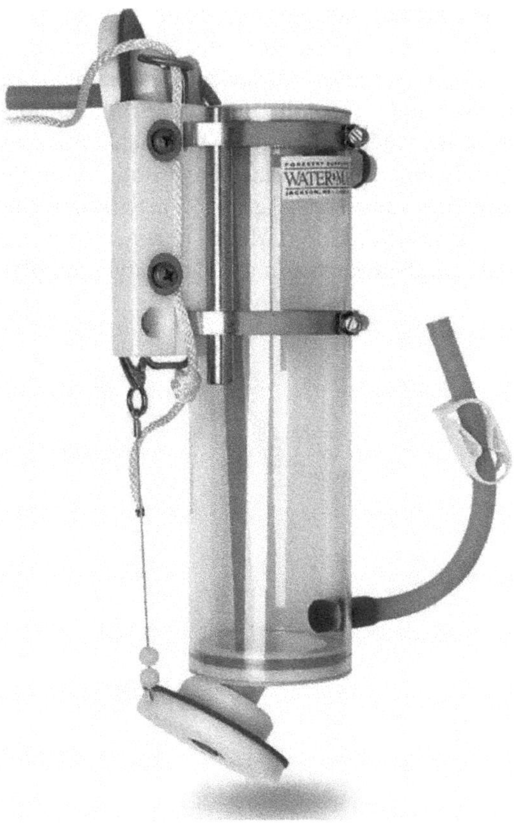

Figure 5.10 Van Dorn style subsurface water sampler Source: Courtesy of Forestry Suppliers, http://www.forestry-suppliers.com/product_pages/Products.asp?mi=50131.

(Figure 5.10). These samplers work by opening the ends of the unit and restraining them by attaching each end of the tubing to a release mechanism. The unit is lowered to the depth of interest and a messenger (a metal weight) is sent down the connecting rope. The messenger hits the release mechanism and both ends of the unit close, trapping the water inside the cylinder. These systems can be used individually or as a series of samplers, with depth, on a single rope, with a series of messenger to close each sampler.

Sediment: Shallow systems can be sampled using hand core sediment samplers such as those shown in Figure 5.11. For deeper water systems a grab surface can be obtained with a Ekman type dredge sampler shown in Figure 5.12. Sediment cores in deep water can be obtained from a wrench-operated boat or ship using an extended version of Figure 5.11, extending several meters long. All coring devices contain a metal or plastic tube containing the sample that can be frozen, sectioned by depth, and extracted for analysis. The sampling of deeper lake systems uses the same type of approach, but the corer is dropped from the boat and retrieved using a rope. Box corers,

Figure 5.11 Hand core sediment sampler. Source: Courtesy of Forestry Suppliers, http://www.forestry-suppliers.com/product_pages/Products.asp?mi=50041&itemnum=77258&title=Wildco%AE%20Hand%20Core%20Sediment%20Sampler.

a combination of a larger version of the Ekman dredge and a sediment coring device, can be used to obtain larger surface area of sediment and extend to considerable depths (1 m plus).

Sampling pore water from benthic layers is also possible with a gradient sampler that features multiple cells that allow water to diffuse into a clean cell and analyzed.

5.6 Important Factors in the Modeling of Lakes: Conceptual Model Development

5.6.1 Definitions of Terms

In order to describe a lake system mathematically, we must first make a list of variables (mathematical symbols) for several terms:

Figure 5.12 Ekman dredge. Source: Courtesy of Forestry Suppliers, http://www.forestry-suppliers.com/product_pages/Products.asp?mi=50071&itemnum=77933&title=Ekman%20Dredge.

V = the volume of the lake (m^3)

Q_i = the inlet flow from the main inlet to the lake (m^3/yr)

Q_e = the outlet, or effluent, flow rate from the lake (m^3/yr) (we usually assume that Q_i is equal to Q_e and we represent both by simply Q)

C_i = the average pollutant concentration in the inlet to the lake (kg/m^3) (this value is zero in many cases)

C_e = the average pollutant concentration in the lake (kg/m^3) and the concentration in the effluent from the lake

k = the first-order rate constant for removal of pollutant from the lake (yr^{-1})

W = the total mass flux of pollutant in the lake, which is equal to the sum ($Q_iC_i + Q_wC_w$)

Note the units used in each of the terms and note that they are compatible with each other. This is important in using the simulator package Fate® (https://educational solutions.github.io/fate/). These terms are used to develop the mass balance of pollutant in the lake and develop equations for the individual components of the mass balance (inflow, outflow, sources, and sinks of pollutant).

5.6.2 Detention Times and Effective Mixing Volumes

In this section we will develop a more conceptual, rather than mathematical, derivation of the chemical and physical processes important in the governing equation for the fate and transport of pollutants in lake systems. In completing this development, we first start with a simple model of a lake and consider it a uniformly mixed basin with one dominate input and one dominant output. Engineers refer to this a continuously stirred tank reactor (CSTR).

In lake systems, it is useful to know or estimate how long water will stay in the system, since this provides an estimate of the minimum time the pollutant will stay in the system. This parameter is called the detention or retention (residence) time, and this brings us to the weakest assumption of commonly used lake models. In order to keep the mathematics relatively simple, we must assume that the lake is completely mixed with respect to pollutant concentrations. In some cases (small to medium sized lakes), this is a valid assumption, but for others (large lakes) it is a weak assumption at best (but still useful in approximations). When we assume that the lake is completely mixed, we can estimate the hydraulic detention time (t_0) by

$$t_0 = V/Q \tag{5.1}$$

which is expressed in years, in accordance with the units specified in Section 5.5.1. Volume (V) is usually expressed in cubic length (m^3) while volumetric flow rate (Q) is expressed in similar cubic length per time (m^3/yr).

Mixing in lakes, an applied form of entropy, is one of the most difficult parameters to estimate. The predominate mixing force is wind blowing across the surface of the lake and is commonly referred to as wind-driven advection (mixing due to the movement of water). The exact extent of mixing can be determined by costly and long-term monitoring projects. One extreme approach would be to release a known mass of dye at the inlet of the lake. Usually a fluorescent dye (Rhodamine WT) is used to enable detection of extremely small concentrations. At low concentrations used in dye studies, the dye itself impacts little toxicity to aquatic species and not long lasting in these environments. A few studies have used the radioactive tracer tritium. Of course, this will not work for very large lakes, since the large volume in these systems will dilute the dye to non-detectable concentrations and large lakes can have detention times of decades. Some recent EPA work, however, fashioned after National Council for Air and Stream Improvement (NCASI) used a radioactive isotope of krypton gas and measured the ratio of Kr/Ar using gas chromatograph/atomic absorption spectrometry to measure atmospheric reaeration very precisely in small and large water bodies (NCASI). After the dye has been placed into the lake, the effluent stream of the lake is monitored with respect to the dye concentration. If the slow increase and subsequent decrease in dye concentration is analyzed, the effective mixing volume can be calculated. Thus, the effective mixing volume is the volume of water actually mixing with the pollutant as opposed to the entire volume of the lake. Of course, if the lake is large the monitoring dye program could take months to years, or even decades to complete. Note that this process is further complicated when stratification of the lake occurs. So, the dye technique for determining mixing is only of use in small ponds and lakes. Usually historical data or "experience" is used to estimate effective mixing volumes for larger lakes.

When the effective mixing volume is determined or estimated, for example 78% of the total volume, this value can easily be used in place of V in Eq. (5.1) to calculate a more accurate estimate of the detention time of water and pollutant in the system. This approach can also be used for stratified lakes, where the depth of the hypolimnion can be measured and the volume of the water body receiving the pollutant can be calculated.

5.6.3 Chemical Reactions

In Chapter 2, a variety of potential degradation schemes were presented, including photochemical, biological, abiotic (chemical), and nuclear reactions. All of these are possible transformation reactions in lake systems. Whatever the type or types of reaction(s), all of these are usually represented by first-order kinetics, and we can add the individual rate constants (k values) together to obtain one overall first-order rate constant. This component of the fate and transport model is of the form

$$C_t = C_0 \cdot e^{(-kt)} \tag{5.2}$$

where

C_t is the pollutant concentration at time t
C_0 is the initial pollutant concentration
e is the exponential function
k is the first-order rate constant, and
t is time.

As noted in Chapter 4, chemists and modelers prefer the first-order rate constant since they more easily fit into our mathematical models. Again, as noted in Chapter 3, first-order rate constants can be easily converted to half-lives, a relatively easily concept to relate to time of degradation or reaction.

Table 5.5 Particle settling velocity as a function of particle size (Lapple 1961).

Classification	Particle diameter range (μm)	Settling velocity in water (cm/s)
Clay	<2	10^{-8} to 2×10^{-4}
Silt	2.0–20	2×10^{-4} to 2×10^{-2}
Fine sand	20–200	2×10^{-2} to 2
Coarse sand	200–2000 (0.2–2 mm)	2–20
Gravel	>2000 (2 mm)	>20

5.6.4 Sedimentation

In addition to wash-out of pollutant in the effluent from lakes, and biological and chemical degradation, pollutants can be removed from a lake system by sorption to particles followed by subsequent settling to the lake bottom. This can be a significant removal mechanism for some pollutants, especially those that do not readily degrade through microbial or chemical means. In order to appreciate how pollutants can thus be removed from the water column of lakes, we will first look at the size of particles that can be present in aqueous systems. Table 5.5 shows the particle settling velocity as a function of particle size. As the particle size decreases, the surface area to the volume ratio of the particle increases, and sorption processes become more important since more pollutant can sorb to the surface. In addition, smaller particles can contain more organic matter on the surface and be even more sorptive reactive. Hence, clay-sized particles can be very important in determining the fate of sorbed pollutants, and as you can see from the data in Table 5.5, they have the smallest settling velocities. This results in the particles and sorbed pollutants settling in the deepest and most quiescent (calmest) regions of the lake. This particle-settling velocity has been validated by monitoring results from lakes that find the highest concentration of polluted sediments in the deepest regions of lakes (referred to as pollutant focusing). In contrast, regions of higher energy flow and thorough mixing in lakes contain larger particles, which generally do not contain high levels of organic matter and therefore do not contain high levels of pollutants.

Settling velocities (ω) given in Table 5.5 were obtained by a relatively simple calculation, defined as Stokes' law

$$\omega = \frac{(2/9)\,g\,(r_s/r_f - 1)\,r^2}{\eta} \tag{5.3}$$

where

g is the acceleration due to gravity (length/time^2),
r_s is the density of the spherical particle (mass/length^3)
r_f is the density of the fluid (mass/length^3)
r is the spherical particle radius (length)
η is kinematic viscosity of the fluid (length^2/time). The kinematic viscosity is the ratio of the dynamic viscosity of the fluid to the density of the fluid. Note that Eq. (5.3) assumes a spherical particle, but average particle radius can be used.

While Eq. (5.3) describes the settling of a particle, it is of little use, since pollutant concentration is not present in this equation. You should recall from Chapters 2 and 3 that sorption behavior of a pollutant is described by the distribution coefficient (K_d) for metals and the partition coefficient (K_p) for hydrophobic pollutants. Thus, we need an expression that incorporates particle removal and pollutant concentration:

$$r_A = -\frac{K_d\,\omega\,S}{H\,(K_d\,S+1)}\,C \tag{5.4}$$

where

r_A is the rate of decrease in pollutant A concentration per unit volume of water (mass/length^3-time),
K_d is the distribution coefficient (or K_p is the partition coefficient),
ω is the particle settling velocity (length/time)
S is the suspended solids concentration (mass/length^3),
H is the water depth (length), and
C is the pollutant concentration in the water (mass/length^3).

Thus, we can account for pollutant removal by sedimentation in calm water. Waters with rapid currents that mix the water and suspended material will result in slower settling rates. Also note the units of the settling rate constant, concentration per time. These are the units of a zero-order, rather than first-order rate expression, whereas our fate and transport models will use first-order expressions. Thus, unfortunately, the rate of pollutant removal cannot be directly substituted into our fate and transport models; still, we can estimate the removal of pollutant through sedimentation. Also note that this is assumed to be a steady-state process, since the concentration of pollutant and the suspended solids concentration in the water are assumed to be relatively constant. This is usually a reasonable assumption, except during storm events where large masses of sediment can enter the lake or re-enter the water column in the form of resuspended sediments.

It is also important to note the rate of sediment accumulation in the bottom of a lake, since the settling sediment can bury previously contaminated sediment. Baker (1994) reports a range from 50 to 600 g of sediment per square meter per year for a variety of lake systems. This translates into an accumulation rate from mm to cm of sediment per year, which can result in significant deposits of sediment. While sediment accumulation and burial of contaminated sediment is important, it is also important to look at sediment resuspension rates. Sediment can be naturally resuspended through bioturbation (the mixing of sediment by bottom feeding fish and organisms [worms and bivalves] living in the sediment) and by storm events. Wetzel (2001, p. 635) reports resuspension rates from 0.5 to 21 g/m^2 d which are very significant.

Considering these sedimentation rates, it is understandable that contaminated sediments can be buried and therefore removed from the aquatic system. In a sense, burial of sediments is a form of natural remediation. An example of this is shown in Figure 5.13 for chromium. This sediment profile is from Upper Mystic Lake, which is the water basin for the Aberjona Watershed, north of Boston, Massachusetts. As you can see from the profile, chrome used in the local tanning industry started in ~1900 and declined after 1925. The spike in chromium in the sediment deposited around 1959 is from rendering operations that utilized chrome-tanned hides in the process of making glue. Note that today the chromium pollution has been isolated and buried by the natural sedimentation of unpolluted, or less polluted, material in the lake. If the sediment cap remains intact then the chromium has been effectively (and inexpensively) removed from the lake system.

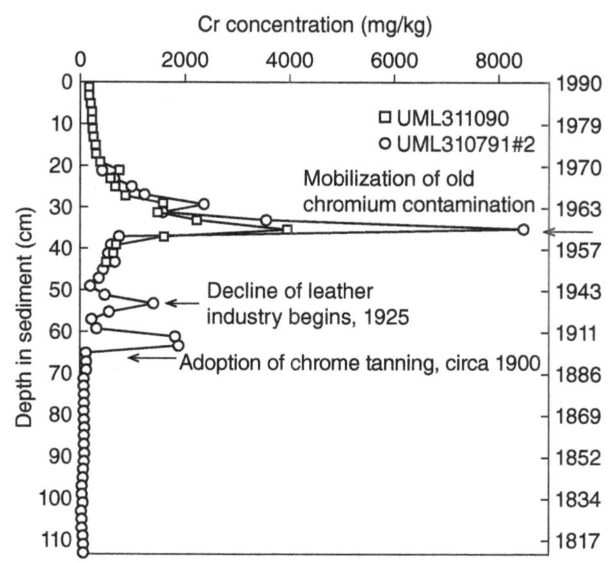

Figure 5.13 An example of natural sedimentation burying chromium-contaminated sediment. Source: Spliethoff and Hemond (1996). Reprinted with permission of American Chemical Society.

5.7 Two Basic Mathematical Models for Lakes (Derivation by John Brooksbank in the Chapter Appendix)

5.7.1 The Basic Model Development

As discussed in Chapter 4, the derivation of the fate and transport equations used in this textbook requires knowledge of calculus, linear algebra, and/or differential equations. In our lake model development, we used the relatively simple assumption of a CSTR.

We will repeat the beginning of the lake equation derivation from Chapter 4. The first step in developing the governing equations for the fate of a pollutant in a lake system is to set up a mass balance for the system. We intuitively should know that mass cannot be created or destroyed only changed in form even if converted to heat energy by combustion. Even in this case mass is conserved. First, quantify all of the mass inputs of pollutant to the system (Eq. (5.5)). This can be expressed as

$$W = Q_w\, C_w + Q_i\, C_i + Q_{trib}\, C_{trib} + P\, A_s\, C_p + V\, C_s \tag{5.5}$$

where

W = the mass input of pollutant to the lake per unit time, kg/time

Q_w = the inflow rate of the wastewater, m^3/time

C_w = the pollutant concentration in the wastewater, kg/m^3

Q_i = the inflow rate of the main river, m^3/time

C_i = the pollutant concentration in the main inlet river, kg/m^3

Q_{trib} = the net inflow rate from all other tributaries, m^3/time

C_{trib} = the net pollutant concentration of the tributaries, kg/m^3

P = annual precipitation, m/time

A_s = mean lake surface area, m^2

C_p = the net pollutant concentration in precipitation, kg/m^3

V = the average lake volume, m^3

C_s = the average pollutant release from suspended lake sediments, kg/m^3-time

A useful exercise is to check the units of each term (area/volume/mass/unit time) or rate is each part of the mass balance. We will see later in this text that rate/time for a given plane is known as flux and has units of mass/time (area) or mg/m^2 day.

In most situations, the mass inputs from the smaller tributaries and precipitation are minor compared to the major input source, and the terms C_{trib} and C_p are ignored. We will further simplify the mass input expression here by assuming that the contribution from contaminated sediments is negligible, although this may not always be the case in certain lakes where benthic or sediment processes are critical. These assumptions simplify the input expression to Eq. (5.6).

$$W = Q_w\, C_w + Q_i\, C_i \tag{5.6}$$

Next, in order to set up a mass balance for the pollutant across the entire system, we need to incorporate outflow into the expression. The equation for the mass outflow from the system is analogous to the input equation, but with only one outlet usually considered. We also assume that there are no additional sources of pollutant and that chemical degradation yields a constant first-order chemical removal rate for the pollutant. Thus the mass balance becomes

$$\text{Change in mass} = \text{inflow} - \text{outflow} + \text{sources} - \text{sinks}$$

$$V\,dC = (Q_w\, C_w\, dt + Q_i\, C_i\, dt) - Q_e\, C\, dt + 0 - V\, C\, k\, dt, \text{ or}$$

$$V\,dC = W\, dt - Q_e\, C\, dt - V\, C\, k\, dt \tag{5.7}$$

where

dC or ΔC = the change in pollutant concentration in the lake

dt or Δt = the incremental change in time, dC/dt or $\Delta C/\Delta t$ is the fundamental theorem of differential calculus meaning a "rate of change"

Q_e = the outlet or effluent flow from the lake (m^3/time)

Q_w = the waste flow to the lake (m^3/time)

Q_i = the inlet or influent flow from the lake (m^3/time)

C = the average lake concentration (kg/m^3)

k = the first-order removal rate for the pollutant (1/time)

Note that since we are looking at changes in concentration and time, we must use an operator to indicate this. The operator is the d or Δ symbol.

The last equation, upon substitution and rearrangement, yields

$$[Q_e\, C - W(t)]\, dt + V dC = -V\, C\, k\, dt \tag{5.8}$$

Upon rearrangement, this reduces to

$$V\frac{dC}{dt} + (Q_e + k\, V)\, C = W(t) \tag{5.9}$$

where Q_e, k, and V of the lake, and thus the quantity $(Q_e + kV)$, are assumed to be constant (usually, they are reasonably constant).

The average detention time (t_0) of water (and thus the pollutant) in the lake, from the time of input to outlet, is defined as

$$t_0 = \frac{V}{Q_e} \tag{5.10}$$

Substitution and further rearrangement into the previous equation, yields the governing non-integrated equation for pollutant transport in a lake (CSTR) system

$$V\frac{dC}{dt} + C\,V\left(\frac{1}{t_0} + k\right) = W \tag{5.11}$$

where dC/dt ($\Delta C/\Delta t$) is the differential part of the equation and represents a change in concentration with change in time, V is lake volume, C is pollutant concentration, t_0 is hydraulic detention time, k is the first-order degradation rate constant, and W is the rate of input of pollutant into the system. It should also be noted that in deriving the model we assume that there are no pollutant sources in the lake or that these internal sources are insignificant (such as from contaminated sediments). Equation (5.11) must now be integrated, based on two sets of boundary conditions, step and pulse, using calculus-based integration factors. Only the basics will be should in the chapter text; an elaborate, step-by-step derivation is shown in the chapter appendix. Equation (5.14) is the net result for the step (continuous) input source and Eq. (5.17) is the new result for the pulse input source.

5.7.2 Continuous (Step) Input Model

To solve for the Continuous (Step) Model, the same approach is used, except in this case rather than setting W equal to 0, we set W as a constant. Thus our starting equation is just Eq. (5.11). An extensive, step-by-step derivation is given in the chapter appendix. Dividing by the volume (V) and substituting in β, yields

$$\frac{dC}{dt} + \beta C = \frac{W}{V} \tag{5.12}$$

where $\beta = (1/t_0 - k)$. Using a variety of mathematical techniques discussed in the chapter appendix and upon integration, one can obtain the governing equation

$$C_{(t)} = \frac{W}{\beta V}(1 - e^{-\beta t}) + C_0 e^{-\beta t} \tag{5.13}$$

If the background concentration is negligible, then $C_0 \approx 0$ and Eq. (5.13) reduces further to

$$C_{(t)} = \frac{W}{\beta V}(1 - e^{-\beta t}) \tag{5.14}$$

These two equations can be used to estimate the concentration of pollutant in a lake that receives a constant input of pollutant and their correct derivation is confirmed by equations in Fjeld et al. (2006). Note that two opposing forces govern the effluent pollutant concentration. The addition of more and more pollutant mass by the source is opposed by washout of the pollutant in the effluent stream and by first-order degradation. Eventually the system will reach an equilibrium pollutant concentration if all parameters remain constant. Thus, as time approaches infinity, the pollutant concentration in the lake approaches

$$C = \frac{W}{\beta\,V} \tag{5.15}$$

where

$$\beta = (1/t_o - k)$$

An physical example of this pollution scenario is shown in Figure 5.14, and a concentration with time profile is shown in Figure 5.15. Note the shape of the plot.

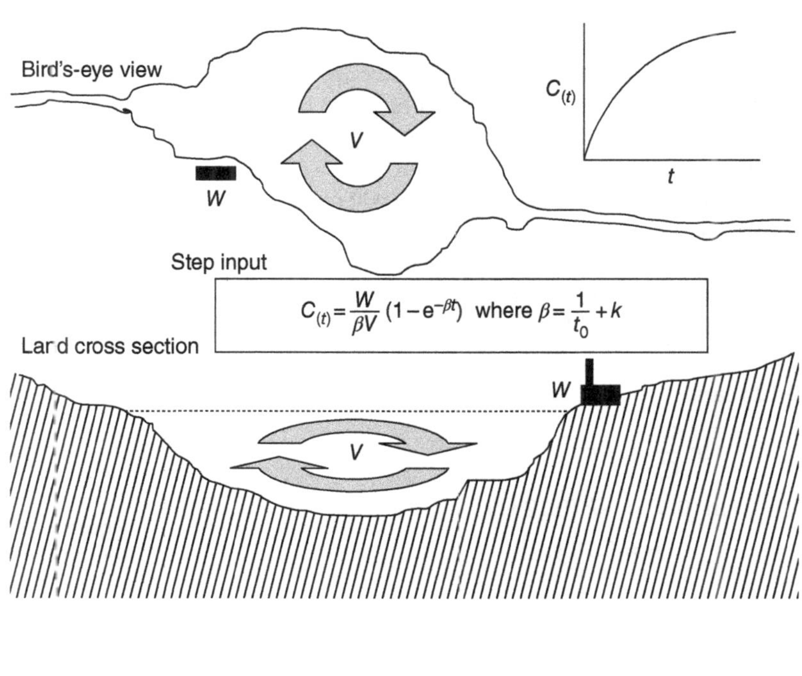

Figure 5.14 Diagram of a step input of pollution to a lake.

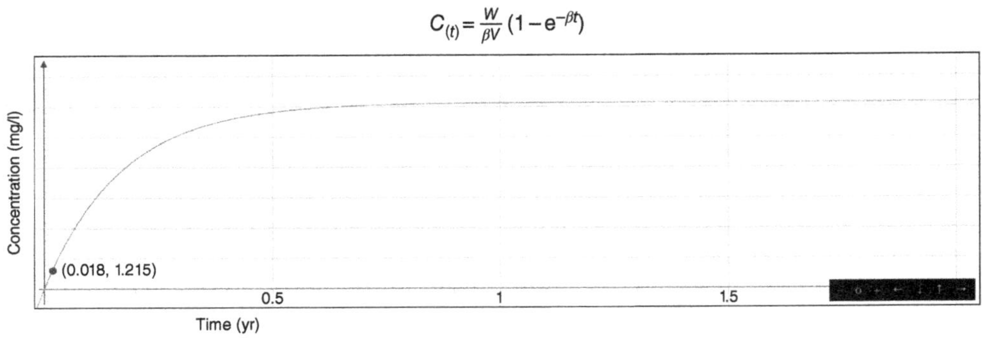

Figure 5.15 The results of a step input of pollutant to a lake (results from Fate). Source: Anders.

Compared to the pulse model, the highest concentration of pollutant for the step model is much later in time. In fact, since the pollutant is emitted slowly into the lake, the concentration slowly builds up in the lake and reaches a maximum.

Problem

A lake in a rural community has an average surface area of 5000 m^2 and a mean depth of 50 m. A stream exits the lake with an average annual flow rate of 45 000 m^3/yr. Aerial application of an insecticide in the area introduces the compound into the lake. The average annual loading of this pollutant to the lake from the atmosphere and from agricultural runoff is estimated at 50 kg/day. Assuming a first-order removal of the insecticide from the lake (half-life = 43.8 days) and that the initial background concentration of insecticide in the lake is negligible, answer the following questions.

What is the detention time of water in the lake?
What is the steady state concentration of insecticide in the lake?
What is the concentration after 0.010 yr?

Solution

The volume of the lake is equal to the average surface area multiplied by the mean depth, and detention time is equal to this volume divided by the average annual flow rate:

Volume = $(5000\ \text{m}^2)\ (50\ \text{m}) = 250\,000\ \text{m}^3$

Detention time = $(250\,000\ \text{m}^3)/(45\,000\ \text{m}^3/\text{yr})$
= 5.56 years.

In order calculate the equilibrium insecticide concentration, we must first convert the first-order half-life to a rate constant, k, expressed in units of reciprocal years. The half-life of 43.8 days is equal to a half-life of 0.12 years.

$$\ln\frac{C_t}{C_0} = -k\,t$$

$$\ln(0.5) = -k\,(0.12\ \text{yr})$$

$$k = 5.78\ \text{yr}^{-1}$$

Now,

$$\beta = \frac{1}{t_0} + k = \frac{1}{5.56\ \text{yr}} + 5.78\ \text{yr}^{-1} = 5.96\ \text{yr}^{-1}$$

$$C = \frac{W}{\beta\,V} = \frac{(50\ \text{kg/day})\ (365\ \text{d/yr})}{5.96\ (250\,000\ \text{m}^3)}$$

$$= 1.22 \times 10^{-2}\ \text{kg/m}^3$$

$$C = (1.22 \times 10^{-2}\text{kg/m}^3)(1000\ \text{g/kg})$$

$$\times\,(10^3\text{mg/g})\left(\frac{1\ \text{m}^3}{1000}\text{l}\right) = 12.5\ \text{mg/l}$$

Determine whether the first-order decay is an important removal process by constructing one plot using the first-order decay and another plot without the first-order decay.

An extension of these results are shown in Figure 5.16 where we plot the effect of having and not having a degradation constant or reaction present. This shows the importance of a chemical being degradable, which will result in a significantly faster removal of the pollutant.

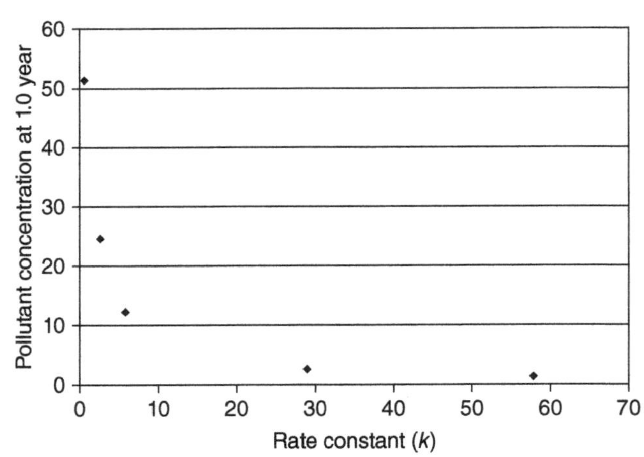

Figure 5.16 The results of the step input from Figure 5.7, showing the effect of degradation rates (results from Fate).

Finally, calculate the concentration after 0.010 years.

$$\beta = \frac{1}{t_0} + k = \frac{1}{5.56} + 5.78 = 5.96$$

$$C = \frac{W}{\beta\, V}\,(1 - e^{-\beta\, t})$$

$$C = \frac{\left(\dfrac{18\,250\ \text{kg}}{\text{yr}}\right)\left(\dfrac{1000\ \text{g}}{\text{kg}}\right)(1000\ \text{mg/g})}{(5.96\ \text{yr}^{-1})(250\,000\ \text{m}^3)(1000\ \text{l/m}^3)} \times (1 - e^{-5.96 * 0.010})$$

$$C = 12.25\,(1 - e^{-5.96 * 0.010})$$

$$= 12.25 - 12.25\,(0.94) = 0.71\,\frac{\text{mg}}{\text{l}}$$

5.7.3 Instantaneous (Pulse) Input Model

The derivation for the governing of the instantaneous lake model again starts with Eq. (5.11) from above:

$$V\,\frac{dC}{dt} + C\,V\left(\frac{1}{t_0} + k\right) = W$$

Now, we will solve this equation under a different set of boundary conditions for the pulse model. Here, we assume that all the pollutant has been added at time = 0, which means that $W = 0$. We will again combine our constants:

$$\frac{1}{t_0} + k = \beta$$

so our equation simplifies to

$$\frac{dC}{dt} + \beta C = 0 \qquad (5.16)$$

Again, using a set of mathematical techniques and upon integration, one can obtain

$$C_{(t)} = C_0 e^{-\left(\frac{1}{t_0}+k\right)t} \qquad (5.17)$$

Equation (5.17) is confirmed by equations in Fjeld et al. (2006). The parameters in this equation were defined earlier. An physical example of such a pollution scenario is shown in Figure 5.17, and a pollutant concentration profile is shown in Figure 5.18.

Note the overall shape of the plot in Figure 5.18. The highest concentration of the pollutant occurs immediately after its addition to the lake, and the pollutant is slowly removed from the lake by generic first-order degradation reactions and by outflow through the effluent of the lake. The student should note the exponential shape of the plot and be able to relate this to Eq. (5.17). A decrease in the first-order degradation rate would prolong the time required for removal of pollutant from the lake. Using Eq. (5.17), note how increasing or decreasing of the other model parameters would affect $C_{(t)}$ and the pollutant removal time.

Problem

We will use the same problem statement used in the step input example (given above), but for this pulse input example we will monitor the fate of the insecticide in the system if the input is ceased after one year. Thus, we can treat the input as a pulse. In this case, we will develop a formula to express the concentration of insecticide as a function of time. Then, we will be able to calculate how long it will take for the insecticide concentration to reach 0.100 mg/l, the detection limit for this compound.

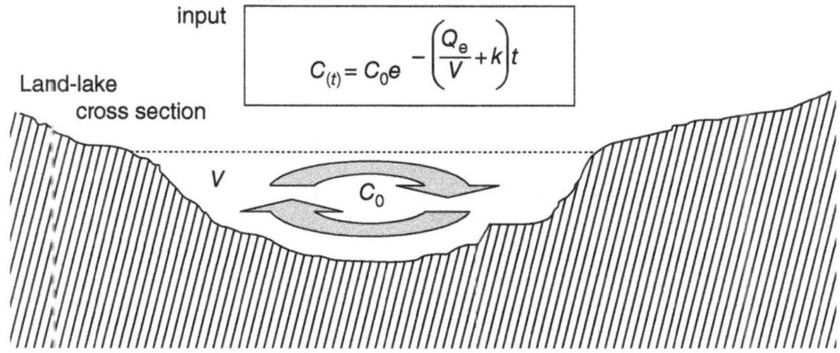

Figure 5.17 Diagram of a pulse input of pollution to a lake.

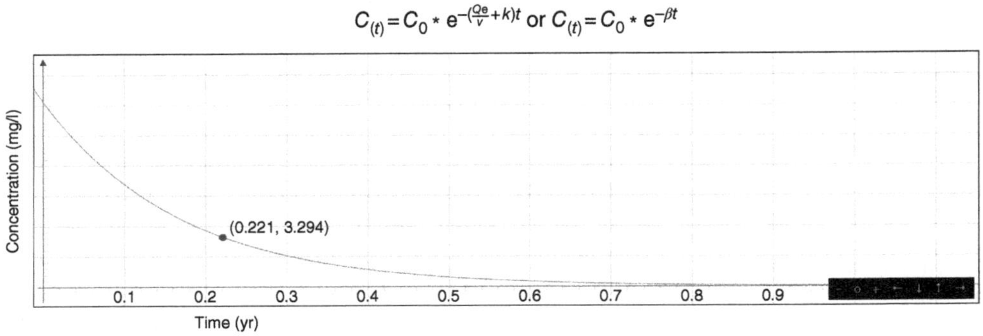

Figure 5.18 The results of a pulse input of pollutant to a lake (results from Fate). Source: Anders.

Solution

The calculations of residence time and the first-order rate constant in the lake are identical to those shown for the step model (Section 5.6.2), since the dynamics of the lake system have not changed.

Now we must calculate the concentration after one year, since this will be the initial concentration for our expression showing the long-term removal of insecticide from the system.

The concentration after one year is

$$C = \frac{W}{\beta V}(e^{-\beta t})$$

$$\text{where } \beta = \frac{1}{t_0} - k = \frac{1}{5.56} + 5.78 = 5.96$$

$$C = \frac{(18\,250\text{ kg/yr})(1000\text{ g/kg})(1000\text{ mg/g})}{(5.96)(250\,000\text{ m}^3)(1000\text{ l/m}^3)} \times (e^{-5.96*1.00})$$

$$C = 12.3\,(e^{-5.96*1.00}) = (12.25)(0.00258)$$

$$= 0.0316\text{ mg/l}$$

Now, for this problem calculate the insecticide concentration as a function of time. This is governed by Eq. (5.17):

$$C_{(t)} = C_0 e^{-\left(\frac{1}{t_0}+k\right)t}$$

where C_0 = 12.2 mg/l, detention time t_0 = 5.56, and k = 5.78 yr^{-1}. A plot of this equation is shown in Figure 5.18.

Next, calculate the time required to reach an insecticide concentration of 0.100 mg/l, again starting with Eq. (5.17).

$$C_{(t)} = C_0 e^{-\left(\frac{1}{t_0}+k\right)t} = \frac{0.100\frac{\text{mg}}{\text{l}}}{12.2\frac{\text{mg}}{\text{l}}} = e^{-\left(\frac{1}{5.56}+5.78\right)t}$$

$$\ln\left(\frac{0.100\text{ mg/l}}{12.2\text{ mg/l}}\right) = -\left(\frac{1}{5.56} + 5.78\right)t$$

$$4.80 = 5.96\,t$$

$$t = 0.81\text{ yr}$$

Thus, it is seen that the lake will recover relatively rapidly after the input of the insecticide is ceased.

5.8 Sensitivity Analysis

The best modelers look critically their modeling results. One common form of determining the accuracy of a model is to perform a sensitivity analysis. In such an analysis, the modeler runs a series of model simulations to determine how much effect an error in a certain parameter will have on the output of the model. To illustrate this, we will perform a sensitivity analysis on first-order degradation constant using the basic (background) step scenario. In the background example problem in Fate, we used a degradation rate constant of 5.78 yr^{-1}. But what if this is in error? How sensitive is the effluent pollutant concentration to the rate constant? We ran a series of calculations using rate constants of 0.578, 2.58, 5.78, 28.9, and 57.8 and a time of 1.0 years. The results of this analysis are shown in Figure 5.19. We

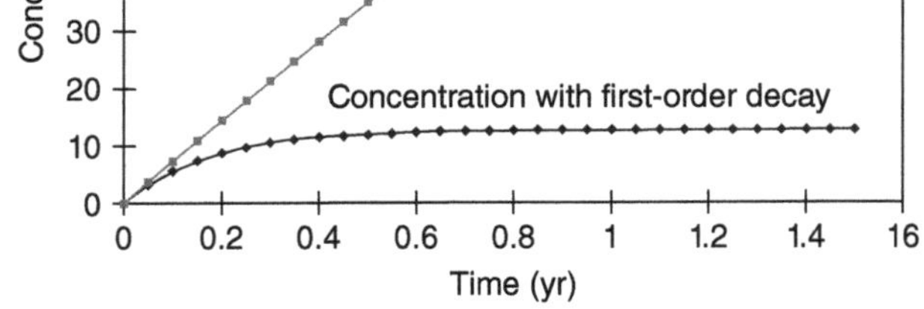

Figure 5.19 The results of a sensitivity analysis on degradation rate constant for the background scenario in Fate.

can now directly see the results are strongly affected by the by the magnitude of the rate constant and that the relationship is exponential as expected. Similar sensitivity analyses can be performed for all of the parameters in each of the basic models.

The reader should be aware of the dangers and limitations of using "canned" programs such as Fate. Always review the input data (boxes) in these programs and be sure you understand the units and how the calculations between the boxes are connected. A good mantra is "always question the results and check everything twice."

5.9 Limitations of Our Models

Incomplete Mixing: We have already discussed the limitations of our models with respect to incomplete mixing, but this can be a limitation of any modeling effort where the true mixing volume of the lake is not known. This unknown quantity will even plague modeling efforts of professionals who use numerical methods of analysis to calculate pollutant concentrations. Thus, the most accurate estimates of effective mixing volumes are needed.

No Internal Sources: If pollutant-sorbed sediment is present in the bottom of the lake, then resuspension of the sediments can act as a source of pollution to lake systems. Normally input from storm events and bioturbation is small or insignificant compared to the overall mass of pollutant in the system. However, it can be large during shipping or harbor dredging operations. This type of re-suspension could be treated as a pulse input since dredging operations are of small time scales compared to the usual time scales of pollutant transport in lake systems.

Human-made reservoirs: These types of reservoirs pose a huge modeling problem, and our simple modeling approach will not work for complicated, multi-basin systems. For example, consider the diagram of Lake Hartwell (Figure 5.20), a human-made reservoir along the Savanna River chain between Georgia and South Carolina (USA). Different mixing basins are separated in Figure 5.20 by the black rectangles which are located at constrictions in the lake. In order to effectively model this reservoir, we would need to develop a series of models, one for each mixing basin. The input for each reservoir would be the output from the previous reservoir. This could easily be handled using the numerical methods approaches described in Chapter 4, and the closing example in this chapter.

5.10 Remediation

No text on pollutant fate and transport would be complete without considering remediation of the contaminated system. Thus, each fate and transport chapter in this book (Chapters 5–9) will close with a section on remediation. Each environmental system discussed in this book presents a unique set of conditions, and thus problems, with respect to remediation. Issues specific to lakes include the hydraulic retention time, t_0, introduced in Section 5.5.2, which is determined by the total volume of the lake and the effluent flow rate. A lake with a high retention time will not naturally or rapidly purge itself of pollutants in the water column, and thus the pollution will be present for extended times. The large volumes of water that most lakes contain makes infeasible any direct *ex situ* (outside of the lake) pump-and-treat process on the lakeshore. While soluble pollutants remain in the water column, other pollutant will concentrate in the sediments. In general, a longer residence time in the lake enables pollutant sorption to particles in the water column, such that the ultimate resting place of the pollution is in the lake sediments. This is especially true for metals and hydrophobic pollutants. Concentration of the pollutants in the sediments has good and bad consequences, as we will see in the following paragraphs.

So, what can we do to remediate a polluted lake? First and foremost is to remove or eliminate the source of pollution to the lake. This is relatively easy for point sources such as the effluent from an industry, but becomes more complicated for effluent from a sewage treatment plant, since treated sewage has to be disposed of somewhere. The problem becomes even greater when we are talking about non-point sources that are not easily identified or eliminated (such as agriculture runoff) or are uncontrollable (such as leachate entering a lake from an abandoned landfill). Nonetheless, identification, isolation, and eliminate of the source of the pollution should be the first step in any remediation plan.

One of the most common remediation efforts undertaken for lakes is to treat eutrophication, the uncontrolled or excessive microbial and algal growth that results from the presence of nutrients from agricultural runoff and sewage treatment facilities. Even the slightest amount of phosphorus can induce eutrophication; thus, elimination of the source is important. In general, the remediation approach for eutrophied lakes is to eliminate the nutrients and let nature take its course (cleansing by removal through burial of nutrients in the sediments and by removing nutrients by outflow from the lake). This approach is usually the only viable option given the large scale of the problem (volume of water and hydraulic detention time in the lake).

Remediation of polluted sediments has also been the focus of many lake remediation efforts. We will discuss remediation strategies from the most to least complicated (and costly). One obvious, but expensive,

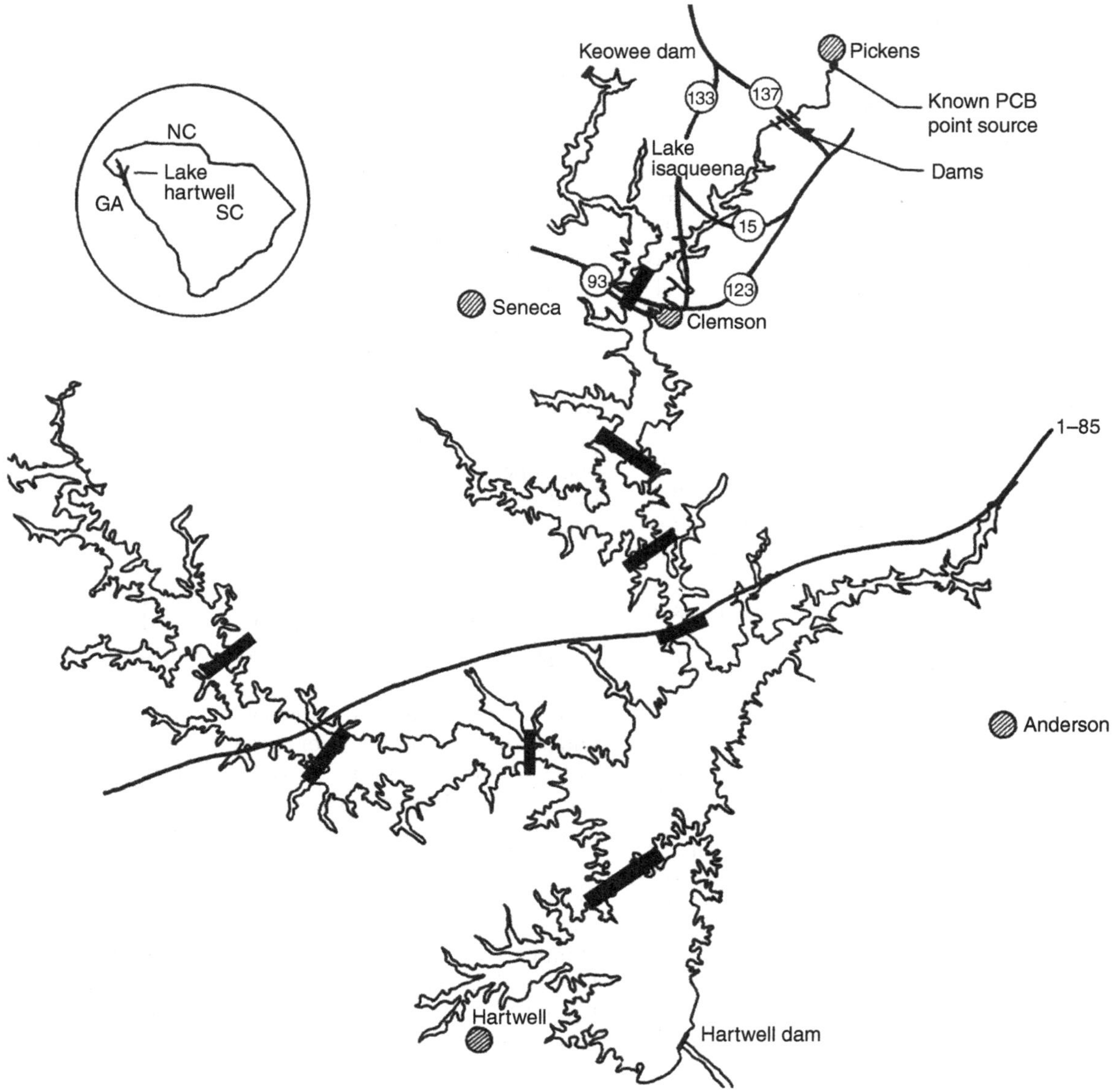

Figure 5.20 A diagram of the shoreline and basins contained in Lake Hartwell (South Carolina, USA). Mixing basins are delineated by solid rectangles. Source: From Dunnivant (1985).

way to remediate contaminated lake sediments is to simply remove the sediment by a technique referred to as dredging. There are many types of dredging, but all involve physically removing the sediment from the lake, placing it on a barge, and transporting it to a treatment or containment facility. But before this "sure fix" is adopted, one must consider the scale of the problem. The cost of dredging uncontaminated sediment is approximately \$5 yd^{-3}, but this cost increases to \$15–20 yd^{-3} for contaminated sediment (National Research Council 1977). And what do you do with the contaminated sediment once you have removed it since it is not a hazardous waste? Treatment costs (thermal degradation, biochemical treatment, extraction) can range from \$100 to 1000 yd^{-3} (National Research Council 1977). Note these sources are dated and costs today are certainly more expensive, depending on your location. You start to see the economic scale of the problem created by this "sure fix" solution. Thus, dredging the sediments of a lake is usually not a viable option. Another issue relates to large potential to resuspend sediments during dredging and re-exposing aquatic life to dangerous chemicals.

Another proposed option is to bioremediate the sediment by adding nutrients and pollutant-degrading microorganisms to the contaminated areas. While this sounds good in theory, it also presents several problems. For example, engineered microbes that effectively degrade pollutants have been raised in the lab, but it is unclear whether these same microbes can effectively compete in the real world of a lake sediment. Furthermore, addition of nutrients to a lake can be a problem in itself, since these nutrients will most certainly migrate to the water column and could promote eutrophication of the lake. In addition, metals do not degrade via microbial action. To date, no human-engineered bioremediation action has been conducted on a lake, although nature is always working through these very same pathways to degrade organic pollutants in lakes.

The next less costly, but also much more viable, approach is to "cap" the contaminated sediment by placing a layer of pollutant-free substrate on the area of interest in the lake. This can be achieved by strategically placing anything from synthetic substrate to clay to sand in the lake. This speeds the process of natural burial and can effectively prevent the contaminated sediments from resuspending and contacting the water column. Of course, considerations that must be made in using this technique include sediment depth from the water's surface, the surface area needing capping, and hydraulic flow rates in the lake that may uncover the contaminated sediments.

The least expensive, and in many cases the most practical, remediation is natural recovery resulting from a combination of biodegradation and burial of the contaminated sediment. Burial occurs by natural sedimentation of suspended lake particles and material that washes into the lake during storm events. As noted in Section 5.5.4, some lakes have natural sedimentation rates of centimeter per square meter per year that can relatively quickly bury a contaminated area of the lake. This remediation approach may not be appropriate for every lake, but does prevent the release of pollutants into the water column during dredging activities. These releases can be significant, as up to 50% of the sorbed mass of pollutant can be released in a period of hours to days (Dunnivant et al. 2005). Sorption and desorption processes were discussed in Sections 3.3 and 3.4.

All of the remediation approaches discussed above must also include some form of long-term monitoring project to document the extent of pollutant occurrence in the lake, future releases of pollutants from the dredged or in-place sediments, and the overall effectiveness of the remediation effort. The extent of this monitoring can range from a few months to decades.

5.11 Numerical Modeling Approaches for Large Lakes

In Chapter 10, on regulatory modeling, we will discuss more complex models for pollutant fate and transport in large lakes. Most notable is WASP used in the Great Lakes and Chesapeake Bay areas. Estuary systems can be effectively modeled using the Dynamic Estuary Model (DEM/DYNHYD) developed for the Sacramento-San Joaquin Delta and San Francisco Bay areas, and EFDC that has been used on the Chesapeake Bay estuary. It should not be surprising that use of these models requires extensive training and experience, thus, these models are only briefly described in Chapter 10. These models would be a possible next step, after this text, to become a professional modeler. WASP, in particular, is currently used to simulate many pollutant types in rivers and lakes because of its versatility and should be on your short list to learn how to use. A sizable user community is developing in which many share their unique experiences with others. Similarly, new modelers should learn how to use crude mass balance algorithms to approximate the desired outcome and/or design better monitoring to facilitate model process understanding.

5.12 Useful Algebraic Model Formulation

There are analyses for lakes and ponds based upon simple mass balance formulations that have seen utility in lakes, and better known as approximations in EPA's model toolkit. One known as simplified method program-variable complexity stream toxins model (SMPTOX) appeared in the early 1990s (LTI, 1992). It is based on the CSTR assumption introduced earlier, and included limited reaction and fate processes. Chapra (2008) developed a useful construct with dissolved and suspended solids fractions and the derivation is summarized in their appendix with an example for Lake Huron and polychlorinated biphenyl (PCB) contamination. With the advent of more sophisticated and reliable modeling platforms (such as WASP) the SMPTOX code was deemed as at the end of its life cycle and formally discontinued, although some state agencies may still have working versions from this time period.

Suggested Papers for Class Discussion

From the hundreds of TMDLs developed for lakes in the U.S., select one to revisit with a crude mass balance only. How does it compare to the more detailed modeling used? Complete as a group project with instructor guidance.

Handlin, M., Molina, A., James, N., McConville, M., and F. Dunnivant. "Effect of suspended uncontaminated sediment on persistent organic pollutant release" Environ. Toxicol. Chem., 2014, 33(2), 375–381.

EIA, (2017), https://www.eia.gov/energyexplained/?page=hydropower_environment.

Eisenreich, S.J., B.B. Looney, and J.D. Thornton. "Airborne organic contaminants in the Great Lakes ecosystem," *Environ. Sci. Technol.*, 1981, 15(1), 30–38.

Holloway, T., A. Feore, M.G. Hastings. "Intercontinental transport of air pollution: Will emerging science lead to a new hemispheric treaty?" *Environ. Sci. Technol.*, 2003, 37(20) 4535–4542.

Engstrom, D.R., E.B. Swain, T.A. Henning, M.E. Brigham, and P.L. Brezonik. "Atmospheric mercury deposition to lakes and watersheds: A quantitative reconstruction from multiple sediment cores," In: *Environmental Chemistry of Lakes and Reservoirs*, Advances in Chemistry Series 237, Ed. L.A. Baker. American Chemical Society, Washington, D.C. Chapter 2, pp. 33–66, 1994.

Michigan, 2017, http://www.michigan.gov/documents/dnr/damsuccess_513764_7.pdf.

5.A Derivation of the two basic forms of fate and transport models for lake system: step (continuous) model and pulse (instantaneous) (derivations by John Brooksbank)

The governing non-integrated equation for pollutant transport in a lake (CSTR) system

$$V\frac{dC}{dt} + C\,V\left(\frac{1}{t_0} + k\right) = W \tag{5.11}$$

where dC/dt ($\Delta C/\Delta t$) is the differential part of the equation and represents a change in concentration with change in time, V is lake volume, C is pollutant concentration, t_0 is hydraulic detention time, k is the first-order degradation rate constant, and W is the rate of input of pollutant into the system. It should also be noted that in deriving the model we assume that there are no pollutant sources in the lake or that these internal sources are insignificant (such as from contaminated sediments). Equation (5.11) must now be integrated, based on two sets of boundary conditions, step and pulse, using calculus-based integration factors. Only the basics will be should in the chapter text; an elaborate, step-by-step derivation is shown in the chapter appendix. We will first discuss integration factors. Equation (5.A.25) is the net result for the step (continuous) input source and Eq. (5.A.36) is the new result for the pulse input source.

5.A.1 Integrating Factors

The technique of using an integrating factor is useful when dealing with differential equations that are first order, meaning that there are no second derivatives or higher, and linear, meaning that all terms involving the function, Y, or any of its derivatives, can be separated. The general form of a linear, first order differential equation is:

$$Y'(t) + p(t)Y(t) = q(t) \tag{5.A.1}$$

where $p(t)$ and $q(t)$ are some functions with respect to t.

Now, we can create some function, $I(t)$, which we define as:

$$I(t) = e^{\int p(t)dt} \tag{5.A.2}$$

The derivative of $u(t)$, $I'(t)$, is thus:

$$I'(t) = p(t) * e^{\int p(t)dt} \tag{5.A.3}$$

If we multiply Eq. (5.A.1) by the integrating factor, $u(t)$, we obtain:

$$e^{\int p(t)dt} * Y'(T) + p(t) * e^{\int p(t)dt} * Y(t) = e^{\int p(t)dt} * q(t) \tag{5.A.4}$$

Notice that, by design, the $Y(t)$ term ends up containing $I'(t)$, allowing us to simplify to

$$I(t)Y'(t) + I'(t)Y(t) = I(t)q(t) \tag{5.A.5}$$

The left hand side of Eq. (5.A.5) is the derivative of $I(t)Y(t)$, allowing us to take the integral:

$$\int I(t)Y'(t) + I'(t)Y(t)dt = \int I(t)q(t)dt \tag{5.A.6}$$

$$I(t)Y(t) = \int I(t)q(t)dt + C \tag{5.A.7}$$

$$Y(t) = \frac{\int I(t)q(t)dt + C}{I(t)} \tag{5.A.8}$$

And, assuming that Eq. (5.A.8) can be solved, we have solved the original differential equation, and have the value of $Y(t)$. Always remember to include the constant of integration (+C).

5.A.2 Continuous (Step) Input Model for a Lake

To solve for the Continuous (Step) Model, the same approach is used, except in this case rather than setting W equal to 0, we set W as a constant. Thus our starting equation is just Eq. (5.11). Dividing by the volume (V) and substituting in β, yields

$$\frac{dC}{dt} + \beta C = \frac{W}{V} \tag{5.A.9}$$

where $\beta = (1/t_0 - k)$.

Now, we multiply by the integrating factor

$$I = e^{\int \beta\, dt} = e^{\beta t} \tag{5.A.10}$$

to obtain

$$\frac{dC}{dt}e^{\beta t} + \beta e^{\beta t}C = \frac{W}{V}e^{\beta t} \tag{5.A.11}$$

And by inverting the Product Rule:

$$e^{\beta t}\frac{dC}{dt} + \frac{d(e^{\beta t})}{dt}C = \frac{d(e^{\beta t}C)}{dt} \tag{5.A.12}$$

By substituting (5.A.23) into (5.A.22), we get

$$\frac{d(e^{\beta t}C)}{dt} = \frac{W}{V}e^{\beta t} \tag{5.A.13}$$

To solve, we integrate:

$$\int \frac{d(e^{\beta t}C)}{dt} = \int \frac{W}{V}e^{\beta t} \tag{5.A.14}$$

$$\int \frac{d(e^{\beta t}C)}{dt} = e^{\beta t}C + c_1 \tag{5.A.15}$$

$$\int \frac{W}{V}e^{\beta t} = \frac{W}{V}\left(\frac{1}{\beta}e^{\beta t} + c_2\right) = \frac{W}{V\beta}e^{\beta t} + \frac{W}{V}c_2 \tag{5.A.16}$$

where c_1 and c_2 are constants of integration

$$e^{\beta t}C + c_1 = \frac{W}{V\beta}e^{\beta t} + \frac{W}{V}c_2 \tag{5.A.17}$$

$$e^{\beta t}C = \frac{W}{V\beta}e^{\beta t} + \frac{W}{V}c_2 - c_1 \tag{5.A.18}$$

$$C = \frac{W}{V\beta} + \frac{Wc_2}{V}e^{-\beta t} - c_1 e^{-\beta t} \tag{5.A.19}$$

$$C = \frac{W}{V}\left(\frac{1}{\beta} + c_2 e^{-\beta t}\right) - c_1 e^{-\beta t} \tag{5.A.20}$$

Now we must find appropriate values for c_1 and c_2. To do this, we set $C = C_0$ at $t = 0$:

$$C_0 = \frac{W}{V}\left(\frac{1}{\beta} + c_2\right) - c_1 \tag{5.A.21}$$

The only way to have this to be true for all values of W, V, and β is to set

$$c_2 = -\frac{1}{\beta} \tag{5.A.22}$$

and

$$c_2 = -C_0 \tag{5.A.23}$$

Giving our solution for the overall concentration as

$$C_{(t)} = \frac{W}{\beta V}(1 - e^{-\beta t}) + C_0 e^{-\beta t} \tag{5.A.24}$$

If the background concentration is negligible, then $C_0 \approx 0$ and Eq. (5.A.24) reduces further to

$$C_{(t)} = \frac{W}{\beta V}(1 - e^{-\beta t}) \tag{5.A.25}$$

Equations (5.A.24) and (5.A.25), can be used to estimate the concentration of pollutant in a lake that receives a constant input of pollutant and their correct derivation is confirmed by equations in Fjeld et al. (2006). Note that two opposing forces govern the effluent pollutant concentration. The addition of more and more pollutant mass by the source is opposed by washout of the pollutant in the effluent stream and by first-order degradation. Eventually the system will reach an equilibrium pollutant concentration if all parameters remain constant. Thus, as time approaches infinity, the pollutant concentration in the lake approaches

$$C = \frac{W}{\beta V} \tag{5.A.26}$$

where

$$\beta = (1/t_0 - k)$$

5.A.3 Instantaneous (Pulse) Input Model for a Lake

The derivation for the governing of the instantaneous lake model again starts with Eq. (5.11) from above:

$$V\frac{dC}{dt} + CV\left(\frac{1}{t_0} + k\right) = W$$

Now, we will solve this equation under a different set of boundary conditions for the pulse model. Here, we assume that all the pollutant has been added at time = 0, which means that $W = 0$. We will again combine our constants:

$$\frac{1}{t_0} + k = \beta$$

so our equation simplifies to

$$\frac{dC}{dt} + \beta C = 0 \tag{5.A.27}$$

Again, we will use the technique of an integrating factor. We multiply Eq. (5.A.27) by

$$I = e^{\int \beta\, dt} = e^{\beta t} \tag{5.A.28}$$

obtaining

$$I\frac{dC}{dt} + \beta IC = 0 \tag{5.A.29}$$

which can also be written as

$$I\frac{dC}{dt} + \frac{dI}{dt}C = 0 \tag{5.A.30}$$

By inverting the Product Rule of differentiation, Eq. (5.A.30) is

$$\frac{d(IC)}{dt} = 0 \tag{5.A.31}$$

To solve, we integrate:

$$\int \mathrm{d}(IC) = \int 0\,\mathrm{d}t \tag{5.A.32}$$

$$IC = c_1 \tag{5.A.33}$$

where c_1 is a constant of integration.

$$C = \frac{c_1}{I} = \frac{c_1}{e^{\beta t}} \tag{5.A.34}$$

$$C = c_1 e^{-\left(\frac{1}{t_0}+k\right)t} \tag{5.A.35}$$

Setting $t = 0$ and $C = C_0$, we see that c_1 is equal to our initial concentration, and our overall equation for the concentration is

$$C_{(t)} = C_0 e^{-\left(\frac{1}{t_0}+k\right)t} \tag{5.A.36}$$

Equation (5.36) is used in Fate to estimate the downstream pollutant concentration resulting from a pulse input.

Concepts

1 List the major lake formation processes.

2 Give three examples of point sources of pollutants to lakes.

3 Give three examples of non-point sources of pollutants to lakes.

4 Draw a basic diagram showing a stratified lake in summer. Label each portion of the lake.

5 List the order of terminal electron acceptors from increasing to decreasing E_H values.

6 Draw a basic concentration versus time plot for a pulse input to a lake.

7 Draw a basic concentration versus time plot for a step input to a lake.

8 For each equation listed in this chapter, verify that the terms are dimensionally consistent. How does this exercise enhance your understanding of the concepts presented?

9 Define each term in Eq. (5.5).

10 Write Eq. (5.7) and define each term.

11 Develop a conceptual model for suspended materials in lakes and their sediment (benthic) layers.

12 What are the chemical cycles in play in most lakes? Why is each important?

13 Why do regulators seek to preserve tree cover around lakes, and seek to restrict projects that remove shoreline areas?

14 With respect to their assimilative capacity, how do rivers and lakes differ?

15 What is the Army Corps of Engineers (COE) role in sediment management. Research programs are discussed on their web site.

16 How are lakes used in providing safe drinking water for the U.S.? What are the minimally required management options?

17 How do nutrient water quality standards impact lake management?

Exercises

Pulse Input Problems (all problems should be worked by hand, but your answers can be checked in Fate; http://www.edusolns.com/fate/2016/lake/pulse/).

1 In the era before 1980 when the chemical was banned, a large container of tetraethyl lead was accidentally released into a lake. The initial concentration of the pollutant, which has a half-life of 50. Years, is measured at 175 mg/l. The lake has a volume of $2.9 \times 106\ \mathrm{m}^3$ and an outlet flow rate of 1.7×10^5. What concentration of the pollutant remains after 10.0 years?
Answer: 84.8 mg/l

2 An accident at a copper processing factory released several barrels of wastewater into the nearby lake. The wastewater contained acidic copper solution that increased the lake Cu levels to 2.5 mg/l. Calculate how many years must pass before the Cu concentration in the lake drops below the actionable level by EPA of 1300 ppb?
Lake specifics: Volume of lake = $4.2 \times 10^6\ \mathrm{m}^3$, Outlet flow rate = $6.5 \times 10^5\ \mathrm{m}^3/\mathrm{yr}$, Cu does not degrade so use a rate constant of 1 000 000 yr^{-1}, Initial Cu concentration = 2.5 mg/l
Answer: 4.2 years

3 A semi-truck carrying a trailer of methyl bromide (a pesticide) falls off the I-90 bridge into Lake Washington, near Seattle. Assume rapid mixing of the relative large lake. What will the concentration in 6.0 months? Lake specifics: detention time of

lake = 2.4 years, volume of lake = 3 km^3, Methyl Bromide water solubility = 9200 ppm, lake outlet flow rate = 1.25×10^9 m^3/yr, methyl bromide removal half-life = 0.575 yr or a $k_{1/2} = 12.055$ yr^{-1}, initial concentration of methyl bromide in the mixed lake = 0.04729 ppm.
Answer: 92.6 ppt

4 5.0 kg of potassium perchlorate is released into a lake with a volume of 4.1×10^6 m^3 and a detention time of 55.6 years. The lake outflow is 73 900 m^3/yr. $KClO_4$ has a first order decay constant of 5.78 years. (a) What is the initial concentration in the well-mixed lake? (b) What is the concentration after 1 years?
Answer: (a) 1.22 ppb; (b) 3.74×10^{-3} ppb

5 A transporting-truck hauling barrels of pesticide precursors containing nitrobenzene crashed along a lake shore in spilling several barrels into the lake. The initial concentration of nitrobenzene in the lake is 2874 mg/l. The lake volume is 1.61×10^8 m^3 and has an annual outflow of 4.79×10^8 m^3/yr. If the half-life of nitrobenzene is 1.08 years, calculate the concentration of nitrobenzene in the well mixed lake after six months, one year, and six years.
Answer: 471 mg/l after 6 months, 77.2 mg/l after 1.00 year, and 1.08×10^{-6} mg/l after 6.0 years.

6 A large fictional industrial producer of aniline is located near the mouth of the Flathead River on Flathead Lake. As a byproduct, the company produces a large quantity of nitrobenzene which they place in a large treatment tank. Early one summer an industrial accident causes a large unknown quantity of nitrobenzene to be released into the lake. Shortly after this release, the EPA stops all fishing and scientific experiments are conducted on the lake to ensure public safety. Scientists find that the concentration of nitrobenzene in the lake is 25 ppb. How many months must pass before the EPA will allow fishing again in the lake if the concentration of nitrobenzene must be less that 17 ppb to ensure the public's safety? Lake specifics are: Flathead Lake is 43.9 km long, an average of 20.9 km wide, and an average depth of 50.2 m; Lake detention time: 3.4 years. Nitrobenzene has a half life of 1.08 years.
Answer: 0.56 years

7 Naphthalene is a hydrocarbon present in many fuels, old moth balls, and is a main component of polyvinyl chloride (PVC) pipe construction. It is quickly evaporated from water and has a half-life of between 3 and 8 hours (use an average of 5.5 hours for this problem). However, a truck transporting naphthalene crashed near a small lake resulting in a uniform concentration of 5.0 mg/l in the lake. The lake has a volume of 305 000 m^3 and an outflow rate of 45 500 m^3/year. The lake is a popular swimming location and will need to be closed until the naphthalene concentration falls below 0.50 ppm. Answer the following questions: (a) What is the detention time of water in the lake? (b) How many days must the lake remain closed to swimmers?
Answers: (a) 6.70 years; (b) ~1.6 days

8 (a) Lake Lukomskoje in Belarus has an average detention of time 7.0 years and a total volume 243 000 m^3. If a fictional plant releases 1000 kg of phenol, what is the initial concentration of phenol? (b) How long will it take to for the lake to reach a safe concentration for its drinking water plant of 0.0006 mg/l? Assume a half-life of 20 days.
Answer: (a) 4.11 mg/l; (b) ~0.5 years

9 A fictional factory located on a lake in Lappi, Finland accidentally spilled 7550 kg of 2-chlorophenol into the lake. The volume of well mixed lake is 1.51×10^7 m^3 and the detention time is 3.4 years. Given that the half life of 2-chlorophenol is 2.5 days, what will be the pollutant concentration after 15.0 years?
Answer: 0.225 mg/l

10 A small lake is surrounded by agricultural fields that are regularly dusted with the pesticide Malathion, which has a half-life of 6.5 days in water. If the initial concentration of Malathion in the lake is 13 mg/l, find the concentration of malathion in the lake one month (30 days) after the use of malathion has been banned.
Lake volume: 20 800 m^3
Outlet flow rate: 3650 m^3/yr.
Answer: 0.52 mg/l

11 Cyanide was a major constituent used in the extraction of gold during the twentieth century and continues to be used in large pit mines in the Global South. At one mining site a few barrels of HCN spilled into a nearby lake, resulting in an HCN concentration of 17.5 mg/l. The degradation/removal half-life for HCN is 334 days. Calculate the HCN concentration in the lake one month and one year later.
Lake volume: 4 000 000 m^3
Outlet flow rate: 29 200 000 m^3/yr.
Answer: 9.0 mg/l for one month; 5.5 μg/l after one year

12 A truck carrying 7.0 U.S. tons (1.00 U.S. Ton = 907 kg) of de-icing salt falls into a lake with a volume of 800 000 m^3. The lake has a single small stream with an outflow of 3000 m^3/day. What is the salt concentration in mg/l after 6.0 years. Note de-icing salt does not degrade or adsorb to sediments, so assume a large half-life is calculating k.
Answer: 2.22 mg/l

13 A lake is surrounded by agricultural fields. The pesticide malathion, used on these crops, has a chemical degradation half-life of 6.5 days. As a result of its location, the lake receives high doses of malathion after aerial spraying of pesticides (by crop-dusting airplanes). If the initial concentration of malathion in the lake is 13 mg/l, find the amount of malathion in the lake eight weeks after the contamination has ended, using the following parameters (check your answers with Fate)
Detention time: 5.70 years
Flow rate: 3.65×10^3 m^3/yr.
Lake volume: 2.08×10^4 m^3
Answer: 0.032 mg/l
Perform a sensitivity analysis using half-lives of twice and half as long.
Step Input Problems (all problems should be working by hand, but your answers can be checked in Fate; http://www.edusolns.com/fate/2016/lake/step/)

14 A chrome plating plant located on a lake has been operating for years without trouble, but one year fishermen notice their catch becoming scarce. Wildlife biologists doing a survey of regional trout populations find that there are significant amounts of chromium ion (Cr^{6+}) in the fish bones, and conclude that the holding tank for the plant must be leaking. If this is true, what concentration of chromium ion species would you expect to find in the lake water if the tank started leaking 1.00 years ago? Use the data from the chart below to do the necessary calculations.
Lake volume = 6.70×10^6 m^3
Outlet flow = 8.9×10^6 m^3/year
Cr^{3+} input per day = 0.150 kg/day
Note: Chromium does not have a rate constant for loss as it does not degrade. Thus, you will need to use a very high half-life value to graph this in Fate-2016 (suggested value: 1 000 000).
Answer: 4.55 μl

15 The drainage area around a lake receives an intense amount of precipitation each year. Precipitation once absorb by a forested watershed has now been replace by urban sprawl that rapidly carry pollutants into the lake. Given the near continuous rain, the lake now receives pollutants as a step input. The volume of the lake is 202 500 000 m^3 with an average outflow of 28 900 000 m^3/yr. the lake receives on average 100 kg/day of total pollutants from the surrounding runoff. The average half-life of the chemicals is 43.8 days. What will be the concentration in 0.20 years, at equilibrium?
Answer: 0.021 mg/l

16 A large lake has received an average of 140 kg DDT every year, for the last 50 years, from the atmosphere inputs. The volume of the lake is 1.64×10^{12} m^3, and the outflow is 7.99×10^9 m^3/yr. The half-life of DDT is 31.3 years. Is the DDT concentration in the lake at equilibrium?
Answer: No, more near 130 years

17 A storage tank of xylene is leaking into a lake at a rate of 11 kg/day. Xylene is very volatile and moderately biodegradable, and it has a half-life of 18 hours when dissolved in water. Using Fate, create a plot of pollutant concentration in the lake over time, assuming the following parameters. Will it take weeks, months, or years for the system to reach equilibrium?
Lake volume: 313 348 796 m^3
Outlet flow rates: 44 650 012 m^3/yr.
Answer: Weeks

18 Lake Managua (Nicaragua) contains many pollutants, one of the main ones being mercury, which is found in high concentrations in the tissue of local fish species. One of the numerous companies that contributed to the mercury pollution was a soda company that released 20.0 kg of mercury daily for 10 years. At the end of the company's ten-year business, what concentration of mercury in the lake was it responsible for?
Lake volume: 7.97×10^9 m^3
Outlet flow rate: 2.52×10^9 m^3/yr.
Answer: 2.90 ppb

19 A lake is being polluted with pentachlorophenol at a rate of 100 kg/day. Pentachlorophenol is a pesticide with a degradation half-life of 48 days. What is the concentration of pentachlorophenol in the lake after 3.0 months?
Lake volume: 20 000 m^3
Outlet flow rate: 6000 m^3/yr.
Answer: 246 ppm

20 A manufacturer of PVC pipe is located along the edge of a lake and is releasing 56 kg/day of vinyl chloride from a leaking storage tank into the lake. After seven weeks, the manufacturer notices the leak and hires a chemist to check the concentration of vinyl chloride in the water. Calculate the concentration of vinyl chloride that he/she should find. The degradation/removal half-life of vinyl chloride in/from water is 7.92 years.
Lake surface area: 1.2×10^6 m^2
Lake depth: 5.2 m
Outlet flow rate: 1.24×0^7 m^3/yr.
Answer: 0.385 ppm

21 Lake Mendota is located in an agricultural area near Madison, Wisconsin, and fertilizer runoff contributes 31 100 kg of phosphates into the lake every year. Phosphates are a major contributor to the algae growth that can lead to eutrophication. The half-life of phosphates is 14.3 days. What is the concentration of phosphates in the lake after this pollution has been occurring for 0.33 years? Does the lake's phosphorus level ever reach 20.0 ppb, the level where algal blooms become a threat?
Lake volume: 481 000 000 m^3
Detention time: 6.21 years
Answer: 3.64 ppb, no algal bloom threat

22 A wood-treating chemical plant on the shores of a lake releases 100 g/day of arsenic (no degradation) into the lake. Assuming the lake is well mixed, determine how many years will pass before the arsenic level in the lake rises above 10.0 ppb. When will the lake approach equilibrium?
Lake volume: 5 00 000 m^3
Outlet flow rate: 20000 m^3/yr.
Answer: in 1.4 years; 80–100 years

23 An old electrical facility is seeping 36.5 kg/yr. of PCBs into a bordering lake. The average half-life of the most toxic PCBs is 7.80 years.
(a) Calculate the concentration of PCBs in the lake after 5.0 years
(b) Calculate the equilibrium concentration of PCBs in the lake and when it will occur
Lake volume: 1.35×10^8 m^3
Detention time: 2.94 years
Answer: (a) 0.011 ppb; (b) 0.030 ppb

Spreadsheet Exercise

Create a spreadsheet that performs the same calculations as Fate for both the step and pulse equation for one problem from above. Construct your spreadsheet so that it is interactive (so you can change numeric values for parameters and the plot automatically updates itself). In each case, create a plot that shows 95% of the decrease in concentration in the case of a pulse release or 95% of the increase in concentration in the case of a step release.

Bibliography

Baker, L.A. (1994). *Environmental Chemistry of Lakes and Reservoirs*, Advances in Chemical Series, vol. 237, 45. Washington, DC: American Chemical Society.

Boyce, W.E. and DiPrima, R.C. (2009). *Elementary Differential Equations*, 9e. New York: Wiley.

Chapra, S.C. (2008). *Surface Water-Quality Modeling*. Long Grove, IL: Waveland Press.

Dunnivant, F.M. (1985). Determination of polychlorinated biphenyl concentrations in selected sediments using a sonication technique, figure 4.1, p. 27. M.S. thesis. Clemson University.

Dunnivant, F.M., Coates, J.T., and Elzerman, A.W. (2005). Labile and Non-labile Desorption Rate Constants for 20 PCB Congeners from Lake Sediment Suspensions. *Chemosphere* **61**: 332–340.

EPA Center for Exposure Assessment Models (2017). Environmental Modeling Community of Practice. https://www.epa.gov/exposure-assessment-models (accessed 7 December 2017).

Fjeld, R.A., Eisenberg, N.A., and Compton, K.L. (2006). *Quantitative Environmental Risk Analysis for Human Health*. New York: Wiley.

Hemond, H.F. and Fechner-Levy, E.J. (2000). *Chemical Fate and Transport in the Environment*, 2e. New York: Academic Press.

Lapple, C.E. (1961). The Little Things in Life. *Stanford Res. Inst. J.* Third Quarter **5**: 95–102.

LTI (1992, September). *Simplified Method Program-Variable Complexity Stream Toxics Model (SMPTOX3) VERSION 2.0 User's Manual*. Washington, DC: EPA.

National Research Council (1977). *Contaminated Sediments in Ports and Waterways: Cleanup Strategies and Technologies*, 10–140. Washington, DC: National Academy Press.

NSCEP (2017), National Service Center for Environmental Publications. https://nepis.epa.gov/Exe/ZyNET.exe/9101WW0R.TXT?ZyActionD=ZyDocument&Client=EPA&Index=1995+Thru+1999&Docs=&Query=&Time=&

EndTime=&SearchMethod=1&TocRestrict=n&Toc=&TocEntry=&QField=&QFieldYear=&QFieldMonth=&QFieldDay=&IntQFieldOp=0&ExtQFieldOp=0&XmlQuery=&File=D%3A%5Czyfiles%5CIndex%20Data%5C95thru99%5CTxt%5C00000036%5C9101WW0R.txt&User=ANONYMOUS&Password=anonymous&SortMethod=h%7C-&MaximumDocuments=1&FuzzyDegree=0&ImageQuality=r75g8/r75g8/x150y150g16/i425&Display=hpfr&DefSeekPage=x&SearchBack=ZyActionL&Back=ZyActionS&BackDesc=Results%20page&MaximumPages=1&ZyEntry=1&SeekPage=x&ZyPURL (accessed 20 November 2018).

Ruttner, F. (1963). *Fundamentals of Limnology*. Toronto: University of Toronto Press.

Shapiro, B.E. (2011) for integral-table.com. This work is licensed under a Creative Commons Attribution-NonCommercial-ShareAlike 3.0 Unported License. http://integral-table.com/downloads/LaplaceTable.pdf (revised 1 July 2016) (accessed 15 June 2018).

Spliefhoff, H.S. and Hemond, H.R. (1996). History of toxic metal discharge to surface waters of the Aberjona watershed. *Environmental Science & Technology* **30** (1): 125.

Wetzel, R.G. (1990). Land-water interfaces: metabolic and limnological regulators. *Verh. Internat. Verein. Limnol.* **24**: 6–24.

Wetzel, R.G. (2001). *Limnology: River and Lake Ecosystems*, 635. New York: Academic Press.

World Almanac (2004). *The World Almanac and Book of Facts*. New York: World Almanac Books.

United States Geological Survey (USGS), https://www.usgs.gov/.

6

Fate and Transport of Pollutants in Rivers and Streams

Case Study: The Rhine River

For our case study for Chapter 6, we return to the Rhine River pollution event (in Western Europe) initially presented in Chapter 1. There are many historical river pollution events to choose from, but the Rhine River incident is classic in the sense that it involved a major river passing through many highly populated areas, and in that the spread of pollution through a system was very well documented. Recall that on 1 November 1986, a storehouse owned by Sandoz Ltd. near Basel, Switzerland, caught fire and released pesticides, solvents, dyes, and various raw and intermediate chemicals (Capel et al. 1988). A map of the route of the Rhine River through Europe was presented in Figure 1.4. A rapid and valuable sampling effort was conducted by the researchers at the University of Zurich and the Swiss Federal Institute for Water Resources and Water Pollution Control to document the movement of pollutants through the river system. After analysis, the researchers used explanatory modeling to reverse fit the field data to a complicated river model to estimate the dispersion and movement of pollutants through the system. For modeling purposes, the release was treated as a pulse (short duration) release. Monitoring points for the various chemicals were set up at the stations labeled in Figure 1.4 (Capel et al. 1988). One of the most abundant pesticides released was disulfoton, a thiophosphoric acid ester insecticide. Figure 1.5 showed the movement and flushing of disulfoton through the Rhine (Capel et al. 1988). Note the bell or Gaussian shape of the concentration profile, which, as we will see in this chapter, is characteristic of a pulse release in a river. As the pulse of disulfoton moved downstream, it was diluted and degraded, as indicated by the broader peaks and lower concentrations shown in Figure 1.5. This was a case in which the model was fit to the data to better understand how pollutants move through the system and in order to better predict downstream concentrations of later accidental releases of pollutants.

In this chapter, we will learn the basic flow properties of river systems; source inputs of pollutants; how we model these systems for pulse and step pollutant inputs; how dispersion, sedimentation, and degradation affect downstream pollutant concentrations; and how rivers can be remediated after pollutant releases.

6.1 Introduction

Rivers and streams make up approximately 0.014% of the terrestrial water on Earth. Human civilizations have always settled near bodies of fresh running surface water, not only for a source of water but also because the water is constantly renewed and wastes can be instantly removed from the area (unless, of course, if one lives immediately downstream from another community!). The cultural importance of rivers in antiquity and in modern-day life is significant and impacts our lives today.

In our modern society, with chemical factories, railways, and highways in close proximity to natural waterways, unintentional releases of hazardous chemicals occur frequently. Once hazardous chemicals are in an aquatic system, they can have a number of detrimental effects that extend for considerable distances downstream from the pollutant source. In this chapter, we will look at the concentration of a pollutant downstream of pulse (instantaneous) and step (continuous) releases. Examples of pulse releases can be as simple as small discrete releases such as pouring a liter of antifreeze off a bridge, or they can be more complex such as an accident that results in the release of acetone from a tanker car. Step releases usually involve a steady input from an industrial process, drainage from non-point sources, or leachate from a landfill located near the stream. Once a pollutant is released to a system, the models we use assume that the pollutant and stream water are completely mixed (i.e. there is no cross-sectional concentration gradient in the stream channel). This is a reasonably good assumption for most systems that are not very wide or impounded by navigation or hydropower dams. The latter can become similar to a lake in hydraulic characteristics. The models used

Pollutant Fate and Transport in Environmental Multimedia, First Edition. Frank M. Dunnivant and Elliot Anders.
© 2019 John Wiley & Sons, Inc. Published 2019 by John Wiley & Sons, Inc.
Companion website: www.wiley.com/go/Dunnivant/pollutant_fate&Transp_EnvironMultimedia

here account for longitudinal dispersion (spreading in the direction of stream flow), advection (transport in the direction of stream flow at the flow velocity of the water), and a first-order removal term (biodegradation, chemical, and/or radioactive decay). They may also simulate nitrification and generally deal with nutrient behavior.

Table 6.1 Drainage area and annual flow of the major rivers of the world.

Rivers by continent	Drainage area (10^3 km^2)	Mean annual flow (m^3/s)
Africa (all rivers)	30 300	136 000
Congo	4 015	40 000
Niger	1 114	6 100
Nile	2 980	2 800
Orange	640	350
Senegal	338	700
Zambezi	1 295	7 000
Asia (all rivers)	45 000	435 000
Brahmaputra	935	20 000
Ganges	1 060	19 000
Indus	927	5 600
Irrawaddy	430	13 600
Mekong	803	11 000
Ob-Irtysh	2 430	12 000
Tigris–Euphrates	541	1 500
Yangtze	1 943	22 000
Huang He (Yellow River)	673	3 300
Europe (all rivers)	9 800	100 000
Danube	817	6 200
Po	70	1 400
Rhine	145	2 200
Rhône	96	1 700
Vistula	197	1 100
North America (all rivers)	20 700	191 000
Colorado	629	580
Mississippi	3 222	17 300
Rio Grande	352	120
Yukon	932	9 100
South America (all rivers)	17 800	336 000
Amazon	5 578	212 000
Magdalena	241	7 500
Orinoco	2 305	14 900
Paraná	2 305	14 900
San Francisco	673	2 800
Tocantins	907	10 000

Source: From Szestay (1982).

Some river systems are even impacted by thermal discharges and are regulated by allowable temperature increases. This is why some industrial facilities have cooling towers for their wastewater discharges.

6.2 Examples of Rivers and Volumetric Flows of Water

Specific physical features pertinent to modeling of a river system are depth, flow rate, and water velocity. Table 6.1 lists the major rivers of the world, and while these constitute most of the flow of fresh water, they represent a very small number of the total streams in the world. Given their size, many of these streams also require special fate and transport models (i.e. 2-D or 3-D), given that the input of pollutant is not instantly and evenly spread across the width of the stream channel and cannot be approximated as such. Students might find interesting some facts about the Amazon River in South America, fresh water from its voluminous discharge extends nearly 50 mi into the Atlantic. To put the size of any given river in perspective, Wetzel and Likens (1991) used the drainage areas of streams to estimate the approximate number of streams of a given length and average water discharge. These data are summarized in Table 6.2. The first column refers to the stream order, which expresses how many streams come together to make up the stream of interest. The more streams join to form the final stream, the higher the order number. Note how few streams achieve the length, drainage area, or discharge rate of the major rivers listed in Table 6.1. A few of major rivers of North American are shown in Figure 6.1.

Table 6.2 The approximate number of streams, average length, and average discharge rate as function of a hypothetical drainage area.

Order	Number of streams	Average length (km)	Average drainage area (km^2)	Average discharge rate (m^3/s)
1	200 000	0.02	0.000 18	0.000 005
2	65 000	0.03	0.000 91	0.000 025
3	20 000	0.06	0.004 14	0.000 12
4	5 500	0.16	0.012 9	0.000 36
5	1 500	0.40	0.090 6	0.002 5
6	400	1.0	0.388	0.011
7	150	2.4	2.20	0.062
8	40	5.6	9.06	0.250

Source: From Wetzel and Likens (2000).

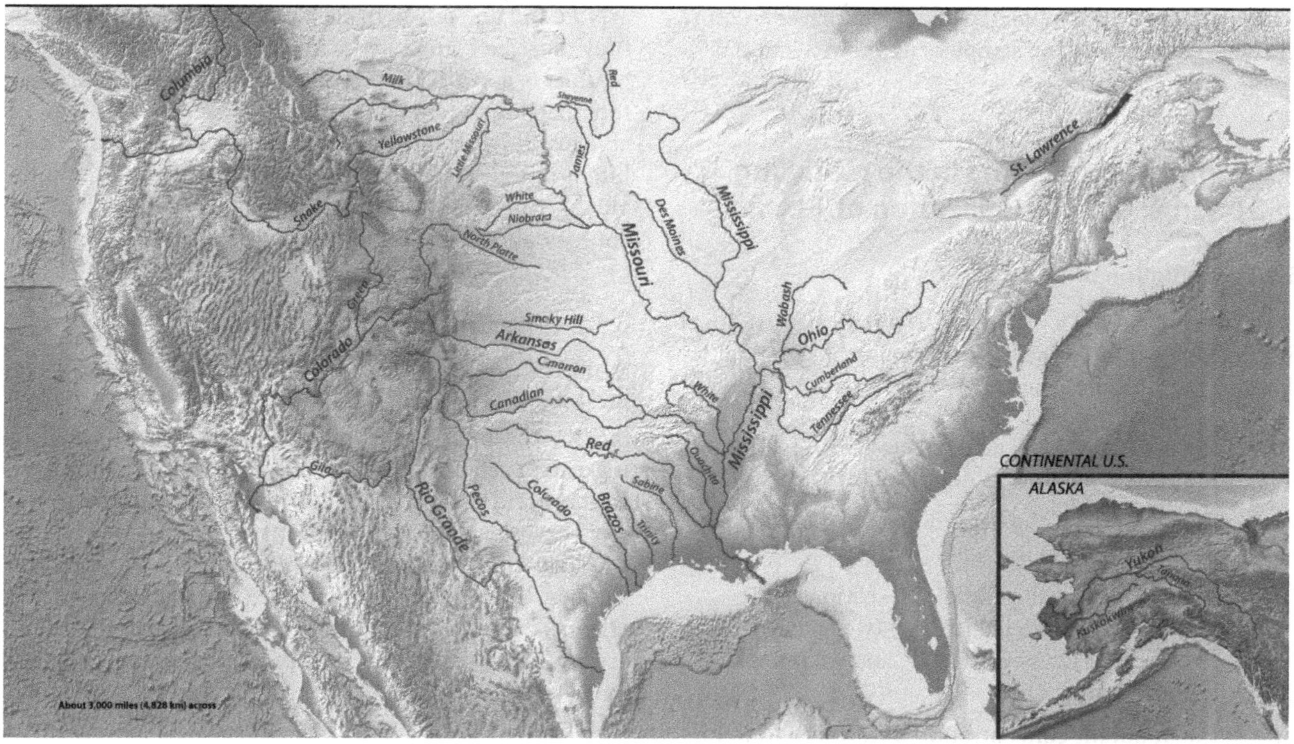

Figure 6.1 Major rivers of North America. Source: https://en.wikipedia.org/wiki/List_of_longest_rivers_of_the_United_States_%28by_main_stem%29.

6.3 Input Sources

Input sources are very similar to those presented in Chapter 1 and discussed for lake systems in Chapter 4. To summarize, we will consider both point and non-point sources. Common point inputs include industrial and feedlot sources, as well as effluent from domestic sewage plants. Non-point sources can include runoff from farming practices and agricultural settings. Underground non-point sources can include leachate from domestic and hazardous waste landfills or storage tanks. Many rivers are used for recreational activities, and outboard motors are notorious for releasing petroleum-related compounds and can be banned in certain pristine waters of the United States. These compounds dissolve into the water and are spread throughout the reach of a river. This is a non-point but continuous source of pollutants.

Another source of pollutants to rivers can be from within the river itself. As we discussed in Chapters 2 and 3, particles in the water attract many inorganic and organic pollutants, and most of these pollutant-laden particles aggregate and settle in quiescent regions of a river. However, if these sediments are resuspended into river water with dilute pollutant concentration, desorption of the pollutants into the water will occur. Resuspension events of importance in most rivers include bioturbation (the mixing of river sediments with water by biota in the river), violent storm events, and dredging of shipping channels. Bioturbation is a constant process and would be treated as a constant non-point input, while the latter two events would be treated as pulse, non-point inputs. As noted, benthic regions of rivers can become highly polluted with carbonaceous and nitrogenous material from point source inputs and respiring algae in certain segments. Water quality modelers call this source *"sediment oxygen demand or SOD"* and this represents a significant DO deficit in some river systems. It is usually expressed as a real demand mg $O_2/m^2/d$. It can be measured by in situ respirometry whereby a known volume of water is enclosed in-stream and the oxygen demand is monitored over time. Sediment oxygen demand (SOD) process models do exist, but currently not often used in regulatory permitting and largely research tools. Typically, a SOD in rivers is of the order of $0.1–1\ mg\ O_2/m^2/d$ and often most significant behind hydropower dams where resuspended sediment and associated contaminants accumulate.

6.4 Sampling of Surface Waters

The sampling of surface waters is generally the same as lakes, presented in Section 5.5. Smaller streams can be sampled with grab samplers, while deeper rivers require the depth discriminating samplers noted in Section 5.5.

6.5 Important Factors in the Modeling of Streams: Conceptualization of Terms

6.5.1 Definition of Terms

In order to describe a stream mathematically, we must first make a list of variables (mathematical symbols) for the important terms. These are

SA = the cross-sectional area of the stream (width, w, multiplied by average depth, d) (m^2)

Q_i = the flow rate at the beginning of the section of the stream to be modeled (m^3/yr)

Q_e = the flow rate at the end of the section of the stream to be modeled (m^3/yr) (we usually assume that Q_i is equal to Q_e and we represent both by simply Q)

$C(x)$ = the pollutant concentration profile as a function of distance (at a fixed time) downstream from the pollutant addition (kg/m^3 or similar units)

$C(t)$ = the pollutant concentration in the stream at a fixed distance as a function of time (kg/m^3 or similar units)

k = the first-order removal rate of pollutant from the river/stream (yr^{-1} or in the Fate® model s^{-1})

t = time (years or in the Fate model, s)

M_o = the total mass of pollutant in the river (kg or Ci)

W = rate of continuous discharge of the waste (mass/time in kg/s or Ci/s)

Note the units used in each of the terms and note that they must be compatible with each other. Compatibility of units is essential in our calculation and is also important in using the simulator package Fate. The terms given above will be used to develop the mass balance of pollutant in the stream and develop equations for the individual components of the mass balance (inflow, outflow, sources, and sinks of pollutant).

6.5.2 The Stream Channel

Water flows in channels created millions of years ago by the glaciers and further carved out of Earth's crust by "weathering" due to moving water currents polishing rock surfaces. Flow is caused by gravity (potential energy) and always goes downhill. The surface topology that defines distinct regions is called "watersheds" through which streams and rivers flow to the oceans and seas. In the United States, there are over 2000 stream channels catalogued by the Interior Department. They are precisely delineated in a public database and used by water resources professionals for water quantity and quality purposes (https://water.usgs.gov/GIS/huc.html).

An illustration of a stream channel in Figure 6.2 shows an instantaneous and pulse input, the volumetric water flow rate (Q), the velocity (v), a first-order degradation (k), and the longitudinal diffusion term (E, although some texts use D to represent this term), which is the subject of Section 6.5.3. Physical characteristics of the stream channel determine the magnitude of each of these terms. Of course, large streams usually have a large flow rate. Streams in steep terrain have higher velocities and more mixing. Streams in gentle sloping areas can have very tortuous flow paths and meander for large distances before reaching the receiving lake or ocean. As we will see in Section 6.5.3, the physical features of the stream

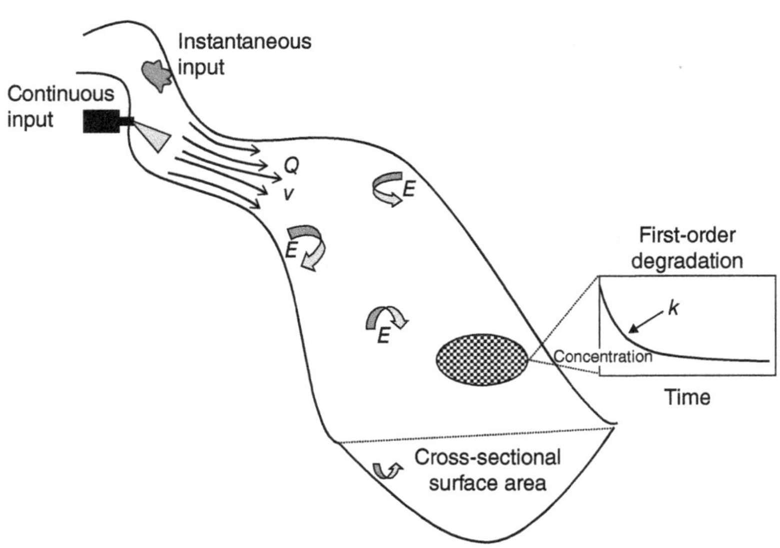

Figure 6.2 Illustration of a stream channel with a pollution release.

channel are responsible for the degree of mixing in a system. On an elevated view, you may have noticed such surface currents attributed to internal mixing. As will be noted later, such mixing is really 3-D and a result of small-scale velocity gradients and the resultant transfer of momentum from high to low velocity.

6.5.3 Mixing and Dispersion in Rivers

All of the transport equations used in flowing media (water and air) are referred to as advective–dispersive. Advection refers to the bulk flow or movement of water (and pollutants). Dispersion is a process primarily resulting from the fact that water velocity will be different in different parts of the river's cross section, which always results in a dilution of the pollutant by unpolluted water. In the one-dimensional river model that we will develop, we are only concerned with advection and dispersion in the longitudinal (x) direction. A chemist can think of dispersion as a form of entropy, since the pollutant always decreases in concentration as the volume of solvent increases (thus, this dilution constitutes an increase of entropy in two ways). This also influences the cost of environmental remediation, since, if the acceptable concentration limit is low, dispersion will cause the pollutant to contaminate a larger volume of water.

There are two fundamental forms of dispersion: hydrodynamic and diffusion based. Turbulent mixing, a macroscale process, results from water velocity gradients along the flow path (represented by the curved arrows in Figure 6.2). Thus, dispersion is greater when higher flow rates are present. Diffusion-based dispersion is a microscale process, and in rivers it occurs at the water–solid interface of suspended particles and at the sediment–water interface. It is based on the movement of molecules in a concentration gradient (Fickian diffusion), which states that compounds move from high to low concentration, and although this process is very important in laboratory studies, it is of little concern in rivers, since the magnitude of the diffusion process is 100- to 1000-fold less than that of turbulent dispersion. Thus, when the equation of fate and transport in streams is derived, diffusion-based dispersion is ignored.

Dispersion is characterized by the longitudinal dispersion (eddy) coefficient, E, (m^2/s) in streams and can be estimated by the method of Fischer et al. (1979)

$$E = 0.011 \frac{v^2 w^2}{du} \tag{6.1}$$

where

$$u = \sqrt{g\, ds} \tag{6.2}$$

where

v = average water velocity (m/s)

w = average stream width (m)

d = average stream depth (m)

$g = 9.81\ m/s^2$ (acceleration due to gravity)

s = the slope or gradient of the stream bed (unitless)

Note that increases in water velocity (influenced by slope of the stream bed) and stream width will increase longitudinal dispersion. Thus, reasonably accurate estimates of longitudinal dispersion can be made for a stream by using a topographic map of the area of interest (for slope) and with a few simple measurements of the stream channel.

Figure 6.3a–c shows the mixing of two river waters in Switzerland, one coming from Lake Geneva and another draining a glacial valley. As these two streams start to mix as shown in Figure 6.3a, they are separated by a concrete barrier just below the water surface. A close-up view of the mixing eddies is shown in Figure 6.3b. The continued mixing is still evident downstream from the observation point (reverse angle), as shown in Figure 6.3c. As illustrated in these photographs, mixing currents in streams can be an important cause of pollutant dilution. Note that for this river, instantaneous mixing does not occur.

Values of E can be determined experimentally by adding a known mass of tracer to the stream and measuring the tracer concentration at various points as a function of time. As mentioned in Chapter 4, a fluorescent tracer is normally used, since it can be detected at extremely low concentrations. An equation defining diffusion (slightly different from Eq. (6.1)) is then fit to the tracer concentration versus time dataset to calculate values of E for each section of the stream. Unlike in lake studies, this method is relatively fast and cost-effective for streams, but is rarely used today except when highly accurate estimates of E are needed. Most stream and river studies simply use Eq. (6.1) to calculate E.

The effect of E on pollutant concentrations in streams is shown in Figure 6.4 by the increase in width of the shaded rectangle. The length of the stream containing the plume of pollution, as a function of time, is illustrated by the dark rectangular bands. Note that as the pollutant is transported downstream, more and more dispersion occurs, and the band becomes longer and longer (and lighter in shading) as the pollutant concentration is diluted.

6.5.4 Removal Mechanisms

As discussed in Chapters 2 and 3, a variety of potential degradation schemes can be present, including

(a) (b) (c)

Figure 6.3 (a) The junction (mixing) of two streams (one from a clear lake and another from a turbid stream), (b) a close-up of the mixing eddies, (c) mixing of the two streams downstream.

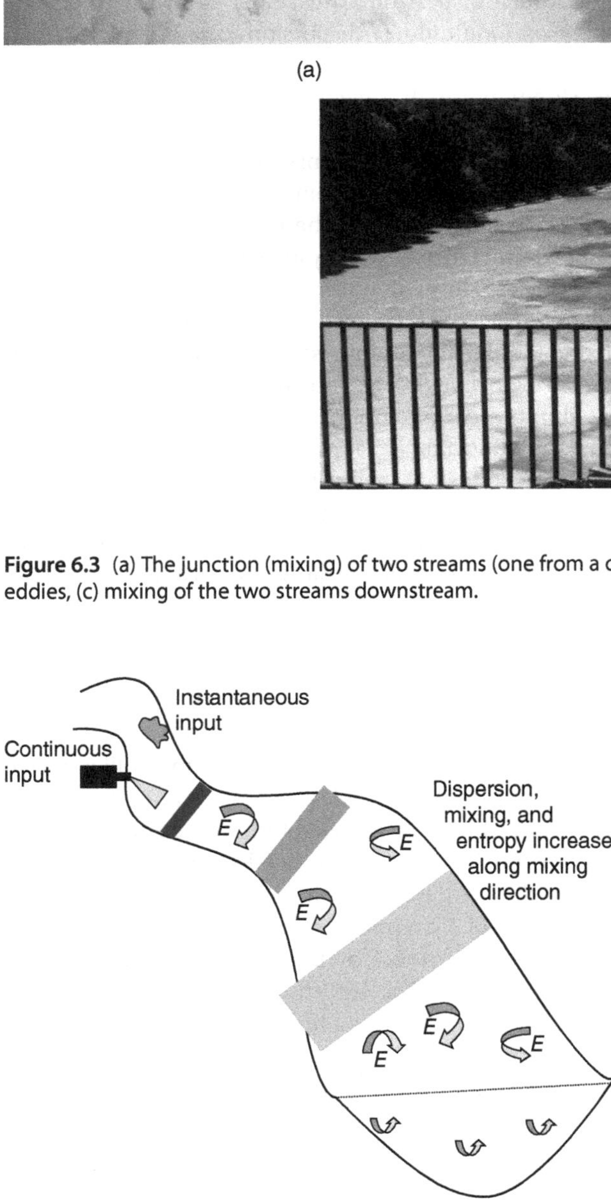

Figure 6.4 A plane view of a stream channel, showing longitudinal dispersion.

photochemical, biological, abiotic (chemical), and nuclear reactions. All these are possible transformation reactions in stream systems. Whatever the type or types of reaction(s), all these are usually represented by first-order reactions, and we can appropriately add the individual rate constants (k values) together to obtain one overall first-order rate constant. This component of the fate and transport model is of the form

$$C_t = C_0\, e^{-kt} \tag{6.3}$$

where

C_t is the pollutant concentration at time t,
C_0 is the initial pollutant concentration,
e is the exponential function,
k is the first-order rate constant (reciprocal time units), and
t is time (same time units as k).

Actual rate constants are usually determined in a laboratory setting via experimentation and in some cases by modeler experience and best judgment. These experiments start with a known initial pollutant concentration, and pollutant concentration is monitored as a function of time. The rate constant of degradation is calculated by plotting the data ($\ln(C/C_0)$ versus time), and if the plot is linear, measuring the slope of the line (k). If the $\ln(C/C_0)$ data plot results in a straight line, then the reaction is said to follow first-order kinetics and the rate constant, k, can be directly used in the fate and transport model. Processes with rates not describable by first-order kinetic expressions require that modelers derive equations that differ from the standard ones presented in this chapter.

Another important removal mechanism is sorption to particles that settle downstream in the river system. This may or may not be important for any given pollution event. Factors that determine the importance of sorption/settling are the aqueous solubility of the pollutant, which in turn is correlated to the distribution/partition coefficient, the presence of suspended matter in the river water, and the rate of turbulence in the river system. As we discussed in Chapters 2 and 3, metal and hydrophobic pollutants tend to sorb to suspended particles, and the more organic matter present in the particle, the more the pollutant will sorb. Most streams have sufficient suspended matter present in the water for this sorption process to be important; however, streams by their basic nature tend to have high turbulence and therefore have less particle settling than lake systems. Storm events increase the ability of a stream to carry suspended particles, and thus place tons of suspended matter in the stream in a short amount of time. As flood waters recede and flow in the stream decreases, particulate matter settles out and this process can take a significant mass of the pollutants with it to the bottom of the stream. But this presents an even more difficult process to understand and model, since sediments in river systems tend not to remain in one place but are redistributed during every storm event. In addition, as we saw in Chapters 2 and 3, resuspension of polluted sediments can result in re-release of pollutants to stream water not originally containing these pollutants. Modeling of these inter-related, potentially very complicated processes is difficult.

We can break the process into individual steps in an effort to understand them. First, we can look at the sorption process that was presented in Section 3.3, where we discussed distribution and partition coefficients. If you do not recall the mathematical meaning of these coefficients, you should review this section. Next, we can relate these sorption reactions to the removal of the particles from the water by settling using the same concepts presented in Eqs. (5.3) and (5.4), where we looked at the Stokes's law settling velocity and rate of removal of pollutants. But Eq. (5.4) was developed for calm water such as that present in lakes. In streams, as the pollutant-laden particles settle, they are also transported downstream. So, we must balance these forces – movement down in the water column and movement down the stream channel – in order to predict how far down the river system the particles will travel before they settle to the bottom of the stream and become incorporated into the sediment. We can estimate the distance downstream by using the same settling velocity calculated from Eq. (5.3) and summarized in Table 5.5, given in centimeters per second or meters per second, and imagining the particle settling in the water column as it is transported downstream. If we divide the average depth of the stream by the settling rate, we obtain a time unit. This is the average time that a particle of specified size will spend in the water column. Thus, if we multiply the stream velocity (in units of meters per second) by the settling time (in seconds), we obtain the distance downstream the particle will travel before it reaches the bottom. We can then estimate where a given particle size will be deposited in a system. You should work through the units of these calculations to confirm our statements and convince yourself of the math behind this concept. Keep in mind that turbulence in the system from mixing eddies will result in the deposition of the sediment at slightly longer distances than those predicted by our calculations. This may require field work to determine more appropriate depositional characteristics.

Although these calculations cannot be readily incorporated into the simple models that we present in this text, you can understand the concepts behind the removal of pollutant-laden particles from stream waters. These same mathematical approaches are used by modelers to incorporate the removal of pollutants by sedimentation in more complicated models based on numerical methods of analysis. The U.S. Army Corps of Engineers has much experience in sediment transport and its effects on chemical fate issues. Similarly, Superfund experiences with remediation of polluted river sediments may also provide valuable insights.

6.6 Mathematical Development of Transport Models (Derivations by John Brooksbank, Here and in Chapter Appendix)

To model a river, we must use an equation that accounts for the two dominant mechanisms of pollution transport within a river: diffusion and advection. Diffusion happens due to the concentration gradient within the river water, and advection is the transport of pollutant due to the natural movement of the river water downstream. To account for both mechanisms, we use the diffusion–advection equation:

$$\frac{\mathrm{d}C}{\mathrm{d}t} = -V\frac{\mathrm{d}C}{\mathrm{d}x} + \frac{\mathrm{d}}{\mathrm{d}x}\left(E\frac{\mathrm{d}C}{\mathrm{d}x}\right) + g(x,t) - d(x,t) \tag{6.4}$$

where C is the concentration of pollutant, V is the water velocity, E is the dispersivity constant, $g(x, t)$ is the constant input function for a step scenario (mass emitted per volumetric flow rate), and $d(x, t)$ is the degradation term of the pollutant. In this equation, the term

$$-V\frac{\mathrm{d}C}{\mathrm{d}x} \tag{6.5}$$

deals with advection or the transfer of pollutant by the actual flow of the water in the river, and the term

$$\frac{\mathrm{d}}{\mathrm{d}x}\left(E\frac{\mathrm{d}C}{\mathrm{d}x}\right) \tag{6.6}$$

deals with dispersion of the pollutant due to differences in concentration throughout the river.

6.6.1 Solution for a Step Pollutant Input

By substituting the step input function (constant mass input with time)

$$g(x,t) = \frac{M_o}{t} \tag{6.7}$$

and the first-order kinetic term

$$d(x,t) = -kt \tag{6.8}$$

into Eq. (6.4), we obtain an equation that can be integrated using Laplace transforms. We show a slightly different solution in end-of-chapter appendix due to more simple boundary conditions. There is a considerable diversity of solutions, similar but different, for this step input scenario in the literature. The result we use in Fate is

$$C(x,t) = \frac{W}{Q\sqrt{1+\frac{4kE}{v^2}}}\,\mathrm{e}^{\frac{vx}{2E}\left(1\pm\sqrt{1+\frac{4kE}{v^2}}\right)} \tag{6.9}$$

where

$C(x, t)$ = the pollutant concentration (in mg/l or μCi/l for radioactive compounds) at distance x and time t (note that t is defined by the water velocity when x is fixed)

W = rate of continuous discharge of the waste (kg/s or Ci/s)

Q = stream flow rate ($\mathrm{m^3/s}$)

E = longitudinal eddy dispersion coefficient ($\mathrm{m^2/s}$)

x = distance downstream from input (m)

v = average water velocity (m/s)

k = first-order decay or degradation rate constant (1/s)

The positive root of the equation refers to the upstream direction ($+x$), and the negative root (what we use in Fate) refers to the downstream direction ($-x$). An example of a concentration–distance plot for the background problem from Fate is shown in Figure 6.5.

Because we are considering a step input, the concentration of pollutant should vary with distance but not with time, and thus time can be dropped from the expression. Note that while time is not directly stated in Eq. (6.9), however, it is present in the combination of the stream velocity, v, and the distance downstream, x.

Example River Problem

An abandoned landfill leaches water into an adjacent stream at a rate of 1500 l/min. The concentration of

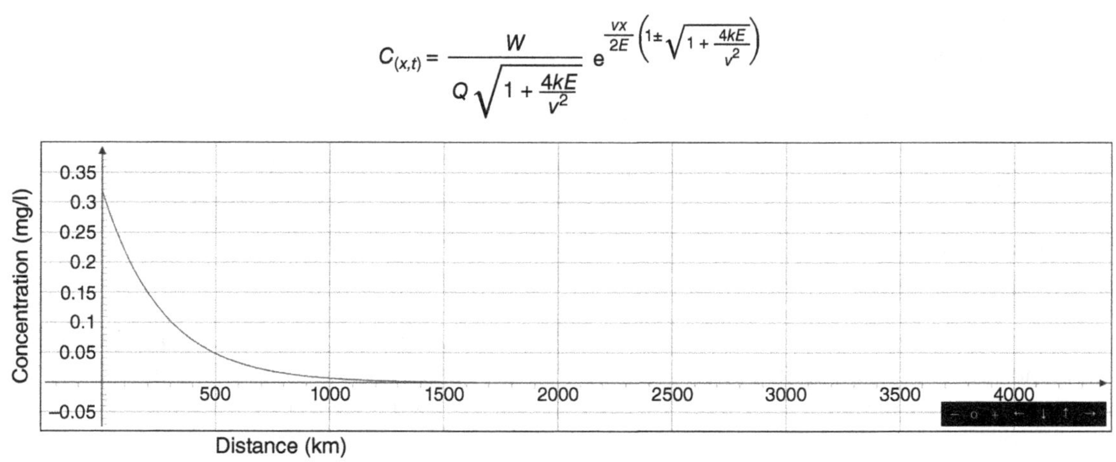

Figure 6.5 Example problem output from Fate® for a continuous (step) pollutant input to a river.

2-chlorophenol in the water is 500 mg/l. The stream is 20 m wide and 2.3 m deep and has a water velocity of 0.85 m/s. The regional slope of the stream channel is 1 ft per 1500 ft distance, and the first-order half-life of 2-chlorophenol is 2.5 days. Construct the concentration profile of 2-chlorophenol in the stream with distance downstream. What is the concentration 25 km downstream from the point source?

Solution

(1) Calculate the mass input to the stream in kg/s.

$$\text{Mass input} = \frac{1500\,\text{l}}{\text{min}}\frac{\text{min}}{60\,\text{s}}\frac{500\,\text{mg}}{\text{l}}\frac{\text{g}}{1000\,\text{mg}} \times \frac{\text{kg}}{1000\,\text{g}} = 0.0125\,\text{kg/s}$$

(2) Calculate the flow rate of the stream in m^3/s.

$$\begin{aligned}\text{Flow rate} &= \text{width}^{*}\ \text{depth}^{*}\ \text{water velocity}\\ &= (20\,\text{m})\,(2.3\,\text{m})\,(0.85\,\text{m/s})\\ &= 39.1\,\text{m}^3/\text{s}\end{aligned}$$

(3) Calculate the rate constant, k, for the first-order decay of 2-chlorophenol.
For a first-order reaction:

$$\ln\frac{C}{C_o} = -kt$$

where

C = the concentration at time t
C_o = the initial concentration of 2-chlorophenol
k = the first-order decay constant, and
t = time in seconds

At the half-life ($t_{1/2}$), one-half of the original concentration remains. Substitution of this into the equation above yields:

$$\ln\frac{\frac{1}{2}C_o}{C_o} = -kt_{\frac{1}{2}}$$

or

$$-\frac{\ln(0.5)}{t_{\frac{1}{2}}} = -\frac{\ln(0.5)}{2.5\,\text{d}} = k = 0.277\,\text{d}^{-1}$$

Thus, the decay rate constant for 2-chlorophenol is $3.21 \times 10^{-6}\,\text{s}^{-1}$.

(4) Calculate the longitudinal dispersion coefficient, E (the coefficient of eddy diffusion).

$$\begin{aligned}\text{Slope} &= \frac{1\,\text{ft}}{1500\,\text{ft}} = 6.67 \times 10^{-4}\\ u &= \sqrt{g\,ds}\\ &= \sqrt{\left(\frac{9.81\text{m}}{\text{s}^2}\right)(2.3\,\text{m})\,(6.67 \times 10^{-4})}\\ &= 0.12\,\text{m/s}\\ E &= 0.011\,\frac{v^2 w^2}{du}\\ &= 0.011\,\frac{\left(\frac{0.85\,\text{m}}{\text{s}^2}\right)(20\,\text{m})^2}{(2.3\,\text{m})\left(\frac{0.123\,\text{m}}{\text{s}}\right)}\\ &= 11.2\,\text{m}^2/\text{s}\end{aligned}$$

(5) Arrange data into proper units:

$Q = 39.1\,\text{m}^3/\text{s}$
$W = 0.0125\,\text{kg/s}$
$w = 20\,\text{m}$
$d = 2.3\,\text{m}$
$v = 0.85\,\text{m/s}$
$E = 11.2\,\text{m}^2/\text{s}$
x = variable (m)
$k = 3.21 \times 10^{-6}\,\text{s}^{-1}$

(6) A plot of concentration versus distance from point source at a given time is shown in Figure 6.5.
(7) Calculate $C(x)$ at 25 km.

$$C(x,t) = \frac{W}{Q\sqrt{1+\frac{4kE}{v^2}}}\,e^{\frac{vx}{2E}\left(1-\sqrt{1+\frac{4kE}{v^2}}\right)}$$

$$C(25\,\text{km}) = \frac{0.0125\,\text{kg/s}}{39.1\,\text{m}^3/\text{s}\sqrt{1+\frac{4(3.21\times10^{-6}\,\text{s}^{-1})\left(\frac{11.24\,\text{m}^2}{\text{s}}\right)}{\left(\frac{0.85\,\text{m}}{\text{s}}\right)^2}}}\,e^{\frac{\left(\frac{0.85\,\text{m}}{\text{s}}\right)(25\,000\,\text{m})}{2\left(\frac{11.24\,\text{m}^2}{\text{s}}\right)}\left(1-\sqrt{1+\frac{4(3.21\times10^{-6}\,\text{s}^{-1})\left(\frac{11.24\,\text{m}^2}{\text{s}}\right)}{\left(\frac{0.85\,\text{m}}{\text{s}}\right)^2}}\right)}$$

$$\begin{aligned}C(25\,\text{km}) &= (3.20 \times 10^{-4})e^{-0.0944}\\ C(25\,\text{km}) &= 2.91 \times 10^{-4}\,\frac{\text{kg}}{\text{m}^3}\\ &= 0.291\,\text{mg/l (or ppm)}.\end{aligned}$$

6.6.2 Solution to a Pulse Pollutant Input

Again we start with the diffusion–advection equation, (6.4 from earlier).

$$\frac{dC}{dt} = -V\frac{dC}{dx} + \frac{d}{dx}\left(E\frac{dC}{dx}\right) + g(x,t) - d(x,t) \tag{6.10}$$

The advection term (Eq. (6.5)), the dispersion term (Eq. (6.6)) and the degradation term (Eq. (6.8)) remain the same, but our pulse (instantaneous) input function now becomes

$$g(x,t) = M\,g(x - x_a) \tag{6.11}$$

where M is the mass emitted and $g(x - x_a)$ is a Heaviside step function term. On integration, using a Laplace transformation, we obtain

$$C(x,t) = \frac{S_0}{A\sqrt{4\pi Et}}\exp\left(\frac{-(x-Vt)^2}{4Et} - kt\right) \tag{6.12}$$

from the end-of-chapter appendix.

An important point in the plot of Eq. (6.12) is the point of maximum concentration. This can be found by taking the derivative of Eq. (6.12), where the slope is equal to zero. At this location, $(x - Vt)$ equals zero and the equation reduces to:

$$C(x,t) = \frac{S_0}{A\sqrt{4\pi Et}}\exp(-kt).$$

Example Surface Water Accidental Spill Problem

One Curie of Cesium-134 (^{134}Cs) is accidentally released into a small stream. The stream channel has an average width of 40 m and an average depth of 2 m. The average water discharge (Q) in the stream is 40 m^3/s, and the stream channel drops 1 m in elevation over a distance of 10 km. Assuming that the ^{134}Cs is evenly distributed across the stream channel, estimate the distribution of ^{134}Cs as a function of distance downstream (using a maximum distance of 30 km) at 1, 3, 6, and 12 hours. Also estimate the ^{134}Cs activity (concentration) at a distance of 10 km at 6 hours after the release. ^{134}Cs has a half-life of 2.07 years.

Solution

(1) Calculate the average stream velocity in m/s.

Cross-sectional area of stream channel

= width*depth = (40 m) (2 m) = 80 m^2

Average velocity

= discharge/cross-sectional area

$= (40\ \text{m}^3/\text{s})/(80\ \text{m}^2) = 0.50\ \text{m/s}$

(2) Calculate the rate constant, k, for ^{134}Cs.
For a first-order reaction:

$$\ln\frac{C_t}{C_o} = -kt \tag{6.13}$$

where

C_t = the concentration (or activity of ^{134}C) at time t

C_o = the initial concentration (or activity) of ^{134}C

k = the decay rate constant

t = time

At the half-life ($t_{1/2}$), one-half of the original concentration remains. Substitution of this into the equation above yields:

$$\ln 0.500 = -kt \text{ or } \frac{\ln 0.500}{2.05\ \text{yr}} = k = 0.338\ \text{yr}^{-1}$$

$$0.338\ \text{yr}^{-1}\left(\frac{\text{yr}}{365\ \text{d}}\right)\left(\frac{\text{d}}{24\ \text{h}}\right)\left(\frac{\text{h}}{60\ \text{min}}\right)\left(\frac{\text{min}}{60\ \text{s}}\right) = 1.07\times10^{-8}\ \text{s}^{-1}$$

Thus, the decay rate constant for ^{134}Cs is 1.07×10^{-8} s^{-1}.

(3) Calculate the longitudinal dispersion coefficient, E (also referred to as the coefficient of eddy diffusion).

$$\text{Slope} = \frac{1\ \text{m}}{10\,000\ \text{m}} = 10^{-4}$$

$$u = \sqrt{g\,ds} = \sqrt{\left(\frac{9.81\ \text{m}}{\text{s}^2}\right)(2\ \text{m})(10^{-4})} = 0.044\ \text{m/s}$$

$$E = 0.011\frac{v^2w^2}{ud} = 0.011\frac{\left(\frac{0.50\ \text{m}}{\text{s}}\right)^2(40\ \text{m})^2}{(0.044\ \text{m})(2\ \text{m})} = 50\ \text{m}^2/\text{s}$$

(4) Arrange data in the proper units:

M_0 = 1 curie = 1×10^6 μCi (in the program, this is entered as 1 Ci)

w = 40 m

d = 2 m

E = 50 m^2/s

t = variable (s)

x = variable (m)

v = 0.50 m/s

$k = 1.07\times10^{-8}$ s^{-1}

A = area (m^2)

(5) The plot showing the concentration profile at 12 hours is shown in Figure 6.6. Using the velocity and

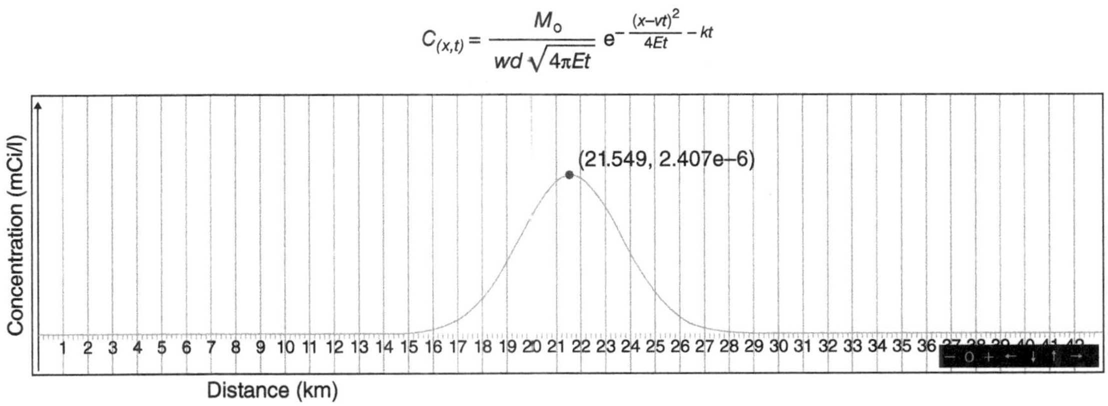

Figure 6.6 Concentration versus distance from point source for the instantaneous (pulse) example problem.

time, we see the peak concentration should be at 21.6 km, which is observed in Figure 6.6. The width of the Gaussian curve depends on the magnitude of E.

(6) Calculate $C(t)$ at 10 km downstream, after six hours ($x = 10\,000$ m and $t = 6$ h $= 21\,600$ s).

$$C(x,t) = \frac{S_0}{A\sqrt{4\pi Et}} \exp\left(\frac{-(x-Vt)^2}{4Et} - kt\right)$$

$$C(x,t) = \frac{1 \times 10^6\ \mu Ci}{(40\ \text{m})(2\ \text{m})\sqrt{4\pi\left(\frac{50\ \text{m}^2}{\text{s}}\right)(21\,600\ \text{s})}}$$

$$\exp\left(\frac{-\left(10000\,\text{m} - \left(\frac{0.50\,\text{m}}{\text{s}} * 21600\,\text{s}\right)\right)^2}{4\left(\frac{50\,\text{m}^2}{\text{s}}\right)(21600\,\text{s})} - \left(1.07 \times \frac{10^{-8}}{\text{s}}\right)(21\,600\ \text{s})\right)$$

$$= (3.39 \times 10^{-6})\,(e^{-0.148})$$

$$= 2.92 \times 10^{-6}\frac{\text{Ci}}{\text{m}^3} = 2.92\ \text{nCi/l}.$$

6.7 Sensitivity Analysis

As in the other fate and transport chapters, we will perform a sensitivity analysis. As an example, we will do this for the water velocity using the basic (background) step scenario in Fate. In the background example of Fate, we used a water velocity of 0.85 m/s. But what if this is in error? How sensitive is the effluent pollutant concentration to the water velocity? The results of a series of calculations using water velocities of 0.25, 0.50, 0.85, 1.00, 1.25, and 1.50 m/s, observing the pollutant concentration at 500 km downstream, are shown in Figure 6.7. We can now directly see that the results are strongly affected by the magnitude of the water velocity and that the relationship is exponential, as expected from the presence of v in the exponential term of the transport equation. The higher water velocity causes pollutants to be transported through the system before they have time to be degraded by the first-order process, which results in higher water concentrations at each distance downstream. Similar sensitivity analyses can be performed for all the parameters in each of the basic models.

As stated in every fate and transport chapter, the reader should be aware of the dangers and limitations of using "canned" programs such as Fate. Always review the input data (boxes) in these programs and be sure you understand the units and how the calculations between the boxes are connected. Once again, remember our mantra for modeling: "Always question the results, and check everything twice."

6.8 Limitations of Our Models

6.8.1 One-Dimensional Versus Two-Dimensional Models and Inputs of Pollutant Plumes in Wide Streams

Neither step nor pulse pollutant inputs are evenly spread across the stream channel, as assumed by the basic models. For most streams, this is not a problem, since the pollutant and stream water are well mixed within a few kilometers downstream of the point of input and most concentration profiles cover tens to hundreds of kilometers. However, for very large streams, the model would be modified to account for mixing (dispersion) in the y direction. This would be accomplished with a two-dimensional approach and could still be modeled using techniques similar to the ones presented in this chapter, but these models are more difficult to implement and appropriately use and often have a steeper learning curve.

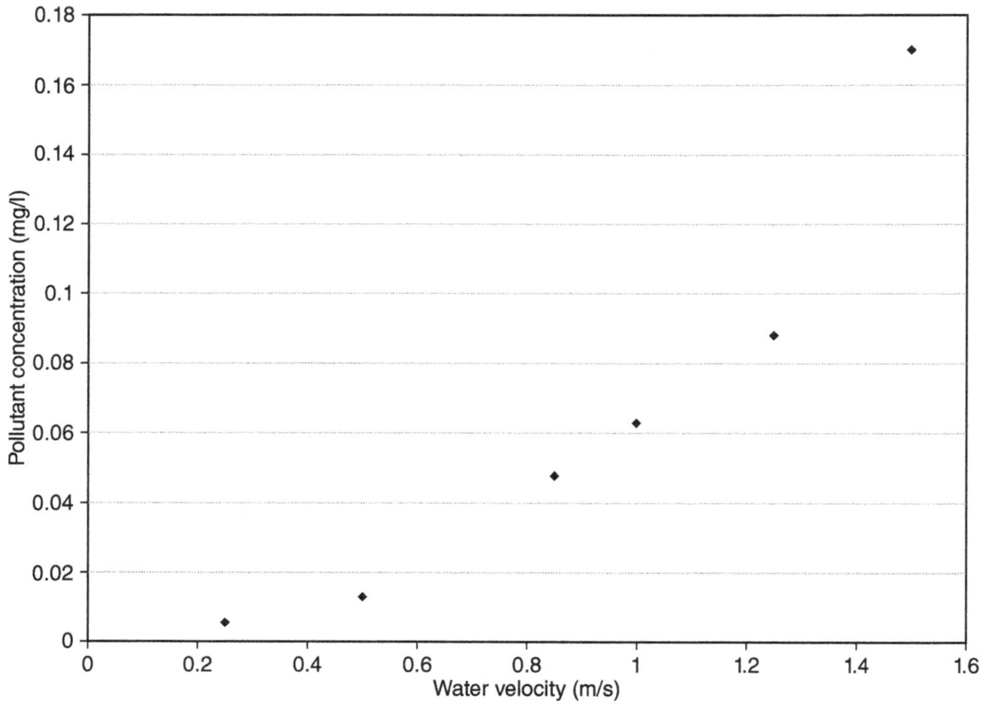

Figure 6.7 Results of a sensitivity analysis of water velocity for a step pollutant input.

6.8.2 Volatilization of Pollutants

When dealing with organic pollutants, volatilization can be an important removal mechanism, since stream waters have considerable contact with the atmosphere. This removal is treated as a Henry's law partitioning process and treated as a first-order removal mechanism. Thus, it is not really a limitation of the model, since it can be included in the kinetic "k" term.

6.9 Remediation of Polluted Stream Systems

As we saw in our case study for this chapter, rivers are highly dynamic systems and pollutants move rapidly through the system. In many cases, the movement is so rapid that we do not have time to respond and clean up the system before the pollution has moved on downstream to an estuary and the ocean. If fact, pollution in a river system is a moving target, and by the time we have adequate technologies in place to respond to a pollution spill, the pollution has moved to another section of the stream. Often, the best we can do is predict the migration of pollution with the water column, warn residents to stay out of the water, and stop the intake of water into their drinking water facilities when the pollution is present in their section of the stream. Meanwhile, the ultimate destination or fate of the pollution is transferred to the lake or estuary system receiving the water from the stream. This presents a common problem for estuary systems – they receive all the point and nonpoint pollution in a watershed. When you consider a large stream, the accumulated pollution can be huge. For example, take the Mississippi River in the United States. It has origins starting far to the north of its Louisiana delta, and it passes through many industrialized and agricultural areas as it flows south. Along the way, it accumulates industrial pollutants and nutrients from sewage effluents and agricultural runoff. It is therefore not surprising that the mouth of the Mississippi outside of New Orleans has been effectively labeled a dead zone due to eutrophication and lack of dissolved oxygen (referred to as hypoxia). Much research is underway at institutions like Louisiana State University (LSU) and Tulane University to find a practical remedy.

There are a few exceptions to the rule of not attempting to treat pollution events in stream systems. For example, leaking barges or oil tankers can be surrounded by booms (floating containment curtains that trap the releases so they can be pumped into containment vessels). In other cases (oil spills), dispersants (surfactants) are used to aid in the effective integration of the spilled oil into the water phase.

Some of the remediation approaches given in Chapter 5 for sediments in lake systems are directly applicable to streams. However, streams are not quiescent like lakes, since the flow in most streams is significant and very dynamic. This prevents the use of remediation technologies such as capping, since even if we can place a protective layer of sediment on a contaminated region of a river, it can easily be removed during the next storm or flood event by turbulent flow. But as rivers make their long journey to their inevitable end, a lake or estuary, they tend to grow in size and width, become more shallow, and decrease in water velocity such that the settling of suspended particles is greatly promoted. This is good and bad. The good is that pollutants are removed from the water column and the biota are no longer exposed to their toxic effects. As more and more sediment is deposited, the pollution is eventually buried and removed from the ecosystem. This process is a natural remediation action in all lakes and estuaries. The downside of this process results from human use of these very same aquatic systems. Shipping channels slowly fill with sediment deposits and must be subsequently dredged to keep the channels open. You should immediately note the problem with this approach, given our previous discussions of pollutant sorption and sedimentation. The streambeds in these quiescent areas are the resting grounds of many toxin-laden particles, and dredging activities always mix some of the removed sediment material with the water column. This creates an ideal environment for the desorption of pollutants and re-release of these pollutants to the water column. Thus, our maintenance of shipping channels and harbors has an inherent environmental consequence associated with it. While our economy depends on these shipping channels, we re-expose ecosystems to buried pollutants when we dredge these channels. As noted previously, the Waterways Experiment Station (WES) of the Corps of Engineers dredges millions of tons per year from the nation's harbors and have much field experience in the fate of chemicals in sediment materials.

It is important to note the scale of the dredging problem. The National Research Council estimates that approximately 14–28 million cubic yards of contaminated sediments must be managed annually (which is actually only 5–10% of the total sediments dredged in the United States) (NRC 1997). This raises a difficult problem: What do we do with this huge volume of sediment that is to be treated as hazardous waste? Recall the costs of dredging and treating these sediments from Chapter 5: \$15–20 per cubic yard to remove and transport contaminated sediments and \$100–1000 per cubic yard to treat the contaminated sediment. You do the math … the maintenance of shipping channels is a very expensive operation.

The more complicated aspects of river sediment cleanup are illustrated in the decades-long process of remediation attempts of the PCB contamination in the Hudson River in New York State (USA). This chapter should close with a reading and discussion of the problem and tentative solution summarized at https://en.wikipedia.org/wiki/Pollution_of_the_Hudson_River.

Suggested Papers for Class Discussion

Capel, P.D., W. Giger, R. Reichert, and O. Warner. "Accidental input of pesticides into the Rhine River", *Environ. Sci. Technol.*, 22(9), 992–997 (1988).

Pollution of the Hudson River, https://en.wikipedia.org/wiki/Pollution_of_the_Hudson_River, accessed 23 Feb., 2018.

Concepts

6.1 Name three point sources of pollution to streams.

6.2 Name three non-point sources of pollution to streams.

6.3 Name three step inputs of pollutants to streams.

6.4 Name three pulse inputs of pollutant to streams.

6.5 Discuss how mixing influences dispersion in streams.

6.6 Discuss the fate in streams for benzene, PCBs, and dioxins. How are they fundamentally different?

Exercises

6.1 (a) Look up the flow rate of a stream in your area using data from your country's geological survey. In the United States, data can be found at www.usgs.gov. Using data in Table 6.2, determine the number of streams of this size. (b) What are some resources available from the USGS site that could be useful to understanding your watershed?

6.2 Manually calculate the eddy dispersion coefficient for this stream using the background data for the stream model in Fate. How would you expect an increase in water velocity, stream width, stream depth, or channel slope to affect E?

6.3 Using the background conditions in the step and pulse models in Fate, perform individual sensitivity analyses on Q, k, E, and v.

Step Input Problems

6.1 In a fictional small town in Wyoming, it is discovered that the local landfill is leaking a small amount, 2.0 g/s, of arsenic into the North Platte River. The communities downriver begin to raise concern about the levels of the arsenic in their drinking water drawn from the river. The EPA determines that arsenic is safe if the concentration is less than 0.01 ppm. Use Fate to determine how far downriver must a person travel to reach safe levels of arsenic in the river.
Flow velocity: 0.876 m/s
River depth: 1.5 m
River width: 18 m
Volumetric flow rate: 23.7 m^3/s
River slope: 0.002 04
Answer: $E = 10.5\ m^2/s$; distance when safe $= \sim 8.5 \times 10^7$ km.

6.2 The Tualatin River located in the Willamette Valley of Oregon is polluted with 0.000 012 5 kg/s of phosphorus (as phosphate) from urban runoff. Scenario 1: Given the following river parameters, use Fate to estimate when the concentration will reach 0.005 ppm? Assume the phosphorous does not have sufficient time to degrade for scenario one, so use a long decay half-life. Scenario 2: Now assume an uptake by algae with a half-life of three days. What difference do you see?
River depth: 6.0 m
River width: 3.0 m
Water velocity: 0.069 m/s
Channel slope: 0.000 92
Answer: Scenario 1: $E = 3.4 \times 10^{-4}\ m^2/s$; distance $= 2.21 \times 10^9$ km (longer than the receiving Willamette River!) Scenario 2: A much higher reduction in free phosphate in the water.

6.3 A pesticide factory located alongside a river has discovered a steady leak in one of its holding tanks. It has slowly been releasing Malathion, a pesticide with a half-life of five days, at a steady rate of 0.117 kg/s. What is the concentration at the site of the factory and 1000 km downstream, given the following conditions? First calculate this scenario manually, and then use Fate to check your answers.
Assuming there is an effective chemical alternative, what should its fate properties look like to ensure minimal downstream impacts?
River depth: 1.7 m
River width: 22.9 m
Water velocity: 2.05 m/s
River slope: 0.000 878
Answer: $E = 118\ m^2/s$; $C = 0.67$ mg/l.

6.4 A stainless steel tank containing 2-chlorophenol is stored next to a disinfectant-producing company that has a river running behind its property. After a car crashed into the tank, a leak in the tank developed and began to release 1.33 kg/s of 2-chlorophenol into the river. The leak was not noticed for 15 days, and 2-chlorophenol has a half-life of 2.5 days. Assume that all the chemical dissolves in the river water. Using the river parameters below, calculate the concentration of 2-chlorophenol in the river 500 km downstream. Use Fate to check your answers.
River depth: 1.4 m
River width: 10.0 m
Water velocity: 2.4 m/s
Channel slope: 0.0005
Answer: $E = 54.6\ m^2/s$; C at 500 km = 20.3 mg/l.

6.5 At a river in Kentucky, there was a 0.0002 kg/s cadmium^{2+} leak from a mining operation into the local river. Determine the concentration of Cd^{2+} in the river 1000 km away from the pollutant source. Include a Fate graph of the concentration of cadmium as it changes over distance. Also, determine the concentration of Cd^{2+} that would occur during drought conditions (assume river depth and velocity are half of their original values).
River depth: 2.0 m
River width: 6.0 m
Water velocity: 5.0 m/s
River slope: 0.001
Answer: Scenario 1. $E = 35.3\ m^2/s$; C at 1000 km $= 1.75 \times 10^{-3}$ mg/l. Scenario 2. $E = 25\ m^2/s$; C at 1000 km $= 9.23 \times 10^{-4}$ mg/l.

6.6 Chemical fertilizer and herbicide are applied to a grass farm and golf course, resulting in a steady runoff of 10.0 kg of soluble 2,4-dichlorophenoxyacetic acid into the river each day. A local brewery located 5.0 km downstream of the chemical source uses water from the river to brew its beer and has found that the water has a bitter taste if it contains more than 20.0 μg/l of 2,4-dichlorophenoxyacetic acid. The half-life of 2,4-dichlorophenoxyacetic acid is 13 days in the river (ignore brewing time). Will brewery customers notice an unpleasant taste in their beer? How far downstream from the pollutant source will the levels of 2,4-dichlorophenoxyacetic acid be low enough that no bitter taste will be noticed?

River depth: 6.0 m
River width: 15.0 m
Flow rate: 10.0 m^3/s
Eddy diffusion: 4.2 m^2/s
Answer: C at 5.0 km will be 0.675 mg/l, well above the taste threshold; yes, they will taste the pesticide.

6.7 A hypothetical landfill is located near Peachtree Creek but has never caused any pollution problems. One year, however, Peachtree Creek floods enough that pollutants from the landfill begin to leach into the stream. One pollutant that leaches into Peachtree Creek is 4-nitrophenol, which enters at a rate of 2.4 kg/s. The half-life of 4-nitrophenol is 9.8 days. Use Fate to determine the distance downstream at which the concentration of 4-nitrophenol is below 1.0 mg/l.
River depth: 6.9 m
River width: 17.0 m
Flow rate: 402.0 m^3/s
River slope: 0.04
Answer: $E = 3.20\ m^2/s$; distance = ~7500 km (a great distance for a river!)

Pulse Input Problems

6.8 The Skagit River flows from the Cascades out to the Puget Sound. Foresters are envisioning a hot summer this year and have set a plan to clear-cut some forest as fire prevention. They also plan to spray herbicides in the area to help clear out some growth. Due to the relentlessly rainy conditions in the North Cascades, the truck delivering their herbicides skids off the road near Concrete, Washington. The driver was not hurt, but the back end of his truck is crushed, leaving a gash that leaks all of the truck's contents into the Skagit. About 10 kg of the transported pesticides contained the endocrine disruptor atrazine. Assume a first-order degradation half-life of five days. Near Concrete, the Skagit River is steep, with a slope of 0.0011 and an average depth of 6 m. It is 150 m across and flows at about 1.10 m/s. In about 40 km, the river flows past the town of Sedro-Woolley. What will the concentration be at Sedro-Woolley in eight hours? In FATE, plot from 0–1.0 × 10^{-6} g/l to 0–40 km downstream. How far down the river will the atrazine be in 24 hours?
Answer: $E = 196\ m^2/s$; In eight hours, $C = 1.32 \times 10^{-3}$ mg/l; In 24 hours, peak will be 95 km downstream.

6.9 A metal plating plant located on a river has an accident in which 102 kg of copper plating solution is released. Assume copper(II) ion is the pollutant in question. The river drops an average of 2.5 m over a distance of 1 km. Given the conditions below, create two plots that show the curve of concentration versus time and distance. Determine where the maximum concentration of copper will be at a time of 20 hours from the time of the spill. Check your answer with Fate.
Flow rate: 96 m^3/s
Depth: 1.97 m
Width: 20 m
Velocity: 1.1 m/s
Half-life: 3000 years (no degradation)
Answer: $E = 12.3\ m^2/s$; the maximum concentration of 0.776 mg/l will occur in 24 hours at a distance downstream of 79 km.

6.10 The Fraser River originates at an elevation of 0.722 km above sea level and flows 1375 km from Central British Columbia to its outlet at the Pacific Ocean. A newly built, fictional, chemical plant at the head of the river dumps 0.100 Ci of Cesium-134 into the river instead of paying to correctly dispose it. Given the following parameters and a Cesium-134 half-life of 2.07 years, what is the maximum activity of Cesium-134 (in mCi/l) to reach the residents living at the mouth of the river?
Average river depth: 12.0 m
Average river width: 55.0 m
Average flow rate: 1.02 m/s
Answer: 1.06 × 10^{-5} mCi.

6.11 Two students returning from an on-site research project crash their car next to a river while driving at night. The accident results in fifty 2.01 jugs of solution containing 50.0 ppm of Cd^{2+} spilling and dissolving into the river. The river runs by a neighborhood 10.0 km downstream from the spill site. How long will it take for the maximum concentration of cadmium to reach the neighborhood? The students decide to take a measurement of the cadmium levels in the river 15.0 km after the neighborhood, but it will take them eight hours to get all their equipment there. Will the cadmium have reached the check point by that time? Cadmium cations do not degrade so assume a half-life of 1.0 × 10^6 years. Use Fate to check your answers.
River depth: 0.8 m
River width: 4.5 m
River slope: 0.0001

Water velocity: 1.0 m/s
Answer: It will take 2.78 hours to reach 10 km downstream. It will take seven hours to reach 25 km, so the students arriving at eight hours will miss the peak.

6.12 A train transporting 1% solution of sodium cyanide was traveling along a river in the countryside when an accident occurred. The railroad tanker car released 250 kg of sodium cyanide (135 kg of cyanide) into the river. The half-life of hydrogen cyanide is 0.9 years. Create a plot of the concentration of the cyanide versus the distance from the point of release using the following parameters. Manually calculate the CN^- concentration every 5 km, from 0.1 to 5.0 km. Research the possible ecological impact on the stream biota given the peak cyanide concentration passing through the river.
Depth of river: 2.0 m
Width of river: 10 m
Water velocity: 0.122 m/s
Stream coefficient of Eddy diffusion: 3.104 m^2/s
Mass of cyanide: 135 kg
It is also important to have a sensitivity analysis for this scenario. What will happen if the degradation rate or half-life is twice as long and twice as short as the value given above? Does the concentration change with varying degradation rate? Use Fate® to check calculations and plots.

6.13 An explosion at a nuclear power plant damages the structure and threatens to cause a meltdown. Workers and an emergency crew shut down the reactor and keep radioactive waste from contaminating the steam released from the cooling towers, but are unable to prevent it from discharging into the nearby river. For purposes of this problem, the waste contains primarily strontium-90, the contaminant of interest. From measurements near the plant, workers estimate that about 17 000 Ci of radioactive material were released into the river. Eight kilometers downstream from the plant is an elementary school right on the water's edge. What will the peak radiation (in Ci/l) at the school be? Use the data from the chart below to perform the necessary calculations and use Fate to check your results.
River depth 2.3 m
River width 35 m
Radioactivity 17 Ci
Water velocity 0.7 m/s
Channel slope 0.0003
Half-life 28.8 years
Answer: The peak of the plume will reach the school in 3.15 hours and have a maximum concentration of 0.094 mCi/l.

6.14 You are an environmental scientist who has been hired to assess the damage caused by an accident in which an eyeglass lens manufacturer, located on a river, released acetone directly into the water. Fortunately, only one 50.0 l container was released. The river has an average depth of 10 m, a width of 45 m, and longitudinal dispersion coefficient of 49.7 m^2/s. The water velocity is estimated to be 3.0 m/s. Acetone has density of 0.786 g/ml and a removal/degradation half-life in water of 20.0 hours. Calculate the concentration of acetone in the river due to this pulse release at distances of 2, 4, and 10 km from the site of the spill. Use Fate to check your answers.
Answer: At 2.0 km, the concentration will be 0.090 mg/l; at 4.0 km, 0.063 mg/l; at 10.0 km, 0.40 mg/l.

Spreadsheet Exercise

Create a spreadsheet that performs the same calculations as Fate for both one step and one pulse problem given above. Construct your spreadsheet so that it is interactive (so you can change numeric values for parameters and the plot automatically updates itself). A programmable calculator might effectively be used for this exercise, as well.

6.A Model Derivations for River and Stream Systems (Derivations by John Brooksbank)

6.A.1 Integrating Factors

The technique of using an integrating factor is useful when dealing with differential equations that are first order, meaning that there are no second derivatives or higher, and linear, meaning that all terms involving the function, Y, or any of its derivatives, can be separated. The general form of a linear, first-order differential equation is:

$$Y'(t) + p(t)Y(t) = q(t) \quad (6.A.1)$$

where $p(t)$ and $q(t)$ are some functions with respect to t.

Now, we can create some function, $I(t)$, which we define as:

$$I(t) = e^{\int p(t)dt} \quad (6.A.2)$$

The derivative of $I(t)$, $I'(t)$, is thus:

$$I'(t) = p(t) * e^{\int p(t)dt} \tag{6.A.3}$$

If we multiply Eq. (6.A.1) by the integrating factor, $u(t)$, we obtain:

$$e^{\int p(t)dt} * Y'(T) + p(t) * e^{\int p(t)dt} * Y(t) = e^{\int p(t)dt} * q(t) \tag{6.A.4}$$

Notice that, by design, the $Y(t)$ term ends up containing $I'(t)$, allowing us to simplify to

$$I(t)Y'(t) + I'(t)Y(t) = I(t)q(t) \tag{6.A.5}$$

The left-hand side of Eq. (6.A.5) is the derivative of $I(t)Y(t)$, allowing us to take the integral:

$$\int I(t)Y'(t) + I'(t)Y(t)dt = \int I(t)q(t)dt \tag{6.A.6}$$

$$I(t)Y(t) = \int I(t)q(t)dt + C \tag{6.A.7}$$

$$Y(t) = \frac{\int I(t)q(t)dt + C}{I(t)} \tag{6.A.8}$$

And, assuming that

$$\int I(t)q(t)dt$$

can be solved, we have solved the original differential equation and have the value of $Y(t)$. Always remember to include the constant of integration (+C).

6.A.2 Laplace Transforms

Another way to solve differential equations is to use the technique of a Laplace transform. A Laplace transform takes an equation in terms of one variable, such as time, and changes it to another variable, such as frequency. The converted equation may then be algebraically solved, before being converted back to the original domain to yield the solution. This technique is used as in many cases the equation is much more easily solved in the converted form than the original.

The Laplace transform, of some function $f(t)$, which we can write as either $F(s)$ or $L(f(t))$, we use the following formula:

$$F(s) = \int_0^\infty e^{-st} f(t)dt \tag{6.A.9}$$

The task of using the transform is made even simpler as there are large tables of Laplace transforms, allowing one to simply find the appropriate entry in the table and apply it. For example, to take the Laplace transform of

$$f(t) = \cos(2\pi t) \tag{6.A.10}$$

we can consult a table and find that:

$$L(\cos(\omega t)) = \frac{s}{s^2 + \omega^2} \tag{6.A.11}$$

as long as ω is a real number. Looking at Eq. (6.A.10), we see that in this case $\omega = 2\pi$, and therefore

$$L(f(t)) = \frac{s}{s^2 + 4\pi^2} \tag{6.A.12}$$

This same process can be applied in reverse, allowing us to find the inverse Laplace transform of quite complex equations fairly easily in many cases.

Of particular use is the knowledge that the Laplace transform of the derivative of a function is not simply the derivative of the Laplace transform of the original function, but is instead:

$$L\left(\frac{df(t)}{dt}\right) = sL(f(t)) + f(0) \tag{6.A.13}$$

6.A.3 River Model – Step

6.A.3.1 Boundary Condition 1: *C* (at infinity) = 0

The governing equation for the step model (continuous input from a point source) is very similar to that of the pulse model. This should make sense, as after all we are modeling the same system, with a different input. The mechanisms for pollutant transport have not changed, so we thus begin with the dispersion/advection equation used in the lake chapter.

$$\frac{dC}{dt} = -V\frac{dC}{dx} + \frac{d}{dx}\left(E\frac{dC}{dx}\right) + g(x,t) - d(x,t) \tag{6.A.14}$$

For the river model, we also can generally assume that the system has reached equilibrium. Most of the sources we will consider, such as industrial plants, will operate for years or even decades. On the other hand, water in a river will flow the entire length of that river quite quickly (even water at the very source of the Mississippi river will make it to the Gulf of Mexico in around 90 days). Therefore, we can assume that enough time has passed that the system has settled into equilibrium. As the system is in equilibrium, the change in concentration at a given point will be zero, and Eq. (6.A.14) becomes

$$0 = -V\frac{dC}{dx} + \frac{d}{dx}\left(E\frac{dC}{dx}\right) + g(x,t) - d(x,t) \tag{6.A.15}$$

For $d(x, t)$, the degradation term, we will again assume only first-order degradation

$$d(x,t) = kC(x,t) \tag{6.A.16}$$

In order to further simplify our equation, we can also use boundary conditions rather than an input function. We will remove the $g(x, t)$ input function from Eq. (6.A.15), giving us the governing equation for the River Model – Step

$$0 = -V\frac{dC}{dx} + \frac{d}{dx}\left(E\frac{dC}{dx}\right) - kC \tag{6.A.17}$$

which can also be written

$$0 = E\frac{d^2C}{dx^2} - V\frac{dC}{dx} - kC \tag{6.A.18}$$

Solving Eq. (6.A.18) without an input function will give us a general form of the solution, but not the final model.

We then reason our way to boundary conditions: certain points of the model at which we know the behavior. For example, at $x = 0$, the concentration will be equal to the amount of pollutant released per second divided by the amount of water that flows past that point per second, as the pollutant will not have had time to degrade or diffuse. We can then assemble a system of equations, with each equation expressing a single boundary condition. Once we have assembled enough boundary conditions, we can solve this system to find the particular version of the general solution to Eq. (6.A.18) that satisfies all the conditions we have set. This final equation is our model.

First, we must solve for the general form. Equations like 6.A.18), which may be written in the form

$$a\frac{d^2C}{dx^2} + b\frac{dC}{dx} + cC = 0 \tag{6.A.19}$$

where a, b, and c are arbitrary constants, are of a type of equation called *Linear* (as each term contains only $C(x)$, $C'(x)$, or $C''(x)$, but no combinations of those terms), *Second Order* (due to the second derivative), and *Homogeneous* (as there is no constant term) *Equations*. The general form of the solution for a Linear Second Order Homogeneous Equation with constant coefficients has been elsewhere derived and is

$$C(x) = k_1 e^{px} + k_2 e^{qx} \tag{6.A.20}$$

where k_1 and k_2 are constants that can be derived from the boundary conditions. The p and q terms in Eq. (6.A.20) are the roots of the *characteristic polynomial*

$$ar^2 + br + c = 0 \tag{6.A.21}$$

where a, b, and c are the coefficients of the original homogeneous Eq. (6.A.19).

For our model as presented in Eq. (6.A.18),

$$a = E \tag{6.A.22}$$

$$b = -V \tag{6.A.23}$$

$$c = -k \tag{6.A.24}$$

Making the characteristic polynomial

$$Er^2 - Vr - k = 0 \tag{6.A.25}$$

We may then find the roots by plugging a, b, and c into the quadratic formula

$$\frac{V \pm \sqrt{V^2 + 4Ek}}{2E} \tag{6.A.26}$$

We can then simplify Eq. (6.A.26) to

$$\frac{V}{2E}\left(1 \pm \sqrt{1 + \frac{4Ek}{V^2}}\right) \tag{6.A.27}$$

Per Eq. (6.A.20), the general solution to our governing equation is thus

$$C(x) = P_1 e^{\left(\frac{Vx}{2E}\left(1+\sqrt{1+\frac{4Ek}{V^2}}\right)\right)} + P_2 e^{\left(\frac{Vx}{2E}\left(1-\sqrt{1+\frac{4Ek}{V^2}}\right)\right)} \tag{6.A.28}$$

where P_1 and P_2 are constants.

From here, we must apply the boundary conditions of our physical model to the general solution to find the particular solution. We may start by noticing that our model includes a degradation term for the pollutant. This means that over time, the amount of pollutant will decrease, or mathematically, that

$$C(\infty) = 0 \tag{6.A.29}$$

This condition can only be met with a negative exponent. Notice that the final term in the exponent of the terms of Eq. (6.A.28),

$$\sqrt{1 + \frac{4Ek}{V^2}}$$

will always be greater than 1 as E, k, and V will always be positive values due to how they are defined. We can then notice that the second term in Eq. (6.A.28) contains

$$\left(1 - \sqrt{1 + \frac{4Ek}{V^2}}\right)$$

which will always be negative. On the contrary, the first term will always be positive. The boundary condition described by Eq. (6.A.29) is thus met by setting $P_1 = 0$. (Note that for the upstream direction, the opposite of Eq. (6.A.29) is true

$$(C(-\infty) = 0)$$

so the logic above will be needed in reverse as well. The model must therefore be broken into two parts, one for the upstream and one for the downstream. We will consider only the downstream from here on, as that is generally the portion we care about, but the math holds just as well for the upstream model.)

Having set P_1 to zero, our model becomes

$$C(x) = P_2 e^{\left(\frac{Vx}{2E}\left(1-\sqrt{1+\frac{4Ek}{V^2}}\right)\right)} \tag{6.A.30}$$

We must then set another boundary condition to determine the value of P_2. We know that at the point the pollutant is released, the concentration will be simply the output of pollutant, W (kg/s), divided by the flow of the stream, Q (m^3/s). We thus substitute the value for $C(0)$ into Eq. (5.17), resulting in Eq. (5.18):

$$\frac{W}{Q} = P_2 e^{\left(\frac{V*0}{2E}\left(1-\sqrt{1+\frac{4Ek}{V^2}}\right)\right)} \tag{6.A.31}$$

This solves very easily to

$$\frac{W}{Q} = P_2 \tag{6.A.32}$$

Substituting this value back into Eq. (6.A.30) yields the final governing equation for a step input of pollutant

$$C(x) = \frac{W}{Q} e^{\left(\frac{Vx}{2E}\left(1-\sqrt{1+\frac{4Ek}{V^2}}\right)\right)} \tag{6.A.33}$$

6.A.3.2 Boundary Condition 2: $C_o = W/Q = 0$

Here we are consistent with the above derivation up to

$$\frac{W}{Q} = P_2 e^{\left(\frac{V*0}{2E}\left(1-\sqrt{1+\frac{4Ek}{V^2}}\right)\right)} \tag{6.A.31}$$

We must then set another boundary condition to determine the value for P_2. But now

$$0 = P_2 \tag{6.A.34}$$

which yields a slightly different expression, most notably the source term (the first term in the equation)

$$C(x) = \frac{W}{Q\left(\frac{V}{2E}\left(1-\sqrt{1+\frac{4Ek}{V^2}}\right)\right)} e^{\left(\frac{Vx}{2E}\left(1-\sqrt{1+\frac{4Ek}{V^2}}\right)\right)} \tag{6.A.35}$$

6.A.3.3 Boundary Condition 3: Unstated Boundary Condition

Yet another governing for a step input of pollutant to a river can be found in Metcalf & Eddy (1991) with variables changed to match our notation:

$$C(x) = \frac{W}{A\sqrt{v^2 + 4kE}} e^{\left(\frac{xv}{2E}\right)\left(1\pm\sqrt{1+\frac{4kE}{v^2}}\right)} \tag{6.A.36}$$

6.A.3.4 Boundary Condition 4: Originally used in EnviroLand, the precursor to Fate

$$C(x,t) = \frac{W}{Q\sqrt{1+\frac{4kE}{v^2}}} e^{\frac{vx}{2E}\left(1\pm\sqrt{1+\frac{4kE}{v^2}}\right)} \tag{6.A.37}$$

This equation is used in Fate 2016 for consistency purposes with our prior publications and software packages.

6.A.4 River Model – Pulse

To model a river, we must use an equation that accounts for the two dominant mechanisms of pollution transport within a river: diffusion and advection. Diffusion happens due to the concentration gradient within the river water, and advection is the transport of pollutant due to the movement of the river water downstream. To account for both mechanisms, we use the diffusion–advection equation:

$$\frac{dC}{dt} = -V\frac{dC}{dx} + \frac{d}{dx}\left(E\frac{dC}{dx}\right) + g(x,t) - d(x,t) \tag{6.A.14}$$

where C is the concentration of pollutant, V is the water velocity, E is the dispersivity, $g(x, t)$ is the input function, and $d(x, t)$ is the degradation term of the pollutant. In this equation, the term

$$-V\frac{dC}{dx}$$

deals with advection, or the transfer of pollutant by the actual flow of the water in the river, and the term

$$\frac{d}{dx}\left(E\frac{dC}{dx}\right)$$

deals with dispersion of the pollutant due to differences in concentration throughout the river. Our first step in solving this equation is to take the Laplace transform of Eq. (6.A.14)

$$L\left(\frac{dC}{dt}\right) = -V * L\left(\frac{dC}{dX}\right) + E * L\left(\frac{d^2C}{dx^2}\right) + L(g(x,t)) - L(d(x,t)) \tag{6.A.38}$$

$$sU - f(x,0) = -V\frac{dU}{dx} + E\frac{d^2U}{dx^2} + G - D \tag{6.A.39}$$

where $U = L(C)$, G is the Laplace transform of the input function, and D is the Laplace transform of the degradation term. $f(x, 0)$ is the initial concentration profile and is therefore equal to C_0.

For a pulse input, we can write $g(x, t)$ using the Dirac Delta Function, $\delta_a(t)$, which is equal to 1 when a equals t, and 0 for all other values of t:

$$g(x,t) = \frac{S_0\delta_0(x)\delta_0(t)}{A} \tag{6.A.40}$$

where S_0 is the mass of the input and A is the cross-sectional area of the river. Equation (6.A.40) is thus equal to

$$\frac{S_0}{A}$$

the input pulse, when

$$(x,t) = (0,0)$$

and equal to 0 for all other values of x and t. We can then take the Laplace transform with respect to t:

$$G = L(g(x,t)) = \frac{S_0\delta_0(x)}{A}L(\delta_0(t)) = \frac{S_0\delta_0(x)}{A} \tag{6.A.41}$$

Now, we can consider the degradation term. Here, we will just use first-order degradation as we have in previous models

$$d(x,t) = kC(x,t) \tag{6.A.42}$$

where k is the first-order rate constant. We then take the Laplace transform:

$$D = L(d(x,t)) = kU \tag{6.A.43}$$

Substituting Eqs. (6.A.41) and (6.A.43) into Eq. (6.A.39) yields:

$$sU - f(x,0) = -V\frac{dU}{dx} + E\frac{d^2U}{dx^2} + \frac{S_0\delta_0(x)}{A} - kU \tag{6.A.44}$$

Applying a starting value of $C_0 = 0$ (since the pollution introduction is included within the $g(x, t)$ term), we can rearrange Eq. (6.A.44) into:

$$\frac{d^2U}{dx^2} - \frac{V}{E}\frac{dU}{dx} - \frac{(s-k)}{E}U = \frac{S_0\delta_0(x)}{EA} \tag{6.A.45}$$

Notice how taking the Laplace transform of Eq. (6.A.14) has transformed it from a partial differential equation, that is a differential equation with partial differentials in multiple dimensions (here, both x and t), into an ordinary differential equation, that is a differential equation with differentials in only one dimension (here, only in x). This changes Eq. (6.A.14) from a form that might not be able to be solved directly to a form about which there are several different methods to solve. This serves as a nice illustration of the power of the Laplace transform.

Now, however, we must use another technique to solve Eq. (6.A.45) in terms of U, before we apply the reverse transformation. To do this, we use a technique called the variation of parameters method. Explaining this method, however, is far beyond the scope of this appendix. For further information on the method itself, consult a differential equations textbook such as *Elementary Differential Equations* by Boyce and DiPrima, and for further detail on applying the method to Eq. (6.A.45), please see Chapter 4 of *Quantitative Environmental Risk Analysis for Human Health*, by Fjeld, Eisenberg, and Compton.

Using the variation of parameters methods, Eq. (6.A.45) can be solved to yield:

$$U = \frac{S_0}{2A\sqrt{E}}\exp\left(\frac{Vx}{2E}\right)\left[\frac{\exp\left(-\frac{x}{\sqrt{E}}\sqrt{\frac{V^2}{4E}+s+k}\right)}{\sqrt{\frac{V^2}{4E}+s+k}}\right] \tag{6.A.46}$$

Now, to find $C(x, t)$, we must take the inverse Laplace transform of Eq. (6.A.46).

$$C(x,t) = L^{-1}(U) = L^{-1}\left(\frac{S_0}{2A\sqrt{E}}\exp\left(\frac{Vx}{2E}\right)\left[\frac{\exp\left(-\frac{x}{\sqrt{E}}\sqrt{\frac{V^2}{4E}+s+k}\right)}{\sqrt{\frac{V^2}{4E}+s+k}}\right]\right) \tag{6.A.47}$$

The inverse transform will only affect terms containing the Laplace variable, s, so we may factor out all terms containing only x and constants:

$$C(x,t) = \frac{S_0}{2A\sqrt{E}}\exp\left(\frac{Vx}{2E}\right)L^{-1} \times\left[\frac{\exp\left(-\frac{x}{\sqrt{E}}\sqrt{\frac{V^2}{4E}+s+k}\right)}{\sqrt{\frac{V^2}{4E}+s+k}}\right] \tag{6.A.48}$$

For convenience, we set

$$\frac{S_0}{2A\sqrt{E}}\exp\left(\frac{Vx}{2E}\right) = W \tag{6.A.49}$$

making the previous equation

$$C(x,t) = WL^{-1}\left[\frac{\exp\left(-\frac{x}{\sqrt{E}}\sqrt{\frac{V^2}{4E}+s+k}\right)}{\sqrt{\frac{V^2}{4E}+s+k}}\right] \tag{6.A.50}$$

Now, we can further rearrange the equation to be in keeping with a form found in the table of Laplace transforms:

$$\frac{C(x,t)}{W} = L^{-1}\left[\frac{\exp\left(-\frac{x}{\sqrt{E}}\sqrt{s-\left(\frac{-V^2}{4E}-k\right)}\right)}{\sqrt{s-\left(\frac{-V^2}{4E}-k\right)}}\right] \tag{6.A.51}$$

Now, from the tables of Laplace transforms, we can find that

$$L^{-1}(F(s-a)) = e^{at}L^{-1}(F(s)) \tag{6.A.52}$$

In Eq. (6.A.51), by setting

$$\frac{-V^2}{4E} - k = a \tag{6.A.53}$$

we can arrive at

$$\frac{C(x,t)}{W} = L^{-1}\left[\frac{\exp\left(-\frac{x}{\sqrt{E}}\sqrt{s-a}\right)}{\sqrt{s-a}}\right] \tag{6.A.54}$$

which we can simplify using Eq. (6.A.52)

$$\frac{C(x,t)}{W} = e^{at}L^{-1}\left[\frac{\exp\left(-\frac{x\sqrt{s}}{\sqrt{E}}\right)}{\sqrt{s}}\right] \tag{6.A.55}$$

Now we turn again to a table of Laplace transforms and find that:

$$L^{-1}\left(\frac{\exp(-b\sqrt{s})}{\sqrt{s}}\right) = \frac{1}{\sqrt{\pi t}}\exp\left(\frac{-b^2}{4t}\right) \tag{6.A.56}$$

In Eq. (6.A.55), we can set

$$\frac{x}{\sqrt{E}} = b \tag{6.A.57}$$

to arrive at

$$\frac{C(x,t)}{W} = e^{at}L^{-1}\left[\frac{\exp(-b\sqrt{s})}{\sqrt{s}}\right] \tag{6.A.58}$$

which we can then find the inverse Laplace transform of using Eq. (6.A.56):

$$\frac{C(x,t)}{W} = e^{at}\frac{1}{\sqrt{\pi t}}\exp\left(\frac{-b^2}{4t}\right) \tag{6.A.59}$$

Now, we substitute back in for w, a, and b from Eqs. (6.A.49), (6.A.53), and (6.A.57) to find

$$\frac{C(x,t)}{\frac{S_0}{2A\sqrt{E}}\exp\left(\frac{Vx}{2E}\right)} = \exp\left[\left(\frac{-V^2}{4E} - k\right)t\right]\frac{1}{\sqrt{\pi t}} \times \exp\left(\frac{-\left(\frac{x}{\sqrt{E}}\right)^2}{4t}\right) \tag{6.A.60}$$

Then solving for $C(x, t)$, we find

$$C(x,t) = \frac{S_0}{2A\sqrt{E}}\exp\left(\frac{Vx}{2E}\right)\exp\left[\left(\frac{-V^2}{4E} - k\right)t\right] \times \frac{1}{\sqrt{\pi t}}\exp\left(\frac{-\left(\frac{x}{\sqrt{E}}\right)^2}{4t}\right) \tag{6.A.61}$$

which can be simplified by combining the exponential terms

$$C(x,t) = \frac{S_0}{A\sqrt{4\pi E t}}\exp\left(\frac{-(x-Vt)^2}{4Et} - kt\right) \tag{6.A.62}$$

the equation used in Fate.

Bibliography

Boyce, W.E. and DiPrima, R.C. (2016). *Elementary Differential Equations and Boundary Value Problems.* New York: Wiley Plus.

Capel, P.D., Giger, W., Reichert, R., and Warner, O. (1988). Accidental input of pesticides into the Rhine River. *Environ. Sci. Technol.* **22** (9): 992–997.

Fischer, H.B., List, E.J., Koh, R.C.Y. et al. (1979). *Mixing in Inland and Coastal Waters.* New York: Academic Press.

Fjeld, R.A., Eisenberg, N.A., and Compton, K.L. (2006). *Quantitative Environmental Risk Analysis for Human Health.* New York: Wiley.

Metcalf & Eddy, Inc. (1991). *Wastewater Engineering: Treatment, Disposal, and Reuse.* New York: McGraw-Hill, Inc.

National Research Council (NRC) (1997). *Contaminated Sediments in Ports and Waterways: Cleanup Strategies and Technologies*, 1. Washington, DC: National Academy Press.

Szestay, K. (1982). River basin development and water management. *Water Qual. Bull.* **7**: 155–162.

An Infobase Company (2004). *The World Almanac and Book of Facts.* New York: World Almanac Books.

Wetzel, R.G. (2001). *Limnology: River and Lake Ecosystems.* New York: Academic Press.

Wetzel, R.G. and Likens, G.E. (1991). *Limnological Analysis*, 2e. New York: Springer Verlag.

Wetzel, R.G. and Likens, G.E. (2000). *Limnological Analysis*, 3e. New York: Springer-Verlag.

7

Dissolved Oxygen Sag Curves in Streams: The Streeter–Phelps Equation

Case Study: Any Stream, Anywhere in the World

The concepts and equations presented in this chapter are ecologically crucial and globally applicable. Virtually every stream in the world has inputs of sewage or other organic wastes, either from humans or from domestic and wild animals. The modeling concepts developed in this chapter are thus applicable to every stream in the world and really speak to the assimilative capacity of all water bodies that perform a self-purification function with ubiquitous bacteria and natural reaeration at the river's surface interface. As we will see in this chapter, the waste presents two problems: the spread of disease caused by microbes and viruses in the waste, and decomposition leading to consumption of valuable dissolved oxygen (DO) in the stream. While the spread of pathogens in the waste is an important problem, in this chapter we will be mostly concerned with modeling the consumption of DO as a consequence of the input of raw domestic sewage. The disposal of sewage has plagued humans since our first permanent settlements. In fact, the accumulation of waste associated with permanent settlements has been suggested as one force promoting a nomadic lifestyle. Unfortunately, the way we overcame this problem through most of history was to build our cities next to streams or rivers, and dump our waste directly into the flowing water. This solved our problem but was not very pleasant for those living downstream. The expression that "*dilution is the solution to pollution*" was the practical remedy in these early days before waste treatment technology evolved in the early 1900s. This crude practice continues today in the rural communities (villages/settlements) of some lesser developed Global South nations. Historically, consider the DO levels as a function of time (from the early 1800s until 1990) in three highly populated areas: New York City, London, and Western Europe (Figure 7.1). As you can see, once the population reached a certain level, the DO level in the streams receiving waste significantly dropped. After living with the problem for decades, we finally installed proper sewage treatment plants, and in the 1970s and 1980s the streams started to recover. Our rivers as open sewers are a practice of our past, fortunately. However, this practice is still used in Global South countries of Africa and Asia.

Chemically, the problem is as follows. When biodegradable waste is added to an aqueous system, two competing kinetics processes are set up – one chemical, by which the microbes in the waste and aquatic system that are ubiquitous consume DO, and the other physical, in which the atmosphere replaces the DO. Thus, we are mostly concerned with the relative rates of DO consumption and reaeration.

In this chapter, we will learn about biochemical oxygen demand (BOD), DO, sewage treatment plants, the kinetics of microbial decomposition, the kinetics of reaeration, and how to remediate a stream or lake system after it has undergone eutrophication.

7.1 Introduction

This chapter is concerned with mathematical modeling the effects of the release of oxidizable organic matter to a flowing body of water. The most common form of organic waste is raw or untreated domestic sewage, but organic waste containing animal waste or from food-processing operations can have the same environmental effects. As we discussed in earlier chapters, the term DO refers to the chemical measurement of how much oxygen is dissolved in a water sample, usually expressed in milligram per liter (mg/l). The BOD is an estimate of how much total DO is required to oxidize the organic matter in a water sample. Thus, the BOD of a water or wastewater is actually calculated as the change in DO from initial DO at saturation to the amount remaining after five or more days, and BOD is expressed in milligram of oxygen per liter (mg O_2/l). Five days was arbitrary and chosen for convenience and the fact that most rivers in England had a flow time to the ocean of five days or less; today, this five-day test decision adds considerable confusion to modeling and understanding the effect of organic

Pollutant Fate and Transport in Environmental Multimedia, First Edition. Frank M. Dunnivant and Elliot Anders.
© 2019 John Wiley & Sons, Inc. Published 2019 by John Wiley & Sons, Inc.
Companion website: www.wiley.com/go/Dunnivant/pollutant_fate&Transp_EnvironMultimedia

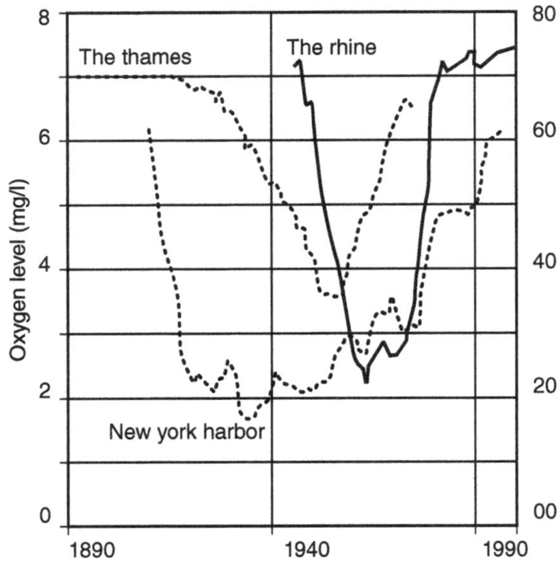

Figure 7.1 The effects of sewage outflow on dissolved oxygen levels in three major rivers. Source: Meadows et al. (2004). Reprinted with permission of Chelsea Green Publishing Company.

waste on stream ecosystems. *Standard Methods for the Examination of Water and Wastewater* (Standard Methods 1998 and more recent editions) includes analytical methods to measure the parameters used in this chapter. In the closing supporting experiment chapter, there is a detailed laboratory exercise to illustrate the BOD_5.

A plot of experimentally consumed oxygen (BOD) versus time is shown in Figure 7.2 for a domestic sewage sample and illustrates the fallacy of using a five-day BOD test. Before we discuss the modeling aspects of BOD in streams, represented by the Streeter–Phelps equation, it is important to gain an appreciation for the extent of the global sewage problem and the environmental issues surrounding wastewater and pathogens contained in the wastewater.

Our standard of living in the United States and other Global North areas of the world is a direct result of having adequate water and wastewater treatment. As early as 1700 BCE, people began to obtain the luxury of running water and then had to deal with the disposal of associated wastes. Though there is evidence of plumbing and sewage systems at many historical sites, including the *Cloaca Maxima* (great sewer) of the ancient Roman Empire, use of sewer and plumbing systems did not become widespread until modern times. Along with providing drinking water and disposing of sewage came the challenge of preventing the rapid spread of disease within populations that utilize a common water source and treatment facility. Many of these advances originated with the Royal Commission for Sanitation in seventeenth- and eighteenth-century England (https://www.uow.edu.au/~sharonb/sewage/history2.html).

Examples of microorganisms and viruses associated with waterborne diseases are (i) bacteria responsible for typhoid fever, cholera, and shigellosis, (ii) viruses causing hepatitis and viral gastroenteritis, and (iii) protozoa that are the agents of the waterborne diseases cryptosporidiosis and giardiasis. These microbes can be

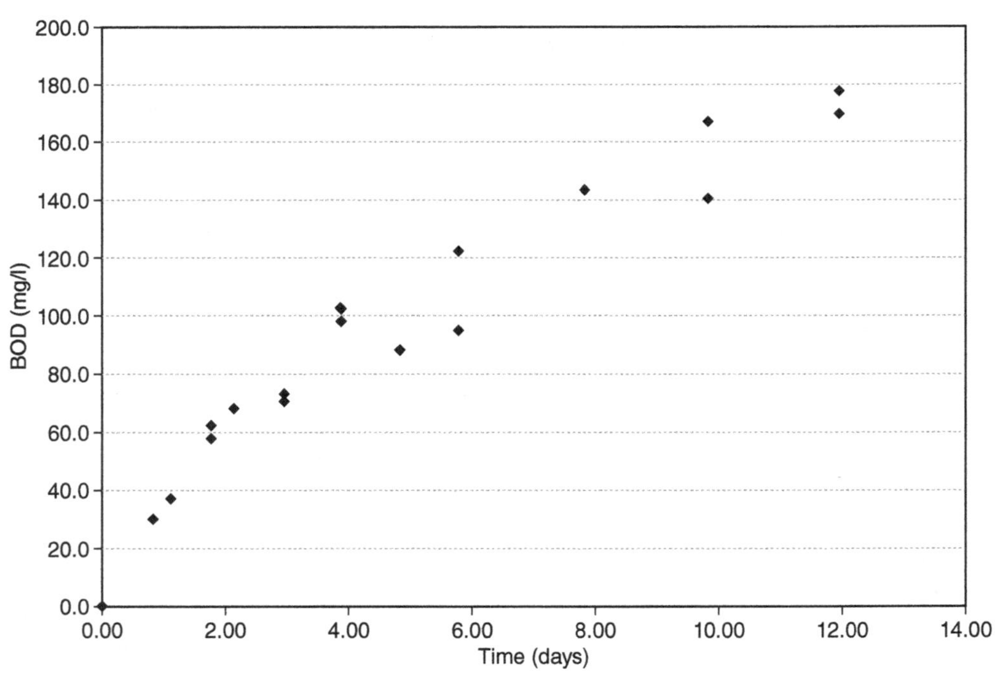

Figure 7.2 Dissolved oxygen demand (BOD) as a function of time in a BOD_L experiment.

killed or removed and their associated diseases can be prevented in domestic water supplies by a combination of sand filtration with chlorination, ozonation, or UV exposure, done at the end of modern sewage treatment processes. While these diseases pose risks to human populations, the release of untreated sewage to waterways can also result in the immediate death of aquatic systems. Surface aquatic systems are aerobic, and the lifeforms contained in these systems are dependent on the constant presence of DO. Most streams are at or near saturation values with respect to DO concentration, with resulting concentrations between 8 and 12 mg/l depending on the temperature and altitude of the stream. When readily oxidizable organic matter, such as domestic sewage, enters the stream, native microorganisms not only rapidly consume DO in the process of oxidizing this organic matter but also consume oxygen faster than it can be replenished through reaeration from the atmosphere. Table 7.1 lists the saturated DO concentration as a function of water temperature and salt content at 1.0 atmospheric pressure. As you can see, increasing temperature and salt content decrease the amount of DO in a water.

A typical plot of the DO concentration as a function of distance from a point source, for a step release of sewage, is shown in Figure 7.3. Note the shape of the curve. Above the entry point of the sewage, the water is near saturation with respect to DO. Where the sewage enters the stream, the DO concentration plummets to near zero, and the lowest DO concentration instream is often called the critical point (located by the dot on the plotted line). It is possible that for a considerable length, downstream of the input anaerobic conditions develop and stress many aquatic life forms. Such incidents may result in putrid odors and slime growth locally. As the organic matter is oxidized and as the stream reaerates, the DO level slowly rises, eventually achieving natural background concentrations and some aquatic life regrow and thrive to some extent as water quality gradually improves and the river recovers.

Figure 7.4 shows the dramatic effect of treating the sewage with 95% of the original BOD removed (for the same model input data used in Figure 7.3) before discharge into the stream. In this model simulation, the stream waters are not significantly affected by the small amount of BOD in the treated wastewater. In this chapter, we will learn to use the Streeter–Phelps equation, the mathematical model used to generate Figures 7.3 and 7.4. This model predicts the effects of sewage input to streams, evaluates the effectiveness of sewage treatment systems on stream quality, and can be used to determine which model parameters are the most important.

Table 7.1 Solubility of oxygen for water in contact with the atmosphere (at 1 atm containing 20.9% oxygen).[a)]

Temperature (°C)	Chloride concentration (mg/l)				
	0	5 000	10 000	15 000	20 000
0	14.6	13.8	13.0	12.1	11.3
1	14.2	13.4	12.6	11.8	11.0
2	13.8	13.1	12.3	11.5	10.8
3	13.5	12.7	12.0	11.2	10.5
4	13.1	12.4	11.7	11.0	10.3
5	12.8	12.1	11.4	10.7	10.0
6	12.5	11.8	11.1	10.5	9.8
7	12.2	11.5	10.9	10.2	9.6
8	11.9	11.2	10.6	10.0	9.4
9	11.6	11.0	10.4	9.8	9.2
10	11.3	10.7	10.1	9.6	9.0
11	11.1	10.5	9.9	9.4	8.8
12	10.8	10.3	9.7	9.2	8.6
13	10.6	10.1	9.5	9.0	8.5
14	10.4	9.9	9.3	8.8	8.3
15	10.2	9.7	9.1	8.6	8.1
16	10.0	9.5	9.0	8.5	8.0
17	9.7	9.3	8.8	8.3	7.8
18	9.5	9.1	8.6	8.2	7.7
19	9.4	8.9	8.5	8.0	7.6
20	9.2	8.7	8.3	7.9	7.4
21	9.0	8.6	8.1	7.7	7.3
22	8.8	8.4	8.0	7.6	7.1
23	8.7	8.3	7.9	7.4	7.0
24	8.5	8.1	7.7	7.3	6.9
25	8.4	8.0	7.6	7.2	6.7
26	8.2	7.8	7.4	7.0	6.6
27	8.1	7.7	7.3	6.9	6.5
28	7.9	7.5	7.1	6.8	6.4
29	7.8	7.4	7.0	6.6	6.3
30	7.6	7.3	6.9	6.5	6.1

a) After G.C. Wipple and M.C. Wipple (1911).

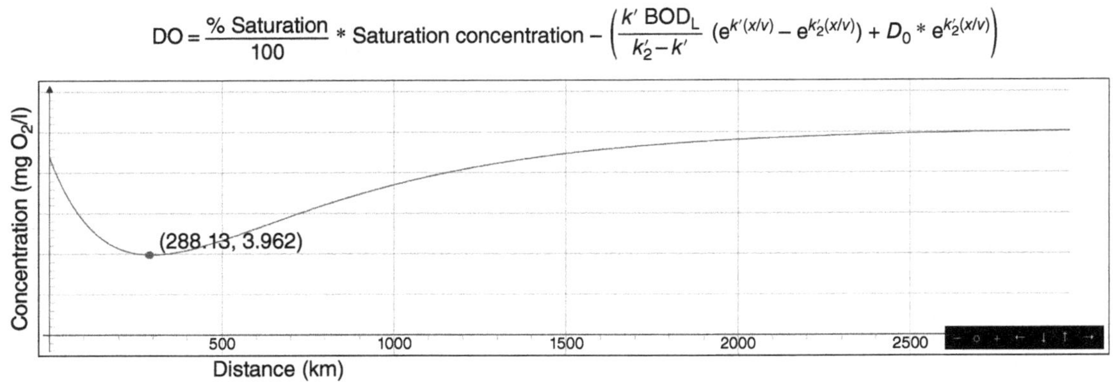

Figure 7.3 DO sag curve plot showing the effects of sewage on DO concentration in a stream (output from Fate®). The dot in the plot is the critical point. Source: Anders.

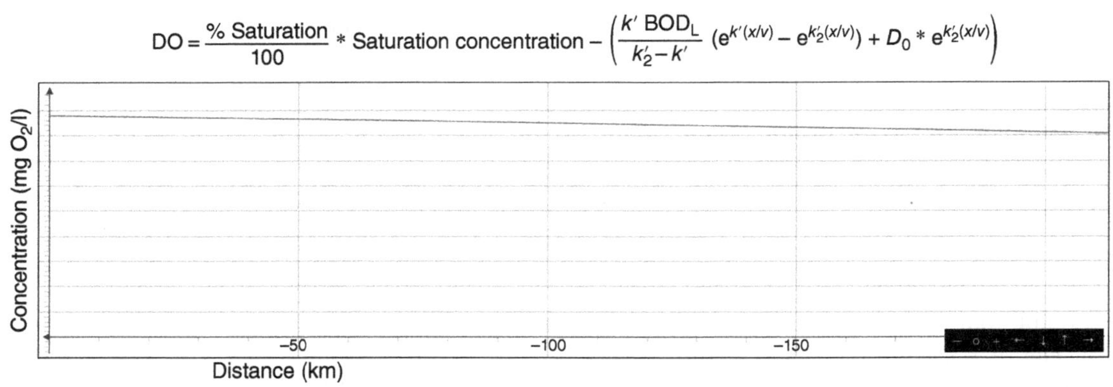

Figure 7.4 DO sag curve plot of the same stream shown in Figure 7.3 but after sewage treatment has been installed (output from Fate®). Notice the absence of a DO sag with a treatment system. Source: Anders.

7.2 Basic Input Sources (Wastewater Flow Rates and BOD Levels)

The Streeter–Phelps equation is designed to model the oxidation of any constant (step) input of oxidizable waste flowing to streams. Thus, the input source can be a variety of point and nonpoint sources, including the waste from domestic sewage plants, food-processing facilities, and inputs from livestock feedlots and agricultural settings. The only wastewater parameters that must be known are the flow rate of wastewater, the temperature of the wastewater, the DO of the wastewater (usually zero), and the concentration of BOD in the wastewater. All these are relatively easy parameters to measure, with the possible exception of BOD. The BOD of a wastewater can be measured in most chemistry laboratories by taking a sample of the wastewater and diluting it so that it does not contain more than approximately 8.0 mg/l of BOD. (The total amount of organic matter in the bottle will not require more than 8 mg DO/l during the incubation period.) The appropriate dilution is determined based on experience with the wastewater or by trial and error in conducting the BOD test. Several 300-ml portions of the diluted wastewater are incubated at a constant temperature (usually 20 °C) for 5–20 days. Twenty days has been assumed by some analysts to be the time to achieve oxidation of the total (ultimate) BOD. In reality, the time to reach its ultimate BOD can be far longer for some calcitrant organic materials like lignin from forested areas or other industrial wastewater effluents (NCASI 1982; Martone 1974) and from the oxidation of nitrogen-containing organic waste.

At predetermined times, the DO of samples is taken and plotted to determine the maximum DO needed to oxidize all the organic matter in the sample as a function of time. A plot of a typical data set was shown in Figure 7.2. The maximum BOD (oxygen consumed) of the diluted sample is the flat portion at the top of the plot (not reached in the experiments used to construct Figure 7.2). Correcting for the dilution factor used to measure the BOD yields the total BOD of the (undiluted) wastewater. Since domestic sewage is the most common source of BOD to most streams, the components of a typical sewage treatment plant will be discussed briefly below.

One of the supporting laboratory experiments given in the last chapter of this book focuses on the determination of DO and the five-day estimate of BOD (the five-day test is used here due to the convenience of a weekly laboratory), but modelers use much longer experiments to determine the ultimate BOD (total carbon and nitrogen chemical demand). A virtual laboratory for the BOD experiment can be viewed at https://sites.google.com/a/whitman.edu/frank-dunnivant-webpage/environmentalsoftware under the *EnviroLab* section.

The focus of modern sewage treatment is to remove turbidity, readily oxidizable organic matter, and pathogenic organisms. These three goals can easily be achieved at a minimal cost. First, we will give an overview of the treatment goals and how the wastewater treatment facility works, and then we will present a diagram of the system. Turbidity (suspended matter, microbes, etc.) is removed in settling clarifiers and in filtration systems. Dissolved organic matter is removed in biological contact units such as trickling filters and activated sludge lagoons. Most pathogens are naturally removed in the various treatment processes, but removal is ensured with the use of sand bed filtration, chlorination, ozonation, or UV exposure. One of the major design criteria for a wastewater treatment plant, and in fact a daily monitoring parameter for facility operators, is the experimentally determined BOD of the incoming and outgoing wastewater.

A diagram of the most basic components of a wastewater treatment plant is shown in Figure 7.5 and is explained in detail in the Quicktime program, *Water*, also available at https://sites.google.com/a/whitman.edu/frank-dunnivant-webpage/environmentalsoftware, under the video *WasteWater*. Section 7.7 discuss the more technical aspects of each component. Water enters the plant from the domestic sewer and first passes through a coarse bar screen to remove large materials that could damage pumps and clog pipes. Next, the water is passed through a finer filtration and grinder unit that removes paper and materials of the size of cigarette butts. Next, the water enters a primary settling tank (clarifier), where most of the remaining particles settle to the bottom of the tank and are pumped to the anaerobic digester (discussed at the end of this section).

The relatively clear water leaving the primary clarifier is next processed in a biological contact unit. The most common biological units are the trickling filter and the activated sludge tank or lagoon. Students are encouraged to view YouTube videos that discuss the basic units of wastewater treatment and its unit processes. A visit to your local facility as a class trip would also serve useful educational needs. This facility and all others like it are obligated to make their monthly National Pollution Discharge Elimination System (NPDES) permit data public

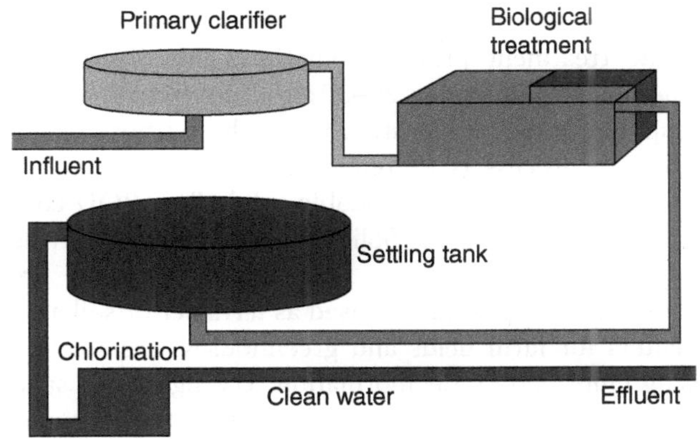

Figure 7.5 Diagram of the basic components of a typical sewage treatment plant.

and located at the State Agency wastewater website (https://video.search.yahoo.com/yhs/search;_ylt=A0LEViVYRHdaaHMAQpEPxQt.;_ylu=X3oDMTEyNGozbWZZ1BGNvbG8DYmYxBHBvcwMxBHZ0aWQDQjQxMjNfMQRzZWMDc2M-?p=activated+sludge+U-tube&fr=yhs-Lkry-SF01&hspart=Lkry&hsimp=yhs-SF01).

Detailed diagrams of each unit process, with tank volumes, flow rates, and retention times, will be given at the end of this chapter.

The goal of both of these treatment process units is to convert dissolved organic matter to particulate matter by allowing microorganisms acclimated to the wastewater organic material to utilize the dissolved organic matter and grow more microbial cells, such that they effectively become particulate matter. The trickling filter unit, a unit not commonly built today, works by spraying the wastewater over rocks that contain mats of bacteria and algae that consume the dissolved BOD. The activated sludge lagoon, a more modern unit, is a highly mixed (aerated), dense slurry of microorganisms that also consume the dissolved BOD. Water leaving the biological contact unit enters settling tanks, where the microbial cells settle out. In the activated sludge lagoon, some of the cells are recycled to keep the microbial cell level high, and in the case of the tricking filter the settled particulate matter is transported to the anaerobic digester. Normally two or three settling tanks in parallel are used to produce high volumes of clear water. However, most governmental regulations require that the water be passed through a coal/sand filter to remove any additional particles and microorganisms. Next, water is chlorinated, contacted with ozone, or irradiated with UV light. Excess chlorine is removed prior to release to a stream. Water retention times in most treatment plants range from 10 to 20 hours, and perform the same cleansing oxidations that occur over kilometers of natural streams with domestic sewage is directly released.

Solids (particulate matter) that have been generated in the treatment process sometimes are digested in anaerobic digesters for 20–30 days, depending on the temperature of the digestion tank. The purpose of the digestion process is to reduce the residual microbial cells to relatively nondegradable solids (low BOD content) and reduce odors. Solids from digesters receiving only domestic sewage or food-processing waste are commonly composted and used as fertilizer or soil augmenters for farm fields and greenhouses. Disinfection using chlorination–dechlorination, UV light, and sand filtration are site specific and depend on the receiving water for the plant effluent. Most of these add considerably to the cost of the treatment plant. The final decision on what type of disinfection to use is by EPA or the State version of EPA.

7.3 Sampling of Wastewater

Sampling of organic waste is relatively easy compared to the sampling of other pollutants and environmental media discussed in this text. Only a few decisions need to be made: where to sample and what time of day to sample. The importance of sampling time is illustrated in Figure 7.6, where the variation of wastewater flow to a treatment plant is given as a function of time of day. In order to estimate the average daily flow into and out of a plant, time-averaged samples are taken every hour over a 24-hour period for BOD_5 measurements. Mostly, any type of clean sampling container will suffice, but samples should be cooled during delivery to a laboratory and processed promptly whenever possible.

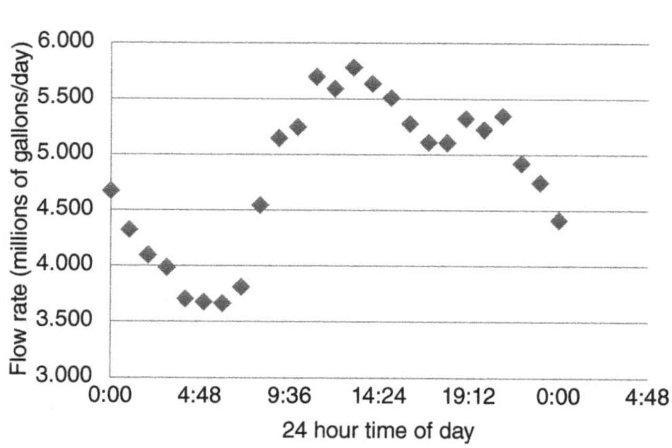

Figure 7.6 Incoming flow rates in millions of gallons over a 24-hour sampling period. Source: Data with complements from the City of Walla Walla (Washington, USA).

7.4 Mass Balance-Based Development of the Basic Streeter–Phelps Model

An early water quality modeling analysis is that for the Ohio River in c. 1920 and addressed the organic contamination from human sources (Streeter and Phelps 1925a,b). Appropriately, it is now known as the Streeter–Phelps (S–P) DO sag equation and has a rich history of use over 100 years throughout the world. It focused on the self-purification potential for streams and included the coupled parameters – BOD and surface reaeration – with point source inputs. An expanded form became popular decades after its first use on the Ohio River. The original 1925 form was one of the first known mathematical models for water quality and sparked a century of water quality advances and started a new field of study in the United States (Fair et al. 1968). Your authors graciously acknowledge and credit the contributions of now Tufts University Professor Steven Chapra (Chapra 1997) and c. 1975 Manhattan College Summer Institute Workshop instructors Professors Donald "Doc" O'Connor and Robert "Bob" Thomann for preparing the material from which the following text was abstracted (Manhattan College 1975). Similarly, business colleagues from Hydroqual, Inc. (Mahwah, NJ) – Thomas Gallagher, Andrew Thuman – shared modeling experiences and caveats from their real-world consulting for clients throughout the United States. They in particular taught engineers to value the utility of simple or "back of the envelope" modeling computations to approximate expected results and anticipated data requirements. The analysis follows below.

An elaborate mathematical derivation for the Streeter–Phelps equation is shown in Appendix 7.A. For a more general conceptual derivation, consider a small volume of water either flowing at a constant rate or stationary in a river or lake. This "control volume" is well mixed such that changes occur instantly throughout. Mass balances for stream DO deficit ($C_{sat} - C$) and BOD can be written as:

$$0 = -v\frac{\mathrm{dBOD_L}}{\mathrm{d}x} - k_\mathrm{d}\mathrm{BOD_L};\quad \text{for } \mathrm{BOD_L} \tag{7.1}$$

$$0 = -v\frac{\mathrm{d}D}{\mathrm{d}x} + k_\mathrm{d}\mathrm{BOD_L} - k_\mathrm{r}D;\quad \text{for DO deficit } D \tag{7.2}$$

where

x is the stream downstream distance, length,
v is the average but constant stream velocity, distance/time,
$\mathrm{BOD_L}$ is the stream ultimate BOD, mass/volume,
k_d is the first-order BOD decay rate, 1/time,

k_r is the DO reaeration rate, 1/time,
t is the travel time (v/x), time,
DO is the dissolved oxygen concentration in mg/l, and
D is the DO deficit in mg/l.

Each mass balance is based on the principle that mass must be conserved unless acted on by chemical reaction or exited from the control volume boundaries.

For the initial (time t equals zero) conditions of the deficit D equals 0.0 mg/l (D_0) and $BOD_{Ultimate}$ equals the ultimate BOD, these equations have a known solution for first-order, independent linear equations as shown subsequently. A more formal solution using integrating factors is provided in this chapter appendix for those students with a math background in differential equation solutions. The example provided later shows how one would calculate the instream deficit and BOD concentration. Hence, these two equations for instream degradation of BOD and the corresponding stream-DO deficit are known as the "classic Streeter–Phelps Equations" or the "DO Sag Equation" by virtue of its overall shape as shown in Figures 7.4 and 7.5.

$$\mathrm{BOD}_{\mathrm{Ultimate}} = \mathrm{BOD}_0\, \mathrm{e}^{-k_\mathrm{d}\frac{x}{v}} \tag{7.3}$$

$$D = \frac{k_\mathrm{d}\mathrm{BOD}_{\mathrm{Ultimate}}}{k_\mathrm{r} - k_\mathrm{d}}\left(\mathrm{e}^{-k_\mathrm{d}\left(\frac{x}{v}\right)} - \mathrm{e}^{-k_\mathrm{r}\left(\frac{x}{v}\right)}\right) + D_0\mathrm{e}^{-k_\mathrm{r}\left(\frac{x}{v}\right)} \tag{7.4}$$

The minimum stream concentration is known as the "sag" point (noted earlier) and the travel time or location of this point of interest in modeling water quality. First noting that any function's minimum value occurs at a point where its first derivative is equal to zero and solving for these values directly determine their values. So, the solution for dD/dt equal 0 (zero) yields a minimum, or D_c, critical deficit and t_c, time to critical minimum concentration shown below:

$$t_\mathrm{c} = \frac{1}{k_\mathrm{r} - k_\mathrm{d}} \ln \frac{k_\mathrm{r}}{k_\mathrm{d}}\left[1 - D_0\frac{(k_\mathrm{r} - k_\mathrm{d})}{k_\mathrm{d}\mathrm{BOD}_{\mathrm{Ultimate}}}\right] \tag{7.5}$$

$$D_\mathrm{c} = \frac{k_\mathrm{d}}{k_\mathrm{r}}\mathrm{BOD}_{\mathrm{Ultimate}} \times \frac{k_\mathrm{r}}{k_\mathrm{d}}\left\{1 - \mathrm{e}^{-k_\mathrm{d}(x_\mathrm{c}/v)}\, D_0\frac{(k_\mathrm{r} - k_\mathrm{d})}{k_\mathrm{d}\mathrm{BOD}_{\mathrm{Ultimate}}}\right\}^{-\frac{k_\mathrm{d}}{k_\mathrm{r}-k_\mathrm{d}}} \tag{7.6}$$

The simplifying equations are messy and the algebra brutal, but the resultant values are highly useful in explaining where a critical deficit appears downstream. In the frequent case of the initial deficit D_0 equals 0, the equations reduce to slightly more manageable forms

$$t_\mathrm{c} = \frac{1}{k_\mathrm{r} - k_\mathrm{d}} \ln \frac{k_\mathrm{r}}{k_\mathrm{d}} \tag{7.7}$$

$$D_\mathrm{c} = \frac{k_\mathrm{d}\mathrm{BOD}_{\mathrm{Ultimate}}}{k_\mathrm{r}}\left(\frac{k_\mathrm{r}}{k_\mathrm{d}}\right)^{-\frac{k_\mathrm{d}}{k_\mathrm{r}-k_\mathrm{d}}} \tag{7.8}$$

Students may ask how does one handle the real case of large BOD loads that cause the water body to become anaerobic or absent DO. In this case, a useful special case was developed by Gundelach and Castillo (1976). It involves a numerical solution to the resulting differential equation and would allow a thorough simulation of a seldom encountered situation in developed countries of the world in the twenty-first century. The specifics are not presented here.

Another special case that some observant students may have noticed is that the classical Streeter–Phelps equation becomes imaginary for the case of k_d equals k_r as division by zero makes the result unstable with no real solution. Practically this does not occur very often as reaction rate correction for both are done with different Arrhenius equation exponents (Tetra-Tech 1985) making such outcomes a low probability. Nevertheless, an alternate derivation adds a correction factor to k_d based on river velocity v, depth H, and stream bed activity coefficient Ω. The latter typically varies from 0.1 to 0.6 and is meant to include benthic surface area effects on bulk water column deoxygenation.

$$k_\mathrm{d} = (k_\mathrm{d})_{\mathrm{old}} + \frac{\Omega}{H} \tag{7.9}$$

Users must be wary of double-counting any sediment oxygen demand (SOD) deficit if it is measured in shallow water rivers, especially those with a rocky substrate that are inclined to accumulate observable slime or benthic growths that serve to increase BOD degradation in the overlying water column.

7.4.1 Expanded Streeter–Phelps Equation Utility

One good feature of the Streeter–Phelps equation for deficits is the utility of the mg/l/d terms. Modelers call these "unit deficits" and represent the addition or depletion of DO by the process they describe degradation or reaeration. Hence, k_{dL} and k_rD represent the depletion of oxygen by degradation of BOD and reaeration, respectively. The relative magnitude of each term relative to others suggests how the model or the prototype behaves. In addition, it might alert the modeler to where more data are required to achieve a more robust simulation. Similarly, information is provided from parameter sensitivity: how a small change in k_d affects computed instream deficits, or how reaeration affects self-purification. One might find that this "sensitivity" changes with stream location – from discharge downstream to the sag point through the recovery phrase. Logically, the student should recognize that sensitivity should change from the initial instream concentrations through the sag point and DO recovery. They could signal that the modeler look more critically at the system under study with available data sources and suggest better or more definitive data. Students who have taken a beginning calculus course should be familiar with this concept and be able to perform this differentiation to obtain these sensitivity functions.

Next we will look at an extension of the Streeter–Phelps analysis to consider other source, and sink terms like nitrification, photosynthesis, SOD, and even distributed contaminant loads are possible, although the latter will not be shown here. All would have a unit deficit that can be compared with others in the simulation for a more accurate simulation. In all these cases, the unit deficits along with sensitivity point to a path to a better model by increasing the number of segments or better characterization of the reaction rates. In short, they reduce uncertainty or better characterize its nature for any management decision made by the managers using the model.

These extensions commonly include nitrogenous BOD (BOD_N) (Thomann 1972), SOD (Mortimer 1942), and the net effect of algal growth and decay (photosynthesis minus respiration or P – R) with mg/l day units (Manhattan College 1975; Tetra-Tech 1985). They are treated with simplistic concepts, although other more detailed and more appropriate models exist and are used that look at nitrification in greater detail (see Chapter 10). Their simplicity does NOT imply inaccuracy or inappropriateness, as they are very useful, as a premodeling tool to determine how detailed a model is required to meet management needs.

The importance of nitrification process and its conversion from organic nitrogen to ammonia, to nitrite and nitrate forms can be significant in the DO resources of rivers and streams and modelers call this the BOD_N component. It was recognized early in the twentieth century that the sum of ammonia and organic nitrogen is the oxidizable portion and called total Kjeldahl nitrogen (TKN). There is an aggressive chemical digestion and analytical test that measures TKN concentrations (AWWA 2012). Further, by the stoichiometry of the oxidation of nitrogen (N) to nitrite and then to nitrate forms, there are 4.57 mg O required for 1 g of N consumed. Hence, the mass balance for BOD_N can be written similarly to that for carbonaceous $BOD_{Ultimate}$ and the solved differential equations become analogous coupled classic Streeter–Phelps equations:

$$BOD_{Ultimate\ N} = BOD_{0N}\, e^{-k_{dN}\frac{x}{v}} \tag{7.10}$$

$$D_N = \frac{k_{dN} BOD_{Ultimate\ N}}{k_r - k_{dN}} \times \left(e^{-k_{dN}\left(\frac{x}{v}\right)} - e^{-k_r\left(\frac{x}{v}\right)}\right) + D_0 e^{-k_r\left(\frac{x}{v}\right)} \tag{7.11}$$

where

D_N = NBOD "unit" deficit, mg/l
BOD_N = NBOD, mg/l
BOD_{0N} = initial or ultimate NBOD, mg/l
$BOD_{Ultimate\ N}$ = ultimate BOD due to nitrogen
k_{dN} = first-order NBOD decay rate, 1/d
t = travel time, d

The unit deficits for sediment oxygen demand (D_{SOD}) are only linked to the DO deficit mass balance, so its magnitude is expressed as:

$$D_{SOD} = \frac{S}{Hk_d}\left(1 - e^{-k_r\left(\frac{x}{v}\right)}\right) \tag{7.12}$$

where

S = measured SOD rate g oxygen/m^2 d
H = stream depth, m

Similarly, algal productivity and respiration $(P - R)_{NET}$ is measured by light and dark bottle technique or estimated empirically by algal chlorophyll a biomass (Tetra-Tech 1985), which can be computed from

$$D_{P-R} = \left(\frac{r_p - r_{sp}}{k_r}\right)\left(1 - e^{-k_r\left(\frac{x}{v}\right)}\right) \tag{7.13}$$

where

D_{P-R} = DO unit deficit from $(P - R)_{NET}$, mg DO/l·day
P = DO production rate, mg DO/l·day
R = algal respiration rate, mg DO/l·day (note DO consumed by respiration of dying cells)

7.4.2 SS Tidal Form for Bays and Estuary

A tidal bay or estuary can be approximated by a steady-state model in which the tidal cycle is uniform and repeating in a sinusoidal pattern. This form allows for a known solution to the linear first-order differential equation as developed by O'Connor (1960) and Thomann and Sobel (1964), and is referenced here.

An early model developed by the Texas Water Development Board (TWDB) with EPA support and subsequent widespread distribution called "DOSAG" performed these exact computations (TWDB 1970). The Georgia Environmental Protection Division provides a link to obtaining working versions of a stream Streeter–Phelps model (GA DOSAG) and its related estuary version GAest (Georgia Environmental Software 2018). Other states and consultants have developed their own versions for simple desktop modeling of water quality. Students are encouraged to pursue finding these information sources on state or university websites. The model could also be developed for a spreadsheet, as the terms require simple numerical computation.

Example Problem

An example is provided to illustrate how the waste concentration is computed and the stream hydraulic data are used:

A city discharges 25 million gallons per day (mgd) of domestic sewage into a stream whose typical rate of flow is 250 cubic feet per second (cfs). The velocity of the stream is approximately 3 mi/h. The temperature of the sewage is 20 °C, while that of the stream is 15 °C. The 20 °C BOD_5 of the sewage is 180 mg/l ($BOD_{Ultimate}$ = 30.6 mg/l), while that of the stream is 1.0 mg/l. The sewage contains no DO, but the stream is 90% saturated upstream of the discharge. At 20 °C, k_d is estimated to be 0.34 d^{-1}, while k_r is 0.65 d^{-1}.

(1) Determine the critical oxygen deficit and its location.
(2) Also estimate the 20 °C BOD_5 of a sample taken at the critical point. Use temperature coefficients of 1.135 for k_d and 1.024 for k_r.
(3) Plot the DO sag curve.
(4) Determine the DO concentration at 1000 km from the point source.

Solution

(a) Determine the DO in the stream before discharge. Saturation concentration at 15 °C (from table on worksheet) = 10.2 mg/l
DO in stream = 0.90 (10.2 mg/l) = 9.2 mg/l
(b) Determine the temperature, DO, and BOD of the mixture using the mass balance approach. Note that units should be compatible.
Flow rate of stream (conversion from cfs to liters per day):

$$(250\ \text{ft}^3/\text{s})(7.48\ \text{gal/ft}^3)(60\ \text{s/min})$$
$$(60\ \text{min/h})(24\ \text{h/d}) = 161.6 \times 10^6\ \text{gal/d}$$
$$(161.1\ \text{gal/d})(3.79\ \text{l/gal}) = 612 \times 10^6\ \text{l/d}$$

Flow rate of sewage effluent (from gallons per day to liters per day):

$$(25 \times 10^6\ \text{gal/d})(3.70\ \text{l/gal})$$
$$= 94.8 \times 10^6\ \text{l/d} = 94.8\ \text{million l/d}$$

Note: Flow rates should be expressed in mgd, million liters/day, or million cubic feet/day.

Temperature of mixture:

$$\begin{aligned}&\text{Net change in temp } (\Delta T)\\&\quad = \text{stream input} + \text{sewage input} - \text{output}\\&0 = (\text{stream flow rate})(\text{stream temp})\\&\qquad + (\text{sewage flow rate})(\text{sewage temp})\\&\qquad - (\text{mixture flow rate})(\text{mixture temp})\\&0 = (612 \times 10^6\ \text{l/d})(15^\circ\text{C})\\&\qquad + (94.8 \times 10^6\ \text{l/d})(20^\circ\text{C})\\&\qquad - (612 \times 10^6\ \text{l/d} + 94.8 \times 10^6\ \text{l/d})\ T_{\text{mixture}}\end{aligned}$$

On rearrangement, yields

$$T_{\text{mixture}} = \frac{(612 \times 10^6\ \text{l/d})(15^\circ\text{C}) + (94.8 \times 10^6\ \text{l/d})(20^\circ\text{C})}{(612 \times 10^6\ \text{l/d} + 94.8 \times 10^6\ \text{l/d})} = 15.7^\circ\text{C}$$

DO of mixture:

$$\begin{aligned}&\text{Net change in DO}\\&\quad = \text{stream input} + \text{sewage input} - \text{output}\\&0 = (\text{stream flow rate})(\text{stream DO})\\&\qquad + (\text{sewage flow rate})(\text{sewage DO})\\&\qquad - (\text{mixture flow rarte})(\text{mixture DO})\\&0 = (612 \times 10^6\ \text{l/d})(9.2\ \text{mg/l})\\&\qquad + (94.8 \times 10^6\ \text{l/d})(0.0\ \text{mg/l})\\&\qquad - (612 \times 10^6\ \text{l/d} + 94.8 \times 10^6\ \text{l/d})\\&\qquad (\text{DO}_{\text{mixture}})\end{aligned}$$

on rearrangement yields

$$\text{DO}_{\text{mixture}} = \frac{(612 \times 10^6\ \text{l/d})(9.2\ \text{mg/l}) + (94.8 \times 10^6\ \text{l/d})(0.0\ \text{mg/l})}{612 \times 10^6\ \text{l/d} + 94.8 \times 10^6\ \text{l/d}} = 7.97\ \text{mg/l}$$

BOD_5 of mixture:

$$\begin{aligned}&\text{Net change in BOD}_5\ (\Delta\text{BOD}_5)\\&\quad = \text{stream input} + \text{sewage input} - \text{output}\\&0 = (\text{stream flow rate})(\text{stream BOD}_5)\\&\qquad (\text{sewage flow rate})(\text{sewage BOD}_5)\\&\qquad - (\text{mixture flow rate})(\text{mixture BOD}_5)\\&0 = (612 \times 10^6\ \text{l/d})(1.0\ \text{mg/l})\\&\qquad + (94.8 \times 10^6\ \text{l/d})(180\ \text{mg/l})\\&\qquad - (612 \times 10^6\ \text{l/d} + 94.8 \times 10^6\ \text{l/d})\\&\text{BOD}_{5\ \text{mixture}}\end{aligned}$$

on rearrangement yields

$$\mathrm{BOD}_{5\ \mathrm{mixture}} = \frac{(612\times10^6\ \mathrm{l/d})(1.0\ \mathrm{mg/l}) + (94.8\times10^6\ \mathrm{l/d})(180\ \mathrm{mg/l})}{612\times10^6\ \mathrm{l/d} + 94.8\times10^6\ \mathrm{l/d}} = 25.0\ \mathrm{mg/l}$$

BOD_L of mixture (at 20 °C):

$$\mathrm{BOD_L} = \frac{\mathrm{BOD_5}}{1-e^{-k_d(x/v)}} = \frac{25.0\ \mathrm{mg/l}}{1-e^{-(0.35\ \mathrm{d^{-1}})(5.00\mathrm{days})}} = 30.6\ \mathrm{mg/l}$$

(c) Correct the rate constants to 15.7 °C:
Rate constants are not linearly related to changes in temperature; therefore, we must correct them using an exponential relationship. Typically, these can be corrected using the two constants and equations given below. Note that 20 °C is used as the reference point since this is where the original data for the k' values were collected.

$$k_d = 0.34(1.135)^{15.7-20} = 0.200\ \mathrm{d^{-1}}$$

$$k_r = 0.65(1.024)^{15.7-20} = 0.588\ \mathrm{d^{-1}}$$

(d) Determine the critical time (t_c) and critical distance (x_c):
In the table note that the saturation value for O_2 at 15.7 °C is 10.1 mg/l; however, the stream is at 90% of the saturation value (9.2 mg/l). Thus, the initial oxygen deficit,

$$D_0 = (\text{initial stream } O_2 \text{value} - O_2 \text{of the mixture}) = (9.2 - 7.97) = 1.23\ \mathrm{mg\ O_2/l}$$

$$t_c = \frac{1}{k_r - k_d}\ln\frac{k_r}{k_d}\left[1-\frac{D_0(k_r-k_d)}{k_d\mathrm{BOD_L}}\right]$$

$$t_c = \frac{1}{\frac{0.587}{\mathrm{d}}-\frac{0.197}{\mathrm{d}}}\ln\frac{0.587\ \mathrm{d^{-1}}}{0.197\ \mathrm{d^{-1}}} \times\left[1-\frac{(1.23\ \mathrm{mg/l})\left(\frac{0.578}{\mathrm{d}}-\frac{0.197}{\mathrm{d}}\right)}{\frac{0.197}{\mathrm{d}}(30.6\ \mathrm{mg/l})}\right] = 2.57\ \mathrm{d}$$

$$x_c = v\,t_c = (3\ \mathrm{mi/h})(24\ \mathrm{h/d})(2.57\ \mathrm{d}) = 185\ \mathrm{mi}$$

or

$$= (3\ \mathrm{mi/h})(1.61\ \mathrm{km/mi})(24\ \mathrm{h/d})(2.57\ \mathrm{d}) = 298\ \mathrm{km}$$

(e) Determine D_c.
To calculate the critical oxygen deficit, D_c, we must first convert the water velocity into units of mi/d: 3 mi/h = 72 mi/d.

$$D_c = \frac{k_d}{k_r}\mathrm{BOD_L}e^{-k_d\frac{x_c}{v}} = \frac{0.197\ \mathrm{d^{-1}}}{0.587\ \mathrm{d^{-1}}}30.6\,e^{-(0.197\ \mathrm{d^{-1}})(185\ \mathrm{mi})(72\ \mathrm{mi/d})} = 6.21\ \mathrm{mg/l}$$

Thus, the DO will be depressed 6.21 mg/l from its saturation value. The minimum O_2 concentration of the stream will be the saturation value minus the D_c, or

$$9.2 - 6.37 = 2.97\ \mathrm{mg\ O_2/l}$$

(f) Determine the BOD_5 of a sample taken at distance x_c.

$$\mathrm{BOD_5} = \mathrm{BOD_L}\,e^{-k'\left(\frac{x}{v}\right)}$$

$$\mathrm{BOD_5} = (30.6\ \mathrm{mg/l})e^{-0.197\ (20\ \mathrm{d})} = 18.3\ \mathrm{mg/l}$$

(g) The oxygen sag curve in kilometer from the point source is shown in Figure 7.8.

First look closely at the overall solution of the derivation, Eq. (7.4). The equation is the net effect of two opposing reactions: the consumption of DO by microorganisms in oxidizing the BOD (represented by k_d) and reaeration of the stream by dissolution of atmospheric oxygen (represented by k_r). Both of these processes can be modeled by first-order reactions, hence the inclusion of exponential terms for the two k values. Of course, the model must incorporate the ultimate BOD (BOD_L) and the stream DO deficit (D_0 in Eq. (7.2)). Note that the purpose of the equation is to calculate the amount of oxygen consumed by the waste (D). It is very important to note that D is *not* the remaining DO content of the stream water, but rather the amount of the original of DO that has been consumed. In order to calculate the resulting DO of the stream, we must subtract D from the original DO of the stream without BOD waste. We use DO of the stream immediately prior to BOD entry for this value. The remaining DO is what is plotted in Fate and most Streeter–Phelps diagrams. The exact equation is given at the top of Figures 7.3 and 7.4.

The term k_r is the first-order rate constant associated with the reaeration of the stream water. Exact measurement of this parameter is difficult since it is dependent on factors such as the stream depth, mixing in the stream, and the degree of water and air contact. For simplification purposes, a set of values has been tabulated by the Engineering Board of Review for Streeter and Phelps (1925b) and can be used based on a qualitative description of the stream. These values have been summarized by Metcalf

Table 7.2 Table of reaeration constants.

Water body	Ranges of k_2' at 20 °C (base 10)	Ranges of k_2' at 20 °C (base e, for calculations)
Small ponds and backwaters	0.05–0.10	0.12–0.23
Sluggish streams and large lakes	0.10–0.15	0.23–0.35
Large streams of low velocity	0.15–0.20	0.35–0.46
Large streams of normal velocity	0.20–0.30	0.46–0.69
Swift streams	0.30–0.50	0.69–1.15
Rapids and waterfalls	>0.50	>1.15

Source: Engineering Board of Review for Streeter and Phelps (1925b), summarized by Metcalf and Eddy (1972).

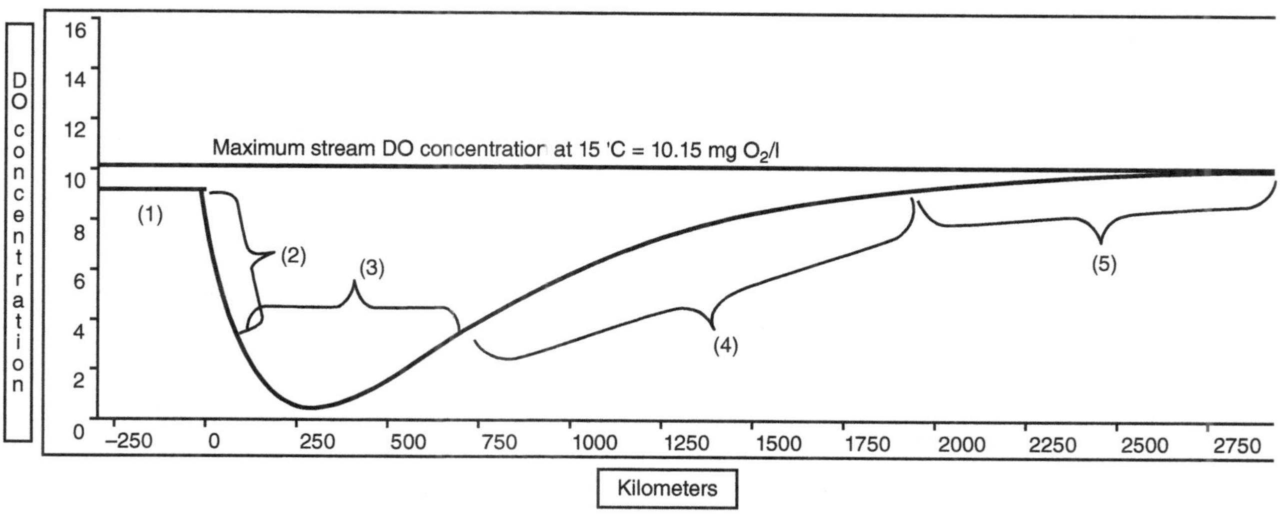

Figure 7.7 A DO sag curve labeled with its various zones of pollution and recovery. Source: Fate®, Anders.

and Eddy (1972) and are given in Table 7.2. Note that the actual k_r values used in Eq. (7.4) are the log to the base e (natural log).

The k_d term in this equation and in Eq. (7.4) is the DO uptake constant, a measure of the rate of the chemical reaction by which microorganisms use DO to degrade organic matter. This is usually obtained from a 20-day BOD experiment, noting limitations above. The data are transformed by what is known as the Thomas slope method, and the slope of the line is equal to the rate constant, k_d, in a day. Such a dataset was shown in Figure 7.2. Decades of data have produced very consistent values for k_d for a variety of wastes, especially domestic sewage.

The average water velocity is represented by "ν." This value is easily measured and is usually given in the problem statement. As noted above, the initial oxygen deficit (D_0) is calculated by subtracting the initial DO content of the stream–waste mixture from the DO level in the stream immediately upstream from the waste input. The plotted value in Fate® is a result of subtracting the oxygen deficit (D) at each point, calculated using Eq. (7.4), from the stream DO concentration above the waste input ($x < 0$). The net result is the remaining DO concentration in the stream.

The DO curve for a BOD-contaminated river can be divided into several zones based on DO content and physical appearance of the stream. These are shown in Figure 7.7. The area of the stream above the entry of sewage is referred to as the Zone of Clean Water (Zone 1 in Figure 7.7) and has the physical appearance of any natural clean system (stable fish, macroinvertebrate, and plankton populations). At the point of sewage entry, microbial consumption of the waste and oxygen begins. This marks the beginning of the Zone of Degradation (Zone 2 in Figure 7.7) that continues down the stream until the oxygen level falls to 40% of the initial value (as compared to the Zone of Clean Water). The Zone of Degradation is also characterized by the presence of more turbid water, an increase in CO_2 levels, and

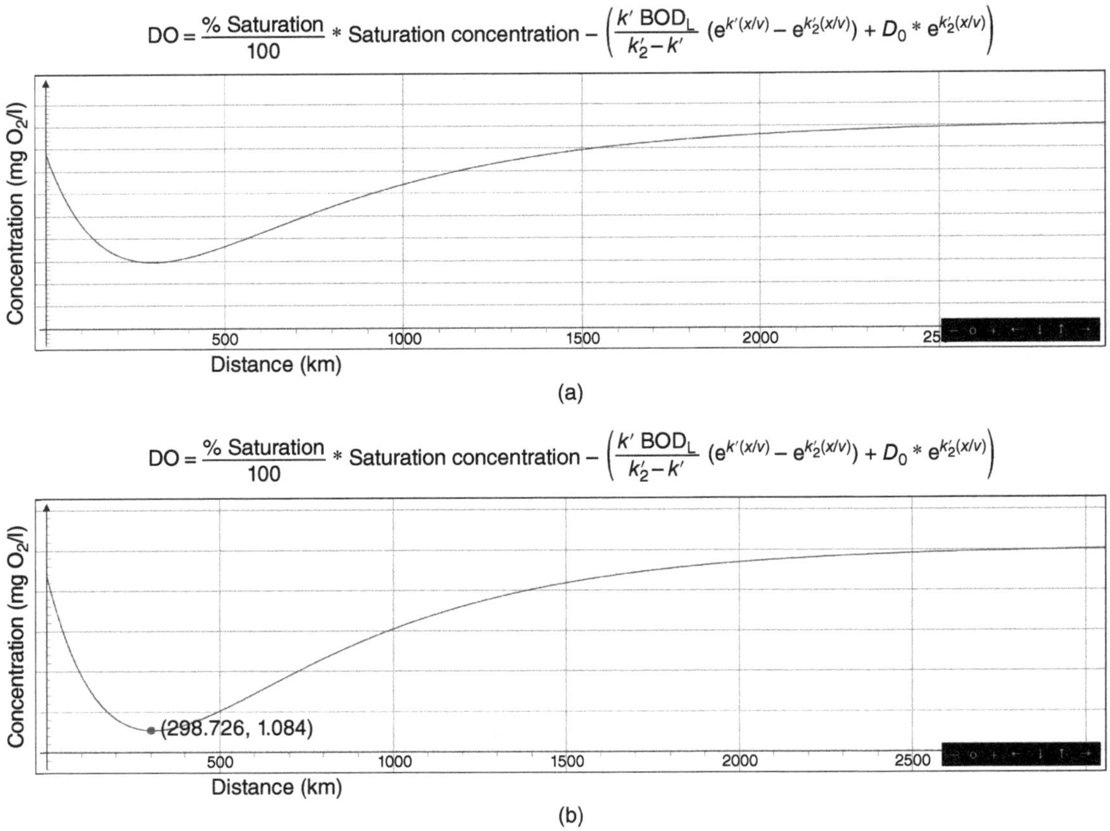

Figure 7.8 (a) A DO sag curve for the example problem. Output from Fate® for the input parameters given in the example problem. (b) DO sag curve for a second input of raw sewage. Source: Anders.

the presence of organically bound nitrogen. Physical characteristics of this zone include a high presence of green and blue-green algae (cyanobacteria), fungi, protozoa, tubifex, and blood worms. Larger aquatic plants die off in this region. Zone 3 in Figure 7.7 begins when the oxygen level falls below 40% of the initial DO concentration and ends where it rises above 40% again. This is the Zone of Active Decomposition. Water in this zone tends to be gray or black and releases gases typical of anoxic environments (H_2S, CH_4, and NH_3). Bacteria and algae are usually the dominant life forms present. As the DO level rises above 40% of the initial value, the Zone of Recovery begins (Zone 4 in Figure 7.7). Carbon dioxide levels decrease, but nitrogen is still present as NH_3 and in organically bound forms. Biological characteristics include a decrease in the number of bacteria and an increase in the presence of protozoa, green and blue-green algae, and tubifex and blood worms. Finally, the Zone of Cleaner Water (Zone 5 in Figure 7.7) is achieved where the chemical and biological characteristics of the stream are similar to those of Zone 1 (Zone of Clean Water).

Multiple sewage inputs:

When there are multiple input point sources with distance downstream, the analyst is required to apply the same equations sequentially at the entry of each point source. Fortunately, the software provided in Fate facilitates this process and the analysts just need to carefully align instream concentrations at each point source location. A simple example is shown here:

Using the default parameters in Fate (the example problem), we see that the time to the critical point is 2.56 days located at 296 km from the point source (Figure7.8a). The BOD level at this point is 15.1 mg/l. But what if a second untreated sewage input occurs here with all other parameters equal to those in the first example problem. The major input parameter that changes is the BOD of the stream (to 15.1 mg/l with an identical waste input flow rate). Now the critical point, critical distance, and BOD at the new point downstream from the second input are: critical point equals 2.64 days, critical distance equals 306 km from second input, and BOD equals 22.2 mg/l (Figure 7.8b). Additional inputs continue to decrease the DO and quality of the stream. Eventually, the DO will approach 0 mg/l as more raw sewage is input to the system.

7.5 Sensitivity Analysis

One utility of the S–P equation, especially with its expanded form, is the ability to compute unit deficits of the source/sink terms – mg/l/d for reaeration, CBOD (chemical biochemical oxygen demand) decay, SOD, etc. to determine which process is having the greater impact on DO at the sag point. The results may dictate collecting further sampling or even switching to a more appropriate model that is more complex (i.e. WASP or Qual-2E versus S–P). In some cases, SOD is the major source of the deficit and reducing wastewater BOD will not necessarily improve overlying DO. This "back of the envelope" type modeling approximation is highly useful in developing credible models in practice. Successful modelers like to use simple models to gain insights into the processes relevant in their current model development. Others seem to know a priori this reality. This is also known as experience!

One form of sensitivity analysis is performed by perturbing (perturbation analysis) one parameter (±5% or 10%) while holding the others constant to determine its potential impact on a result. Those parameters that are deemed sensitive may warrant attention to laboratory/field sampling during field monitoring or may determine model discretization as well as how these processes are even modeled.

Another type of sensitivity analysis is based on the first derivative of the model outcome – say DO in a Streeter–Phelps model. Recall that the first derivative of any function is the rate of change of the function with respect to the variable selected. In our example, the derivative of the DO with respect to its degradation rate k may point to its relative sensitivity in the model.

Each modeling chapter has discussed the need to question the model and input parameters. Model parameters for a sensitivity analysis of the Streeter–Phelps equation include the stream flow rate (Q), the biological oxygen consumption rate (k_d), and the stream reaeration rate (k_r). This will be the focus of a homework problem.

7.6 Limitations of Our Basic Model and More Elaborate Models

7.6.1 Average Reaeration Rates for Streams

The basic Streeter–Phelps equation used in this chapter only allows for an average reaeration constant to be used. This is appropriate for many large streams, but smaller streams alternate between riffle areas (small rapids) and calm pools. In order to account for these varying reaeration areas, one would need to use a numerical methods approach. However, by using an average reaeration constant, adequate estimates of the DO curve for most streams can be obtained.

There are several other approaches (Table 7.1) for estimating reaeration rate. One promoted by the International Hydrological Program is

$$k_\mathrm{r} = 2.148\, v^{0.878} H^{-1.48} \tag{7.14}$$

where

k_r is the reaeration rate constant,
v is the average flow velocity, in m/s, and
H is the average depth of flow in the river, in m.

Obviously from the exponents of the equation, this is not an equation derived from fundamental physics, but a model fit to field data.

7.7 Remediation

The first step in remediating a system that has been contaminated with biodegradable organic matter is to remove (or limit) the source of the pollution. We have been successful in doing this in developed countries by installing relatively simple sewage treatment plants, even in highly populated areas where our waste generation is excessive. But in order to completely restore a system to pristine conditions, additional treatment is usually needed. The typical sewage treatment plant has sufficient technology to remove essentially all the degradable organic matter and most of the nitrogen (ammonium, nitrite, and nitrate). However, these two nutrients are usually not the limiting nutrient in aquatic systems. A limiting nutrient is simply the microbial nutrient that is in limiting supply (limits the growth) in a given system. For freshwater aquatic systems, this is usually phosphorus in the form of phosphate. Therefore, by adding phosphate, we increase the growth potential, and DO consumption, of microbes in the system. Most sewage treatment systems today do not remove phosphate, although Europe is making considerable headway at doing this. In the next 5–10 years, the US EPA is expected to impose phosphate removal requirements on sewage treatment plants. Although it is amazing the EPA has no federal phosphate emission limit, a few states have created them. Phosphate removal is relatively simple and low cost, as iron phosphate can be precipitated in the settling basins of treatment plants by adding iron chlorine to the influent waters to the plants. Iron phosphate is highly insoluble and is incorporated in the waste sludge. This is actually an advantage when the sludge is used as a soil augmentation (fertilizer). Sewage treatment plants in locations prone to eutrophication often have taken such steps.

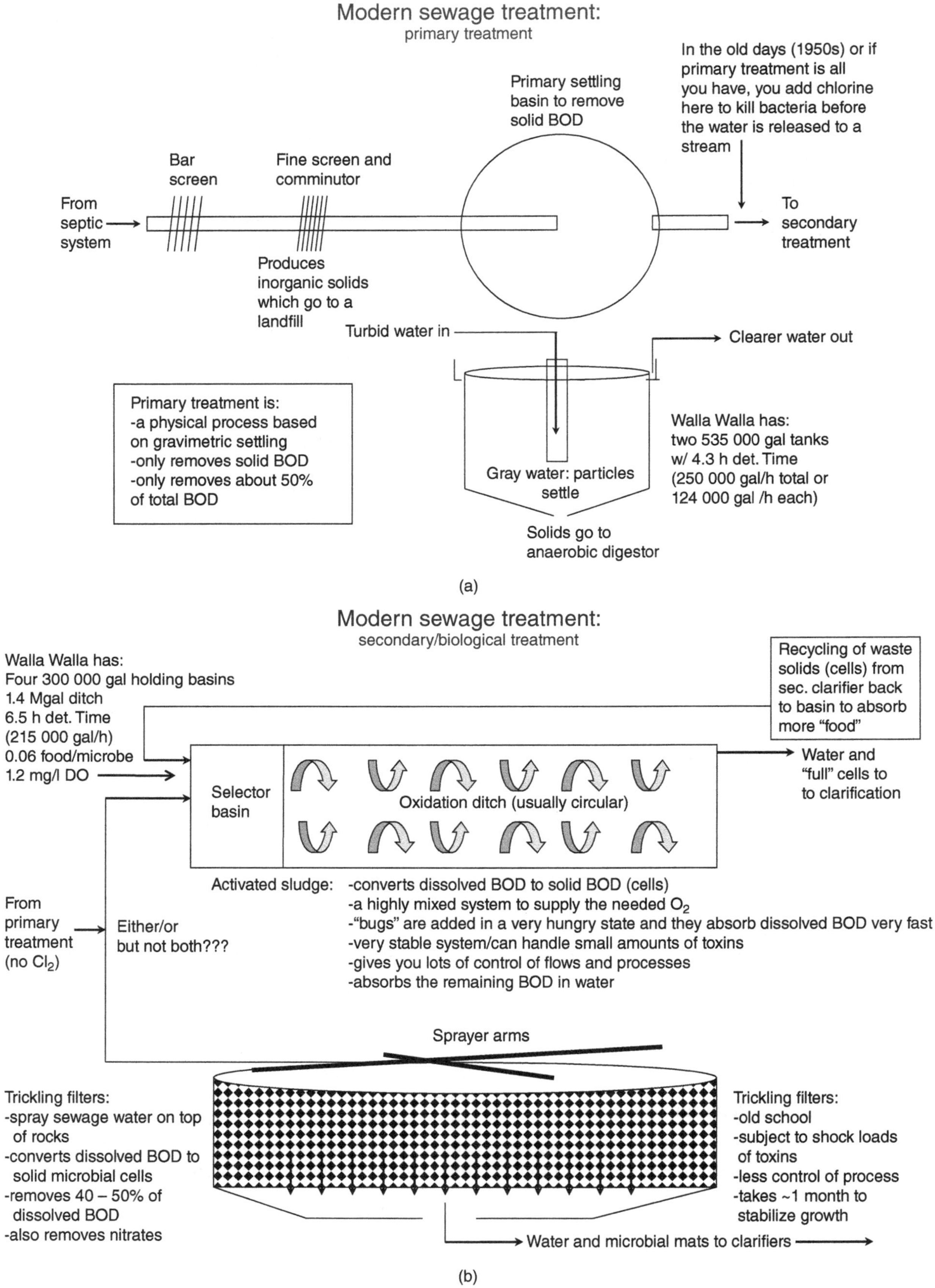

Figure 7.9 (a–c). A diagram of the City of Walla Walla (Washington, USA) wastewater treatment plant.

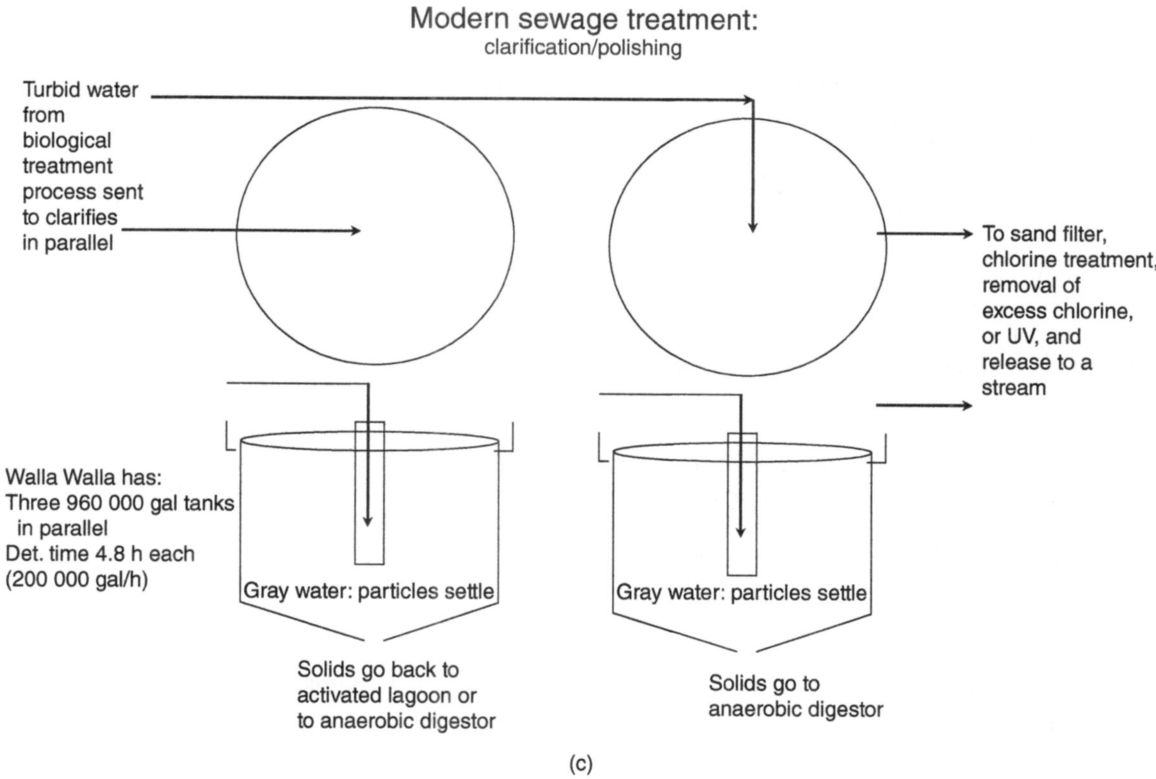

Figure 7.9 (*Continued*)

A schematic of a topical domestic wastewater treatment plant is shown in Figure 7.9a–c. This is an expansion of Figure 7.5, with added dimensions, flow rates, and detention times for each component of the plant. The governing formula for designing such a time is based on the volumetric wastewater flow, the volume of each unit (tank), and the detention time needed for adequate treatment, expressed as

$$V = Q\,t_r \tag{7.15}$$

where

V is the volume of each unit (tank) needed (in m^3),
Q is the volumetric flow rate of wastewater (in m^3/d), and
t_r is necessary retention time of each unit process.

Once the source of the waste has been removed, we simply must wait for the system to recover. In most cases, because of the enormous volumes of water that would need to be treated for most contaminated aquatic system, we cannot directly treat the water in the lake or river. In theory, small lakes could be aerated using mechanical aerators or diffusers, but this can be expensive over extended time periods. Streams can recover very rapidly, since they naturally reaerate and flush the nutrients out of the system. But this can create a even greater problem for the downstream receiving system, usually estuaries. Many estuaries are the spawning grounds for fishery industries, and the constant input of nutrients, especially limiting nutrients, can result in eutrophication and destruction of the ecosystem. Such a case is the dead zone at the mouth of the Mississippi River in the United States. As more advanced sewage treatment facilities are installed and as we better handle nonpoint sources of organic waste, streams and estuaries will continue to recover.

7.8 One Last Note on Estuaries

An estuary is otherwise a river that is physically linked to an ocean or sea, and as such, experiences tidal activity depending on geographical location (Nedwell and Raffaelli 1999). In both hemispheres, they are normally enclosed bodies of water along coastlines where fresh water and salt water meet and mix and serve as a transition zone between oceans and continents. They are known by different names, such as bays (Chesapeake Bay), sounds (Pamlico Sound), inlets (Lake Washington in Puget Sound), harbors (Boston and New York Harbor) and sloughs (South in the Coos in Oregon). They can be small or large and have a huge impact on a region's economy (as in the case of Chesapeake Bay) or much less in many others. In this transition zone, they serve to dilute the content of the incoming tidal ocean waters.

These estuarine waters are extremely productive and generate more organic material than equally sized grassland, forest or agricultural areas, and unlike many rivers in this regard (EPA.gov/NEP). Unique communities of animals and plants are developed and supported by the sheltered tidal waters of estuaries. Because of their biological productivity, estuaries provide an ideal location for migratory birds to rest and feed during their fall and spring journeys. Wildlife and fish species rely on the shelter of estuaries as protected spawning areas. Human physical and social infrastructure can also benefit from the presence of estuaries in the case of sheltered ports and harbors that enable improved shipping and navigation. Look at the ports of the original 13 colonies and trace their evolution from the seventeenth century through today to illustrate this fact about our history and estuarine development (Penn 2001; Johnston 2004).

Many fluvial sediments as well as pollutants coming from upstream rivers and feeder streams flow through estuaries where they are potentially filtered out, creating cleaner water and benefiting both human and marine life. So, we see the soils and plant life found in estuaries also form a natural buffer zone between sea and land, and help to absorb the impact of storm surges and floodwaters (Nedwell and Raffaelli 1999).

The combination of inflowing waters also enable an estuary to contain a high level of nitrogen and phosphorous nutrients in both the sediment and the water column, and create a unique and highly supportive environment for a wide variety of plant and animal life to thrive. The US EPA has established the National Estuary Program (http://EPA.gov/NEP) to aid the public in the Agency's programs and in providing information sources. From a modeling prospective, estuaries have been the laboratories for advances in state of the art. We will allude to some of these in Chapter 10 during the discussion of the WASP model development in the Chesapeake Bay Program. Simulation of nutrient kinetics is but one advance worthy of noting.

Suggested Reading for Discussion

Chapters 1 and 2 of *Environmental Success Stories: Solving Major Ecological Problems and Confronting Climate Change*. Frank M. Dunnivant. Columbia University Press. March 2017.

Concepts

7.1 Explain the difference between DO and BOD. What are the C and N forms of BOD? Why do they matter practically?

7.2 Draw a basic BOD plot (DO consumed versus time).

7.3 What is the difference between BOD_5, BOD_{20}, and BOD_L?

7.4 Name three sources of BOD to a stream.

7.5 There are two main reasons for installing sewage treatment plants, one human and one ecological. Name these.

7.6 Draw a basic sewage treatment plant, label each treatment unit, and explain what it does.

7.7 Draw the basic shape of the Streeter–Phelps (DO sag) curve and label the defined zones.

7.8 Describe each zone of the DO sag curve.

7.9 Perform a sensitivity analysis on v, k_2', and k' using the background example in Fate.

7.10 What is the basic design equation for a wastewater treatment plant?

Exercises

7.1 A cattle stockyard borders a small creek. Due to its close proximity to the creek, there is considerable seasonal runoff. With the given stream parameters below, and the rate constants for the BOD of the runoff, manually create a Streeter–Phelps graph of the change in DO concentration versus distance (BOD_5 = 2600 mg/l). Next, create a plot of the same situation but with treated runoff (BOD_5 = 65 mg/l). Assume that the treatment plant is 98% efficient. Use Fate to check your answers.
k_d: 0.52 d^{-1}
k_r: 0.71 d^{-1}
Temperature of the stream: 12 °C
Temperature of the waste: 18 °C
Flow rate in (stream): 500 m^3/min
Flow rate out (waste): 83.6 m^3/min
Stream velocity: 3.96 km/h.

7.2 A flood event forces a sewage treatment plant out of commission. Thus, all the domestic sewage flows directly into the large stream running adjacent to town. Create a plot showing the concentration of DO in mg/l versus distance from the plant using the following parameters:

BOD rate constant (k_d): 0.34 d^{-1}
Reaeration rate constant (k_r): 0.65 d^{-1}
Stream temperature: 17 °C
% saturation: 100%
Waste temperature: 22 °C
Stream flow: 1.2 mgd
Waste flow: 0.19 mgd
Initial waste DO: 0 mg/l
BOD_5 of stream: 1 mg/l
BOD_5 of waste: 180 mg O_2/l
Stream velocity: 1.9 km/h
k_r temperature coefficient: 1.024
k_d temperature coefficient: 1.135.
Now, for your sensitivity analysis, plot the previous graph assuming the plant removes 98% of the initial BOD_5. Use Fate to check your answers.

7.3 A vegetable-processing plant is assessing the risk of a pond breach, for a waste holding pond that is adjacent to a river. They hire you to determine the concentration of DO (in mg/l) in the stream if a breach were to occur. They wish to know the effects on DO given two cases: (i) no BOD removal performed within the plant and (ii) removal of 98% of the BOD. The effluent from the pond has a BOD of 50 mg/l and a flow of 175 m^3/s. The stream has a BOD of 1 mg/l, 100% DO saturation, a flow of 300 m^3/s and a velocity of 3 m/s. The initial concentration DO of waste is zero. The temperature of the stream is 15 °C and the temperature of the waste is 20 °C. The k' temperature coefficient is 1.135; the k_r temperature coefficient is 1.024. Use $k_d = 0.12\ d^{-1}$ and $k_r = 0.25\ d^{-1}$ in order to calculate the DO at 500 km downstream from the spill and 3000 km downstream. Use Fate to plot the entire graph of distance (km) versus concentration O_2 (mg/l).

7.4 Dairy product-processing plants produce waste that is extremely rich in organic matter. Consider a cheese-processing center in the Midwest that releases its waste into a large stream without treatment. The cheese waste is released at 21 °C, has a BOD_5 of 12 000 mg/l, and a flow of 480 000 gallons per day. There is no DO in the waste released from the plant. The large stream that the waste is released into is at a temperature of 15 °C, has a flow of 90 mgd, a velocity of 10 km/h, a BOD_5 of 1 mg/l, and a DO of 9.0 mg/l. At 21 °C, k_d is 0.4 d^{-1} and k_r is 0.75 d^{-1}. (The temperature coefficients for k' and k_2' are 1.135 and 1.024, respectively.) Plot the DO curve in kilometer from point source. Does the system become anaerobic? Where?

Now, consider that the cheese plant installs a new cleaning system that eliminates 95% of the BOD from its waste. Make a new plot showing the DO curve after treatment.
Use Fate to assist you in making the plots and to check your work. The data from this problem is organized in the table below.

	Stream	Waste
Temperature	15 °C	21 °C
BOD_5	1.0 mg/l	12 000 mg/l
Flow	90 mgd	480 000 gal/day
DO	9.0 mg/l	0 mg/l
Velocity	10 km/h	—

	Value (d^{-1})	Temperature coefficients
k_d	0.4	1.135
k_r	0.75	1.024

7.5 Until the late 1900s, domestic sewage treatment was basically nonexistent. Presently, wastewater treatment in industrialized countries may be at its height; almost nothing gets through the rigorous treatment, except when accidents happen. Yet accidents do occur: on one river with a flow of 216 ft^3/s and a velocity of 2.2 km/h, the wastewater treatment facility fails. Normally, the treatment plant has a 98% removal rate of BOD, but on this day everything passes through without treatment. The plant treats, on average, 25 million gallons of water per day. The river has a reaeration constant of 0.47 d^{-1} and a BOD rate constant of 0.103 d^{-1}. The water coming into the plant has a temperature of 16 °C and the treated water leaving the plant is 20 °C. The temperature coefficients for the stream and the waste are 1.135 and 1.024, respectively. The BOD_5 of the stream is 1 mg/l and the BOD_5 of the treated waste is 4 mg/l. The stream has a DO saturation of 90%. Using the Streeter–Phelps equation, compare the DO of the stream with and without treatment at 100, 400, 600, 1000, 1500, and 2500 km downstream. Use Fate to check your answers.

7.6 A city of 250 000 people in a developing country traditionally releases its sewage directly into a river. Each resident produces an average of 4.0 l of sewage per day. The local government is considering two options to deal with the sewage problem:

(a) Promote home recycling technologies such as reusing water from showers, baths, and clothes washing, which reduces the sewage produced by the city by 50%.

(b) Establish a biomass nutrient recycling technology, which has a 70% BOD removal for all waste.

The budget for each project is about the same. What is the best option? To help you make a decision, use Fate to plot distance versus DO concentration in the stream for each situation. Use the following parameters:

BOD rate constant (k_d): 0.4 d^{-1}
Reaeration rate constant (k_r): 0.69 d^{-1}
Sewage temperature: 22 °C
Stream temperature: 15 °C
Stream % saturation: 100%
Stream flow: 4 million l/d
Stream velocity: 5 km/h
Initial waste DO: 0 mg/l
BOD_5 stream: 2 mg/l
BOD_5 waste: 200 mg/l
k_r temperature coefficient: 1.024
k_d temperature coefficient: 1.135.

7.7 An earthquake causes the rupture of a town's secondary waste water pipe and releases all the domestic waste into a small creek that runs near the perimeter of town. Given the state of disaster across the region, the pipe is not fixed for 30 days. Use the following parameters and Fate to create a plot that shows the concentration of DO in mg O_2/l in the stream versus distance downstream from the location of the pipe rupture.

BOD rate constant (k_d): 0.48 d^{-1}
Reaeration rate constant (k_r): 0.69 d^{-1}
Stream temperature: 12 °C
Waste temperature: 24 °C
Stream flow: 347 000 gal/d
Waste flow: 103 700 gal/d
Initial waste DO: 0 mg O_2/l
BOD_5 of stream: 5 mg/l
BOD_5 of waste: 123 mg/O_2/l
Stream velocity: 5.36 km/h
k_d temperature coefficient: 1.135
k_r temperature coefficient: 1.024.

7.8 Due to a technical glitch, a waste treatment center's ability to reduce the BOD levels of waste decreases from 90% efficiency to 65% efficiency. You have been hired to monitor the DO levels of the river that the plant empties into. The BOD_5 level of the sewage being treated by the plant is 628 mg O_2/l. The river has a flow of 1456 mgd and a temperature of 17 °C with a velocity of 4.2 million/h. The plant's wastewater has a flow of 32.2 mgd and a temperature of 22 °C. The BOD_5 of the stream is measured at 2.3 mg/l and has a BOD rate constant of 0.23 d^{-1} with a reaeration rate constant of 0.38 d^{-1}. The DO saturation of the stream is 100%, and the temperature coefficients are $k_1 = 1.135$ and $k_2 = 1.024$. Use Fate to create two graphs of the DO level in the stream versus distance from the plant: one for each percent efficiency.

7.9 A large metropolitan area slightly east of Cahora Bassa on the Zambezi River in Mozambique does not possess a sewage treatment facility. The city's untreated sewer system creates 255 m^3/s of sewage with a BOD_5 of 215 mg/l. Calculate the DO at a major city 200 km down river and when the river enters the ocean 650 km downstream from the initial point source. Calculate the DO at the same points if a treatment facility is constructed that removes 90% of the organic material.

k_d: 0.45 d^{-1}
k_r: 0.18 d^{-1}
Temperature of the stream: 20 °C
% DO saturation: 100
Temperature of the waste: 18 °C
Flow rate in (stream): 3000 m^3/s
Flow rate out (waste): 235 m^3/s
Initial waste DO: 0 mg/l
BOD_5 of stream: 2 mg/l
Stream velocity: 2.46 km/h
k_r temperature coefficient: 1.024
k_d temperature coefficient: 1.135
Stream velocity: 1.23 km/h
Use Fate to check your answers.

7.10 A river has a temperature of 25 °C, an average velocity of 4 km/h, a 75% oxygen saturation, and a BOD_5 of 1.5 mg/l. A wastewater treatment facility is installed on the river that reduces the BOD_5 of the sewage to 10% of its untreated value. The untreated waste has a BOD_5 of 71 mg/l. In both scenarios, the waste entering the river has a velocity of 2.5 km/h, a temperature of 28 °C, and a DO level of 0 mg/l. Using the following parameters, calculate the river's concentration of DO 550 km downstream from the waste treatment center (i) when the facility is working and (ii) when the facility is not working.

k_d: 0.35 d^{-1}
k_r: 0.65 d^{-1}
k_1' temperature coefficient: 1.024

k_2' temperature coefficient: 1.135
Use Fate to check your answers.

7.11 A fictional paper mill on the Snake River treats and releases their waste directly into the river. The river flows at a rate of 152 m^3/s, has a velocity of 9.2 km/h, and has a temperature of 17 °C. The BOD_5 of the river is 1 mg/l. The flow rate of waste into the river is 35 m^3/s, and the waste has a temperature of 25 °C. The BOD_5 of the untreated waste is 6000 mg/l and is reduced to 25 mg/l after treatment. The BOD rate constant k_d is 0.46 d^{-1} and the reaeration rate constant k_r is 0.83 d^{-1}. The k_d temperature coefficient is 1.024 and the k_r temperature coefficient is 1.135. One day, the waste from the mill is accidentally released into a river without being treated. Calculate the DO in the river 3 km downstream from the plant both before and after the accidental release. Use Fate to check your answers.

7.12 A water treatment plant processes roughly 1.5 mgd of wastewater and releases it into the neighboring river. Usually the plant removes 95% of the BOD from the stream. However, the trickling filter is broken resulting in only 70% removal of BOD. The effluent from the plant is discharged into a river which has a flow rate of 100 ft^3/s and has a velocity of 2 mph. The temperature of the stream is 19 °C, and the temperature of the treated wastewater is 22 °C. The 20 °C BOD_5 of the sewage and of the stream is 155 and 1.5 mg/l, respectively. The wastewater contains no DO and upstream from the plant, and the river is 95% saturated with oxygen. At 20 °C, k_d is 0.15 d^{-1} and k_r is 0.35 d^{-1}.
Use the information above to answer the following questions:
Plot both DO sag curves of the plant operating at 95% efficiency and at 70% efficiency.
Calculate the critical oxygen deficit and its location for both removal efficiencies.
A particular fish needs water with DO concentrations of at least 6.7 mg/l and is generally found 500 km downstream – will they be able to remain there?
Use Fate to check your calculations.

7.13 A sewage treatment plant empties its treated waste into a nearby river. Suppose the average domestic sewage has a BOD_5 of 300 mg/l and 85% of this is removed by the treatment plant. The temperature of the river is 19.0 °C, and the river has a flow rate of 13.2 m^3/s and a velocity of 0.069 m/s. The sewage plant has a flow rate of 6.96 m^3/s and a temperature of 21 °C. Since the river is large and slow moving, let k_r = 0.40 and k_d = 0.34. At what distance downstream from the treatment plant, does the concentration of DO return to 5 mg/l, the minimum level acceptable for aquatic life? Use Fate to check your answers.

7.14 A town sits on a stream that feeds into a larger river 42 km downstream. The BOD rate constant for the town's waste is 0.31 d^{-1}, the reaeration rate is 0.67 d^{-1}, and the temperature of the stream is 18 °C. Waste enters the stream at a temperature of 23 °C. The flow rates of the stream and the sewage are 623 and 95 million liters/d, respectively. Assuming the stream, which flows at 3.2 km/h, already has 1 mg/l BOD, and the waste is very potent (200 mg/l BOD) with all other values as given in Fate, what is the DO sag curve of the water entering the large river if the town has no waste treatment plant? What if the town installs a plant that removes 95% of the BOD? Use Fate to create graphs of both scenarios.

7.15 The flooding of a creek that neighbors a waste treatment center causes the release of untreated waste into the stream. The creek's flooded channel has a depth of 6.90 m, a width of 137 m, and a flow rate of 402 m^3/s. The temperature of the stream is 20 °C when the release occurs, while the temperature of the sewage is 26 °C. For this scenario, the BOD rate constant for oxidation is 0.34 $days^{-1}$ and the reaeration constant is 0.80 $days^{-1}$. The DO concentration in the waste is 0 mg/l and the BOD_5 of the waste is 100 mg/l. The BOD_5 of the creek is 3 mg/l and the stream is 95% saturated with oxygen. If the flow rate of the untreated waste is 94.8 m^3/s, what is the system's critical distance and what is the DO concentration at this value? Use Fate to check your answers.

7.16 Sewage from a wastewater lagoon runs into a nearby creek at a rate of 0.009 91 m/s. The slow-moving lagoon water has a temperature of 20 °C. The creek flows at a rate of 0.0959 km/h and already contains 1.00 mg/l of BOD. The creek temperature is 15 °C. The BOD rate constant is 0.340 d^{-1}, and the reaeration constant is 0.650 d^{-1}. If the sewage has a BOD concentration of 100.00 mg/l, how far from the lagoon will the lowest concentration of DO be found? What distance downstream from the treatment center has less than 5.00 mg/l DO, the minimum requirement to meet water quality standards? Use Fate to check your answers.

7.17 A wastewater treatment plant experiences an unexpected pipe leak and releases 15 ft^3/s of sewage into the nearby stream. The average flow of the stream is 3 ft^3/s. The temperature of the sewage is 25 °C, and the temperature of the stream is 22 °C. The BOD_5 of the sewage is 200 mg/l and that of the stream is 1 mg/l. The sewage does not contain any DO, but the stream is 100% saturated. The constant k_d is 0.34 d^{-1} and k_r is 0.65 d^{-1}. Determine the location of the critical DO concentration. Determine the location where there is 50% saturation in the stream. You may use Fate to help obtain your answers.

7.18 You discover that sewage has been leaking into a local stream and decide to calculate how this sewage leak will affect the DO in the stream. Your goal is to find out how far downstream the DO will be affected by the leaked sewage. The constants k_1' and k_2' are 0.050 and 0.200 d^{-1}, respectively. You also know that the average temperature of the stream is 20.0 °C, and the temperature of the waste is 25.0 °C. The flow rate of the stream is 700 m/s and the flow rate of the waste is 60.0 m/s. The stream's velocity is 3.00 km/h. You assume that the sewage is mixed uniformly throughout the stream. The tested BOD_5 for the sewage is 70.0 mg O_2/l. The stream was 100% saturated with oxygen. Use Fate to create a graph of the Streeter–Phelps/DO sag curve and determine how far downstream the concentration of DO in the stream returns to 90% of its original value.

7.19 A sewage plant has an unnoticed malfunction and begins releasing some untreated waste into a small nearby river each day. Using Fate and the following parameters, determine how far downstream, if ever, the river will become anaerobic.
BOD rate constant (k_d): 0.34 d^{-1}
Reaeration constant (k_r): 0.22 d^{-1}
Stream temperature: 15 °C
Waste temperature: 20 °C
DO saturation of river: 100%
k' temperature coefficient: 1.135
Stream flow rate: 500 m^3/s
Waste flow: 75 m^3/s
Waste DO = 0 mg/l
BOD_5 stream = 1 mg/l
BOD_5 waste = 180 mg/l
Stream velocity: 2.5 km/h
k_d temperature coefficient: 1.135
k_r temperature coefficient: 1.024.

7.20 With no sewage treatment facility, a large town is releasing 2.12 million gallons of sewage into a river every day. The sewage has an average temperature of 21 °C and contains no DO. The river is 75% saturated with oxygen upstream of the discharge. The river flows at a rate of 20.0 ft^3/s, has an average temperature of 16 °C, a velocity of 1.30 m/s, and a BOD_5 of 10.0 mg/l. The BOD_5 of the sewage is 175 mg/l. The rate constants and temperature coefficients for this relatively swift stream are $k_d = 0.35$ and $k_r = 0.75$.
Using Fate:
(a) Determine the critical oxygen deficit T_C and its location D_c.
(b) Determine how far downstream from the town the river becomes anaerobic.
(c) Plot the DO sag curve and determine the BOD_L.
(d) Determine the DO concentration of the river 1200 km from the point source.

7.21 A city's waste treatment center is incapacitated during a flood, and untreated wastewater flows into the nearby river at a rate of 140 l/min. The river flows at a rate of 550 l/min. The velocity of the river is 15 km/h. The temperature of the sewage is 20 °C, while the temperature of the river is 12 °C. The 20 °C BOD_5 of the wastewater is 200 mg/l, while that of the stream is 1.00 mg/l. The sewage contains no DO, while the river is 100% saturated with DO before coming in contact with the waste. At 20 °C, k_d is determined to be 0.30 and k_r is 0.45.
Using Fate, determine where, if ever, the river turns anaerobic.

Spreadsheet Exercise

Create a spreadsheet that performs the same calculations as Fate for one or more of the earlier problems. Construct your spreadsheet so that it is interactive (so that you can change numeric values for parameters and the plot automatically updates itself).

7.A Derivation of the Streeter–Phelps (DO Sag Curve) Equation (By John Brooksbank)

The derivation of the Streeter–Phelps equation requires the use of an integrator factor, presented in earlier derivations in Chapters 5 and 6. This will be presented first and then the actual equation derivation.

7.A.1 Integrating Factors

The technique of using an integrating factor is useful when dealing with differential equations that are first order, meaning that there are no second derivatives or higher, and linear, meaning that all terms involving the function, Y, or any of its derivatives, can be separated. The general form of a linear, first-order differential equation is:

$$Y'(t) + p(t)Y(t) = q(t) \tag{7.A.1}$$

where $p(t)$ and $q(t)$ are some functions with respect to t.

Now, we can create some function, $I(t)$, which we define as:

$$I(t) = e^{\int p(t)dt} \tag{7.A.2}$$

The derivative of $u(t)$, $I'(t)$, is thus:

$$I'(t) = p(t) * e^{\int p(t)dt} \tag{7.A.3}$$

If we multiply Eq. (7.A.1) by the integrating factor, $u(t)$, we obtain:

$$e^{\int p(t)dt} * Y'(T) + p(t) * e^{\int p(t)dt} * Y(t) = e^{\int p(t)dt} * q(t) \tag{7.A.4}$$

Notice that, by design, the $Y(t)$ term ends up containing $I'(t)$, allowing us to simplify to

$$I(t)Y'(t) + I'(t)Y(t) = I(t)q(t) \tag{7.A.5}$$

The left-hand side of Eq. (7.A.5) is the derivative of $I(t)Y(t)$, allowing us to take the integral:

$$\int I(t)Y'(t) + I'(t)Y(t)dt = \int I(t)q(t)dt \tag{7.A.6}$$

$$I(t)Y(t) = \int I(t)q(t)dt + C \tag{7.A.7}$$

$$Y(t) = \frac{\int I(t)q(t)dt + C}{I(t)} \tag{7.A.8}$$

And assuming that $\int I(t)q(t)dt$ can be solved, we have solved the original differential equation and have the value of $Y(t)$. Always remember to include the constant of integration ($+C$).

7.A.2 BOD Model (Streeter–Phelps Equation)

The Streeter–Phelps equation describes the DO deficit in a stream due to the consumption of oxygen by biological processes. The equation arises from two competing first-order processes: the consumption of oxygen, which is first order with respect to the concentration of BOD, and the dissolution of oxygen from the atmosphere into the water, which is first order with respect to the difference in DO concentration between the theoretical saturation limit and the actual DO level within the water. The overall process can thus be written as:

$$\frac{dC(t)}{dt} = \text{reaeration} - \text{consumption} \tag{7.A.9}$$

where C is the concentration of DO in the stream. The rate of consumption can be written as a general first-order process:

$$\text{rate of consumption} = -k_d[\text{BOD}] \tag{7.A.10}$$

where k_d is the first-order rate constant, and [BOD] is the concentration of BOD. Likewise, since the concentration of BOD is determined with a first-order process, and we know that the general solution of a first-order process with initial concentration C_0 and rate constant k is:

$$C(t) = C_0 e^{-kt} \tag{7.A.11}$$

we can write the concentration of BOD as

$$[\text{BOD}] = \text{BOD}_L e^{-k_d t} \tag{7.A.12}$$

where BOD_L is the initial concentration of BOD in the solution. By substituting Eq. (7.A.10) into Eq. (7.A.12), we can write the rate of consumption as:

$$\text{rate of consumption} = -k_d \text{BOD}_L e^{-k_d t} \tag{7.A.13}$$

The rate of reaeration can be written as:

$$\text{rate of reaeration} = k_r(C_s - C(t)) \tag{7.A.14}$$

where k_r is the first-order rate constant, C_s is the saturation concentration of oxygen for the given temperature, and C is the current oxygen concentration in the water.

By substituting Eqs. (7.A.13) and (7.A.14) into Eq. (7.A.9), we can write the overall governing equation as

$$\frac{dC}{dt} = -k_d \text{BOD}_L e^{-k_d t} + k_r(C_s - C) \tag{7.A.15}$$

We can simplify this by defining the oxygen deficit, D, as:

$$D(t) = C_s - C(t) \tag{7.A.16}$$

$$\frac{dD(t)}{dt} = 0 - \frac{dC(t)}{dt} \tag{7.A.17}$$

Equation (7.A.15) then becomes:

$$\frac{dD(t)}{dt} = k_d \text{BOD}_L e^{-k_d t} - k_r D(t) \tag{7.A.18}$$

We can collect all terms involving $D(t)$ to one side to put Eq. (7.A.18) into standard form:

$$\frac{dD(t)}{dt} + k_r D(t) = k_d \text{BOD}_L e^{-k_d t} \tag{7.A.19}$$

We then multiply by an integrating factor, described above. For a generic first-order linear differential equation of the form

$$\frac{dY(t)}{dt} + f(t)Y(t) = g(t) \tag{7.A.20}$$

the integrating factor takes the form of

$$u(t) = e^{\int f(t)dt} \tag{7.A.21}$$

For Eq. (7.A.19), the integrating factor is thus

$$u(t) = e^{\int k_r dt} \tag{7.A.22}$$

$$u(t) = e^{k_r t} \tag{7.A.23}$$

Equation (7.A.19) becomes:

$$e^{k_r t}\frac{dD(t)}{dt} + k_r e^{k_r t}D(t) = k_d BOD_L e^{-k_d t} * e^{k_r t} \tag{7.A.24}$$

$$e^{k_r t}\frac{dD(t)}{dt} + k_r e^{k_r t}D(t) = k_d BOD_L e^{(k_r - k_d)t} \tag{7.A.25}$$

We integrate both sides:

$$\int e^{k_r t}\frac{dD(t)}{dt} + k_r e^{k_r t}D(t) = \int k_d BOD_L e^{(k_r - k_d)t} \tag{7.A.26}$$

$$e^{k_r t}D(t) = \frac{k_d BOD_L}{(k_r - k_d)} e^{(k_r - k_d)t} + C \tag{7.A.27}$$

Now to determine C, we set our initial conditions:

$$D(0) = D_0 \tag{7.A.28}$$

where D_0 is the initial deficit of the stream. Given this, Eq. (7.A.19) becomes:

$$D_0 = \frac{k_d BOD_L}{(k_r - k_d)} + C \tag{7.A.29}$$

$$C = D_0 - \frac{k_d BOD_L}{(k_r - k_d)} \tag{7.A.30}$$

So, the final solution of the equation becomes:

$$e^{k_r t}D(t) = \frac{k_d BOD_L}{(k_r - k_d)} e^{(k_r - k_d)t} + D_0 - \frac{k_d BOD_L}{(k_r - k_d)} \tag{7.A.31}$$

which simplifies to:

$$D(t) = \frac{k_d BOD_L}{(k_r - k_d)} e^{-k_d t} + D_0 e^{-k_r t} - \frac{k_d BOD_L}{(k_r - k_d)} e^{-k_r t} \tag{7.A.32}$$

$$D(t) = \frac{k_d BOD_L}{(k_r - k_d)} (e^{-k_d t} - e^{-k_r t}) + D_0 e^{-k_r t} \tag{7.A.33}$$

Bibliography

AWWA (2012). https://www.ecfr.gov/cgi-bin/text-idx?SID=a6bb8a02b6d783f9356758b5ff0ed106&mc=true&node=pt40.25.136&rgn=div5#se40.25.136_13 (accessed 10 August 2018).

Baity, H.G. (1938). Some factors affecting the decomposition of sewage sludge deposits. *Sewage Works J.* **10**: 539–568.

Boyce, W.E. and DiPrima, R.C. (2009). *Elementary Differential Equations*, 9e. New York: Wiley.

Chapra, S. (1997). *Surface Water Quality Modeling*. Loveland, CO (USA): Loveland Press.

Craun, G. (1986). *Waterborne Diseases in the United States*. Boca Raton, FL: CRC Press.

Environmental Protection Agency. National service center for environmental publications. https://www.epa.gov/nscep (accessed 10 August 2018).

EPA (2018). https://www.epa.gov/sites/production/files/2015-10/documents/sediment_oxygen_demand507_af.r4.pdf (accessed 11 December 2018).

Fair, G. et al. (1968). *Water and Wastewater Engineering, Vol. 2, Water Purification and Wastewater Treatment and Disposal*. New York: Wiley.

Georgia Environmental Division (2018). Environmental software. http://epdsoftware.wileng.com (accessed 10 August 2018).

Gundelach, J. and Castillo, J. (1976). Natural stream purification under anaerobic conditions. *J. Water Pollut. Control Fed.* **47** (7): 1758–1765.

Johnston, R. (2004). *A Journey into an Estuary (Biomes of North America)*. Carol Rhoda Books, Inc.

Jolánkai, G. (1997). *Basic river water quality models, computer aided learning (CAL) programme on water quality modelling* (WQMCAL version 1.1), International Hydrological Programme, Technical Documents in Hydrology, No. 13

Manhattan College (1975). Manhattan College Summer Institute on Water Quality Modeling (organized by Professors Donald J. O'Connor & Robert V. Thomann).

Martone, C. (1974). On the Measurement of Ultimate BOD and its Kinetic Formulation for Pulp & Paper Effluents. MS. dissertation. Department of Civil Engineering, Tufts University, Medford, MA.

Meadows, D., Randers, J., and Meadows, D. (2004). *Limits to Growth: The 30-Year Update*. White River Junction, VT, USA: Chelsea Green Publishing Company.

Metcalf & Eddy, Inc. (1972). *Wastewater Engineering: Treatment, Disposal, and Reuse*, 3e. NY: McGraw-Hill, Inc.

Metcalf & Eddy, Inc. (2013). *Wastewater Engineering: Treatment Disposal and Use*, 5e. New York: McGraw Hill.

Mortimer, C.H. (1942). The exchange of dissolved substances between mud and water I and II. *J. Ecol.* **29**: 280–329.

Nauman, Hal. (1983), A comparison of the prevalence of infectious gastrointestinal disease and the quality of individual, surface source water system in Lincoln County, OR. MS thesis. Oregon State University.

NCASI (1978). Interfacial Velocity Effects on Sediment Oxygen Demand, National Council for Air and Stream Improvement, Inc. Technical Bulletin No. 317.

NCASI (1982). National Council for Air and Stream Improvement, Inc. A review of ultimate BOD estimation and its kinetic formulation for pulp and paper mill effluents. Technical Bulletin No. 382. http://www.ncasi.org/Members/Login.aspx?returnTo=/Downloads/Download.ashx?id=3545.

NCASI (1985). A Study of Uncertainty Analysis Techniques and Their Application to the Mathematical Water Quality Model SNSIME. Technical Bulletin No. 0463.

Nedwell, D. and Raffaelli, D. (1999). *Advances in Ecological Research*, vol. 29. London: Academic Press.

O, Connor, D.J. (1960). Oxygen balance of an estuary. *J. San. Engr. Div. ASCE* **86** (SA3): 35–55.

O, Connor, D.J. (1967). The temporal and spatial distribution of dissolved oxygen in streams. *Water Resour. Res.* **3** (1): 65–79.

Penn, J. (2001). *Rivers of the World: A Social, Geographic, and Environmental Sourcebook*. ABC-CLIO.

Sawyer, C.N. and McCarty, P.L. (1978). *Chemistry for Environmental Engineering*. New York: McGraw-Hill Book Company.

Shapiro, B.E. (2011). for integral-table.com. This work is licensed under a Creative Commons Attribution-NonCommercial-ShareAlike 3.0 Unported License. Revised with corrections 1 July 2016. http://integral-table.com/downloads/LaplaceTable.pdf (accessed 11 December 2018).

Snoeyink, V.L. and Jenkins, D. (1980). *Water Chemistry*. New York: Wiley.

Standard Methods (1998). *Standard Methods for the Examination of Water and Wastewater*, 20e. American Water Works Association.

Streeter, H.W. and Phelps, E.B. (1925a). A study of the pollution and natural purification of the Ohio river. United States Public Health Service, U.S. Department of Health, Education, and Welfare.

Streeter, H.W. and Phelps, E.B. (1925b). A study of the pollution and natural purification of the Ohio river. III. Factors concerned in the phenomena of oxidation and reaeration. Public Health Bulletin no. 146, Reprinted by U.S. Department of Health, Education and Welfare, Public Health Service, 1958, ISBN B001BP4GZI.

Tchobanoglous, G. and Burton, F.L. (1991) (in Metcalf and Eddy, Inc.)). *Wastewater Engineering: Treatment, Disposal, and Reuse*. New York: McGraw-Hill.

Tetra-Tech (1985). *Rates, Constants, and Kinetics Formulations in Surface Water Quality Models*, 2e. US Environmental Protection Agency, EPA, EPA/600/3-85/040.

Thomann, R.V. (1972). *Systems Analysis and Water Quality Management*, 110. New York: McGraw Hill.

Thomann, R.V. and Sobel, M. (1964). Estuarine water quality management and forecasting. *J. San. Engr. Div. ASCE* **90** (SA5): 9–36.

TWDB (Texas Water Development Board) (1970). Simulation of Water Quality in Streams and Canals, Program Documentation and Guidance for Its Use, PB-202 974.

Wikipedia (2017). https://en.wikipedia.org/wiki/Estuary (accessed 5 February 2018).

Wipple, G.C. and Wipple, M.C. (1911). Solubility of oxygen in sea water. *J. Am. Chem. Soc.* **3**: 362.

8

Fate and Transport Concepts for Groundwater Systems

Case Study: The Test Area North Deep Well Injection Site at the Idaho National Environmental and Engineering Laboratory (INEEL)

There are an endless number of polluted soils and aquifers that could be used as case studies for this chapter, but we will use the one described in Chapter 1 located at the Idaho National Environmental and Engineering Laboratory (INEEL), a Department of Energy (DOE) site in the State of Idaho (the United States). This US government owned and operated hazardous waste site was placed on the US EPA's National Priorities List (NPL) on 21 November 1989. The injection well (TSN-05) was drilled in 1953 to a depth of 93 m (305 ft) with a diameter of 30.5 cm (12 in.). The groundwater surface at the site is approximately 63 m (206 ft) below the land surface. During its years of operation, this well received approximately 133 000 l (35 000 gal; 193 000 kg) of liquid and dissolved trichloroethylene (TCE), organic sludges, treated sanitary sewage, metal filing process waters, and low-level radioactive waste streams. Basically, the well was the means of disposal for any liquid or semiliquid waste (domestic and hazardous) produced at the remote and isolated site. Although several pollutants of concern were disposed of and detected in the groundwater at Test Area North (TAN), our main concern here is the TCE groundwater plume.

Figure 1.4 showed isoconcentration circles (isopleths; lines of equal TCE concentration) for the area surrounding the disposal site that were obtained by fitting a step model to field measurements of TCE concentrations in the groundwater (explanatory modeling). Assuming, as expected, that TCE is present in the aquifer as a dense nonaqueous phase liquid (DNAPL), it solubilizes slowly, and the input of TCE can therefore be considered a continuous source (step input over an extended period of time). You will note that the isopleth near TSN-05 injection well represents a concentration of 1000 ppb, while the lowest concentration shown in this figure is 3 ppb (the maximum allowed drinking water concentration). If no remedial action was to be taken in 1994 (the proposed year of remediation if any was to be attempted) and the DNAPL continued to release TCE to the groundwater, another step model predicted that the TCE plume would expand as shown in Figure 1.5 by the year 2044.

In this chapter, we will learn the chemical and physical processes responsible for pollutant fate and transport in groundwater systems such as the one in Idaho. As we will see, special approaches have been developed specifically for groundwater in order to sample and study the sites. Polluted groundwater systems present special challenges to the environmental chemist and engineer and have accounted for a significant portion of the Superfund budget. Students can learn more about active Superfund sites at EPA national map showing sites and proximity to populated areas in Chapter 12 when Superfund (Comprehensive Environmental Response, Compensation, and Liability Act, CERCLA) is introduced. Not surprisingly, many are located close to populated cities and industrial complexes in the northeast and southeast.

8.1 Introduction

Groundwater is the most prominent reservoir of liquid freshwater, although glaciers constitute the most common reservoir of freshwater overall. Data given in Figure 2.1 showed that globally there are approximately 9.5×10^7 km^3 of groundwater, which accounts

Pollutant Fate and Transport in Environmental Multimedia, First Edition. Frank M. Dunnivant and Elliot Anders.
© 2019 John Wiley & Sons, Inc. Published 2019 by John Wiley & Sons, Inc.
Companion website: www.wiley.com/go/Dunnivant/pollutant_fate&Transp_EnvironMultimedia

for 98% of the liquid freshwater on the Earth. Yet groundwater is somewhat more difficult to access, with the exception of flow from natural springs, than is water in lakes and rivers. The residence time of water in the ground is also very long, from 200 to 1000 years, with some waters dating millions of years old. Thus, a contaminated aquifer can take much longer to recover by natural means than would relatively rapidly flowing surface waters. In addition, contaminated groundwater systems are much more difficult to access for treatment and remediation efforts than other aquatic systems for obvious access reasons.

Water in the ground can be grouped into two categories, based on whether it is present in saturated or unsaturated media. In general, unsaturated areas are those where the soil or fractured rock does not contain air spaces and freely drains water, whereas saturated soil or fractured rock contains air spaces and retains most or all of its water. Water flows in both saturated and unsaturated media. Geologic media can also be classified into two basic categories: consolidated (solid or fractured rock) and unconsolidated (sediments). Figure 8.1 shows a cross-section of the Earth's surficial crust. The source of all groundwater is from the land surface (the recharge area on the left-hand side of the figure), and it flows downward until it contacts an impermeable layer of soil/rock (the aquiclude in Figure 8.1). Water can enter the soil rapidly in major recharge areas, such as the one shown in Figure 8.1 or by slow percolation through the soil. Rivers and lakes also contribute water to the subsurface. Water-saturated layers of the subsurface may be separated by layers of permeable or impermeable soil or rock deposits. Saturated layers between two impermeable soil layers are referred to as confined aquifers. The uppermost unconfined layer of water is referred to as the surface aquifer and is the most likely portion of the subsurface to be contaminated, since it is located closest to the land surface.

There are many laws that regulate the purity or quality of groundwater. Drinking water in the United States is largely derived from groundwater aquifers rather than surface water locations. Accordingly, these resources were addressed in the United States, the major law is the Safe Drinking Water Act (SDWA), originally passed by Congress in 1974 and amended in 1986 and 1996. As noted in the title, it is primarily concerned with the use of water removed from the ground for drinking water purposes. There are other sections of the SDWA and other laws regulating the disposal of waste into the ground and the use of water for irrigation purposes. Many areas of the world use groundwater as a source of drinking water, and while much of the water is pristine, there are also many areas that are contaminated. One major natural form of contamination is arsenic, which is leached into the water from mineral deposits. Of course, there are many man-made pollutants present in groundwater in urban areas or near industrial complexes (discussed in the next Section 8.2). The SDWA establishes several "barriers" against the pollution of water. These include regulations to protect the water supplies and for treatment of drinking water, ways to maintain distribution systems, and guidelines for providing information to the public. Under the SDWA, the US Environmental Protection Agency sets limits on the amount of each pollutant or microorganism that can be present in drinking water. These are referred to as primary pollutants and have maximum contaminant limits (MCLs). For each pollutant, the concentration deemed safe for consumption is determined based on the risk assessment principles discussed in Chapter 11.

8.2 Input Sources

Sources of pollution to groundwater may include any liquid that can drain into the unsaturated or saturated zone (direct spills of liquid or pollutants washed into the subsurface by rain). Examples are shown in Figure 8.1b,c and include liquids from leaking surface or subsurface storage tanks, material storage (coal, road salt, and ore leachate), drainage and leachate from farming operations, contamination entering the aquifer from rivers and lakes, subsurface mining operations, pollutants disposed of in waste injection wells, saltwater intrusion, and leachate from domestic and hazardous waste landfills. Two groundwater pollutants that are in the news today and will be in the news for years to come are perchlorate (from solid rocket fuel) and methyl tertiary butyl ether (MTBE, an additive to gasoline). As noted in Section 8.1, groundwater systems are much more complicated than river or lake systems. In rivers and lakes, we have one distinct advantage in pollution treatment or remediation efforts: the flushing of the system and removal of the polluted water by rapid flow through the system. In comparison, flow rates in groundwater are very slow, and access to the system for treatment of the water or soil is highly limited.

Many sources of pollution produce one or more biodegradable pollutants that can result in the consumption of available dissolved oxygen. Thus, the same set of microbial communities and degradation schemes discussed in the lake chapter for anaerobic waters can, and often do, occur in groundwater systems. This results in a changing oxidation state (E_H) of the water. Near sources such as landfills or hazardous waste injection wells, organic waste is oxidized and the dissolved oxygen is consumed. This results in an anaerobic zone in the aquifer, which will proceed through the usual order of terminal electron acceptors (O_2, NO_3^-, SO_4^{2-}, and finally CO_2). As unpolluted and oxygenated water mixes with the polluted water, the aquifer will establish a facultative zone (an area containing only a small amount of oxygen). As more and more clean water mixes with the polluted water, dissolved oxygen levels will slowly return to normal. However, this sequence of mixing and terminal electron acceptors can occur over a great distance in the subsurface if a large amount of contamination is present, and the water in this region will be unfit for consumption. Thus, the waste entering the system must be characterized or a history of the waste inputs must be known in order to predict the chemical reactions occurring in the aquifer. In addition, as we illustrated in Chapter 3, the pH and E_H of the system (as controlled by microbial degradation of pollutants) can greatly affect sorption (K_d or K_p) and degradation reactions (k). As we will see in this chapter, each of these parameters can be the governing factor determining the fate and transport of pollutants in groundwater systems.

8.3 Monitoring Wells

Taking environmental samples to monitor the concentration of a pollutant in river, lake, and atmospheric

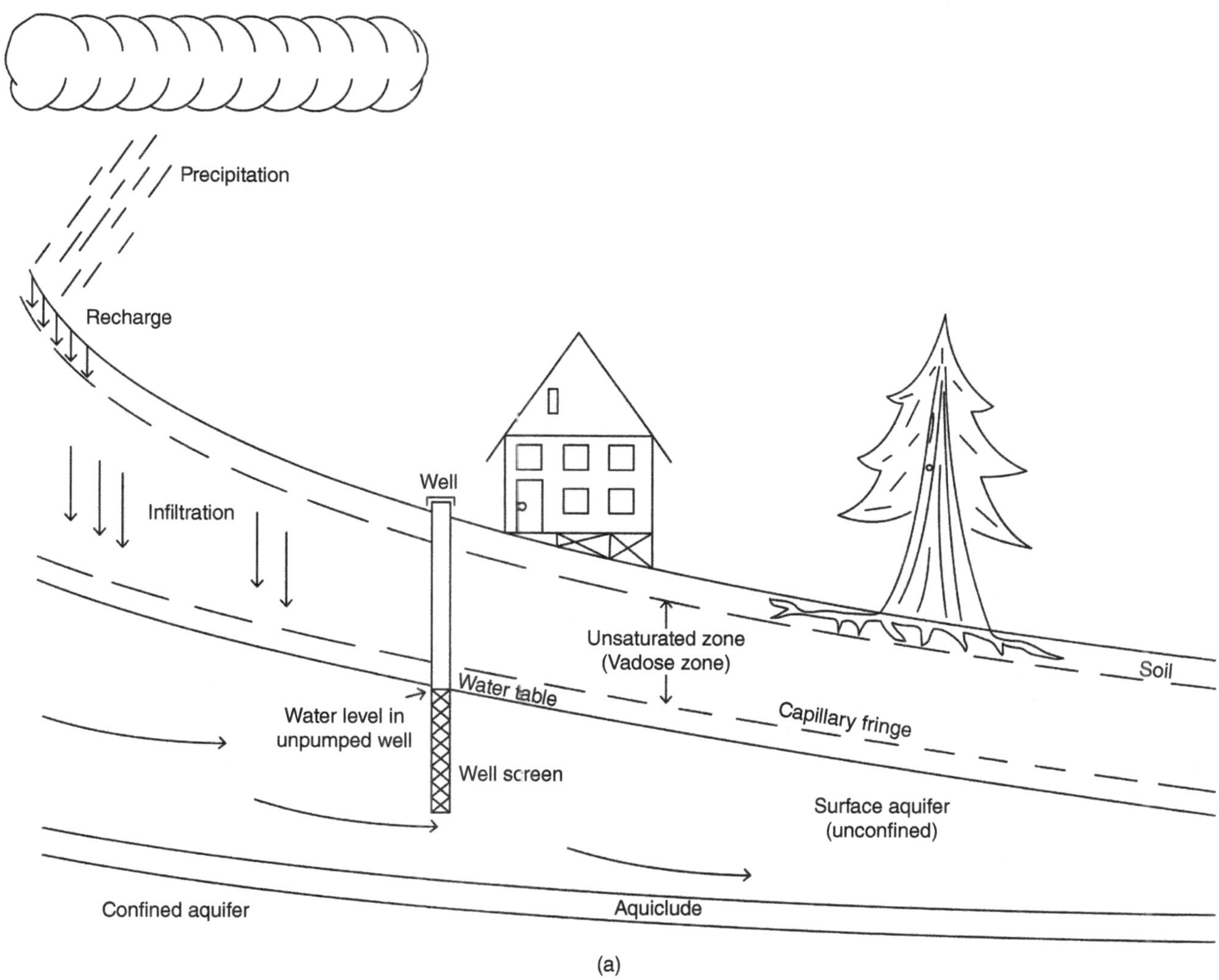

Figure 8.1 (a–c) Cross section of an aquifer.

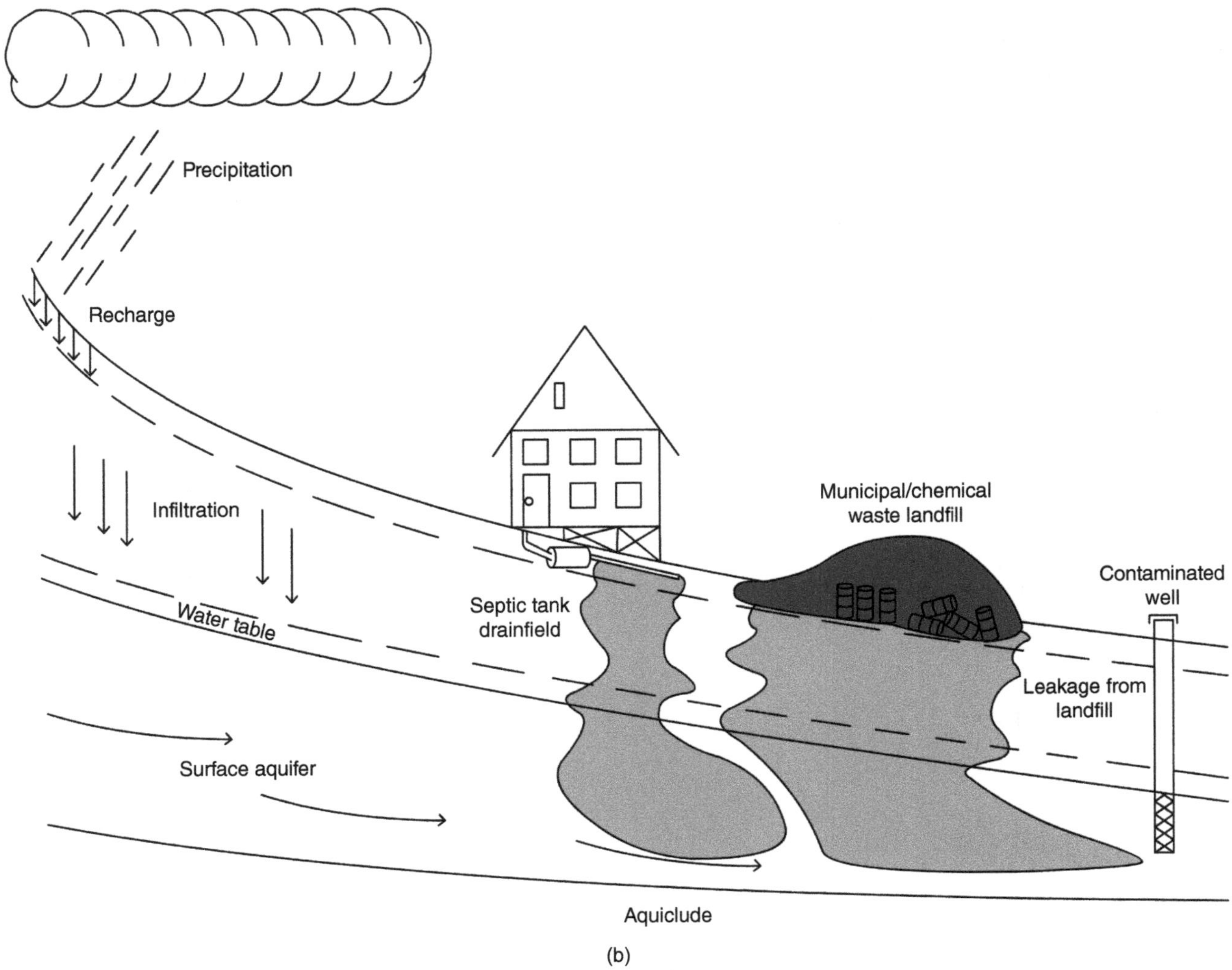

Figure 8.1 *(Continued)*

systems is relatively easy, although there are strict procedural guidelines for sampling and analysis. In these systems, you can see and feel the sample matrix. However, the sampling of groundwater is very different. For simplification purposes, we will limit our discussions primarily to saturated groundwater systems, but remember that contamination also can occur between the ground surface and the flowing aquifer in the unsaturated zone.

First, one has to determine the depth of the water table (boundary of the saturated zone; Figure 8.1), and then which way the water is flowing. Although today there are nonintrusive techniques that do this, we usually resort to the very intrusive and potentially destructive method of installing a monitoring well, since this also enables us to obtain water samples to analyze for pollutants. A monitoring well consists of a metal or polyvinyl chloride (PVC) casing (tube), installed from the land surface to a given depth in the aquifer. There is no perfect good way to install such a well without significantly disrupting the nature of the aquifer. In environmental sampling, we always want to take a "representative and unbiased" sample, so that the water we obtain is characteristic of any water in the system. Obviously, if you violently drill a hole into a system, grinding and mixing dirt and rock fragments with the water, it is difficult to

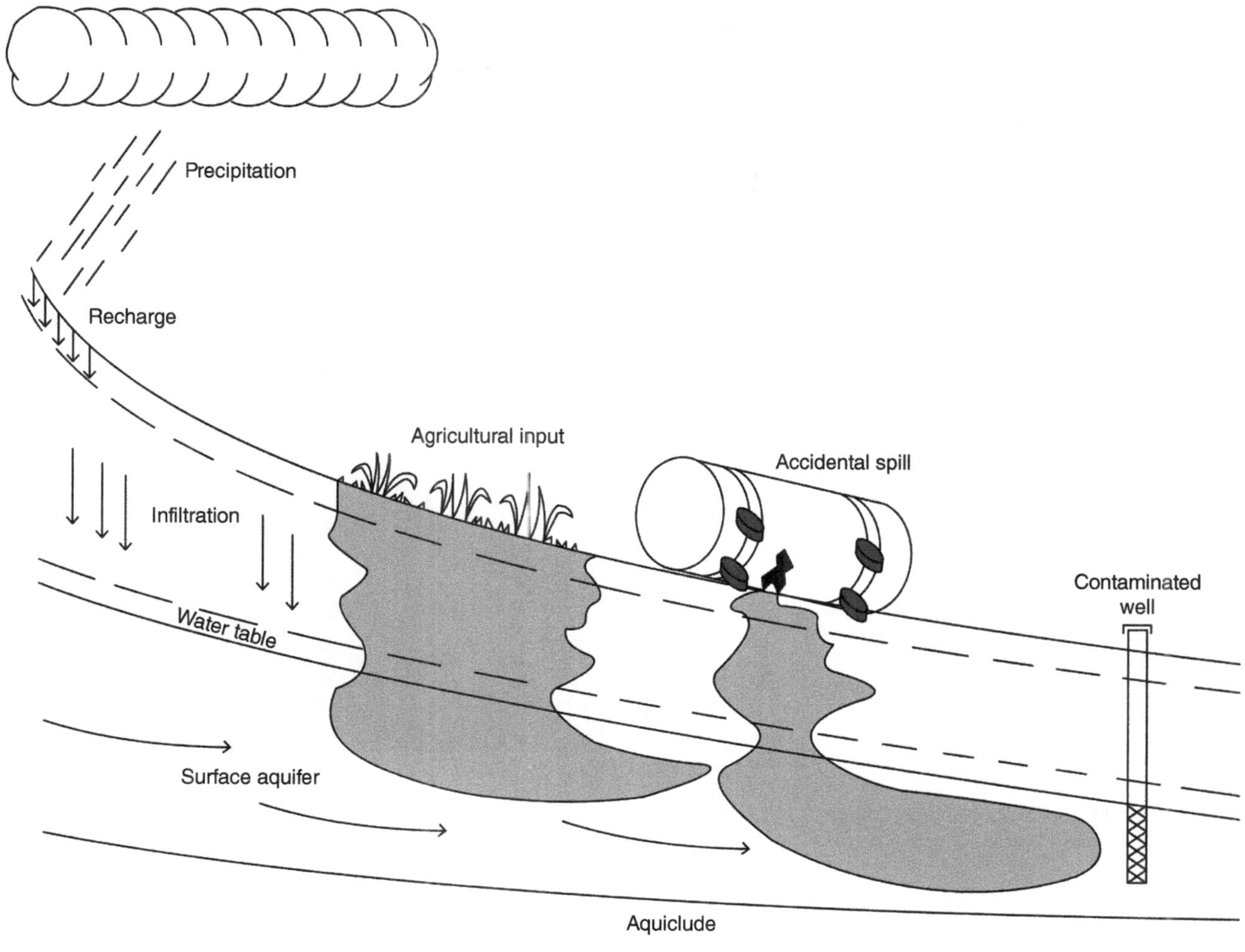

Figure 8.1 (*Continued*)

obtain a representative sample. But for now, this is the only technically feasible and economical method we have of obtaining groundwater samples.

We will next present a few of the well-installation techniques used today, since installing groundwater monitoring wells is often an expensive part of a characterization process and since our modeling results depend on obtaining samples to validate our modeling efforts. Also, as noted earlier, the sampling of groundwater requires special considerations. The types of drilling techniques we will cover include the cable tool, direct rotary drills, and auger and coring systems.

8.3.1 Cable Tool Percussion Method

The cable tool percussion method of drilling wells was probably used in the first attempts at installing a deep well. The technique goes back at least 4000 years and was developed by the Chinese, who successfully drilled wells to approximately 915 m (3000 ft) at this time. A photograph of a modern system is shown in Figure 8.2. The drilling process works by repeated lifting and dropping of the drill bit (shown in Figure 8.3). Water is usually used to suspend the broken rocks and media, but this is not always necessary. Periodically, when sufficient debris has accumulated at the bottom of the well being drilled, a drill bit is replaced with a bailer and the solids

Figure 8.2 A cable tool drill rig. Source: Courtesy of US DOE-INEL photograph.

Figure 8.3 A cable tool drill bit and bailer. Source: Courtesy of US DOE-INEL photograph.

are removed. The cable tool technique can be used in almost any geologic media, but works especially well in coarse glacial sediments (till), boulder deposits, or rock strata that are highly disturbed, broken, fractured, or cavernous. Well depths typically run from 90 m (~300 ft) to 1520 m (5000 ft). Major advantages of the cable tool technique include the following (Johnson Filtration Systems 1986):

- drill rigs are relatively inexpensive
- drill rigs are simple in design and use
- machines have low energy requirements
- recovery of samples of geologic media is possible
- wells can be drilled in areas with little water supply
- wells are not contaminated by drilling equipment (muds, discussed later)
- drill rigs can be operated in all temperature regimes
- wells can be drilled in media where loss of circulation is a problem (discussed in the Section 8.3.2, rotary drill rigs), and
- water yield from the formation can be determined at any point in the drilling operation.

Disadvantages include relatively slow drilling rates and high casing costs resulting from heavier diameter wall casing needed.

8.3.2 Direct Rotary Drill Method

A much faster drilling technique is the direct rotary drill method. In this approach, the drill bit is directly spun (motor driven) to break the geologic media. The drill shaft is hollow, and some form of fluid is forced down the inside of the shaft, removing the cuttings from the well by forcing them up the well between the drill shaft and the borehole. You may have seen this type of drilling technique in use, or in movies on television, since it is used in wells for water, gas, and petroleum exploration. A typical rotary drill rig is shown in Figure 8.4. Drilling fluids must be used to remove cutting, in order to prevent the drill bit from becoming lodged in the borehole. Air, water, and synthetic "mud" (a clay mineral, bentonite) have been used for this purpose. Synthetic mud contains clay in suspension and, in some cases, surfactants. Air and clean water are preferable for installing monitoring wells for environmental sampling, since synthetic mud is transported into the geologic medium during the installation process. Once there, the mud cannot be completely removed on well development (Section 8.3.5) and affects the movement of pollutants. Water and synthetic mud are necessary, however, in traditional direct rotary drilling in unconsolidated media, since they

Figure 8.4 A direct rotary drill rig. Source: Courtesy of US DOE-INEL photograph.

Figure 8.5 A direct rotary drill rig being used to drill a diagonal well. Source: Courtesy of US DOE-INEL photograph.

also serve to hold the formation open (prevent collapsing) around the drill bit and borehole. With this in mind, it follows that air can only be used in semiconsolidated or consolidated media, which will not be subject to collapse, as would unconsolidated material. One added feature of the rotary drill technique is that it can drill monitoring wells under permanent facilities by drilling at steep angles, as illustrated in Figure 8.5.

The major advantages of the direct rotary drill method include the following (Johnson Filtration Systems 1986): relatively high drill rates in all types of media, minimal casing required during the actual drilling process, and rapid rig mobilization and demobilization.

Major disadvantages of this technique include high cost of drilling rigs, high level of maintenance required by drilling rigs, special procedures required for collection of media samples, and possible plugging of the drill bit and borehole caused by loss of pressure of the drilling fluid in fractured formations.

8.3.3 Augers

The most common type of drilling method for installation of shallow monitoring wells is the hollow stem auger. A small, hand-operated system is shown in Figure 8.6, but larger truck-mounted systems are common. These systems are only used in unconsolidated media. The hollow stem allows soil samples to be collected and analyzed for pollutant contamination. This technique is used both for soil sampling and for installation of monitoring wells. These systems allow relatively rapid installation but are limited to shallow depths.

Figure 8.6 A handheld auger drilling system. Source: Courtesy of Forestry Suppliers, Inc. http://www.forestry-suppliers.com.

8.3.4 Well Casing, Grouting, and Sealing the Well Casing

In most cases, once a borehole is drilled it must be lined with casing to prevent collapse of the geologic media. Commonly available casing materials include aluminum, carbon steel, stainless steel, and PVC. For monitoring wells, the type of casing used depends on the type of contamination. For example, PVC casing would not be used for monitoring the subsurface around a leaking underground organic chemical tank, since the liquid chemical might dissolve the PVC casing. PVC casing would also sorb any hydrophobic pollutant from the water. Similarly, metal casing would unlikely be used in wells for monitoring metal contamination.

Grouting is a term used to describe the filling of the space between the well casing and the borehole (referred to as the annular space). This filling is needed to prevent water from entering the well from the surface and in order to isolate different regions of the well. Some monitoring operations, especially in the unsaturated zone of a formation, use lysimeters for obtaining water samples. A lysimeter is a porous cylinder (usually ceramic) that contains two tubes that lead back to the land surface. A vacuum can be pulled on the lysimeters, allowing the collection of water over long periods of time. When sufficient time has passed, one tube of the lysimeter is pressurized, which lifts the waters in the lysimeter to the surface through the second tube for sampling. Lysimeters are placed at areas in the unsaturated subsurface where water is expected to collect during infiltration from rain or flooding. Collection areas can be located at different depths in unconsolidated media or at fractures in consolidated media. In any event, water must be prevented from entering at the land surface or flowing between different lysimeters. To prevent the inflow of water from another region of the subsurface, a clay medium, usually bentonite (one form of grout), is placed above and below the lysimeter, and sand or silica is placed immediately around the lysimeter to allow water to freely pass to the collection point. The addition of grouting materials to a borehole is illustrated in Figure 8.7. Above the lysimeter, more bentonite is placed to seal the lower lysimeter from water flowing above. Another lysimeter is placed at the next collection point and so on until each fracture or sampling locale has a lysimeter and the well is full. Many wells/boreholes are sealed at the surface with cement, but given the high pH associated with the $Ca(OH)_2$, cement would not be used near any sampling location.

Wells at monitoring stations are drilled to the desired depth and then typically a porous section of pipe (referred to as a screen) is placed in the borehole to allow collection of aquifer water. Screen depths, where water is expected to be collected, can range from a few centimeters to several meters, depending on the section of the aquifer that you desire to sample. Sand or silica is placed between the screened casing and the borehole, and grout, usually bentonite, is placed above the screened area. Again, cement is avoided except at the land surface.

Figure 8.7 The addition of grouting materials to a borehole. Source: Courtesy of US DOE-INEL photograph.

8.3.5 Well Development

As noted at the beginning of this section, the installation of monitoring wells is an intrusive and destructive process, and the area immediately adjacent to the borehole/well needs to be returned to its original condition or as close to it as possible. Thus, well development refers to (i) the repair of damage to the subsurface (geologic) formation and (ii) the alteration of the aquifer to allow water to flow freely to the well for collection. First and foremost, it includes removal of any drilling fluid (other than air) that was used in the drilling process. As noted earlier, the installation of most environmental monitoring wells does not allow the use of synthetic muds or surfactants. Yet it is necessary to remove any drill cutting (rock chips and disrupted material) that may be present in the formation and contaminate subsequent samples. In order to restore the aquifer to near-normal conditions, the well can be "over-pumped" (pumped at a high water removal rate for hours or days). Another technique is to reverse the flow several times to dislodge particles in the formation. In general, states, federal agencies, or concerned clients have an established procedure for developing the well. It should be noted that you do not simply walk up to a monitoring well and take a sample. Usually, three to five well casing volumes of water must be removed before a water sample is taken for chemical analysis. The purpose of this is to ensure that you have a representative sample from the aquifer and not stagnant water from the well casing.

8.3.6 But How Good Is a Groundwater Monitoring Well?

Groundwater monitoring wells have been used for decades, but as amazing as it may seem, the effectiveness of well grouting was only recently evaluated. Dunnivant et al. (1997) found that standard well installation procedures for consolidated media were highly effective in isolating different productive zones within a borehole. An example of this effectiveness is shown in Figure 8.8. The dark areas of the borehole represent the location of bentonite used to seal the borehole and prevent water from flowing from above or below. This lysimeter installation was later flooded with at least 1 m of water for 50 days, yet no water passed directly down the borehole. Water and tracer (^{75}Se) did migrate through the subsurface via fractures in the basalt media and arrived at some lysimeter locations but not at others. The first lysimeter placement at 6.4 m shows a typical tracer profile, with time, for a pulse input. The next two lysimeters in the borehole at 10.4 and 18.9 m did not receive water or tracer over the entire experiment (50 days of water application). The next lysimeter at 37.8 m did receive water, apparently from the beginning of the flooding when no tracer was present. The next two lysimeters at 42.7 and 52.7 m received water and tracer. Again, these and other experimental measurements confirm that water did not flow in the borehole between lysimeter placements, which validates the grouting technique used. These and other borehole studies during the experiment by Dunnivant et al. (1997) found that the standard grouting technique was effective in sealing the borehole in consolidated media. Another study has found standard grouting techniques to also be effective in unconsolidated media (Christman et al. 2002).

8.4 Groundwater Sampling Equipment

Once a well or borehole has been installed and pumped for development, it will need to be tested for water purity or any pollutants present. After well development, the general rule for a new well or an older monitoring well is to remove the water present in the well casing prior to taking a sample of analysis. The purpose is to remove the stagnant water in the well casing. The purging of three well volumes is generally required. Decades ago, this purging was completed by hand with a water bailer – a cylindrical tube that would fit into the well casing, lowered by an inert rope, that would fill with water and allow the water to be removed. Such a bailer is shown in Figure 8.9. Bailers can be made out of a variety of materials (Teflon®, metal, plastic, etc.), depending on the analyte of interest.

In the last decade, portable battery-powered pumps have been economically manufactured that can be lowered into at least 2-in. wells to remove water from the wells. This has greatly aided and speeded up the sampling efforts in groundwater monitoring. Such a device is shown in Figure 8.10.

In situ, monitoring of several common analyses can be performed by a set of probes lowered into the well, allowing monitoring a one depth or over the entire depth of the monitoring well. One common, in situ probe is the Hydrolab shown in Figure 8.11.

8.5 Chemistry Experiments Used to Support Modeling Efforts

As we discussed in Chapters 2 and 3, chemistry can play important roles in fate and transport processes. In our previous discussions on river and lake systems, we mostly noted these roles in a qualitative manner, with the exception of the degradation rate constant, k. However, in groundwater systems, pollutants dissolved in the water (the mobile phase) are in constant contact with the soil/rock matrix (the stationary phase). Hence, all the chemistry related to sorption phenomena plays a very important role. Recall that we summarize sorption reactions in terms of the distribution coefficient (K_d) for metals or partition coefficient (K_p) for hydrophobic pollutants. Also recall that for metal pollutants, the pH, E_H, salinity, complexation, and content of natural organic matter (NOM) determine the degree of adsorption. For hydrophobic pollutants, the most important factors are the presence of NOM and clay particles. However, we will see in a few paragraphs that the type of NOM, either dissolved or sorbed, is also important.

8.5.1 K_d and K_p Values

The sorption coefficient is perhaps the most important factor determining the fate and transport of a pollutant in groundwater systems. We will most often use a generic

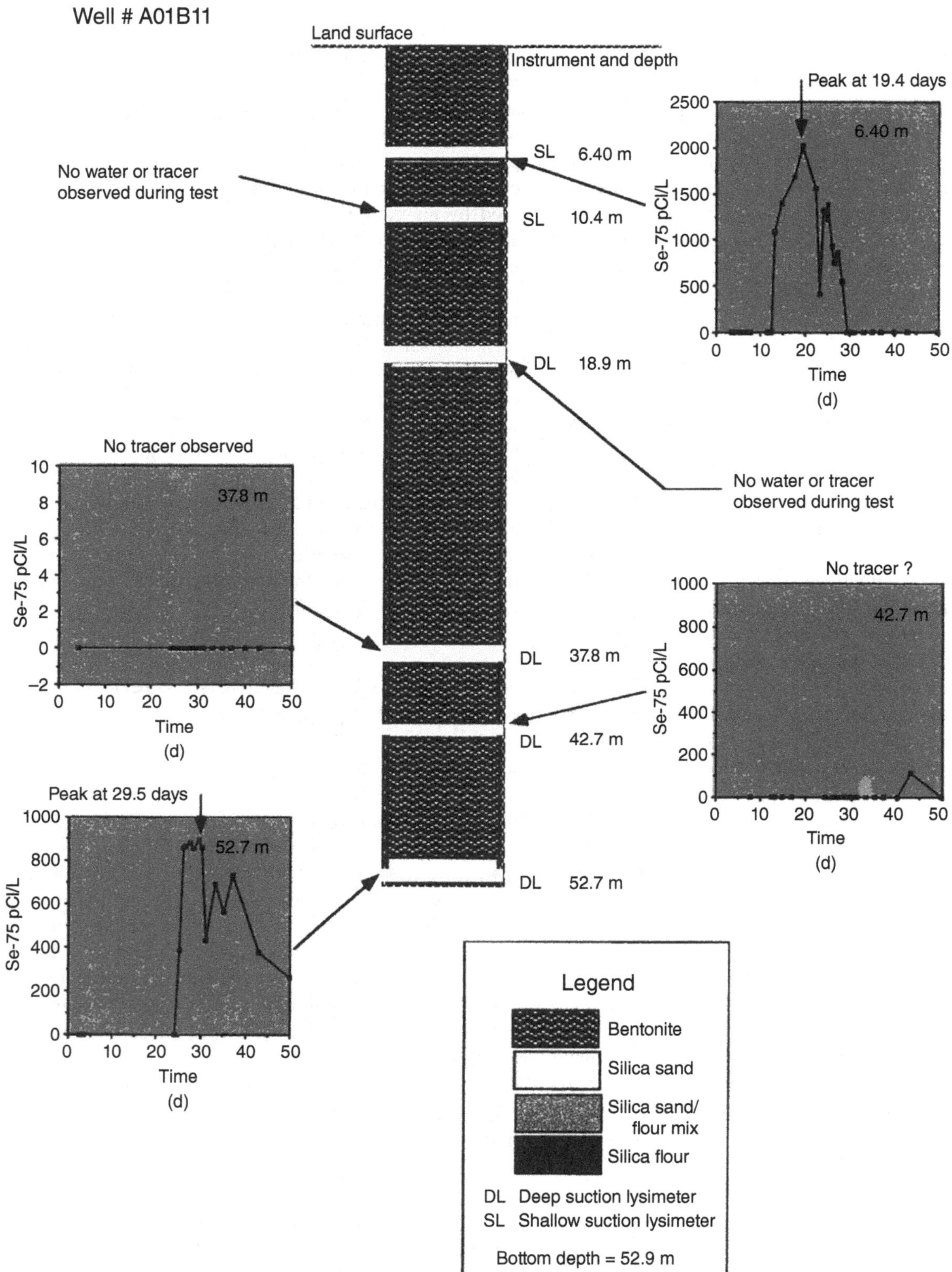

Figure 8.8 Illustration of the effectiveness of grouting in lysimeter placements. Source: US DOE; Wood and Norrell (1996).

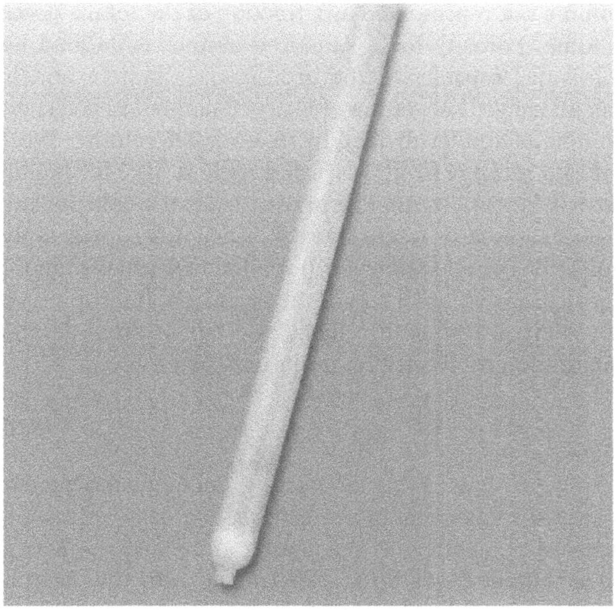

Figure 8.9 A standard bailer. WaterMark® Dome-End 36″ Disposable Polyethylene Single Check Valve Bailer. http://www.forestry-suppliers.com/product_pages/products.php?mi=32421&itemnum=78015&title=WaterMark%C2%AE%20Dome-End%2036%E2%80%9D%20Disposable%20Polyethylene%20Single%20Check%20Valve%20Bailer. Source: Courtesy of Forestry Suppliers, Inc. http://www.forestry-suppliers.com.

K, since we will be discussing both the adsorption of metals and the partitioning of hydrophobic pollutants. For metal pollutants, there is no one equation that can be used to estimate the distribution coefficient. The extent of adsorption for metal pollutants must be evaluated experimentally. For hydrophobic pollutants, predictive equations correlating adsorption with pollutant solubility have been developed independently by a number of researchers. Three of these are shown below (Eq. (8.1), Chiou et al. 1979; Eq. (8.2), Karickhoff et al. 1979; Eq. (8.3), Means et al. 1980):

$$\log(K_{oc}) = 3.803 - 0.557 \log(S_1) \tag{8.1}$$

$$\log(K_{oc}) = 0.440 - 0.540 \log(S_2) \tag{8.2}$$

$$\log(K_{oc}) = 4.273 - 0.686 \log(S_3) \tag{8.3}$$

where

K_{oc} is the organic carbon partition coefficient between the soil organic matter and the water,
S_1 is the pollutant aqueous solubility in μmol/l,
S_2 is the pollutant aqueous solubility in mole fraction, and
S_3 is the pollutant aqueous solubility in mg/l.

Recall that K_p can be calculated from K_{oc} by

$$K_p = K_{oc}^{*} \text{ fraction of organic carbon} \tag{8.4}$$

Figure 8.10 A battery-operated pump of purging and sampling groundwater monitoring well. Proactive Environmental Products™ Tornado Sampling Pump. http://www.forestry-suppliers.com/product_pages/products.php?mi=26217&itemnum=76183&title=Proactive%20Environmental%20Products%E2%84%A2%20Tornado%20Sampling%20Pump. Source: Courtesy of Forestry Suppliers, Inc. http://www.forestry-suppliers.com.

Thus, if you know the aqueous solubility of the pollutant, either an experimental value or an estimated value from SPARC (https://www3.epa.gov/ceampubl/learn2model/part-two/onsite/sparcproperties.html; see Chapter 2), you can estimate the partition coefficient. But the user of these equations should note that each equation was developed for a particular set of compounds and experimental conditions, and minor or major errors may be present in your estimate of K. A much more accurate method is to measure K_d or K_p using the experimental setup discussed at the end of Chapter 3 or using the column study approach discussed below.

8.5.2 Relationship Between *K* and the Groundwater Fate and Transport Equation

No matter how you obtain a value for K, it is not directly entered into the fate and transport equation. Groundwater fate and transport equations calculate movement of water and pollutants in terms of how much water passes through the system or unit volume of media. Thus, we need the K term expressed in terms of water volume. To begin understanding how this works, we will only discuss the movement of water through a unit volume of soil or

Figure 8.11 An in situ monitoring device for groundwater monitoring wells: Hydrolab MS5 – Mulitparameter Mini Sonde for 2-in. wells. http://www.ott.com/en-us/products/water-quality-2/hydrolab-ms5-mulitparameter-mini-sonde-57/. Parameters that can be measured include: temperature, conductivity, water depth, pH, oxidation–reduction potential, dissolved oxygen, turbidity, chlorophyll *a*, blue-green algae, rhodamine WT, ammonium, nitrate, and chloride. Source: Photo courtesy of OTT Hydromet.

fractured rock. Unlike in rivers and lakes, where a unit volume of the system was essentially all water, in groundwater systems, most of the unit volume is soil or rock. In groundwater flow and pollutant movement, we talk in terms of the number of pore volumes that need to be passed through a unit volume of media to move the pollutant. One pore volume is equal to the volume of water contained in the pores of one unit volume. In theory, it should take one pore volume for a tracer that does not react with the soil or rock media (referred to as a conservative tracer) to pass through the system. For a sorbing pollutant, the volume of water needed to push the pollutant completely through the system will be greater than one pore volume. We can mathematically define R, the retardation coefficient, as

$$R = \frac{V}{V_1} \tag{8.5}$$

where V is the volume of water needed to pass the tracer or pollutant completely through the system, and V_1 is the volume of water in the unit volume of media (the pore volume).

This approach for determining R works fine for laboratory settings, but is of little use in the real world, where we would have to collect and measure the volume of all water passing through a polluted aquifer. For these real-world applications, we can define R in a different manner using properties of the aquifer itself. This new approach requires a few more terms. First, we need the porosity (n) of the aquifer, which is similar to the pore volume but represented as a fraction of the total aquifer volume. Porosity for a saturated system is defined by the water capacity of the groundwater system and is calculated as the volume of water contained in a certain volume of aquifer divided by that aquifer volume. Typical values range from 0 for solid rock to ~0.4 for highly porous media. Another term we need is the bulk density (ρ_b) of the soil or fractured rock, since it is considerably different from that of water (for which we usually ignore density) and in order to use the units of K (liter of water per kilogram of soil). Finally, we can express R, the retardation coefficient, in terms of K by

$$R = 1 + \frac{\rho_b K}{n} \tag{8.6}$$

Now, we have a way of relating pollutant movement to water movement by measuring two easily obtained physical parameters of the aquifer and K. Again, K can be determined by a direct experiment or by the column experiment, described next, which more accurately reflects the aquifer properties and conditions.

8.5.3 Column Studies for Evaluating Pollutant Transport in Subsurface Media

In designing a column study for a polluted aquifer, it is important to use soil/rock media characteristic of the site, and, as we will see, water collected from the site. If variations in geologic media are present, several column experiments may be necessary. As an example of a column study, we will present the work of Dunnivant et al. (1992a,b), who investigated the transport of cadmium and a polychlorinated biphenyl in a sandy aquifer material containing dissolved and sand-adsorbed NOM. The goal of this investigation was to evaluate the effect of mobile (dissolved) NOM on pollutant transport through the soil columns. We will use pore volumes and R to describe the movement of conservative tracers and pollutants through the soil columns.

In soil column experiments, step or pulse inputs of pollutants can be used. In the study by Dunnivant et al. (1992a), a step (continuous) input was used. Before a soil column can be used to study pollutant movement, it must be characterized with a conservative tracer to make sure that dispersion is low. Excessive dispersion in the column, resulting from poor column packing or construction, will negate the interpretation of the sorption data. Bromide is commonly used as a conservative tracer, and the results from this experiment are used to estimate the pore volume of the column and dispersion. Dilute bromide solution is added as a step input, and small measured volumes of elution waters (water exiting the column) are collected and analyzed for bromide. The resulting dataset, referred to as a breakthrough curve

(BTC) since it shows the breakthrough of chemical from the column, is tested using a computer program and an estimate of dispersion is obtained. If the column is acceptable, it is then used to test the transport of a pollutant through the column.

After an acceptable column has been constructed and tested with the bromide tracer, the column is saturated with NOM at a specified dissolved NOM level. This is achieved by pumping a NOM solution through a column of soil identical to the soil used in the pollutant experiment but not containing pollutant. This step is necessary in order to simplify the experimental design, since NOM and the pollutants sorb to the soil at different rates and we want to have only one variable in the experiment (the sorption and desorption of pollutants in the soil column). In general, NOM saturation experiments take three to five days (Dunnivant et al. 1992b).

Finally, water containing the cadmium, and in some cases NOM, is passed through the column. Discrete volumes of eluent water are again collected, the volumes measured, and the samples analyzed for Cd^{2+}. Figure 8.12 contains several BTCs for Cd^{2+} as a function of dissolved NOM. First, note the BTC to the right. This is the BTC for Cd^{2+} dissolved in water not containing NOM and is our reference point. Note the axes of the plot. The y-axis is the fraction of Cd^{2+} collected and the x-axis is expressed not in time, but in pore volumes. The midpoint of this plot (0.50 on the y-axis) corresponds to R for Cd^{2+} (corresponding to an R value of 68 and from which we calculate a K_d of 14.1). We can also calculate K_d since we can also measure the porosity and bulk density of the soil column. Many studies, including the one by Dunnivant et al. (1992a) also measure K_d using the procedures described in Chapter 3 and compare it to the column-calculated value. Obviously, agreement between the two techniques gives more credence to the approach and results. These concepts will become clearer if you complete the K_d and column study laboratory experiments in Chapter 16.

We noted in the introduction to this section that it is important to use soil and water from the site under investigation. The importance of this is illustrated in Figures 8.12 and 8.13. The site under investigation contained varying amounts of dissolved NOM; thus, it was important to determine the effect of the presence of NOM in the dissolved phase since NOM is known to bind to metal pollutants. In Figure 8.12, note that as the level of NOM is increased from 5.2 to 20.4 to 58.1 mg/l, the cadmium moves faster and faster through the column, as indicated by the fewer pore volumes of water needed to remove the Cd^{2+} from the column. The R values for these BTCs are 47, 36, and 26, respectively. Note the error that would have been introduced to the laboratory study if no NOM had been included in the

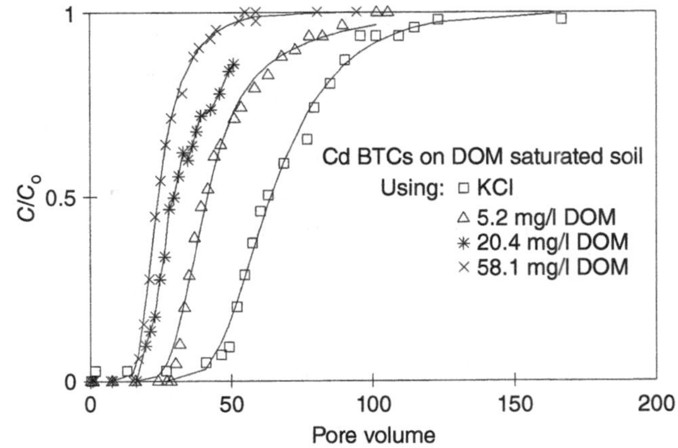

Figure 8.12 Breakthrough curves for Cd^{2+} on a NOM-saturated soil column as a function of dissolved NOM concentration. Source: Dunnivant et al. (1992c). Copyright 1992. Reprinted with permission of American Chemical Society.

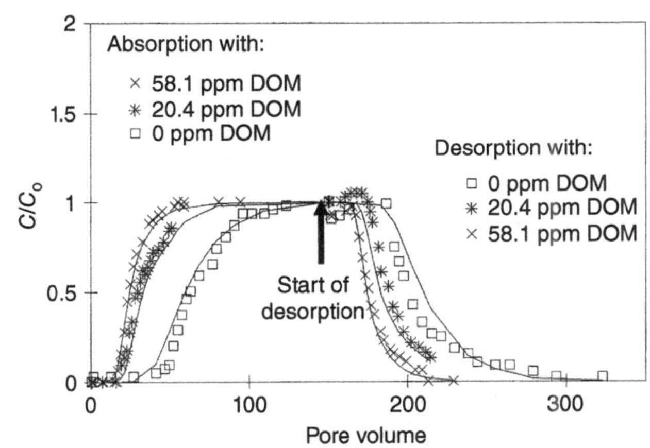

Figure 8.13 Breakthrough curves for the adsorption and desorption process for Cd^{2+} as a function of NOM concentrations. Source: Dunnivant et al. (1992c). Copyright 1992. Reprinted with permission of American Chemical Society.

water used in this study, while NOM was present in the aquifer. An R value of 68 would have been assumed, whereas in reality an R value as low as 26 could have been accurate for behavior of cadmium in the aquifer. Hence, the researchers would have significantly underestimated the mobility of cadmium in the system.

Recall again what the BTCs represent. Initially, as water containing pollutant is passed through the column, pollutants sorb onto the soil and no detectable pollutant exits the column. Eventually, the sorption sites on the soil are saturated and pollutant starts to exit the column. In theory, the BTC should be a perfectly vertical line starting at a C/C_0 value of 0.0 and going immediately to 100.0. However, this does not occur in laboratory experiments or in the environment due to two factors: the slow kinetics of sorption to the soil and the dispersion that occurs in the column. Finally, all the sorption sites are covered and the effluent pollutant from the column reaches the inlet concentration. But what happens if we remove the input of pollutant? This is illustrated in Figure 8.10, where the inlet water to the pollutant-saturated column is changed to NOM-containing water without pollutant. This point is indicated by the upward arrow in Figure 8.13. Note the predictable trend. NOM increases the mobility of cadmium during the adsorption process by shifting the BTC to the left as compared to the eluent water not containing NOM and increases mobility in the desorption process by shifting the BTC to the left. Increasing the NOM results in a greater leftward shift and increased mobility. These types of experiments illustrate the need to characterize the water and soil at the site under investigation in order to closely match the chemistry of the systems.

The transport of the polychlorinated biphenyl 2,2′,4,4′,5,5′-tetrachlorobiphenyl (TCB) was also investigated in the soil columns. These results are shown in Figure 8.14. The transport of the pollutant (TCB) in water not containing NOM is indicated by the stars (the right-most plot). Note the much larger pore volume scale, as compared to the Cd^{2+} experiment, which indicates a much higher R value, and in turn, a much higher K_p value for the TCB. This is common; hydrophobic pollutants tend to have higher K values as compared to metal pollutants, especially in organic-rich soils or sediments. Again, as mobile (dissolved) NOM is added to the elution water, the hydrophobic pollutant becomes more mobile. R values were 1018 (0.0 mg/l NOM; K_p = 245 l/kg), 625 (5.2 mg/l NOM), 456 (10.2 mg/l NOM), and 293 (20.4 mg/l NOM). Note that the length of experiments measuring pore volumes of 1000 require weeks to months if they are to yield representative transport predictions.

One final point should be made concerning column studies and the need to use laboratory conditions that are representative of the field site. Some chemical parameters, such as pH and E_H, can significantly affect the fate of a pollutant. Recall the Swiss study mentioned in Chapter 2 documenting the reduction of substituted nitrobenzenes by reduced NOM. If reducing E_H conditions are present in the field, you must incorporate these into your K determination or column study in order to obtain representative results. And in regard to transformation reaction rates, surface-catalyzed degradation rates in the environment tend to be orders of magnitude faster than those by dissolved catalysts. Hence, these types of reactions can be extremely important.

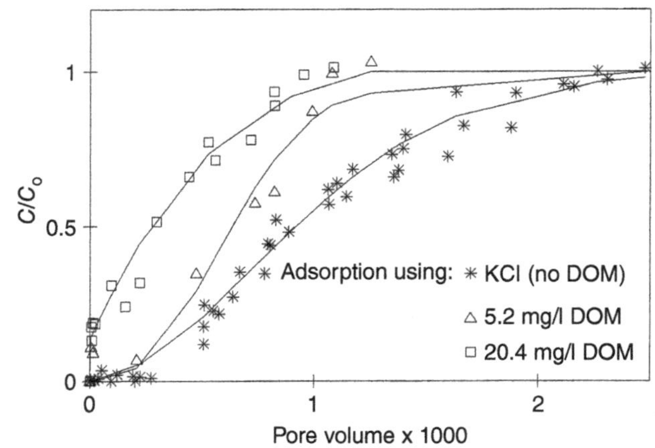

Figure 8.14 Breakthrough curves for a polychlorinated biphenyl (PCB) as a function of NOM on a soil column. Source: Dunnivant et al. 1992c. Copyright 1992. Reprinted with permission of American Chemical Society.

8.6 Direction of Water Flow (The Three-Point Problem)

Determining the direction of pollutant transport is easy in river and atmospheric systems but more complicated in groundwater systems. Yet, the direction of flow must be known in order to calculate the gradient (magnitude of water flow), which is defined as the difference in water elevation (head) between two points along the flow line, divided by the longitudinal distance between the two points. Ideally you would only need two wells, one upgradient and one downgradient, but this would require that the wells' locations define a line parallel to water flow. Such a representation is shown in Figure 8.15a in which the wells are located on the same water flow line. Rarely, if ever, will you come across such an ideal situation in the real world. Typically, you need at least three monitoring wells of known location (on an x–y grid) to yield measurements of depth to the water table, as shown in Figure 8.15b. The depth to the water table values are converted to depth above sea level or related to another reference point. With this information, you have sufficient data to solve a problem commonly referred to as the "three-point problem." There are two ways to solve this problem, graphically and with an exact mathematical calculation. We will illustrate the graphical method and refer you to Fate for the mathematical method.

Figure 8.15b gives the general layout of three wells and heads in each well. Well A has a location of (152, 870) with a head of 21 m, Well B has a location of (174, 370) with a head of 20 m, and Well C has a location of (826, 804) with a head of 19 m. To determine the direction of groundwater flow, you arrange the wells on graph paper as shown in Figure 8.12c and draw the following lines between them.

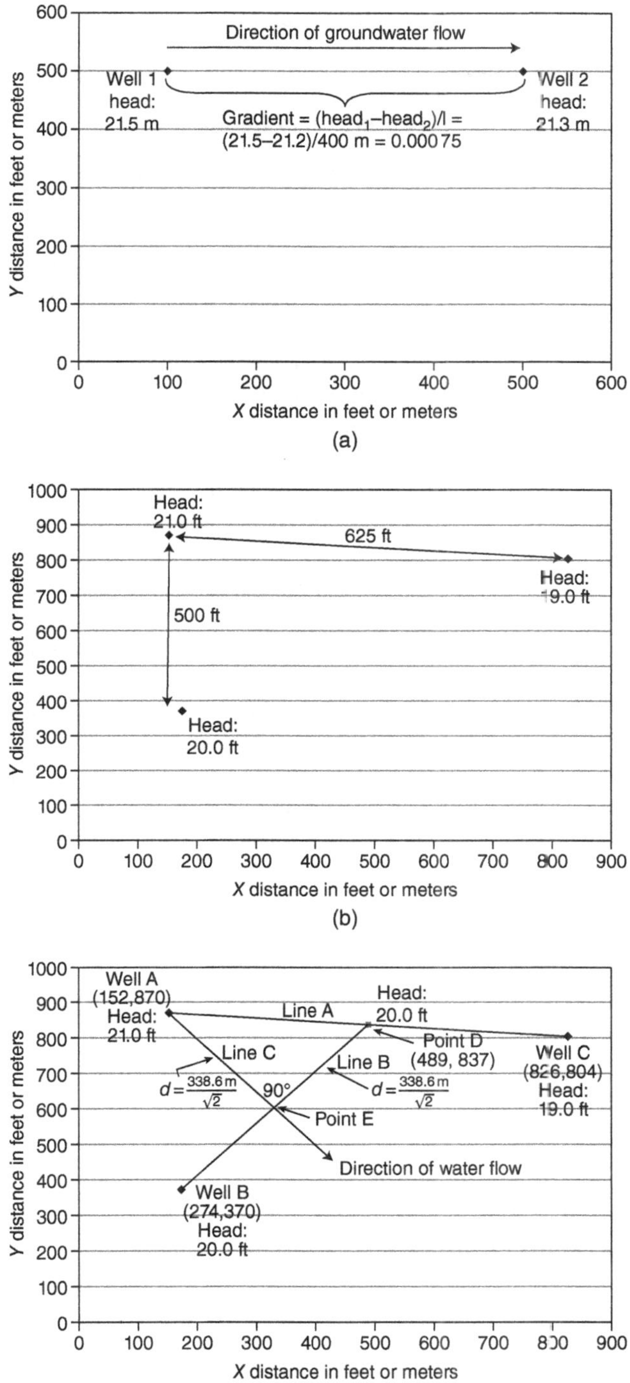

Figure 8.15 (a–c) The three-point problem.

First, a line is drawn between the highest and the lowest wells (Line A in Figure 8.12c). Next, you must find the point (Point D) along this line A that has the same head as the intermediate-head well (Well B). This point is located by assuming that the water table between the two wells has constant gradient. For our example, therefore, this point is halfway between the two wells (20.0 m is halfway between 21.0 and 19.0 m) and is at (Point D: 489, 837). Next, a line is drawn from the well with the highest head (Well A) to Line B (the line between Well B and Point D) such that these two lines intersect at a 90° angle (refer to Figure 8.15c). (In some cases, you will draw the line from the lowest well. It will be clear when to do this.) This line gives the direction of water flow and, therefore, the direction of pollutant migration.

In order to find the hydraulic gradient, we must first find the distance between Well A and Point D. To accomplish this, we use the Pythagorean theorem, where the distance d is calculated by

$$\begin{aligned} d^2 &= (x_D - x_A)^2 + (y_D - y_A)^2 \\ &= (489 - 152)^2 + (837 - 870)^2 \\ &= 113569 + 1089 \\ &= 114658 \\ d &= 338.6\ \text{m} \end{aligned} \tag{8.7}$$

Using $a^2 + b^2 = c^2$, we can obtain the distance of the short legs of the triangle, where the distance from Well A to Point E (the intersection of Lines B and C), or the distance from Point D to E is

$$d = \frac{338.6\ \text{m}}{\sqrt{2}}.$$

Thus, the hydraulic gradient (dh/dL) is

$$\text{Gradient} = \frac{(21.0 - 20.0\ \text{m})}{\frac{338.6\ \text{m}}{\sqrt{2}}} = 0.004\,18.$$

The direction of water flow can also be determined using the Three-Point Module in Fate® (http://edusolns.com/fate/2016/welcome/home/). This is found under the Section 8.5.1, where an identical problem to the one worked above is illustrated. Fate uses a vector analysis approach to solve the three-point problem. The hydraulic gradient (dh/dL) is used in Darcy's law to determine the flow rate of water through porous media:

$$Q = K_{HC} A \left(\frac{dh}{dL}\right) \tag{8.8}$$

where

Q = the flow rate (m^3/day),
K_{HC} = the hydraulic conductivity or coefficient of permeability (m/day),
A = the cross-sectional area (m^2), and
dh/dL = the hydraulic gradient.

As a point of reference, groundwater velocities range from 0.01 to 42 m/d for sand and gravel to 150 to 200 m/d for gravel with cobbles.

8.7 Physical Parameters Important in Pollutant Fate and Transport

In Section 8.4, we mentioned the importance of bulk density and porosity in relating K_d and K_p to retention (R) in an aquifer or column. These physical measurements are also important in estimating water velocity. In rivers and lakes, logically, the water flows downhill, and it is also generally easy to tell how fast the water is flowing with a tool as simple as a tennis ball. But in aquifers, where we cannot see the slope of the water table, it is difficult and in many cases impossible to predict how fast the water is moving. This is further complicated by the fact that water is not the major space-occupying substance in the subsurface, where soil or rock is the predominate media. Thus, when measuring the volume of water, you must account for the media porosity. There are three major factors that contribute to the movement of pollutants in the subsurface environment: sorption phenomena; chemical, biological, and nuclear degradation; and dilution of the pollutants by mixing and dispersion. We have already discussed sorption phenomena and degradation rates. In this section, we will concentrate on dispersion, a highly complicated and very unpredictable phenomenon that presents a major challenge for the prediction of fate and transport in groundwater.

8.7.1 Sources of Dispersion in Geologic Media

Dispersion occurs primarily because of small- and large-scale mixing phenomena. These are illustrated in Figure 8.16 for porous media. As polluted water flows through the porous media, it does not travel as a unified volume, but mixes with unpolluted water. The mixing is primarily random (Figure 8.16a), although dispersion also is affected by the heterogeneity of the subsurface, with respect to mineralogy, size of particles, or diameter or curvature of fractures. This is illustrated in Figure 8.16b for porous media containing "lenses" of different media (usually clay sedimentary deposits) dispersed through the aquifer. We account for these unpredictable mixing events with a "fudge factor" in our modeling efforts, defined as E_x or D_x, the longitudinal dispersion coefficient in square meter per year (m^2/yr) (or the dispersivity in meters also represented in some text by α_x).

Diffusion results in dilution of the polluted water. More importantly from a modeling standpoint, it also increases the extent of the zone of polluted water. To understand how dispersion and the dispersion coefficient are used, we will use the following case study.

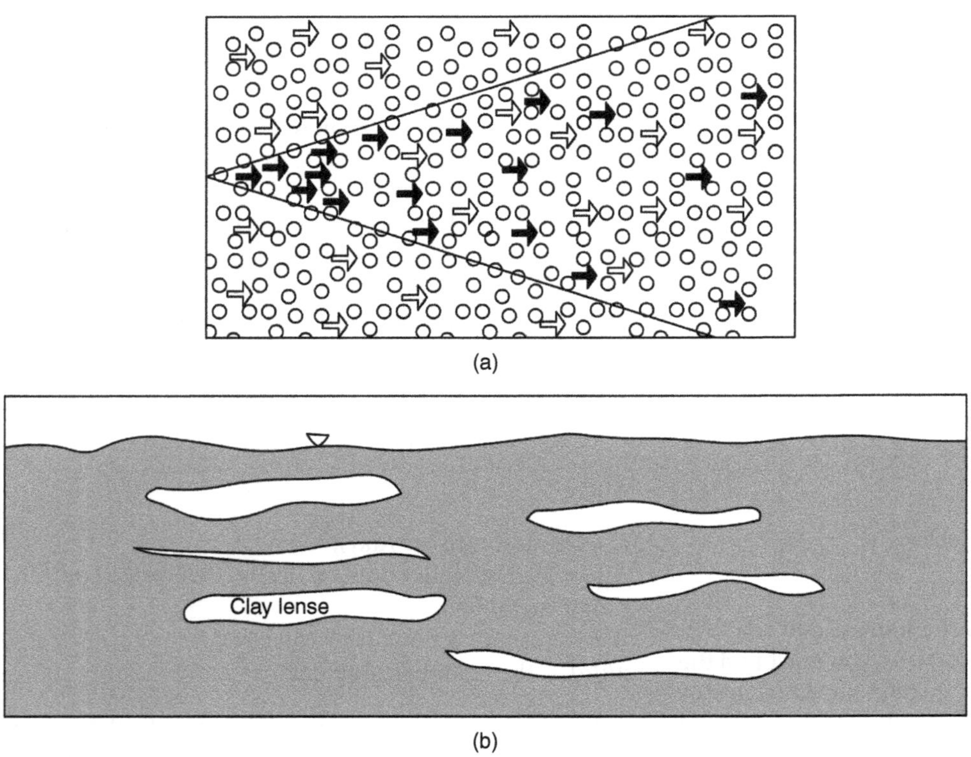

Figure 8.16 Sources of dispersion in geologic media.

8.7.2 Return to the Case Study: The INEEL Water and Tracer Infiltration Experiment

The following case study was conducted at the INEEL located near Idaho Falls, ID, USA. This facility is operated by subcontractors for the US DOE. Its primary historical function was to design and test nuclear reactors as well as to conduct secret projects of the US Department of Defense. These operations generated hazardous and radioactive wastes that had to be disposed of on-site. In order to understand the hazardous waste situation that exists today, we must try to place ourselves in the shoes of the scientists who worked at the INEEL during the Cold War. The scenario presented next does not in any way excuse the actions taken by the DOE during these times, but does shed light on their thinking at that time.

In the midst of the Cold War, nearly everything at national laboratories was secret and everyone was under suspicion. The problem facing workers at the INEEL was that they needed a secure, easily accessible place to dispose of hazardous and radioactive waste that was located near their construction and testing facilities and could not be seen from the highways crossing the INEEL (for security reasons at the time). Given that the INEEL is located on the very flat Snake River Plain, the only potential hidden places are low spots. This did not seem to be a problem at the time, since the average precipitation rate was so low as to not create flooding events; however, as it turned out, the disposal site that was finally selected was a very bad choice.

Waste was buried in barrels and containers, as shown in Figure 8.17, in unlined trenches and covered with local soil (unlined cover). For many years, this did not present a problem, but in 1962, 1969, and 1982, heavy late spring snowfalls in the nearby mountains were followed by warm winds and rapid snow melts that resulted in the flooding of a river (the Big Lost River) adjacent to the burial site. This resulted in the burial site being flooded three times for periods of days to weeks. The flooding event in 1969 is shown in Figure 8.18. The obvious concern regarding these flooding events was the release and movement of radioactive materials, both into the groundwater and out of the disposal site. The Snake River Plain Aquifer, a major pristine source of water to the region, is located below the site, but fortunately at a depth of ~550 ft. The goal of the following experiment was to determine whether contamination could reach the aquifer during such a flooding event and to estimate water velocities, chemical tracer velocities, and dispersion values at the site for future pollutant modeling efforts. Before we discuss the experiment, we should mention that the disposal site at the INEEL has now been made more resistant to flooding by an earthen berm placed around the site, and by more control and diversion of the Big Lost River.

Figure 8.17 Disposal of radioactive waste at the INEEL. Source: Courtesy of US DOE-INEL photograph.

Figure 8.18 Flooding of the waste site in 1969. Source: Courtesy of US DOE-INEL photograph.

A site for the experiment, approximately 1.5 km south of the disposal site and with similar geology, was chosen so as not to interfere with current operations at the burial site. An 183-m diameter basin was constructed and instrumented with lysimeters and wells for monitoring and collecting water and chemical tracers (refer to Figure 8.16). Tracers included a semipulse addition of ^{75}Se, ^{85}Sr, ^{160}Tb, which have a sufficiently short decay half-life such that no detectable radiation would be present at the site after approximately two years. The basin was flooded with tracer-free water for 6 days, followed by ~11 days of tracer input, followed by ~19 days of tracer-free water to flush the tracers through the system. There were 101 monitoring locations for water and tracer.

An experiment of this size is rare, to say the least, but was necessary due to the scale of potential contamination of a major pristine water source. Many smaller experiments were conducted to answer fundamental research questions as well as to address the obvious problem, the transport of radioactive waste to the aquifer. These

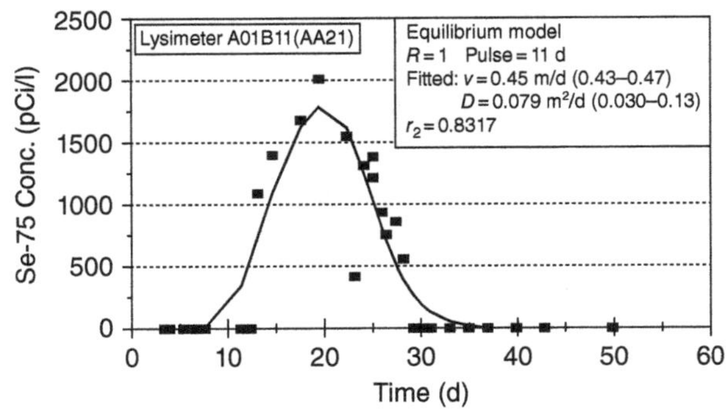

Figure 8.19 A BTC for ^{75}Se at the INEEL experiment. Source: From Dunnivant and Newman (1995); US DOE; public domain.

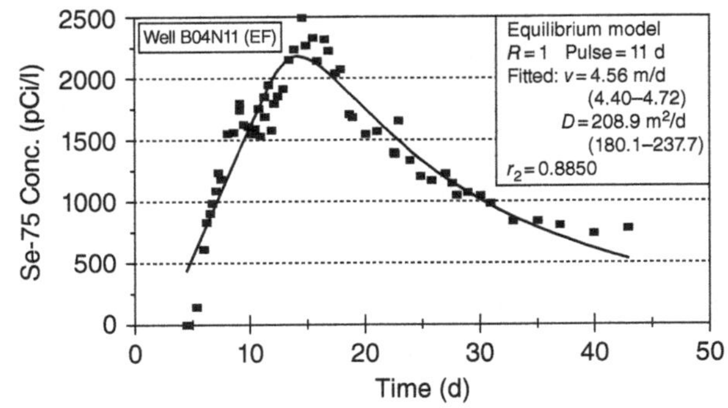

Figure 8.20 A BTC for ^{75}Se at the INEEL experiment. Source: From Dunnivant and Newman (1995); US DOE.

experiments are summarized in Dunnivant et al. (1998). The geology at the burial area is primarily composed of undulating, fractured basalt flows with some soil covering. Fortunately, although unknown to the scientists working at the burial site during the Cold War, there were several sedimentary interbeds (buried soils) located between the land surface and the Snake River Plain Aquifer. The scientists conducting the infiltration experiment (and the US DOE) now hoped that these interbeds had impeded the flow of water and pollutants to the aquifer during the three mentioned flooding events. But first let us look at some results from the experiment, specifically the BTCs for the conservative tracer ^{75}Se.

Of the 101 sampling sites, water was found and recovered from only 30, while the conservative tracer (^{75}Se) was detected in only 26 of these. In fact, during the two-year planning of the experiment, some scientists had expected that no water or tracer would be recovered from any monitoring site due to the complex, fractured nature of the geologic media underlying the site. Figures 8.19 and 8.20 contain two of the BTCs for ^{75}Se. The lines in these BTCs represent the model fit (explanatory modeling from Chapter 1) from which estimates of water velocity and dispersion were obtained, fulfilling one of the major goals of the experiment.

In an experiment conducted in the natural environment and at this scale, it is not uncommon to obtain unexpected results. It has often been proposed that unconnected and dead-end fractures exist in fractured media, but to our knowledge no one has extensively demonstrated their flow characteristics during a tracer study. Researchers found a variety of BTCs, suggesting the complexity of groundwater flow at the site. A summary of the results are shown in Figure 8.21.

As mentioned, one goal of the experiment was to obtain BTCs to estimate water velocity and dispersion in the subsurface at the INEEL. These estimates were successful from BTCs such as the one shown in Figure 8.22a, where water apparently flowed directly from the basin to the fracture being monitored. Unexpectedly, water was also observed during the experiment in some fractures in which tracer never appeared at the monitoring location (Figure 8.22b). This may be due to the monitoring of a dead-end fracture that initially filled with tracer-free water and was never drained to receive tracer. Another unexpected finding was the lack of water and tracer in lysimeter locations immediately below the water basin. These fractures were apparently isolated within an otherwise solid basalt flow section. In addition, nonclassical BTCs (non-bell shaped) were observed, in which the first water observed contained the highest concentration of tracer (Figure 8.22c). This may be explained by overflow of a dead-end fracture into a fracture where a monitoring station was located. The complexity of the BTCs strongly illustrates the need to characterize polluted sites individually, although costs of such extensive characterization may be prohibitive.

Thirteen estimates of dispersivity (in meter) were obtained from the experiment that can now be used to estimate past and future pollution transport at the site. In order to understand dispersion in the subsurface environment, we must distinguish between two terms, dispersion coefficient and dispersion. The goal of estimating a dispersion coefficient is to provide a method of estimating the dispersivity (given in meters) for any distance along the flow path. In theory, the degree of dispersion or mixing and dilution of pollutants should increase as you move downstream. To obtain values of dispersion, we simply divide the dispersion coefficient by the water velocity. The units of dispersivity are meters, so, for the example, the dispersivity represents the length of the spread of pollutants.

The explanatory modeling effort used to model the transport of water and conservative tracers in the INEEL experiment fits the general transport equation (Section 8.7) for groundwater to the BTCs. This yielded

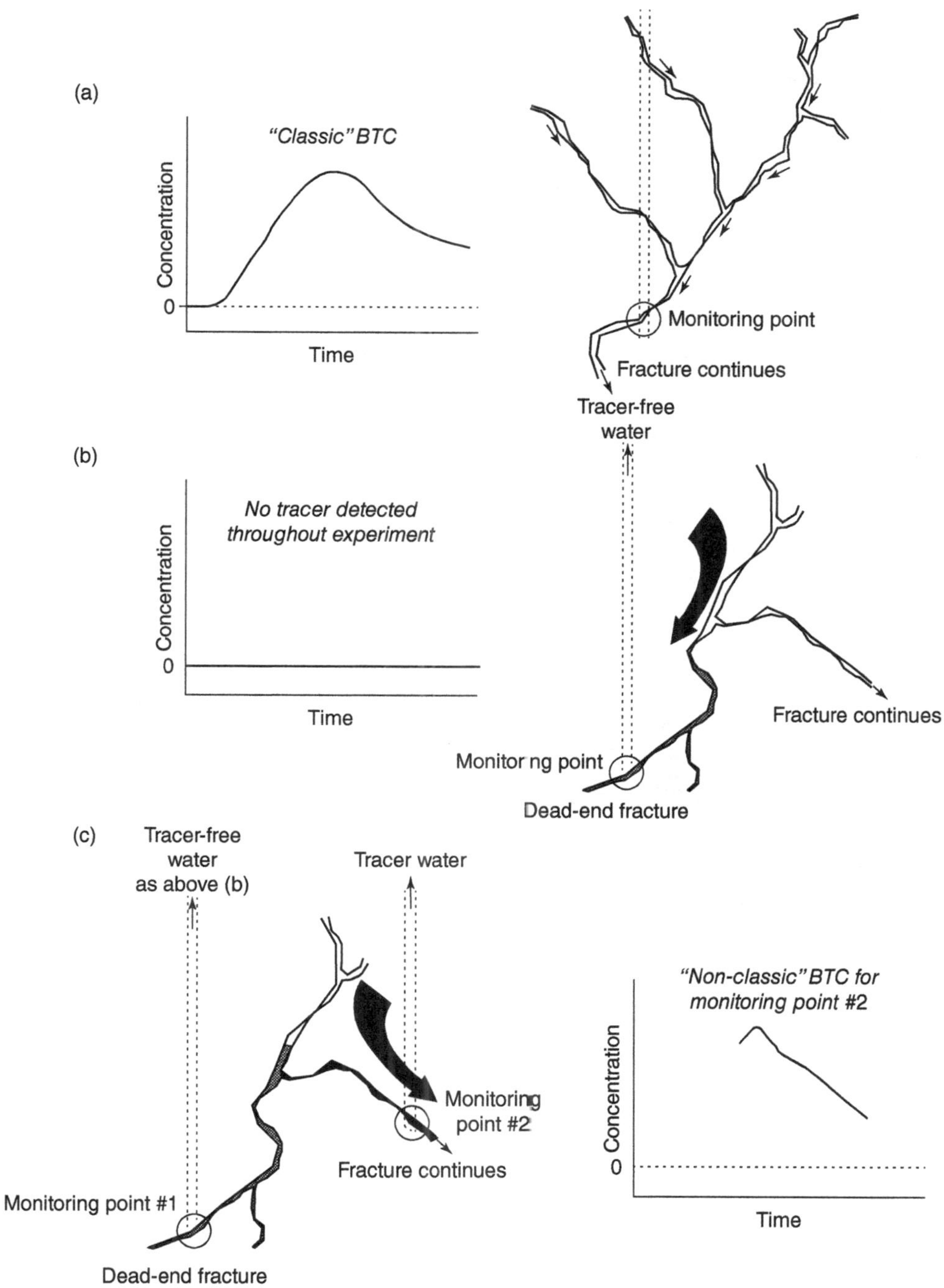

Figure 8.21 An illustrative summary of results from the INEEL experiment. Source: From Dunnivant and Newman (1995); US DOE.

13 estimates of water velocity, with units of meters per day (m/d) and dispersion coefficients with units of square meters per day. However, experimental data do not always confirm the theory. Consider the data from the INEEL experiment plotted in Figure 8.23. In general, the lysimeter data follow a trend: increasing dispersivity with increasing distance from the point source. But this trend falls apart when the interbed (a semi-impermeable layer) is reached at 68.7 m (a log value of 1.84 in Figure 8.23). The dispersivity along the interbed ranges from 0.10 to 45.8 m, over a factor of 15. This is a common observation in field experiments, where a large range of dispersivity values are often obtained for similar distances from the point source. Most researchers average the values for future modeling efforts. The data from the INEEL experiment were averaged to yield one estimate for the lysimeters (2.16 m) and another for the interbed locations (16.2 m).

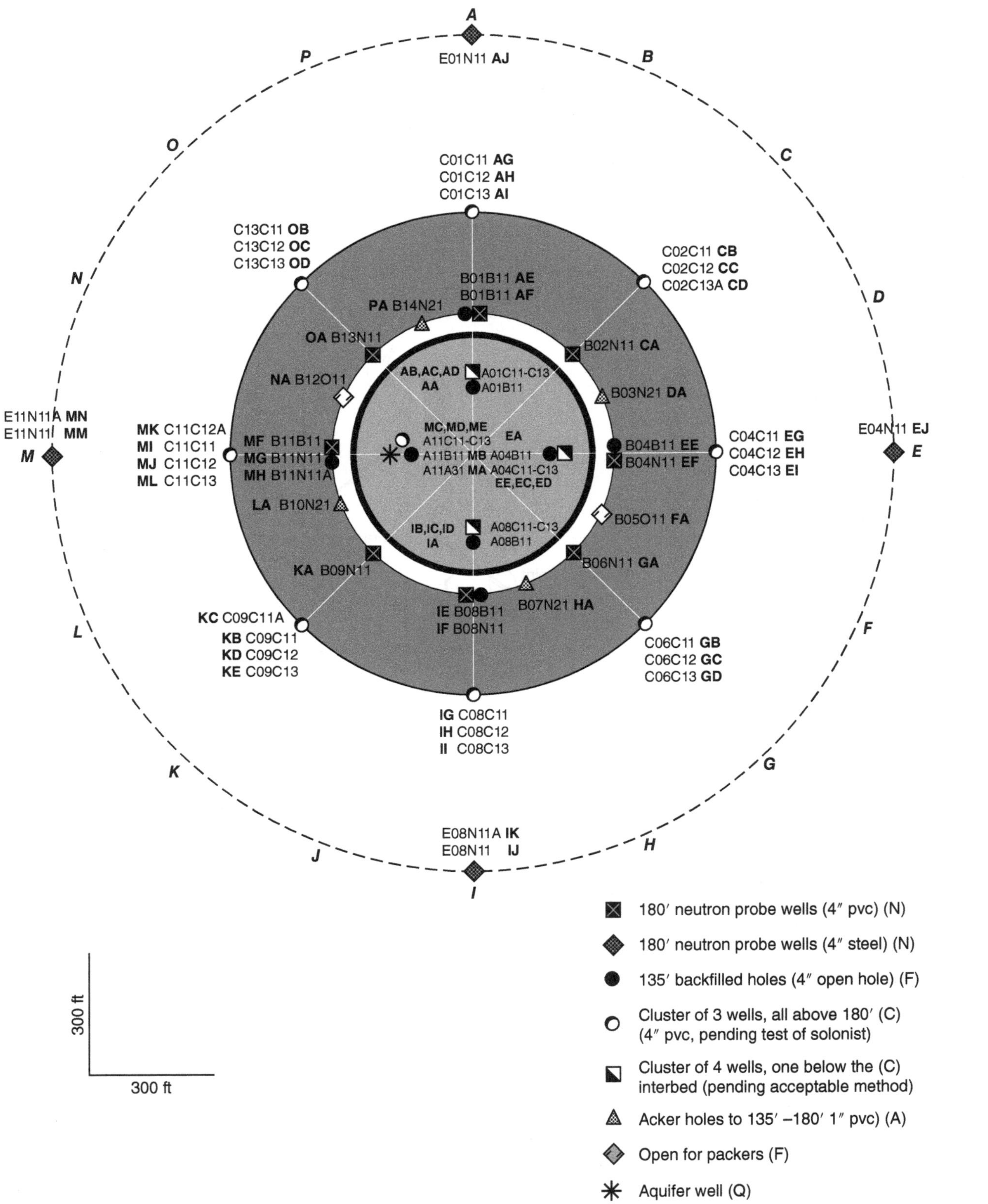

Figure 8.22 An overview of the infiltration experiment at the INEEL. Source: From Dunnivant and Newman (1995); US DOE; public domain.

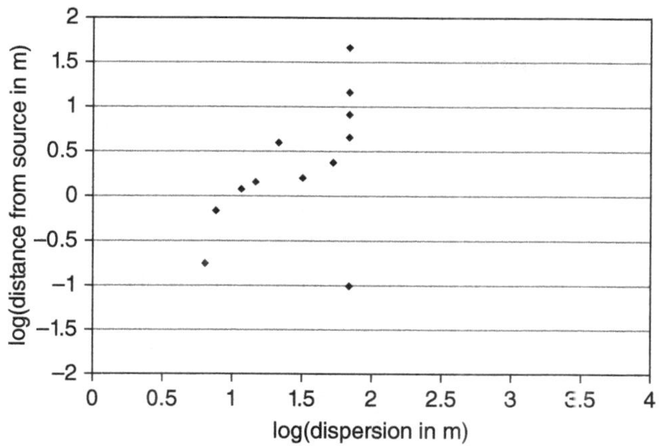

Figure 8.23 Summary of dispersivity values for the INEEL experiment. Units are log(*m*) on both axes. Source: From Dunnivant and Newman (1995); US DOE.

Regarding the main question of the INEEL experiment, "Did pollutants travel to the Snake River Plain Aquifer during the flooding events?," researchers of the experiment concluded that this was unlikely. Neither water nor tracer in the experiment appeared to penetrate the first continuous sedimentary interbed located beneath the basin. The Cold War burial site is also underlain by at least two semicontinuous, relatively thick, sedimentary interbeds. And it should be noted that the experiment flooded the basin for 50 days while the flooding events of the burial sites lasted a few weeks at the longest. Scientists rarely have an absolute answer, and results of the infiltration experiment are still being debated, but most scientists feel that pollutants have not reached the aquifer from the burial site. Again, the point of considering this experiment in such detail was to note the complexity of determining dispersion at a site.

8.7.3 Toward a Universal Estimate Technique for Dispersion Coefficient and Dispersivity

Hydrologists and modelers have long sought a way of estimating dispersivity for a given aquifer based on data from other aquifers. The desire for such a generalized model is understandable, given the huge effort and expense of characterizing every polluted site, as done in the INEEL experiment. However, very limited success has been achieved in developing such a predictive method. Gelhar et al. (1992) was one of the first to attempt to use this approach, when they compiled dispersivity data (in meters as a function of distance from a point source) from 59 field studies of dispersivity into one data plot. An more extensive and problematic plot can be found in Fetter (1993, figure 2.27, p. 90). These dispersivity data are shown in Figure 8.24, with the solid diamonds representing data from porous media

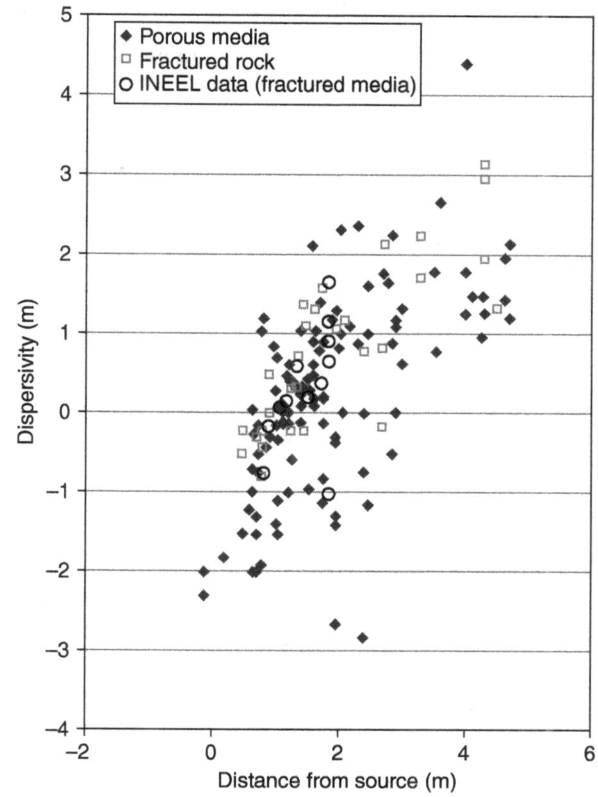

Figure 8.24 A summary of dispersivity data from various field investigations. Units are log(*m*) on both axes. Source: Data from Gelhar et al. (1992).

(sand, etc.), the hollow squares representing data from fractured rock, and the hollow circles representing data from the INEEL experiment (fractured media). Several points should be made concerning this plot as compared to other published versions of the Gelhar et al. (1992) data. First, we have intentionally chosen to use the same scale on the x- and y-axes. Some other versions of this plot have extended the x-axis to give the standard elongated rectangle along the x-axis, but this seriously biases the interpretation of the data. After plotting the data in the elongated rectangle, researchers often then draw a regression line through the data to "create" a trend of increasing dispersivity with increasing distance from the point source. As the reader can clearly see from our plot in Figure 8.24, there is little to no evidence to support such a linear trend in either the porous or fractured media data sets. Although dispersivity does seem to generally increase with increasing distance from the point source, we would not call this a predictable trend and definitely should not draw a regression line through it as many hydrologists have done. For example, say you want to predict the dispersivity at your site at a distance of 100 m (a log value of 2 on the x-axis in Figure 8.20). You could choose a dispersivity from approximately

0.001 78 m (a log value of −2.72) to approximately 316 (a log value of +2.5). The "acceptable" values from such a "correlation" vary over 5 orders of magnitude! A child with a dartboard could produce more accurate estimates of dispersivity. But here we see the problem: there is a great need to develop such a "correlation" in order to save money and resources on site characterization, and even when we spend large amounts of money to characterize a site (i.e. the INEEL experiment), we still obtain a range of dispersivity values (a factor of 15).

The lack of correlation in Figure 8.24 raises another issue. When fate and transport modeling is performed, its goal is to feed risk assessment calculations, and the most important result from the fate and transport modeling is the pollutant concentration in the water. So, which dispersivity value would you select taking into consideration this ultimate goal? If you are from the party responsible for the cost of cleanup (remediation), you would be financially wise to use the high estimate of dispersivity, since this will result in lower pollutant concentrations and less needed cleanup. However, if you are the person who might drink the water in the future, you would be wise to choose the lower dispersivity estimate in the fate and transport calculations, since it will result in higher pollutant concentrations and thus increase the chance that the pollutant will be cleaned up. Who is correct? There is no single, correct answer to this question, but it is best to be conservative and err on the safe side (use low estimates of dispersivity) in most cases.

So, how does one obtain estimates of dispersivity for a polluted site? There is not one absolute answer to this. Unfortunately, you cannot usually perform the extensive characterization experiments at the polluted site, due to cost and fear of increasing the environmental damage from the contamination by pumping pollutants off site. But when possible (when sufficient accuracy is needed to warrant the expense), dispersivity and water velocity estimates can be obtained through experimentation at a site with geology similar to that of the polluted area (generally near the polluted site). As a worst case scenario, one could use data from Figure 8.24 or similar plots, but fate and transport prediction must use the complete range of dispersivity values present in the plot at each distance from the point source. One of the best, and perhaps most common, ways of estimating site-specific dispersivity coefficients is to perform explanatory modeling on the contaminated site based on historical input source data and results from an extensive groundwater monitoring program. Then the estimated dispersivities or dispersivity coefficients, water velocities, and K values can be used to perform future fate and transport modeling.

8.8 Derivation of Mathematical Models for Groundwater

8.8.1 Step Pollutant Input into a Groundwater System

Once again, this model starts very similarly to the step model for the river system discussed above, with the addition of the retardation factor (R). Just like the river system and the pulse model, we start with the previous advection/dispersion Eq. (8.10)

$$\frac{dC}{dt} = -\frac{v}{R}\frac{dC}{dx} + \frac{d}{dx}\left(\frac{E}{R}\frac{dC}{dx}\right) + g(x,t) - d(x,t) \tag{8.9}$$

However, unlike the river model, here we cannot assume steady-state conditions. Due to how slowly pollutants can travel through groundwater systems, it can take years or even decades for the system to fully reach equilibrium conditions. To develop a useful model, therefore, we must include how the conditions evolve over time. As usual, we will assume first-order degradation for the $d(x, t)$ term. The input term here is again most easily dealt with through solving a general form of the equation, then imposing boundary conditions on that solution to solve for the final form of the equation. This gives us a governing equation of

$$\frac{dC}{dt} = \frac{-v}{R}\frac{dC}{dx} + \frac{d}{dx}\left(\frac{E}{R}\frac{dC}{dx}\right) - kC \tag{8.10}$$

The solution to which is far beyond what may comfortably be discussed in this chapter. There is one detail that is worth discussing, however. During the course of the derivation, it becomes necessary to solve an integral of the form

$$\int a\,e^{-bt^2}\,dt \tag{8.11}$$

where a and b are constants. However, there is no general, analytical solution to integrals of the form presented in Eq. (8.11). How then can the derivation proceed? Luckily, this form of integral is found over and over in applied math problems, especially those dealing with heat diffusion or the normal distribution. Due to the number of times this integral is needed, mathematicians have developed a workaround for this integral. A specific integral has been simply defined as the *error function*, erf(x)

$$\mathrm{erf}(x) = \frac{2}{\sqrt{\pi}}\int_0^x e^{-x^2}\,dx \tag{8.12}$$

A number of ways have been developed to numerically approximate the error function. Thus, when integrals such as Eq. (8.11) are encountered, one can define the integral in terms of the error function. Then, when solving for the answer to a specific problem, the value of the error function can be looked up in a table (see Kaye 1955). This allows us to develop models for which there is no exact, analytical solution. The solution to the one-dimensional groundwater – step model uses the complementary error function, erfc(x), which is defined simply as

$$\operatorname{erfc}(x) = 1 - \operatorname{erf}(x) \tag{8.13}$$

The full solution to Eq. (8.11) is in Ogata and Banks (1961) which was derived using dispersivity (in m), not the dispersion coefficient (in m^2/yr) as in the pulse model in the Section 8.8.2. Refer to Eq. (8.9) for conversion between these two parameters.

$$C(x,t) = \left(\frac{C_0}{2}\right)\left(e^{\left(\frac{x}{2\alpha_x}\right)\left(1-\left(1+\frac{R4k\alpha_x}{v}\right)^{\frac{1}{2}}\right)}\right)$$
$$\left(\operatorname{erfc}\left(\frac{x-\left(\frac{v}{R}\right)(t)\left(1+4k\frac{R\alpha_x}{v}\right)^{\frac{1}{2}}}{2\left(\alpha_x\left(\frac{v}{R}\right)t\right)^{\frac{1}{2}}}\right)\right.$$
$$\left.+e^{\left(\frac{x}{\alpha_x}\right)}\operatorname{erfc}\left(\frac{x+\left(\frac{v}{R}\right)(t)\left(1+4K\frac{R\alpha_x}{v}\right)^{\frac{1}{2}}}{2\left(\alpha_x\left(\frac{v}{R}\right)t\right)^{\frac{1}{2}}}\right)\right) \tag{8.14}$$

where

C_0 = the initial concentration of the contaminant,
x = distance from the source,
α_x = longitudinal dispersivity in m (not dispersivity in m^2/time),
k = the first-order reaction rate,
v = velocity,
t = time, and
erfc = the complementary error function.

The final term in the above Eq. (8.14),

$$+e^{\left(\frac{x}{\alpha_x}\right)}\operatorname{erfc}\left(\frac{x+\left(\frac{v}{R}\right)(t)\left(1+4k\frac{R\alpha_x}{v}\right)^{\frac{1}{2}}}{2\left(\alpha_x\left(\frac{v}{R}\right)t\right)^{\frac{1}{2}}}\right),$$

is generally considered insignificant and is ignored. This term is ignored in Fate (http://edusolns.com/fate/2016/welcome/home/). There are also a number of complementary error function tables and calculators on the Internet. The calculation below used is found in https://www.miniwebtool.com/complementary-error-function-calculator/?number=0.249 (accessed 30 March 2018).

Example Problem

A waste lagoon containing 100 μg/l benzene is contaminating the underlying aquifer. The aquifer has a porosity of 30%, a solid density of 2.29 g/cm^3, a bulk density of 1.6 g/cm^3, a velocity of 10 m/yr, and a dispersion coefficient (α_x) of 100 m^2/yr. The distribution coefficient of benzene for this aquifer material has been measured to be 5 ml/g. Benzene biodegrades through a first-order reaction at a first-order rate constant of 0.025 yr^{-1}. Calculate the concentration 10 m downgradient from the lagoon 10 years after the input.

Solution

1) Calculate the retardation factor from the distribution coefficient.

$$R = 1 + \frac{\rho_b K_d}{n}$$

$$R = 1 + \frac{1.6\,\text{g/cm}^3\ 5\,\text{ml/g}}{0.30} = 27.7.$$

2) If necessary, calculate the dispersivity.
   For a velocity of 10 m/yr, and a dispersion coefficient of 100 m^2/yr, the dispersivity is equal to 10 m.
3) If necessary, calculate the first-order degradation rate from the half-life.

$$-\frac{\ln 0.50}{t_{\frac{1}{2}}} = k = -\frac{-0.693}{27.7\,\text{yr}} = 0.025\,\text{yr}^{-1}.$$

4) Convert data into proper units
   C_0 = 100 μg/l,
   v = 10 m/yr,
   α_x (dispersivity) = 10 m,
   R = 27.67, and
   k = 0.025 yr^{-1}.
5) Input data into the Fate (http://edusolns.com/fate/2016/welcome/home/) program or governing equation and obtain/draw a graph.
   Results are shown in Figure 8.25.
6) Calculate the concentration 10 m downgradient from the lagoon 10 years after the input.

$$C(x,t) = \left(\frac{C_0}{2}\right)\left(e^{\left(\frac{x}{2\alpha_x}\right)\left(1-\left(1+\frac{R4k\alpha_x}{v}\right)^{\frac{1}{2}}\right)}\right)$$
$$\left(\operatorname{erfc}\left(\frac{x-\left(\frac{v}{R}\right)(t)\left(1+4k\frac{R\alpha_x}{v}\right)^{\frac{1}{2}}}{2\left(\alpha_x\left(\frac{v}{R}\right)t\right)^{\frac{1}{2}}}\right)\right)$$

$$C(x,t) = \left(\frac{100\ \mu\text{g/l}}{2}\right)$$

$$\left(\mathrm{e}^{\left(\frac{10\ \text{m}}{2(10\ \text{m})}\right)\left(1-\left(1+\frac{(27.3)4\left(\frac{0.025}{\text{yr}}\right)(10\ \text{m})}{10\ \text{m/yr}}\right)^{\frac{1}{2}}\right)}\right)$$

$$\left(\text{erfc}\left(\frac{10\ \text{m} - \left(\frac{10\ \text{m/yr}}{27.7}\right)(10\ \text{yr})\left(1+\frac{4\left(\frac{0.025}{\text{yr}}\right)(27.7)(10\ \text{m})}{10\ \text{m/yr}}\right)^{\frac{1}{2}}}{2\left(10\ \text{m}\left(\frac{10\ \text{m/yr}}{27.7}\right)10\ \text{yr}\right)^{\frac{1}{2}}}\right)\right)$$

$$C(x,t) = \left(\frac{100\ \text{mg/l}}{2}\right)\left(\mathrm{e}^{(0.500)(-0.931)}\right)$$

$$\left(\text{erfc}\left(\frac{10\ \text{m} - \left(\frac{10\ \text{m/yr}}{27.7}\right)(10\ \text{yr})1.94}{12}\right)\right)$$

$$C(x,t) = \left(\frac{100\ \text{mg/l}}{2}\right)0.628\ (0.725)$$

$$C(x,t) = 22.8\ \mu\text{g/l benzene.}$$

8.8.2 Instantaneous Pollutant Input to a Groundwater System

The derivation of the one-dimensional groundwater model is very similar to that of the river model. After all, both are modeling the pollutant concentration over time from water moving in a one-dimensional line. The only mathematical difference is from the retardation factor (R). Due to the adsorption of pollutant onto the soil particles, both advection and dispersion will be slowed by a factor of R. This gives a governing equation of (8.15)

$$\frac{\mathrm{d}C}{\mathrm{d}t} = -\frac{v}{R}\frac{\mathrm{d}C}{\mathrm{d}x} + \frac{\mathrm{d}}{\mathrm{d}x}\left(\frac{E}{R}\frac{\mathrm{d}C}{\mathrm{d}x}\right) + g(x,t) - d(x,t) \tag{8.15}$$

where

C is the concentration of pollutant,
v is the water velocity,
E is the dispersion coefficient (m^2/time) (some books use the letter D), not dispersion (m) as used in the pulse ground water (m),
$g(x,t)$ is the input function, and
$d(x,t)$ is the degradation term of the pollutant.

In this equation, the term

$$-v\frac{\mathrm{d}C}{\mathrm{d}x} \tag{8.16}$$

deals with advection, or the transfer of pollutant by the actual flow of the water in the ground, and the term

$$\frac{\mathrm{d}}{\mathrm{d}x}\left(E\frac{\mathrm{d}C}{\mathrm{d}x}\right) \tag{8.17}$$

deals with dispersion of the pollutant due to differences in concentration throughout the groundwater.

This is solved by essentially the same derivation presented in Chapter 6: River Model – Pulse yielding the final form (8.18)

$$C(x,t) = \frac{S_0}{A\sqrt{4\pi\frac{E}{R}t}}\exp\left(-\frac{\left(x-\frac{v}{R}t\right)^2}{4\frac{E}{R}t} - kt\right) \tag{8.18}$$

where S_0 is the mass of the input (sometimes represented by M, mass) and A is the cross-sectional area (m^2).

Except in groundwater applications E, sometimes represented by D, the dispersion coefficient, depending on the governing equation derivation, is used.

Note that in Eq. (8.12), we use the dispersion coefficient (in m^2/time), not dispersivity (in m), in transport equations and model, where it is a function of distance from the point source, where

$$D = \alpha_x * v \tag{8.19}$$

$$C_{(x,t)} = \frac{C_0}{2}\left(\mathrm{e}^{\left(\frac{x}{2\alpha_x}\right)\left(1-\left(1+\frac{4kR\alpha_x}{v}\right)^{\frac{1}{2}}\right)}\right)\left(\text{erfc}\left(\frac{x-\left(\frac{v}{R}t\left(1+\frac{4kR\alpha_x}{v}\right)^{\frac{1}{2}}\right)}{2\left(\alpha_x\frac{v}{R}t\right)^{\frac{1}{2}}}\right)\right)$$

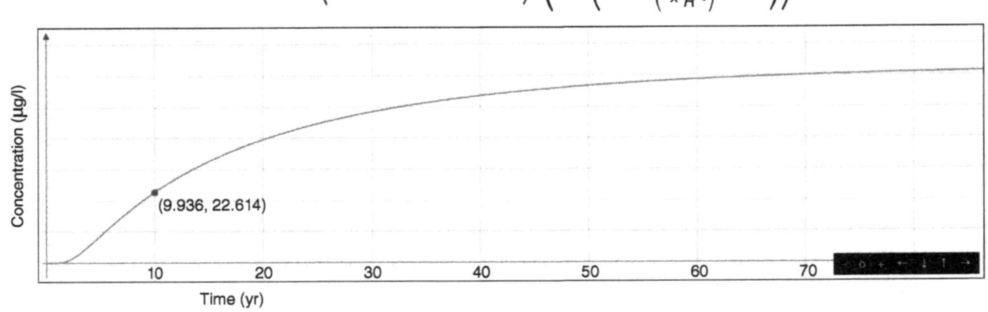

Figure 8.25 Example step problem output from Fate®. Source: Anders.

where

D is the dispersion coefficient (m^2/yr),
α_x is the called the longitudinal dispersivity (m), and
v is the water velocity (m/yr).

This equation is confirmed in Fjeld et al. (2006, table 4.1, p. 84). Indeed, we could imagine the river model (Chapter 6), as merely a special case of Eq. (8.19), where due to the nature of a river $R = 1$.

An important point in the plot of Eq. (8.8) is the location of the apex of the bell-shaped concentration curve. This point can be found where the exponent term

$$x - \frac{v}{R}t = 0 \tag{8.20}$$

Example Problem

A $100\,m^3$ tanker containing a $100\,mg/m^3$ solution of 2,4-dinitrotoluene crashes and empties its entire contents above an aquifer. The cross-sectional area of the spill is $50\,m^2$. The aquifer has a porosity of 30%, a bulk density of $1.6\,g/cm^3$, a velocity of 10 m/yr, and a dispersion coefficient (D_x) of $10\,m^2$/yr. The distribution coefficient of 2,4-dinitrotoluene for this aquifer material has been measured to be 2.5 ml/g. 2,4-Dinitrotoluene biodegrades through a first-order reaction at a rate of $0.693\,yr^{-1}$. Calculate the concentration 10 m downgradient from the lagoon 10 years after the input.

Solution

A $100\,m^3$ tanker containing a $100\,mg/m^3$ solution of 2,4-dinitrotoluene crashes and empties its entire contents above an aquifer. The cross-sectional area of the spill is $50\,m^2$. The aquifer has a porosity of 30%, a bulk density of $1.6\,g/cm^3$, a velocity of 10 m/yr, and a dispersion coefficient (E_x or D_x) of $10\,m^2$/yr. The distribution coefficient of 2,4-dinitrotoluene for this aquifer material has been measured to be 2.5 ml/g. 2,4-Dinitrotoluene biodegrades through a first-order reaction at a first-order rate constant of $0.693\,yr^{-1}$. Calculate the concentration 10 m downgradient from the lagoon 10 years after the input.

1) Calculate the retardation factor from the distribution coefficient:

$$R = 1 + \frac{\rho_b K_d}{n}$$

$$R = 1 + \frac{1.6\,g/cm^3\;2.5\,ml/g}{0.30} = 14.3.$$

2) Correct the cross-sectional area of the spill site to the void volume (rather than the total volume):

$$\text{Cross-sectional area}^* \text{ porosity} = 50\,m^2 \times 0.30 = 15\,m^2.$$

3) Calculate the total mass of dinitrotoluene spilled:

$$\text{Volume spilled} * \text{concentration of solution}$$
$$= (100\,m^3)\left(1000\frac{mg}{m^3}\right) = 100\,000\text{ mg or }100\text{ g}.$$

4) If necessary, calculate the first-order degradation rate from the half-life:

$$k \text{ given} = 0.693\,yr^{-1}.$$

5) Arrange data into proper units
groundwater velocity, $v = 10$ m/yr,
the retardation factor, $R = 14.33$,
the mass of contaminant, $M = 100\,000$ mg
the dispersion coefficient, $D = 10\,m^2$/yr,
the reaction rate constant, $k = 0.693\,yr^{-1}$, and
the cross-sectional area, $A = 15\,m^2$.

6) Input data into the Fate program (http://edusolns.com/fate/2016/welcome/home/) or governing equation and obtain/draw a graph.
Results are shown in Figure 8.26.
An additional important distance point in Figure 8.26 is the apex of pollutant concentration. This occurs in Eq. (8.18) when

$$\frac{\partial C}{\partial x} = 0, \text{and } R \neq 0 \text{ and } ADt \neq 0$$

or

$$x = \frac{tv}{R} \tag{8.21}$$

where x is analogous to the critical point in the other fate and transport modules previously discussed in this book. At this point, the pre-exponential and exponential terms in Eq. (8.8) reduce to zero. This occurs in this example problem at 6.9 m. But note, as with the river pulse model, the Gaussian-shaped peak moves from the input well or source through the aquifer as x (distance) and t (time) increase resulting in a shifting peak moving downgradient. For every point where x and t results in a zero change in concentration with change in distance, there is a peak.

7) Calculate the concentration 10 m downgradient from the lagoon 10 years after the input.

$$C(x,t) = \frac{S_0 \text{ or } M}{A\sqrt{4\pi\frac{E}{R}t}} e\left(-\frac{\left(x - \frac{v}{R}t\right)^2}{4\frac{E}{R}t} - kt\right)$$

$$C(x,t) = \frac{100\,000\text{ mg}}{15\,m^2\sqrt{4\pi\frac{10\,m^2/yr}{14.33} * 10\,yr}}$$

$$e\left(-\frac{\left(10\,m - \left(\frac{10\,m/yr}{14.3}\right)(10\,yr)\right)^2}{4\frac{10\,m^2/yr}{14.3} * 10\,yr} - 0.693\frac{1}{yr} * 10\,yr\right)$$

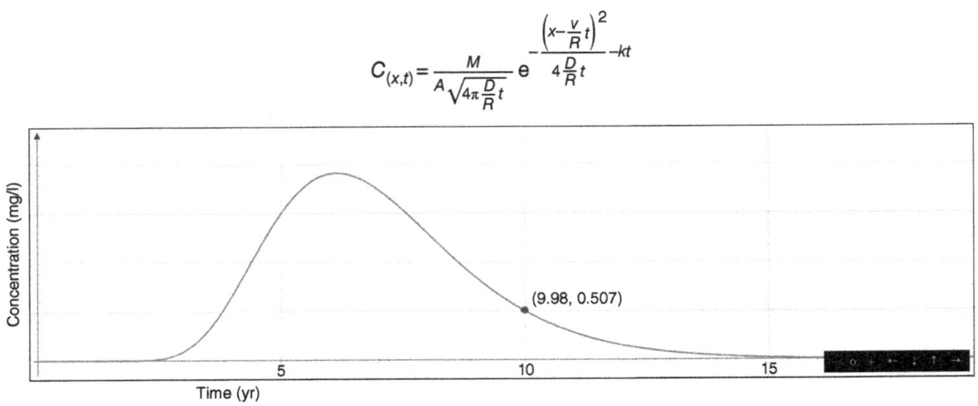

Figure 8.26 Example instantaneous problem output from Fate®. Source: Anders.

$$C(x,t) = \frac{100\,000\ \text{mg}}{15\ \text{m}^2 * 9.36}\text{e}\left(-\frac{9.13}{27.9} - 6.93\right)$$

$$C(x,t) = 712 * 0.000\,708$$

$$= 0.502\ \text{mg/m}^3 \text{ or } 502\ \mu\text{g/l}\ 2,4-\text{dinitrotoluene}.$$

8.8.3 More Realistic but Complex Models: 2-D and 3-D Models

The governing pulse and step pollutant fate and transport models in this chapter are the most complex covered yet in this book, but they are gross simplifications of reality. Take the pulse example above that can be illustrated in Figure 8.27. In the one-dimensional equation above, we assumed that the pollutant source was introduced over the height of the receiving aquifer and only dispersed and reacted in the x direction. In reality, from a point source in the well, it will be diluted by dispersion in the y direction (width of the plume) and in the z direction (height and depth of the aquifer). The net result is an overestimation of concentration by the one-dimensional model. The three-dimensional models thus require, in Eq. (8.17), an additional erfc term in the y direction with a_y and an additional erfc term in the z direction with an a_z term, making this extremely complex.

In contrast, take the more realistic representation of a step pollutant input shown in Figure 8.28. Here the pollutant source is introduced at one point in the well, and as shown is dispersed in the x direction with the flow of ground water. Not shown is the dispersion in the y and z directions. Again, the one-dimensional model overestimates the downgradient pollutant concentration.

While the one-dimensional models developed above can be helpful in certain circumstances, especially for teaching purposes and when the aquifer is constrained by impervious layers to the point that it starts to resemble an underground river, in most cases more complexity is required to gain an accurate understanding of pollutant behavior. The models can be extended to cover additional dimensions by modification of the governing equation. The one-dimensional models began with the equation

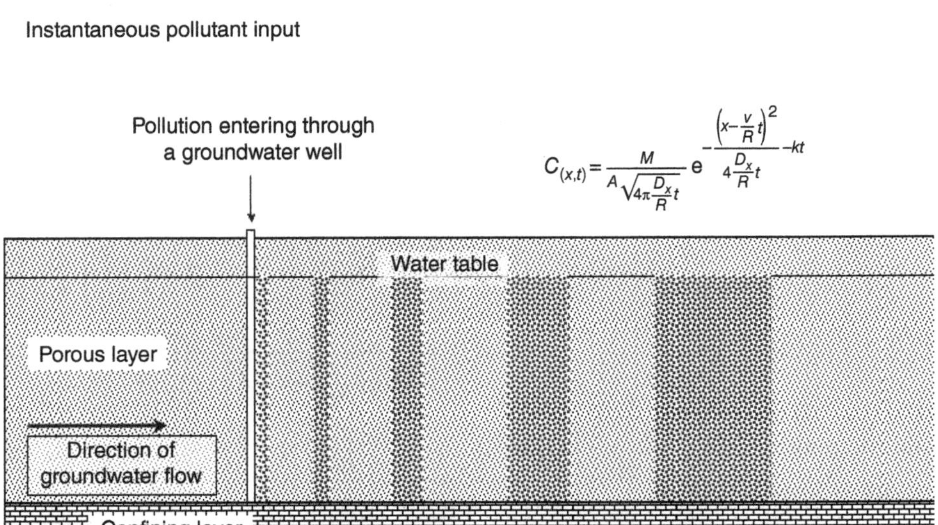

Figure 8.27 An illustration of an instantaneous (pulse) input of pollution to and one-dimensional dispersivity in an aquifer. Source: Fate®, Anders.

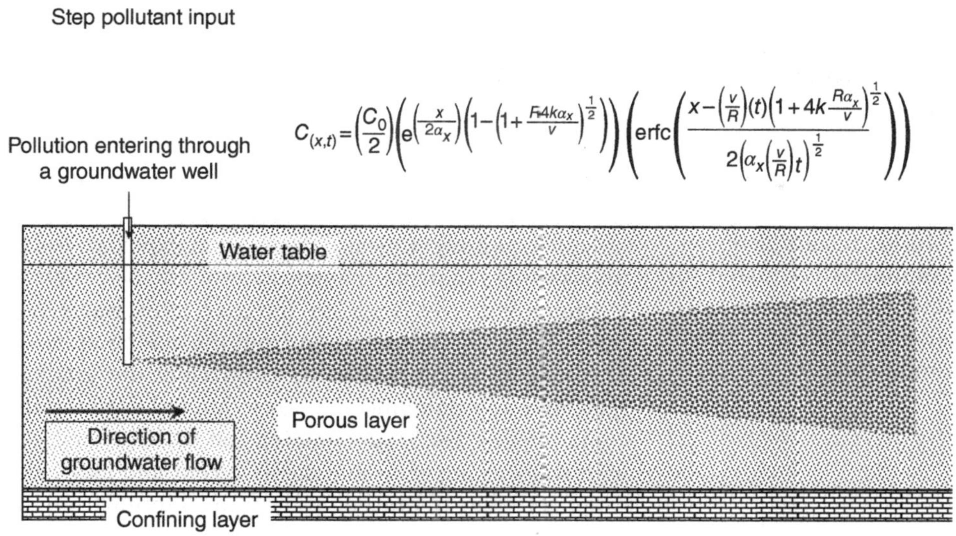

Figure 8.28 An illustration of a step (continuous) input of pollution to and two-dimensional dispersivity in an aquifer. Note the model shown is one dimensional.

mass balance equation

$$\frac{dC}{dt} = -\frac{V}{R}\frac{dC}{dx} + \frac{d}{dx}\left(\frac{E}{R}\frac{dC}{dx}\right) + g(x,t) - d(x,t),$$

which contains terms for advection and dispersion only in the x direction. In order to describe additional dimensions, we simply add terms to the governing equation that deal with dispersion and advection in the y and z dimensions. One three-dimensional treatment of the mass balance is thus

$$\frac{dC}{dt} = -\frac{V}{R}\nabla C + \frac{E}{R}\nabla^2 C + g(x,y,z,t) - d(x,y,z,t) \tag{8.22}$$

where ∇ is an operator known as the gradient and is the partial derivative of a function in all dimensions

$$\nabla C = \frac{\delta C}{\delta x} + \frac{\delta C}{\delta y} + \frac{\delta C}{\delta z} \tag{8.23}$$

As can be imagined, the solutions to equations such as Eq. (8.23) become extremely complex, comparable to a master-level thesis in mathematics, and are thus rarely if ever used in teaching or as an algebraic model. Numerical models are generally used instead, due to their relative simplicity. Additionally, the use of numerical models can allow for more accurate results through combining models. Due to the inhomogeneous nature of the Earth, flow in aquifers is not simple linear flow, as the one-dimensional equations assume. Instead, flow is changed by impermeable barriers, changes in soil composition, and even soil particle size. To take these into account, a region-specific numerical model is often developed. The pollutant transport model can then be built on top of this flow model, using the first model for the advection term, instead of assuming constant flow. This "nesting" of models can only be accomplished numerically, via computer modeling.

Such a solution is shown in Fjeld et al. (2006, equation 4.15, p. 97) for an instantaneous (pulse) input of a conservative tracer

$$C(x,y,z,t) = S_0 \frac{e^{-\frac{(x-vt)^2}{4D_x t}}}{\sqrt{4\pi D_x t}} \frac{e^{-\frac{y^2}{4D_y t}}}{\sqrt{4\pi D_y t}} \frac{e^{-\frac{z^2}{4D_z t}}}{\sqrt{4\pi D_z t}} \tag{8.24}$$

8.9 Sensitivity Analysis

While chemical factors, such as sorption and degradation, can determine the downgradient pollutant concentration, the most difficult parameter to estimate is hydrodynamic dispersion. Sensitivity analysis should include a variety of values for distribution and partition coefficients and chemical, biological, and nuclear degradation rates. In terms of dispersion estimates, Figure 8.24 should be used for your initial estimate and should be combined with water velocity to obtain estimates of dispersion coefficients. Note that your sensitivity analysis should use large ranges of dispersion coefficients, in some cases from 2 to 4 orders of magnitude.

8.10 Limitations of Our Models

In this chapter, we presented the classic one-dimensional general transport equation in order to keep the mathematics simple and concentrate on the important processes. It should be noted that professional modelers use two- and three-dimensional models, and use numerical methods of analysis, which allow for variations in aquifer conditions such as water velocity, sorption phenomena, degradation rates, and, most importantly, dispersion. The use of extra dimensions in the modeling (and the real-world) results in lower downgradient pollutant concentrations as compared to one-dimensional models. Obtaining an accurate estimate of dispersion or the dispersion coefficient is the major limitation of groundwater modeling.

8.11 Remediation

Contaminated groundwater systems are one of the most difficult environmental media to remediate, due to their location below the land surface (out of direct observation), the large volumes of soil and water involved, and the slow water velocities inherent in these aquifers. Many technologies and approaches have been developed specifically for groundwater remediation with mixed and site-specific successes and failures. There are numerous books on groundwater remediation, and given the expansive research in this area we will only briefly summarize some of the common approaches to this problem. In general, we will begin with the least expensive and complicated and progress to the more expensive.

A summary of suggested and common remediation actions are given in Figure 8.21 (EPA 1998), and this list has been expanded on since its original publication. These actions are divided into natural attenuation, containment, and active restoration. The first two categories, while similar, are two distinct approaches adopted by EPA. Both of these can be undertaken in a variety of cases and for a variety of reasons. For example, even when a risk assessment shows no immediate risk, governing bodies (EPA, local government, or the public) sometimes decide that some control of the site is necessary. Natural attenuation or containment may also be employed when there is no acceptable remediation plan at present and the site needs to be removed from possible human contact, or when remediation of the site using one of the acceptable technologies may create a worse problem than waiting for the development of a whole new remediation approach. Notice that all remediation plans involve long-term monitoring of the site.

Today, natural attenuation may be expanded to a form of active restoration, by employing the addition of nutrients to aid in the microbial removal of pollutants. These projects still take years to decades to complete, but operating costs are relatively inexpensive compared to other approaches such as removal and treatment of the soil. Institutional controls, a common listing in Figure 8.29, are basically easements placed on the land deed specifying and limiting how the land can be used or developed in the future.

In some cases, containment may be the only viable and affordable option. Containment does nothing to treat the site, but limits the further spread of pollution and is intended to protect local citizens from exposure to the toxins while a more permanent solution can be designed. Containment uses ways to limit or stop water and therefore pollutant migration from the site boundaries. This can be accomplished by controlling the gradient (placing water pumping and extraction wells around the contaminated area), surrounding the underground site with walls of impermeable media (grout and clay), and capping the site to prevent water from entering from the ground surface and spreading the pollution to new areas.

There are a seemingly endless number of active treatment technologies for polluted groundwater and soil sites. These are divided into extraction of polluted groundwater (termed "pump and treat"), direct treatment options that may involve the removal of the contaminated soil and water, and institutional controls that limit use of the water from the site. The first approach, pump and treat, was originally thought of as the cure-all of polluted groundwater. The concept is simple: polluted groundwater is pumped out relatively slowly, treated at the ground surface, and then the "clean or cleaner" groundwater is pumped back into the polluted aquifer in order to displace more pollution. The water extraction and injection wells are placed so as to contain the polluted groundwater plume. By using partition and distribution coefficients, it is relatively easy to estimate how many aquifer volumes of groundwater must be removed and treated for the pollution to be effectively removed. *However, after decades of use, this technology has proven ineffective in many, though not all, cases.* Sites that were once thought to be remediated have been later found to still contain unsafe levels of pollutants in their water. This is due to the slow desorption of pollutants from the soils. After the pump and treat system has been shut off, the pollution reoccurs by desorbing from the polluted soil. Thus, the pump and treat approach has been augmented with the addition of in situ biological and chemical treatment processes to increase pollutant degradation and/or removal.

There are many ex-site treatment technologies that can be used on the polluted groundwater after it has been brought to the surface. These are too numerous to even mention here but include physical, biological, and chemical treatment. After treatment, the water may be reinjected or disposed of in other manners depending on the local restrictions.

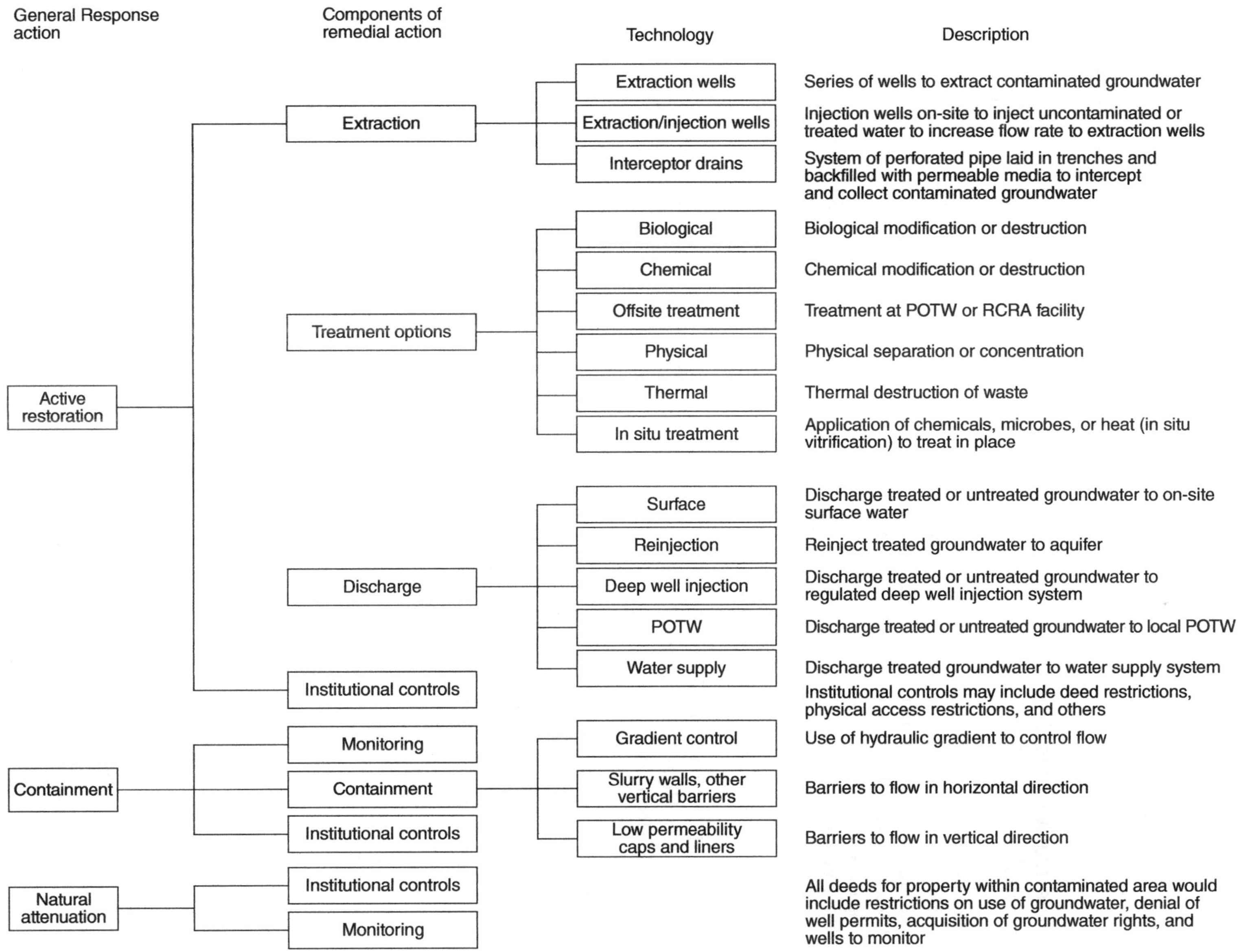

Figure 8.29 A summary of technologies used in groundwater remediation. Source: From EPA (1998), pp. 5–10).

In situ treatments include reductive cements and zero-valent iron for reductive dehalogenation of chlorinated solvent plumes.

Polluted aquifers are one of the most difficult and costly sites to remediate. In fact, most of the US Superfund effort has focused on remediating groundwater. Yet, there is much to be accomplished in the future given the difficult nature of these systems.

8.12 Numerical Models Used by Professionals

There have been numerous numerical models designed for a variety of aquifer and media conditions, and include both the modeling of water flow and pollutant fate and transport. A few of these will be discussed in Chapter 10, but US Geological Survey and EPA are at the forefront of developing these models.

Suggested Papers for Class Discussion

Dunnivant, F.M., M.E. Newman, C.W. Bishop, D. Burgess, J.R. Giles, B.D. Higgs, J.M. Hubbell, E. Neher, G.T. Norrell, M.C. Pfiefer, I. Porro, R.C. Starr, and A.H. Wylie. Water and radioactive tracer flow in a heterogeneous field-scale system, *Groundwater*, 1998, 36(6), pp. 949–958.

Dunnivant, F.M., P.M. Jardine, D.L. Taylor, and J.F. McCarthy. Cotransport of cadmium and hexachlorobiphenyl by dissolved organic carbon through columns containing aquifer material, *Environ. Sci. Technol.*, 1992a, 26(2), pp. 360–368.

Concepts

8.1 Contrast the following pairs of terms:
Saturated versus unsaturated media
Porous versus fractured media
Dispersivity versus dispersion coefficient
Consolidated versus nonconsolidated media

8.2 Research the Safe Drinking Water Act (SDWA) on the Internet. Write a one-page summary of the act with respect to groundwater, priority pollutants, and maximum contaminant levels.

8.3 Describe three examples of step inputs and three examples of pulse inputs of pollutants to a groundwater system in your area.

8.4 Discuss the difficulties in installing a monitoring well with respect to obtaining a representative groundwater sample.

8.5 Describe the major ways of installing a groundwater monitoring well. List the advantages and disadvantages of each.

8.6 Describe each of the three ways of obtaining a K_d or K_p for a pollutant.

8.7 Select two pollutants from Table 2.6 and calculate K_p value for them using two of the predictive equations in this chapter. Compare the values between the two equations.

8.8 Explain the relationship between R and K.

8.9 Draw typical breakthrough curves (BTC) for pulse and step inputs of pollutant.

8.10 Use Figure 8.19 to explain the difficulties in estimating dispersion for an untested system.

8.11 Summarize the three ways of estimating/determining dispersion in a groundwater system.

8.12 You work a problem using a one-dimensional groundwater model and obtain a pollutant concentration at a specific time and point in an aquifer. How would using a 2-D and 3-D model change your answer, increase or decrease pollutant concentration?

Exercises

Solutions are available using the Fate software at http://edusolns.com/fate/2016/welcome/home/.

Pulse (instantaneous) problems

8.1 A storage tank breaks during an earthquake event, spilling gasoline treated with MTBE into the surrounding soil (glacial till) and contaminating the groundwater. Create a plot (calculations every 4 m downgradient) of concentration of MTBE versus distance (0–22 m) 10 years after the spill, using the following parameters:
Cross-sectional area: 50 m^2
Groundwater velocity: 5 m/yr
Initial concentration of MTBE resulting from spill: 35 mg/m^3
Volume spilled: 100 m^3

Half-life of MTBE: 1.0 year
Dispersion coefficient: 10 m^2/yr
Bulk density: 1.6 g/cm^3
K_d: 1.072 ml/g
Porosity: 0.23
Now, for your sensitivity analysis, prepare the previous plots using 1 m^2/yr as your dispersion coefficient.
Answer: The maximum MTBE concentration will occur at 5.0 years at 0.56 ppm.

8.2 A tanker car from a mining operation leaks and spills 100 g HCN in water over a sand aquifer with an area of 20 m. The EPA asks you to assess the damage and to determine the probable concentration of HCN in the groundwater. The velocity of the groundwater is 180 m/yr, the dispersion coefficient is 2 m^2/yr, and the retardation factor is 1.050. The half-life of HCN under the conditions at the site is 1.00 year. The bulk density of the sand is 1.5 g/cm^3 and the porosity is 0.30. The K_d for HCN is 0.010. Calculate the concentration of HCN 0.04, 0.055, and 0.065 years after the spill. Use a monitoring well at 10 m downgradient from the spill.
Answer: At 0.04 years, the HCN concentration equals below the detection limit; at 0.055 years, HCN equals 1924 mg/l; at 0.065 years, HCN equals 274 mg/l.

8.3 A manufacturing company stores perfluorooctanoic acid (PFOA), a compound used in polymerization reactions, in 10 kg tanks in a warehouse. One day, one of the tanks ruptures, spilling its contains in a floor drain and into the subsurface below the building. The warehouse is built on a relatively sandy aquifer, with a porosity of 0.40 and a density of 1.6 g/cm^3, so the groundwater flows at a rate of 300 m^2/yr. This also results in a high dispersivity, which is estimated at 10 m^2/yr. The cross-sectional area of the spill is 1000 m^2. The K_d for PFOA is 251 cm^3/g, and it degrades with a half-life of 19.8 years. The EPA hires you to examine what the downgradient profile of the contamination will look like in 50 years. As a point calculation, what will be the PFOA concentration at 10 years and 4 m?
Answer: 474 mg/l.

8.4 Cadmium nitrate waste was spontaneously released into an aquifer from a tank. Assume no degradation of the metal. Calculate the concentration of Cd^{2+} in a groundwater well located at 10 m and 50 years later, using the conditions below.
Cd concentration in the storage tank: 1500 mg/m^3
Tank size: 300 m^3
Cross-sectional area: 100 m^2
Groundwater velocity: 8 m/yr
Dispersion coefficient: 8 m^2/yr
Porosity: 0.20
Bulk density: 1.8 g/cm^3
K_d: 10 cm^3/g
Answer: 101 mg/l.

Step (Continuous) Inputs

8.5 Arsenic in the ground contaminates an aquifer by a naturally occurring, continuous geologic process. The sandy aquifer has a bulk density of 1.08 g/cm^3 and a solid density of 2.63 g/cm^3. The distribution coefficient (K_d) in this scenario is 0.027 cm^3/g. Consider that the initial concentration of arsenic at the source is 21 μg/l, the longitudinal dispersion is 10 m^2/d (3650 m^2/yr), and the linear water velocity is 100 m/yr (confirm that the dispersivity equals 3650 m). Using this information, determine what the equilibrium As concentration (in μg/l) will be at a distance of 100 m. The data from this problem is organized in the table below.
Note: Arsenic does not have a rate constant since it is stable (does not degrade). However, you can use a very high value to graph this in Fate (suggested value: 60 000).
Use Fate to create a plot of the scenario and to check your results.

Bulk density	1.08 g/cm^3
Solids density	2.63 g/cm^3
K_d	0.027 cm^3/g
Initial concentration	21 μg/l
Dispersion coefficient	10 m^2/day
Dispersivity	3650 m
Water velocity	100 m/yr

Answer: 18.3 μg/l at 100 m.

8.6 Xylene is one of the major components of gasoline. It can be toxic to humans when released into groundwater drinking sources. Some of the harsher effects of xylene on the human body include brain hemorrhaging or death. One Summer in 1979, before gas stations had started using double-layered underground storage tanks

(USTs), a UST at a gas station began corroding. Gasoline had been spilling out of the UST for 10 years before any remediation efforts were undertaken. Xylene has a half-life of 30 days. The ground beneath the UST was fairly porous: it had a bulk density of 1.44 g/cm^3 and a solid density of 2.94 g/cm^3. The initial concentration of xylene in the groundwater was 37 ppm Below the UST, the ground water moved at a velocity of 15 m/yr and dispersed longitudinally at 50 m^2/yr (with a dispersivity of 3.3 m). The K_d of xylene is 1.22 cm^3/g. For this system, what was the concentration of xylene 20 m from the UST 1, 2, 5, and 10 years after the spill began?
Answer: 7.93 ppb at 1 year, 873 ppb at 2 years, 14 200 at 5 years, and 30 400 at 10 years.

8.7 You are a chemist who has been hired to analyze the step release of a DNAPL (1,1,1-trichloroethane, TCE) into a coarse sand aquifer. The average linear groundwater velocity for this aquifer is 10 m/yr with a longitudinal dispersivity of 19 m. Use a bulk density of 1.44 g/cm^3 and a solid density of 2.40 g/cm^3. The sand has a retardation factor for TCE of 28 (K_d equals 7.5). The pollutant has a half-life of 56 days and is initially present at a concentration of 5.50 ppb as a constant input from the DNAPL source. How much pollutant (what concentration) is present at distances of 20.0 m from the release point at times of 5, 10, 20, and 50 years after the accident?
Answer: 0.074 ppb at 5 years, 0.463 ppb at 10 years, 1.20 ppb at 20 years, and 2.57 ppb at 50 years.

8.8 Chromium(VI) is leaking into the groundwater from an abandoned wood treatment facility with an initial concentration of 20 µg/l. The subsurface is fractured basalt with a bulk density of 1.99 g/cm^3 and a solid density of 2.4 g/cm^3. Chromium(VI) as the dichromate anion has a K_d of 0.030 ml/g and does not degrade. The groundwater velocity is 8 m/yr, with a dispersivity of 30 m. What is the Cr ion concentration 10 m from the leakage site in 10 years?
Answer: 15.9 ppb.

8.9 Nitrates are a common contaminate in groundwater systems because they are present in fertilizer runoff, manure, and septic system leachate. Ingestion of nitrate can inhibit the ability of red blood cells to carry oxygen; infants are the most at risk and can develop "blue baby syndrome" due to lack of oxygen. As a result, EPA has set the maximum nitrate concentration (MCL) in drinking water at 10 mg/l. In one instance, the nitrate concentration leaching into a sandy aquifer is 13 000 µg/l. A well is located 100 m downgradient, when will it reach the MCL for nitrate?
Aquifer/pollutant characteristics:
Bulk density equals 1.99 g/cm^3; solid density equals 2.4 g/cm^3
Partition coefficient equals 0.017 cm^3/g for an ion
Nitrate degradation half-life equals 120 years in aerated soils
Dispersivity equals 5 m
Water velocity equals 18 m/yr
Answer: ~1.7 years to reach MCL at 100 m downgradient.

8.10 A nuclear waste site discovered that one of its USTs was seeping into the groundwater below. The tanks were seeping plutonium at a rate of 10 950 µg/l. The porosity is 25%, has a bulk density of 1.8 g/cm^3, a solid density of 2.4 g/cm^3, and a K_d value of 0.9 cm^3/g. The groundwater velocity is 8.6 m/yr. The dispersivity is 25 m. Assume plutonium does not decay at an appreciable rate. Based on this information, calculate the concentration of plutonium after 2 years 25 m away from the source.
Answer: 188 µg/l at 2 years and 25 m downgradient.

8.11 In one of your inspections as an EPA official, you discover that a company's underground acetone storage tanks has been leaking relatively pure acetone into the groundwater for the past 6 months. Knowing the parameters of the aquifer (given below), what will the concentration of acetone be 20 m downgradient from the tank? Treat the input as a step, continuous input.
Time = 6 months
$x = 20$ m
initial acetone concentration = 10 000 µg/l
bulk density = 1.8 g/cm^3; solid density = 2.4 g/cm^3
$K_d = 0.5$ cm^3/g
$v = 800$ m/yr
$k = 0.23$/day
dispersivity = 10.5 m
Answer: 9410 µg/l.

Capstone Exercise

Research an aquifer or groundwater pollution event and collect all necessary data to use Fate in order to model a pollution scenario. Develop a problem statement similar to those given earlier.

Spreadsheet Exercise

Create a spreadsheet that performs the same calculations as Fate for a pulse input of pollutant to a groundwater system. Construct your spreadsheet so that it is interactive (so that you can change numeric values for parameters and the plot automatically updates itself).

Bibliography

Chiou, C.T., Peters, L.F., and Freed, V.H. (1979). A physical concept to soil-water equilibria for nonionic compounds. *Science* **206**: 831–832.

Christman, M.C., Benson, C.H., and Edil, T.G.B. (2002). Geophysical study of annular well seals. *Ground Water* **22** (3): 104–112.

Dunnivant, F.M., Porro, I., Bishop, C. et al. (1997). Verifying the integrity of annular and back-filled seals for vadose-zone monitoring wells. *Groundwater* **35** (1): 140–148.

Dunnivant, F.M., Jardine, P.M., Taylor, D.L., and McCarthy, J.F. (1992a). Cotransport of cadmium and hexachlorobiphenyl by dissolved organic carbon through columns containing aquifer material. *Environ. Sci. Technol.* **26** (2): 360–368.

Dunnivant, F.M., Jardine, P.M., Taylor, D.L., and McCarthy, J.F. (1992b). Transport of naturally occurring dissolved organic matter in laboratory columns containing aquifer material. *Soil Sci.* **56** (2): 437–444.

Dunnivant, F.M., Jardine, P.M., Taylor, D.L., and McCarthy, J.F. (1992c). The cotransport of cadmium, 2,2',4,4',5,5'-hexachlorobiphenyl, and 2,2',4,4',6,6'-hexachlorobiphenyl by dissolved organic carbon in laboratory columns containing aquifer material. *Environ. Sci. Technol.* **26**: 360–368.

Dunnivant, F.M., Newman, M.E., Bishop, C.W. et al. (1998). Water and radioactive tracer flow in a heterogeneous field-scale system. *Groundwater* **36** (6): 949–958.

Dunnivant, F.M. and Newman, N.E. (1995). Migration of Radionuclide Tracers through Fractured Media: Preliminary Modeling Results of Breakthrough Curves Obtained during the Large-Scale Aquifer Pumping and Infiltration Tests. U.S. Department of Energy, Idaho National Engineering Laboratory, INEL-95/288, ER-WAG7-84 (November 1995).

EPA (1998). *Guidance on Remedial Actions for Contaminated Ground Water at Superfund Sites.* Office of Emergency and Remedial Response, EPA 540 G-88 003, OSWER Directive 9283 (1–2 December 1988).

Fetter, C.W. (1993). *Contaminant Hydrogeology*. New York: Macmillan Publishing Company.

Fjeld, R.A., Eisenberg, N.A., and Compton, D. (2006). *Quantitative Environmental Risk Analysis for Human Health*. New York: Wiley Interscience.

Gelhar, L.W., Welty, C., and Renfeldt, K.R. (1992). A critical review of data on field-scale dispersion in aquifers. *Water Resour. Res.* **28** (7): 1955–1974.

Johnson Filtration Systems, Inc (1986). *Groundwater and Wells*. St. Paul, MN.

Karickhoff, S.W., Brown, D.S., and Scott, T.A. (1979). Sorption of hydrophobic pollutants on natural sediments. *Water Res.* **13**: 241–248.

Kaye, J. (1955). A table of the first eleven repeated integrals of the error function. *J. Math. Phys.* **34**: 119–125. https://doi.org/10.1002/sapm1955341119.

Means, J.C., Wood, S.G., Hassett, J.J., and Banwart, W.L. (1980). Sorption of polynuclear aromatic hydrocarbons by sediments and soils. *Environ. Sci. Technol.* **14**: 1524–1528.

Ogata, A. and Banks, R.B. (1961). A solution of the differential equation of longitudinal dispersion in porous media. Geological Survey Professional Paper 411-A.

van Genuchten, M.T. and Alves, W.J. (1982). *Analytical Solutions of the One-dimensional Convective-Dispersive Solute Transport Equation*. Technical bulletin no. 1661. Washington, DC: US Department of Agriculture.

Wood, T.R. and Norrell, G.T. (1996). Integrated Large-Scale Aquifer Pumping and Infiltration Tests. Groundwater Report OU 7-06, Summary Report, U.S. Department of Energy, Idaho National Engineering Laboratory, INEL-96/0256 (October 1996).

USGS (2018) United States Geological Survey. www.usgs.gov (accessed 23 May 2018).

EPA (2018) United State Environmental Protection Agency. www.epa.gov (accessed 23 May 2018).

9

Fate and Transport Concepts Atmospheric Systems

Case Study: The Union Carbide-Bhopal Accident

One of the worst industrial accidents in history occurred overnight on 2–3 December 1984 at a chemical plant in Bhopal, India. The plant, owned and operated by Union Carbide India Limited (UCIL), was primarily a battery company, but the owners also ventured into the manufacture of carbaryl, the active ingredient in the pesticide Sevin (Shrivastava 1987, p. 42). Many reasons for the accident have been proposed, but cutbacks in operations and the fact that the refrigeration units for the chemical storage tanks were off-line weigh heavily in the cause of the accident (Bogard 1989, p. 3). It has been estimated that as much as 42 tons of methyl isocyanate (MIC) escaped over a matter of minutes due to an uncontrolled reaction in storage tank 610 (Lapierre and Moro 2002, p 270). Thermal degradation of the MIC during the chemical reaction may have resulted in the formation of cyanide gas (Kurtzman 1987, p. 101; Shrivastava 1987, p. 70), but this is disputed by Union Carbide. MIC is almost twice as dense as air; therefore, the gas blanketed the ground and did not readily mix with the surrounding air. The cloud, approximately 100 yd wide (Lapierre and Moro 2002, p. 299), spread with the prevailing southerly wind into the impoverished neighborhoods around the plant. Death estimates vary widely, from 1754 to 15 000, while estimates of the number injured range from 200 000 to 300 000 (Shrivastava 1987, p. 65; Bogard 1989, p. 341).

In this chapter, we will learn about factors important in the fate and transport of airborne pollutants and how to model pollutant releases such as this one, using simple concepts included in the simulator Fate®. As in our other fate and transport models, we will use instantaneous (puff) inputs and continuous (step) inputs to the atmosphere. But, as we will see in this chapter, atmospheric modeling requires knowledge of atmospheric and wind conditions. Also, transport and dilution of the pollution occur so quickly that modelers do not usually include degradation reactions in their model estimates.

After you finish reading this chapter, we will ask you to return to the Bhopal accident and model the event using Fate. As input for your modeling approach, you will use the puff model in fate. Your input conditions are 42 tons of MIC (Lapierre and Moro 2002, p. 270), a stack height of 0.00, a clear night with a temperature inversion, and a light wind (2–3 m/s). For comparison purposes and concentration threshold limits, we will use an MIC threshold limit value of 0.02 parts per million (Kumar and Mukerjee 1985, p. 131). Plot the concentration of MIC as a function of distance from the source at 20.0 minutes after the release. Note that this is an iterative approach, similar to that shown later in Figure 9.7, which will require you to calculate the concentration many times using Fate.

Consequences of this incident in the United States were severe and led to the Community Right to Know provisions in a regulation known as "Right to Know" that applies equally to public and private facilities. Late in 1986, the US Congress passed the Emergency Planning and Community Right to Know Act (EPCRA). EPCRA was designed to help America's communities deal safely and effectively with the many hazardous substances used throughout society, and the Act had two purposes: (i) to encourage and support emergency planning for responding to chemical accidents and (ii) to provide local governments and the pubic with information about possible chemical hazards in their community and allow "first responders" at the local level to respond efficiently and effectively. The regulation does not place limits on which chemicals can be stored, used, released, disposed, or transferred at a facility. It only requires a facility to document, notify, and report information. A threshold level is established for chemicals based upon the Chemical Abstract Service Number (CAS), and such amounts at or in excess of this threshold become reportable. Each section of the law applies different requirement, has different deadlines, and covers a different group of chemicals. A chemical safety officer at each facility is typically trained for most large manufacturing operations.

Pollutant Fate and Transport in Environmental Multimedia, First Edition. Frank M. Dunnivant and Elliot Anders.
© 2019 John Wiley & Sons, Inc. Published 2019 by John Wiley & Sons, Inc.
Companion website: www.wiley.com/go/Dunnivant/pollutant_fate&Transp_EnvironMultimedia

9.1 Introduction

The atmosphere is the environmental medium (compartment) where we live and breathe, and therefore, we are very sensitive to inputs of pollution to the air. Modeling of atmospheric pollution can be used to determine human exposure to existing pollution sources and to predict future exposures from industrial accidents. There are many sources of atmospheric pollution, including industrial smokestacks, fugitive (or nonpoint) industrial emissions, gasoline stations, industrial accidents, automotive and railroad accidents, volcanoes, and forest fires. In this chapter, we will develop relatively simple models to predict the fate and transport of pollution released from these and similar sources. One major difference between the atmosphere and the other environmental media is that the atmospheric models must allow mixing in three dimensions since air and pollutants readily mix. Thus, in contrast to the one-dimensional modeling equations presented elsewhere in this textbook, we will be using a three-dimensional model for our atmospheric studies.

9.2 Input Sources

Input sources of pollution to the atmosphere are highly variable, and classification of these as either pulse or step depends on the time scale of observation. Pulse and step inputs are defined in the previous chapters as short-term and long-term, respectively. But many pollution events, both natural and man-made, do not neatly fit into these two categories. The complexities of the timing of these pollution events could be incorporated into a more complicated model, but this would require a numerical method of analysis approach. Here, we want only a basic understanding of atmospheric modeling. Therefore, as we have for other systems, we will break the pollution events into pulse (referred to as puff in atmospheric scenarios) and step (plume) categories by selecting time scales that allow modeling by our two simple scenarios. For example, if a railroad tanker of acetone (a very volatile organic compound) derails and spills its contents, this could be considered a pulse release over a period of months if we are looking at long-range transport, but we would rarely be concerned with the downwind concentrations of acetone at these distances because the concentrations would be extremely dilute. Therefore, a more realistic modeling approach would be to model the release as a step release using the volatilization rate of acetone as our input source. An example that would fit into the pulse scenario for both the short- and long-term time scales is the release of radiation from a catastrophic nuclear accident. For example, the release of radioactivity from Chernobyl could have been modeled as a pulse release in considering short-term radioactive dose rates for nearby countries. Scientists also monitored the spread of radiation around the world for the Chernobyl release, and this could also have been modeled as a pulse release but on a much longer time scale. Therefore, in using the atmospheric models, the modeler must set the time duration (a boundary condition), which, in turn defines the source input function.

9.3 Atmospheric Sampling Equipment and Efforts

Water samples (rain, snow, and ice) can be obtained using a sampling system as simple as a plastic or stainless steel bucket or as sophisticated as the automated sampler shown in Figure 9.1. Other types of atmospheric samplers actually have sensors to detect if it is precipitating or sunny and take wet or dry (particulate) samples. For sampling in remote areas, solar-powered units are available (Figure 9.2). Strictly, dry particulate samples

Figure 9.1 The Model 200 Wet-Only Rainwater Sampler designed by Ecotech Pty Ltd., Blackburn, Victoria: http://www.ecotech.com.au/rainwat.htm. Source: Reproduced with permission from ECOTECH Pty Ltd.

Figure 9.2 The MicroVol 1100 Particulate Sampler designed by Ecotech Pty Ltd., Blackburn, Victoria: http://www.ecotech.com.au/uvol1100.htm. Source: Reproduced with permission from ECOTECH Pty Ltd.

Figure 9.3 The HV3000 High Volume Air Sampler designed by Ecotech Pty Ltd, Blackburn, Victoria: http://www.ecotech.com.au/hv3000.htm. Source: Reproduced with permission from ECOTECH Pty Ltd.

can be obtained using a low-volume atmospheric sampler similar to the one shown in Figure 9.3. Air enters the unit at the top and is pulled through a large weighed filter (typically the size of an 8.5 × 11 in. piece of note paper). The mesh or pore size of the filter paper can be selected to collect a specific particle size. This approach allows for the total mass of particles to be determined as well as for laboratory analysis of the particles. Isokinetic sampling is required to obtain unbiased samples for laboratory analysis. In any technique for collecting airborne particle matter, the collector is so designed such that the airstream entering it has a velocity equal to that of the air passing around and outside of the collector (McGraw Hill 2003).

Sampling of indoor and outdoor gases is relatively easy using a portable personnel pump similar to the one shown in Figure 9.4. In this system, the flow rate

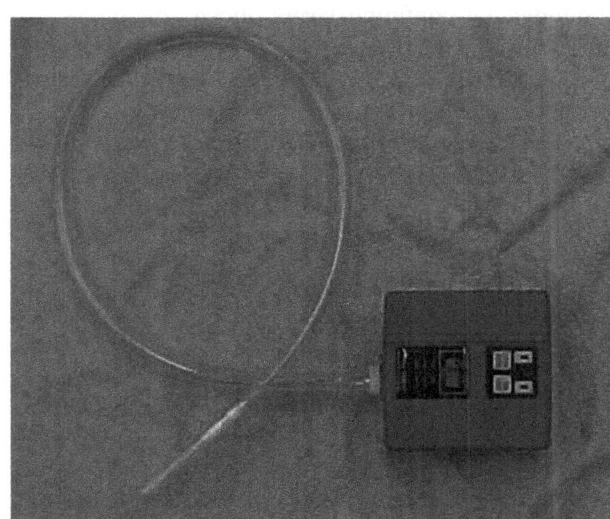

Figure 9.4 The Supelco Q-MAX Pump for taking small samples of inorganic compounds.

of the pump is calibrated to a specified value (typically 2.0 l/min). A sampling tube containing a resin that is specifically designed to sample a compound or set of compounds is attached to the pump. The pump is actually a vacuum pump that pulls air first through the sample collection tube and then into the pump, thus not contaminating the air by the pumping system. The resin tubes are returned to the laboratory, broken open, extracted into a solvent that effectively desorbs the analytes, and analyzed (usually be GC or HPLC). These types of systems are used in industrial workplace settings for monitoring the exposure of volatile solvents.

9.4 Important Factors in the Modeling of Atmospheric Pollution: Conceptual Model Development

9.4.1 One- Versus Two- Versus Three-Dimensional Models

First, we will compare other fate and transport models to the general atmospheric model. The aquatic models in this textbook were given only for one dimension, the x or longitudinal direction. Streams and lakes usually can be adequately modeled using one-dimensional models since most of the dispersion is in the longitudinal direction, whereas groundwater systems can be modeled in one, two, or three dimensions (x, y, and z). Two dimensions are normally required for accurate modeling because the groundwater is not constrained by a river or lake bank, and dispersion can occur in all directions. Vertical dispersion, although important near a pollution source, becomes less important when the groundwater system is bounded by confining layers above and below the aquifer of interest, which is why we used the simpler one-dimensional model in considering both the instantaneous and pulse groundwater releases. However, deep aquifer systems require at least two-dimensional models. Virtually, all atmospheric models, including the one used here and in Fate, are three-dimensional by necessity.

9.4.2 Mixing and Dispersion in Atmospheric Systems

Although the aquatic models may have seemed complicated, they are simple compared to most atmospheric models. To account for complex wind currents and mixing, atmospheric models have to incorporate three dimensions, which automatically makes the governing equations more complex to solve. As usual, we make many assumptions so that our model will be more manageable. For example, the models given later in this chapter ignore buoyancy effects, and thus, they are designed for gases with the same density as the atmosphere. As previously discussed, the models distinguish between step and instantaneous sources, although many atmospheric pollution episodes can lie between these two extremes. Unlike the aquatic models that allow first-order decay processes, our atmospheric models do not allow degradation of pollutants. This assumption is justified for modeling of pollutants over relatively short distances (under 10 km or 7 mi). With a 5-mi per hour wind, it will take less than 1.5 hours for the pollutant to be dispersed this distance (our model limit, explained later), and most of the atmospheric photochemical reactions (except for the production of smog) require the pollutant to be in the atmosphere over a much longer time frame (hours to days). The dominant force resulting in the decrease in pollutant concentration is dispersion, which can rapidly dilute pollutant concentrations in the atmosphere. However, understanding and accounting for dispersion can be very complicated. First, we will look at the movement of atmospheric gases over the Earth's surface.

A profile of the wind's velocity with increasing height shows a steep increasing parabolic shape, with low velocity at the Earth's surface due to friction between the moving air and the ground. The surface wind velocity, however, is also subject to many complex variables. For example, the roughness of the Earth's surface can significantly impact the shape or steepness of the wind velocity–height profile. The wind velocity profile over an open grassland is illustrated in right-hand side of Figure 9.5, showing that wind speed rapidly approaches its maximum as height above the surface increases. Compare this to an urban setting where tall buildings impede the path of the wind and slow its speed near the land surface. This expands the velocity–height gradient

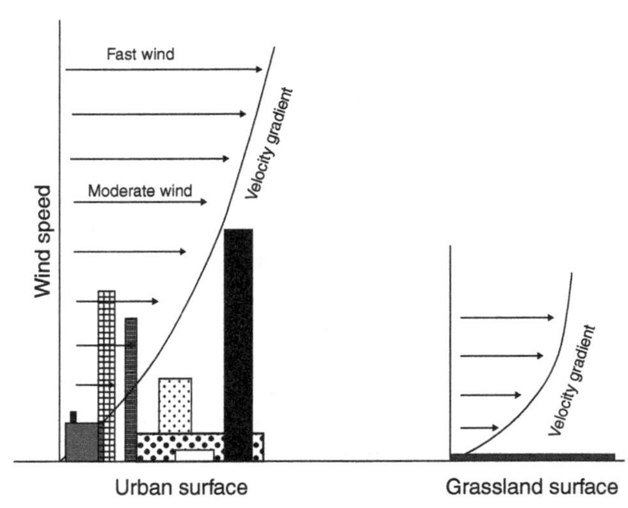

Figure 9.5 The effect of surface roughness on wind speed.

well above the Earth's surface. The resulting lower wind velocity could decrease the turbulence and subsequent dispersion and may result in stagnant pockets of the atmosphere that can contain clear or polluted air. The increase in the roughness of the surface due to the presence of buildings will greatly affect flow patterns and ground-level pollutant concentrations. The wind velocity and direction are also highly variable in urban settings. For example, in Chicago in winter, the cold wind may actually be blowing into your face every time you turn the corner of a building. Observations such as this demonstrate that atmospheric processes are too complicated even for our most sophisticated models. In our brief treatment here, we will simplify our model by assuming that an average wind speed can be used, and in general, we will not account for differences in surface roughness or small-scale changes in wind direction.

Although surface roughness can greatly affect turbulence and mixing, the magnitude of wind speed can also increase mixing. Recall that the movement of any fluid or gas is referred to as advection, and wherever gradients exist, this movement results in the mixing of gases. We will refer to this mixing as dispersion because the net result is a dilution of pollutant concentrations. Considering the combined effects of wind velocity and the atmospheric temperature as a function of height above the surface, we obtain the three basic turbulence scenarios shown in Figure 9.6. First, consider an isolated pocket of atmosphere at nighttime temperatures (shown in Figure 9.6a). This type of condition occurs where a thick cloud layer prevents the Earth from radiating its heat to space as it cools during the night. Under these conditions, an emission from an industrial stack will take the shape of the plume shown in Figure 9.6a. The released gases will rise or sink until their density (temperature) matches that of the surrounding (diluting) atmospheric gases. Then, the plume will take the shape of a thin layer.

Under daytime heating conditions, the temperature–height profile will be similar to the one shown in Figure 9.6b. The warmest point in the atmosphere is at the Earth's surface, and as the heated air rises it cools, thus producing the temperature profile shown in Figure 9.6b. In a steady wind, the plume will spread in all directions, but mostly in the longitudinal direction. With a lower temperature–height gradient and a higher wind velocity, extreme turbulence will be observed (Figure 9.6c). In order to attempt the modeling of these conditions, we must greatly simplify the temperature and wind relationships.

We begin our simplification process by attempting to combine the effects of wind velocity, temperature–height profiles, and cloud cover into a set of atmospheric stability categories. As we do this, remember that our goal is to come up with a way to characterize dispersion (mixing) of the pollutant with the atmospheric gases. Table 9.1 shows a qualitative approach to the combined effects of wind speed and cloud cover collected for rural settings in England. Cloud cover is a good reflector of heat back to the Earth. The categories range from strongly unstable (category A reflected in Figure 9.6c) to very stable (category G) and distinguish between day and night conditions.

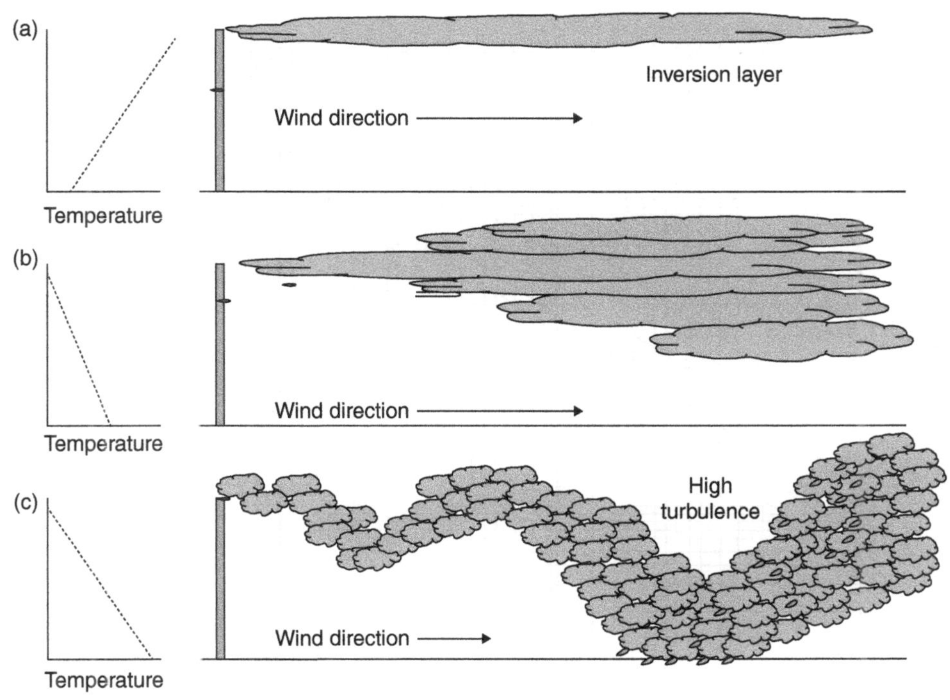

Figure 9.6 Three basic turbulence scenarios for plumes. (a) mixing in a inversion layer, (b) mixing in a neutral atmosphere, (c) mixing in a turbulent atmosphere.

Table 9.1 Pasquill stability categories (collected for rural settings in England).

	Day, degree of cloud insulation			Night	
Wind speed (at 10 m elevation m/s)	Strong	Moderate	Slight	Thinly overcast or greater than 50% low clouds	Less than 50% cloud cover
<2	A	A–B	B	G	G
2–3	A–B	B	C	E	F
3–5	B	B–C	D	D	E
5–6	C	C–D	D	D	D
>6	C	D	D	D	D

Turner (1994) adds the following notes on selecting the appropriate category:

1. Strong insolation corresponds to sunny midday in midsummer in England; slight isolation to similar conditions in midwinter.
2. Night refers to the period from 1 h before sunset to 1 h after sunrise.
3. The neutral category D should also be used, regardless of wind speed, for overcast conditions during day or night and for any sky condition during the hour preceding or following night as defined in note number 2.

Source: Turner (1970, 1994) and Pasquill (1961) (EPA: public domain).

Next, the somewhat qualitative categories in Table 9.1 are used to mathematically predict values for horizontal dispersion coefficients (Table 9.2), which are estimates of mixing in the x and y directions. We do not have a way to mathematically predict these values accurately, and the data in Tables 9.1 and 9.2 are empirical (based on experimental observations). We usually assume that dispersion (σ in Table 9.2) in the x and y directions is the same; thus, Table 9.2 can be used to estimate σ_x and σ_y simultaneously. The equations given in Table 9.2 were used to draw the lines in Figure 9.7 for the six stability categories. Note that dispersion increases as you move away from the point source of pollution. This should be intuitive because mixing continues as you move away from the point source. Therefore, for every pollutant concentration you attempt to estimate, you must select a distance from the point source. The limitation of this is that in Fate, you can only plot a slice of the concentration profile in the y and z plane. You therefore have

Table 9.2 Pasquill–Gifford horizontal dispersion parameters.

$\sigma_y = 1000^{*} \tan(T)/2.15$

where x is the downwind distance (in km) from the point source and T is one-half Pasquill's q in degrees. T, as a function of x, is determined by each stability category from Table 9.1.

Stability	Equation for T
A	$T = 24.167 - 2.5334\ \ln(x)$
B	$T = 18.333 - 1.8096\ \ln(x)$
C	$T = 12.5 - 1.0857\ \ln(x)$
D	$T = 8.333 - 0.7238\ \ln(x)$
E	$T = 6.25 - 0.5429\ \ln(x)$
F	$T = 4.167 - 0.3619\ \ln(x)$

Source: From Turner (1994, 1970).

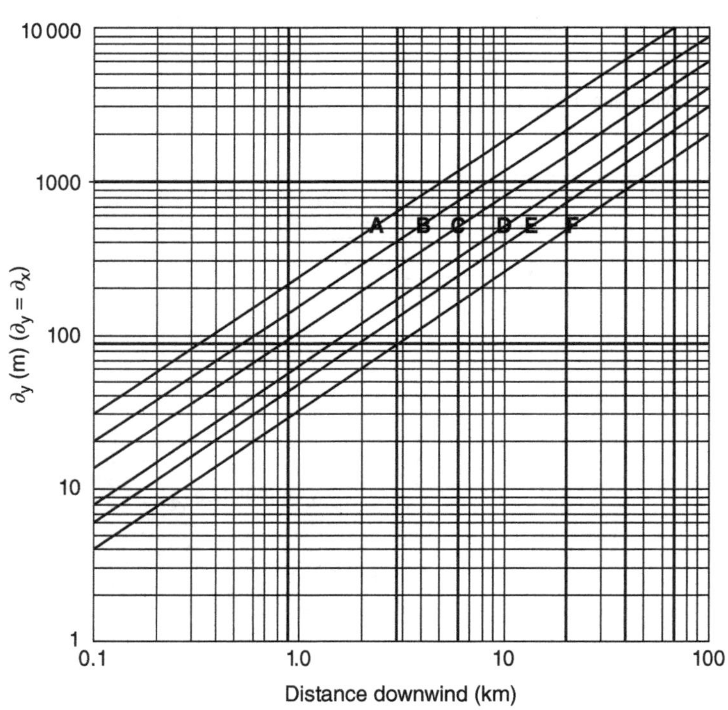

Figure 9.7 Pasquill–Gifford horizontal dispersion parameters for rural and step inputs. Source: Turner (1970) and Pasquill (1961); Graph: Anders in Fate®.

Table 9.3 Pasquill–Gifford vertical dispersion parameter for rural step inputs.

	Vertical dispersion parameter: $\sigma_z = a x^b$, where x is in km			
Stability	**Distance (km)**	a	b	σ_z **at upper boundary**
A	>3.11			5000
	0.5–3.11	453.85	2.1166	
	0.4–0.5	346.75	1.7283	104.7
	0.3–0.4	258.89	1.4094	71.2
	0.25–0.3	217.41	1.2644	47.4
	0.2–0.25	179.52	1.1262	37.7
	0.15–0.2	170.22	1.0932	29.3
	0.1–0.15	158.08	1.0542	21.4
	<0.1	122.8	0.9447	14.0
B	>0.35			5000
	0.4–35	109.30	1.0971	
	0.2–0.4	98.483	0.9833	40.0
	>0.2	90.673	0.93198	20.2
C	All values of x		61.141	0.91465
D	>30.0	44.053	0.51179	
	10.0–30.0	36.650	0.56589	251.2
	3–10.0	33.504	0.60486	134.9
	1–3	32.093	0.64403	65.1
	0.3–1	32.093	0.81066	32.1
	<0.3	34.459	0.86974	12.1
E	>40.0	47.618	0.29592	
	20.0–40.0	35.420	0.37615	141.9
	10.0–20.0	26.970	0.46713	109.3
	4–10.0	24.703	0.50527	79.1
	2–4	22.534	0.57154	49.8
	1–2	21.628	0.63077	33.5
	0.3–1	21.628	0.75660	21.6
	0.1–0.3	23.331	0.81956	8.7
	<0.1	24.260	0.83660	3.5
F	>60	34.219	0.21716	
	30.0–60.0	27.074	0.27436	83.3
	15–30.0	22.651	0.32681	68.8
	7–15	17.836	0.4150	54.9
	3–7	16.187	0.4649	40.0
	2–3	14.823	0.54503	27.0
	1–2	13.953	0.63227	21.6
	0.7–1	13.953	0.68465	14.0
	0.2–0.7	14.457	0.78407	10.9
	<0.2	15.209	0.81558	4.1

Source: Turner (1970) and Pasquill (1961) (EPA: public domain.)

to manually plot the concentration gradient in the x, or longitudinal, direction of wind flow in a spreadsheet. Fate uses Figure 9.3 as input for rural step, horizontal dispersion parameters.

Dispersion in the vertical (z) direction is somewhat more complicated to predict and again is based on experimental observations. We can estimate the vertical dispersion coefficient, σ_z, by using the same atmospheric stability categories from Table 9.1 but with a more precise treatment of the wind speed. The equation governing the estimate of vertical dispersion is the following:

$$\sigma_z = ax^b$$

where

x is the distance in km and
a and b are fitting parameters obtained from Table 9.3.

Plots showing the dependence of vertical dispersion coefficients (in meters) on distance from the point source for the different stability categories are shown in Figure 9.8. Values from Figure 9.8 are used in Fate. We have been describing dispersion, but what exactly is it? As we have noted, dispersion is a physical process and is a function of the distance from the point source. Dispersion can be mathematically described as the mixing between the pollutant plume and the natural atmospheric gases. The values you read from the graph or calculate using the equations are given in meters or kilometers. Thus, the values given represent the width of the pollutant plume at the specified distance from the point source and reflect the amount of atmosphere the pollution has mixed with. You can visualize the spread of the plume downwind as a narrow funnel, flaring out as you move away from the point source, and becoming more diluted in pollutant concentration.

9.5 Mathematical Development of Model

We will start with the derivation of a one-dimensional system. The derivations of both the one-dimensional atmospheric models, step and instantaneous, are very similar to that of the river and groundwater models. After all, both are modeling the pollutant concentration over time from water or air moving in a one-dimensional line. But, there are several mathematical differences. First given the relatively fast speed of airflow, we ignore any reaction term. Second, there is no significant adsorption to particles in the air; therefore, we ignore the retardation factor (R). We start again with the basic $\mathrm{d}C/\mathrm{d}t$ mass balance equation as before.

$$\frac{\mathrm{d}C}{\mathrm{d}t} = -\frac{v}{R}\frac{\mathrm{d}C}{\mathrm{d}x} + \frac{\mathrm{d}}{\mathrm{d}x}\left(\frac{E}{R}\frac{\mathrm{d}C}{\mathrm{d}x}\right) + g(x,t) - d(x,t) \tag{9.1}$$

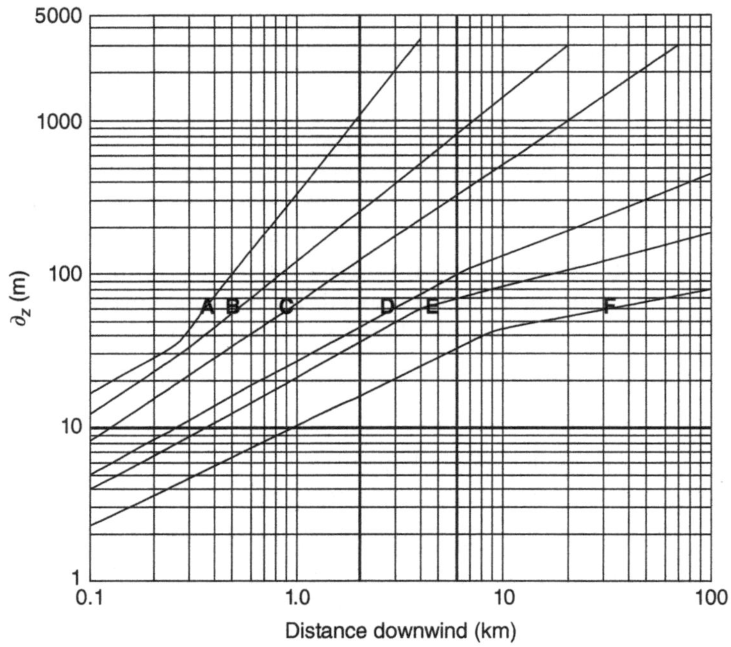

Figure 9.8 Pasquill–Gifford vertical dispersion parameters for rural and step inputs. Source: Turner (1970); Graph: Anders.

where

C is the concentration of pollutant,
v is the water velocity,
E is the dispersivity constant,
$g(x, t)$ is the input function, and
$d(x, t)$ is the degradation term of the pollutant.

In this equation, the term

$$-V\frac{dC}{dx} \tag{9.2}$$

deals with advection, or the transfer of pollutant by the actual flow of the air in the system, and the term

$$\frac{d}{dx}\left(E\frac{dC}{dx}\right) \tag{9.3}$$

deals with dispersion of the pollutant due to differences in concentration throughout the atmospheric system.

This reduces the one-dimensional equation to

$$\frac{dC}{dt} = -v\frac{dC}{dx} + \frac{d}{dx}\left(E\frac{dC}{dx}\right) + g(x,t) \tag{9.4}$$

9.5.1 Step Input (Plume Model) of Pollutant

The only additional term needed for the plume model is the source term

$$\frac{Q_m}{u} \tag{9.5}$$

where

Q_m is the pollutant source (mass/time) and
u is the wind velocity (length/time).

Giving us an overall one-dimensional governing equation as follows:

$$\frac{dC}{dt} = -v\frac{dC}{dx} + \frac{d}{dx}\left(E\frac{dC}{dx}\right) + \frac{Q_m}{u} \tag{9.6}$$

Now, we must extend Eq. (9.1) to a three-dimensional model. Using the many assumptions stated earlier (rapid mixing and no pollutant degradation) and the estimated horizontal and vertical dispersion coefficients, we may expand the one-dimensional governing equation developed previously to include terms accounting for the extra dimension. For example, the dispersion coefficient, E, will be broken into separate coefficients for each dimension. A model (Eq. (9.7)) for the steady-state plume can then be derived from the expanded equation, using differential equation techniques to estimate the pollutant concentration at any point (x, y, and z) downwind from the continuous source. This is referred to as the steady-state plume model generally derived and confirmed in Fjeld et al. (2006, p. 162).

$$C(x,y,z) = \frac{Q_m}{2\pi\sigma_y\sigma_z u}\left(e^{-\frac{1}{2}\left(\frac{y}{\sigma_y}\right)^2}\right) \times \left[e^{-\frac{1}{2}\left(\frac{z-H_r}{\sigma_z}\right)^2} + e^{-\frac{1}{2}\left(\frac{z+H_r}{\sigma_z}\right)^2}\right] \tag{9.7}$$

where

$C(x, y, z)$ is the concentration of pollutant in the plume as a function of x, y, and z (mass/length^3),
x, y, and z are distances from the source (length) (see Figures 9.7 and 9.8),
Q_m is the pollutant source (mass/time),

$\sigma_x = \sigma_y$ is the horizontal dispersion coefficient (length),
σ_z is the vertical dispersion coefficient (length),
u is the wind velocity (length/time), and
H_r is the height of the release (length).

Notice that the model is divided into two basic parts: the mass of pollutant released, represented by the first term in the equation, and dispersion, characterized by the wind speed, the y and z coordinates that yield estimates of dispersion (mixing), and the height of the release above the Earth's surface. Input values for all these parameters are relatively simple to estimate or measure using the techniques described earlier. Note that the longitudinal distance, x, is a part of the equation since you must select the x distance to obtain estimates of σ_x and σ_z.

The equation can be simplified if the point of interest (at the receptor) is immediately downwind of the pollution source. For the concentration along this three-dimensional centerline of the plume ($z = 0$, and $H_r = 0$), we can use a simplification of Eq. (9.7) (generally derived and confirmed in Fjeld et al. 2006, p. 162):

$$C(x, y, 0) = \frac{Q_m}{\pi \sigma_y \sigma_z u} \left(e^{-\frac{1}{2}\left(\frac{y}{\sigma_y}\right)^2} \right) \left(e^{-\frac{1}{2}\left(\frac{H}{\sigma_z}\right)^2} \right) \tag{9.8}$$

A typical simulation of downwind pollutant concentration is shown in Figure 9.9 for a z value of 1.0 m (height above ground level), a y distance of 0.0 km (along the x–z axis), and an x value of 1.5 km (distance downwind). In Figure 9.9, the peak of the Gaussian-shaped plot is along the center x axis (y = 0 m) and 1 m above ground or about nose level for a child. The pollutant concentration declines as you go to the left or right of the centerline (positive or negative y values).

A similar output would be obtained by plotting a y value of 0.0 (along the center line), an x distance of 1.5 km, and calculating the pollutant concentration as you move up above the Earth's surface. This is illustrated in Figure 9.10. In this plot, as you go from left to right on the x-axis, you are moving up away from the Earth's surface.

A useful function of Fate is to evaluate the pollutant concentration as a function of distance from the point source along the centerline of the wind flow. Fate cannot plot this directly because dispersion in the x, y, and z directions are functions of the distance from the point source. In order to accomplish this, we must repeatedly use steps 5 and 6 in manual calculations or in the Fate simulator. Systematically change the x distance, increasing it incrementally, and record the pollutant concentration given in step 6 (in Fate). A plot like the one shown in Figure 9.11 can be obtained by summarizing the output data. Note that the pollutant concentration decreases, as expected, as you move away from the point source. This type of plot is similar to those used to illustrate the other fate and transport models in this text.

Example Problem

For the following data (which would be measured at the source), manually calculate the downwind concentration profile over a distance of 1.5 km (i) for varying distance from the center line (varying y coordinate) where z = 4

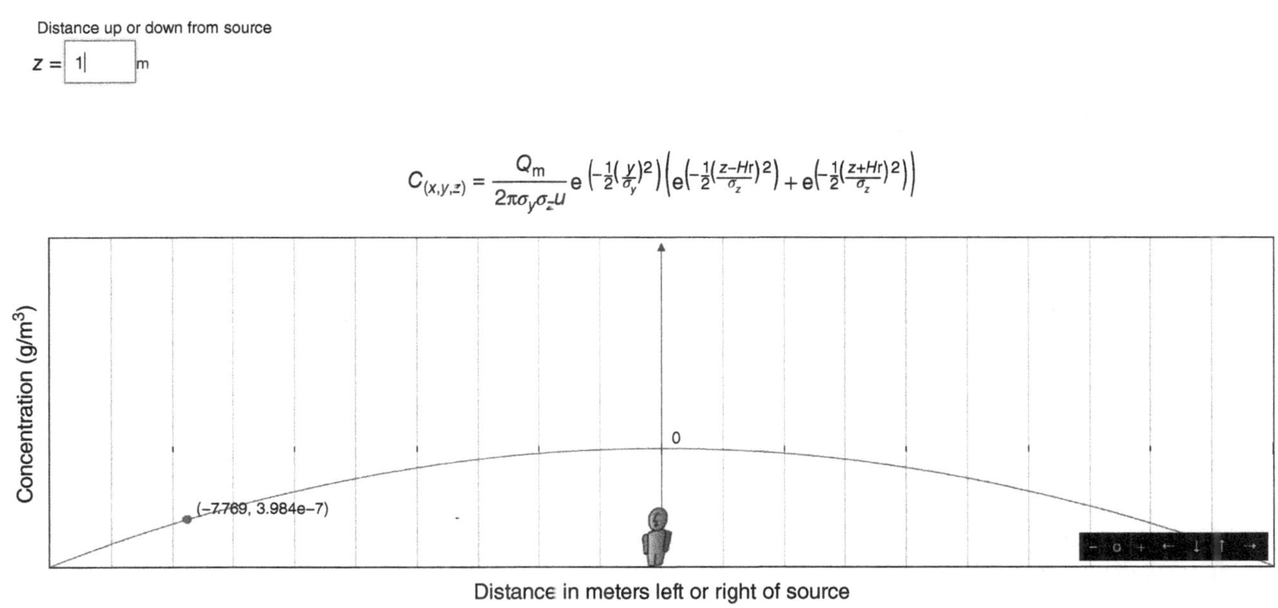

Figure 9.9 Output from Fate© for a continuous release (plume) of pollutant into the atmosphere at an x value of 1.5 km. Source: Anders in Fate®.

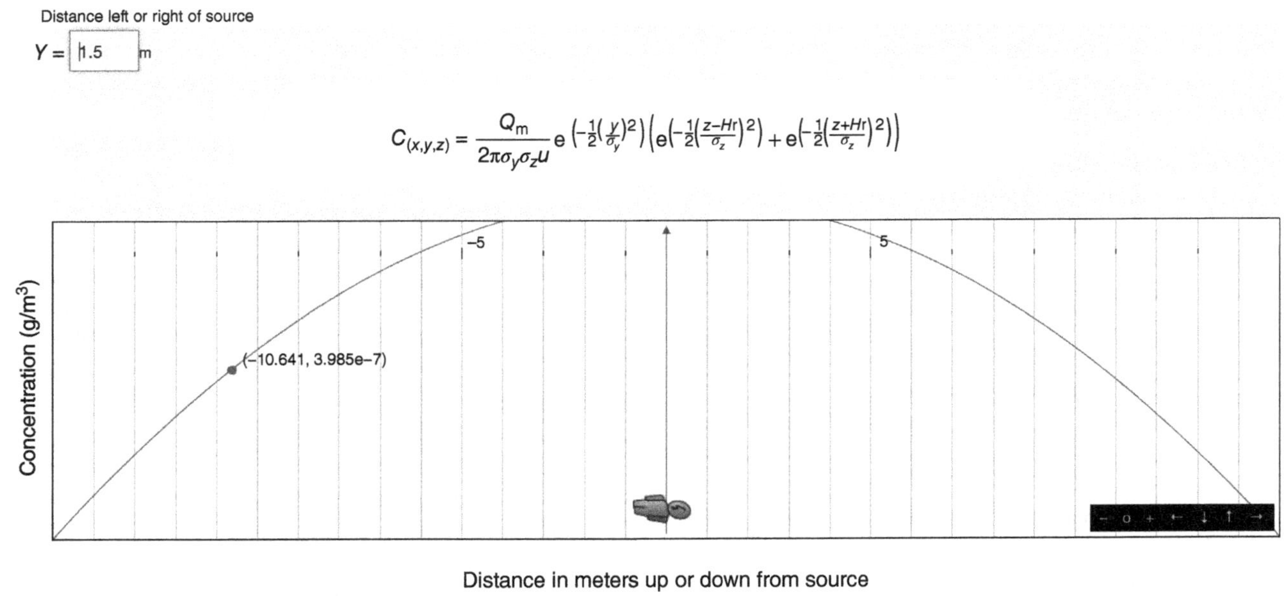

Figure 9.10 Output from Fate© for a continuous release (plume) of pollutant into the atmosphere showing variations in plume concentration with changing vertical position in relation to the source ($x = 1.5$ km). Source: Anders in Fate®.

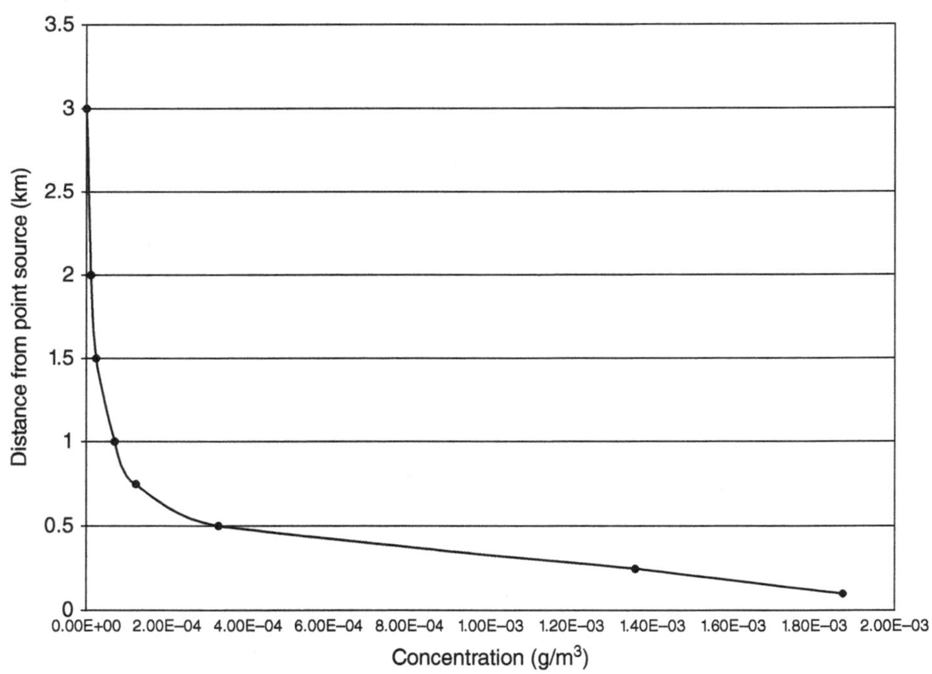

Figure 9.11 An evaluation of the pollutant concentration as you move away from the point source (plume model). Source: Output from Fate®, Dunnivant.

and (ii) for varying distance above the ground surface (z coordinate) where $y = 1.5$.

H_r: Height of pollutant release is 30.0 m
u_s: Stack exit velocity of gases is 0.80 m/s
u: Wind speed is 0.80 m/s
d: Inside stack diameter is 0.80 m
P: Atmospheric pressure is 1010 mbar
T_s: Stack gas temperature is 285 K
T_a: Atmospheric temperature is 280 K

The source rate of emission is 20.0 g/s.

Use an atmospheric stability condition of A (low wind speed, day time hours, and strong cloud cover).

Solution

The calculated effective height of release is calculated by

$$\Delta H_r = \frac{u_s d}{u}\left(1.5 + 0.002\,68\,P\,d\left(\frac{T_s - T_a}{T_s}\right)\right) \tag{9.9}$$

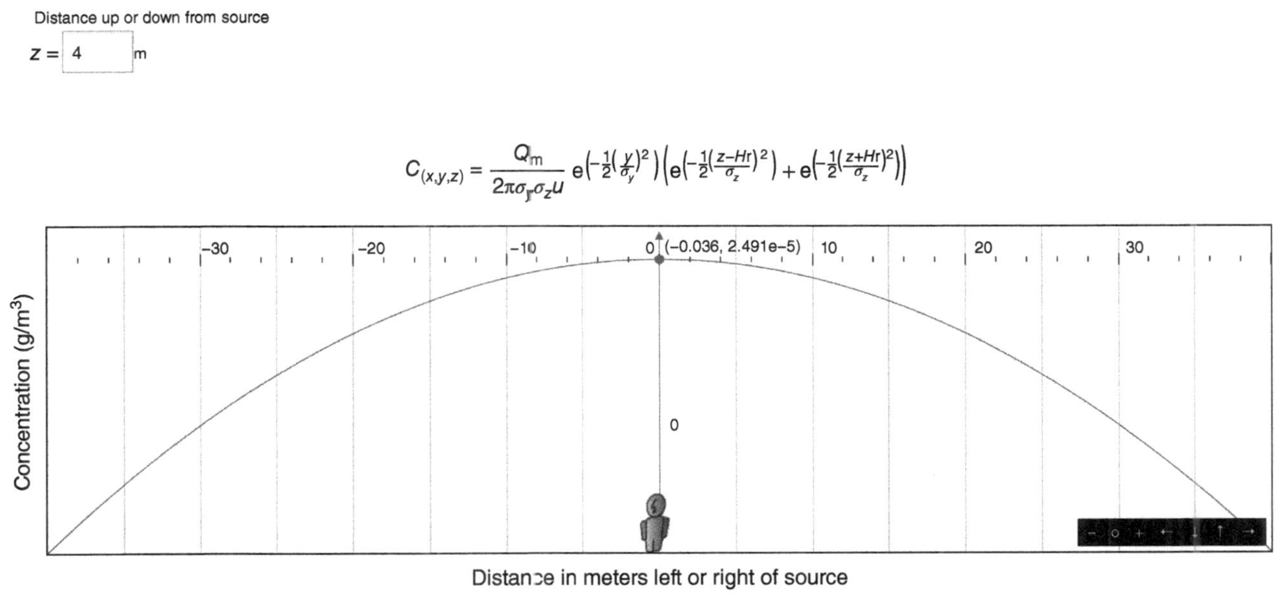

Figure 9.12 Concentration profile in the positive and negative y direction for the example problem (plume model). Source: Anders.

and results in an effective stack height of 31.2 m (the height the gaseous plume rises to). Note that temperature must be in K.

From the data for dispersion given in Figures 9.7 and 9.8 and an atmospheric stability in category A, the horizontal dispersion is 298.2 m, and the vertical dispersion is 1.071×10^3 m.

$$C(x, y, z) = \frac{Q_m}{2\pi\sigma_y\sigma_z u}\left(e^{-\frac{1}{2}\left(\frac{y}{\sigma_y}\right)^2}\right) \times \left[e^{-\frac{1}{2}\left(\frac{z-H_r}{\sigma_z}\right)^2} + e^{-\frac{1}{2}\left(\frac{z-H_r}{\sigma_z}\right)^2}\right] \quad (9.10)$$

Using the continuous fate and transport equation (Eq. (9.7)), the concentration profile (in g/m^3 versus m) for the positive and negative y directions is shown in Figure 9.12.

The concentration profile for the z (height) direction is shown in Figure 9.13.

9.5.2 Instantaneous Input (Pulse or Puff Model) of Pollution

The only additional term needed for the pulse model is the source term, where we will use the Dirac Delta function,

$$\delta_x(t) \quad (9.11)$$

equal to 1 for $t = x$ and equal to 0 for all other values of t. The input function for a 1D model will be the following:

$$\frac{Q_m\delta_0(x)\delta_0(t)}{u} \quad (9.12)$$

where

Q_m is the pollutant source (mass) and
u is the wind velocity (length/time).

Again, this input function will be expanded to include terms accounting for the y and z dimensions, before being used to solve for a 3D model of the pollutant.

For a pulse rather than a step input, dispersion is handled a little differently. In the step (plume) model, we can use either rural or urban dispersion estimates, whereas urban dispersion parameters are usually used for the pulse (puff) model. The "puff" dispersion estimates are derived from experimental observations made by McElroy and Pooler (1968) near St. Louis and from Briggs (1972). Calculations for estimating the horizontal and vertical dispersion coefficients are shown in Table 9.4. We will again assume that dispersivity (σ in the x and y directions is the same). Atmospheric stability categories are the same as those described in Table 9.1. Vertical and horizontal dispersion coefficients are shown in Figures 9.14 and 9.15.

Using the stability categories, wind speed, and the equations shown in Table 9.4, we can now estimate the atmospheric pollutant concentration downwind from an instantaneous source (also referred to as pulse or puff) by the following equation:

$$C(x, y, z, t) = \frac{Q_m}{(2\pi)^{3/2}\sigma_y\sigma_z u}\left(e^{-\frac{1}{2}\left(\frac{y}{\sigma_y}\right)^2}\right) \times \left[e^{-\frac{1}{2}\left(\frac{z-H_r}{\sigma_z}\right)^2} + e^{-\frac{1}{2}\left(\frac{z+H_r}{\sigma_z}\right)^2}\right] \quad (9.13)$$

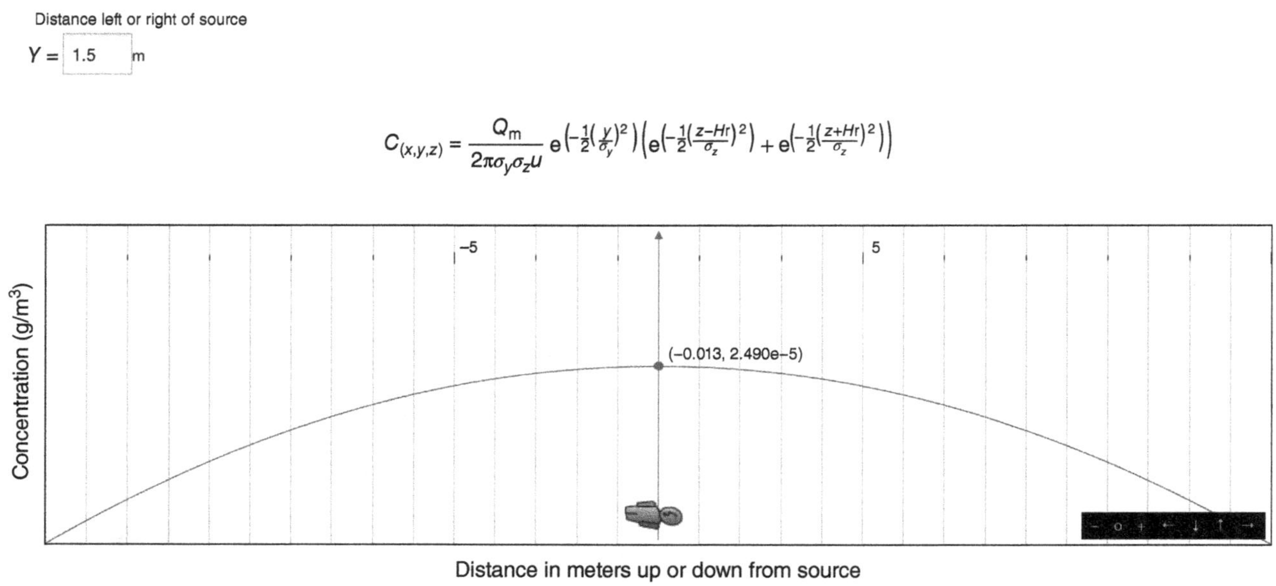

Figure 9.13 Concentration profile in the positive and negative z direction for the example problem (plume model). Source: Anders.

Table 9.4 Urban dispersion parameters for step and pulse inputs.

Pasquill type of stability	σ_y (m)	σ_z (m)
A–B	$0.32/(0.0004x)^{-0.5}$	$0.24/(0.001x)^{0.5}$
C	$0.22/(0.0004x)^{-0.5}$	$0.20x$
D	$0.16/(0.0004x)^{-0.5}$	$0.14/(0.0003x)^{-0.5}$
E–F	$0.11/(0.0004x)^{-0.5}$	$0.08/(0.0015x)^{-0.5}$

For distances, x, between 100 and 10000 m.
Source: Turner (1994), Briggs (1972), and McElroy and Pooler (1968).

where

$C(x, y, z, t)$ is the concentration of pollutant in the plume as a function of x, y, and z (mass/length^3) and time. A similar equation is given by Fjeld et al. (2006),
x, y, and z are distances from the source (length); see Figures 9.10 and 9.11,
t is time,
Q_m is the pollutant source (mass/time),
σ_x and σ_y are the horizontal dispersion coefficients (length),
σ_z is the vertical dispersion coefficient (length), and
H_r is the height of the release (length).

Note that time is included here because the distance traveled (x) since the instantaneous pollutant release is a function of wind velocity (u) and time (t), where

$$x = ut \tag{9.14}$$

For the concentration along the centerline ($y = 0$, $z = 0$, and $H_r = 0$), we can use a simplification of Eq. (9.10) to yield the following equation:

$$C(x, 0, 0, t) = C(ut, 0, 0, t) = \frac{Q_m}{(\sqrt{2\pi})^{3/2}\, \sigma_x \sigma_y \sigma_z} \tag{9.15}$$

Simulation outputs (for Eq. (9.13)) from Fate are shown in the example problem below.

Example Problem

For the following data (which would be measured on-site), manually calculate the downwind concentration profile for a distance of 10 km (i) for varying distance from the center line ($y = 0$) where $z = 30$ and (ii) for varying distance above the ground surface ($z = 0$) where $y = 1.5$.

Height of pollutant release is 30.0 m,
Wind speed is 4.0 m/s, and
Total source mass is 2000 g.

The given wind speed results in an atmospheric stability of condition D for clear daytime conditions.

Solution

Using the equations for dispersion or the figures, the horizontal dispersion is 0.7517 m and the vertical dispersion is 0.4990 m.

$$C(x, y, z, t) = \frac{Q_m}{(2\pi)^{3/2}\sigma_y \sigma_z u}\left(e^{-\frac{1}{2}\left(\frac{y}{\sigma_y}\right)^2}\right) \times \left[e^{-\frac{1}{2}\left(\frac{z-H_r}{\sigma_z}\right)^2} + e^{-\frac{1}{2}\left(\frac{z+H_r}{\sigma_z}\right)^2}\right] \tag{9.16}$$

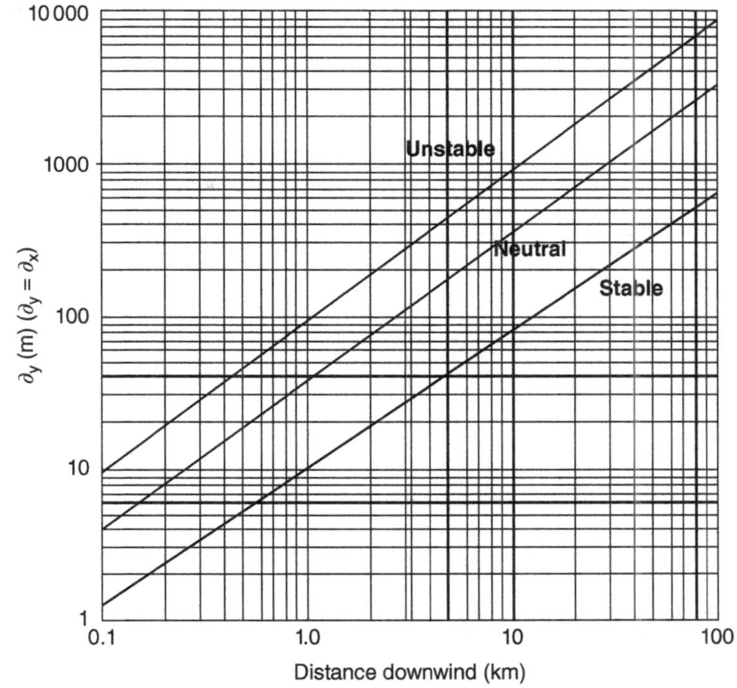

Figure 9.14 Pasquill–Gifford horizontal dispersion parameters for urban settings. Source: From Turner (1970); public domain. (Graph: Fate® and Anders.)

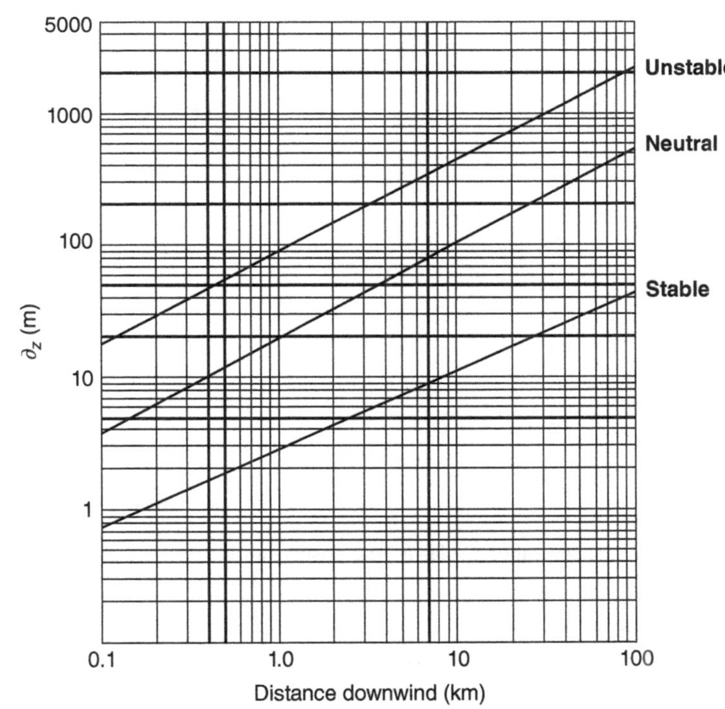

Figure 9.15 Pasquill–Gifford vertical dispersion parameters for urban settings. Source: Turner (1970); public domain. (Graph: Fate® Anders.)

Using Eq. (9.7), $z = 30$ m, and $y = 1.5$ m, the concentration profile (in g/m^3 versus m) for the positive and negative y directions is shown in Figure 9.16.

The concentration profile for the z (height) direction is shown in Figure 9.17.

9.6 Sensitivity Analysis

Each chapter has discussed the need to question the model and input parameters. Model parameters for a sensitivity analysis of the two atmospheric equations

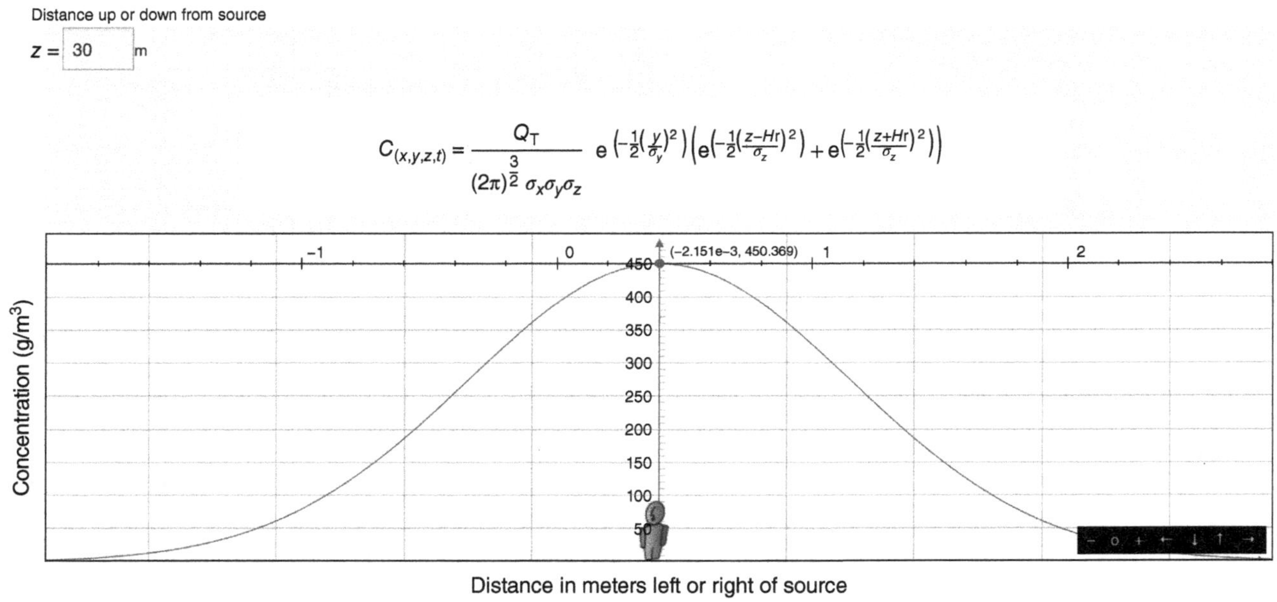

Figure 9.16 Concentration profile for the positive and negative *y* direction for the example problem (puff model). Source: Anders.

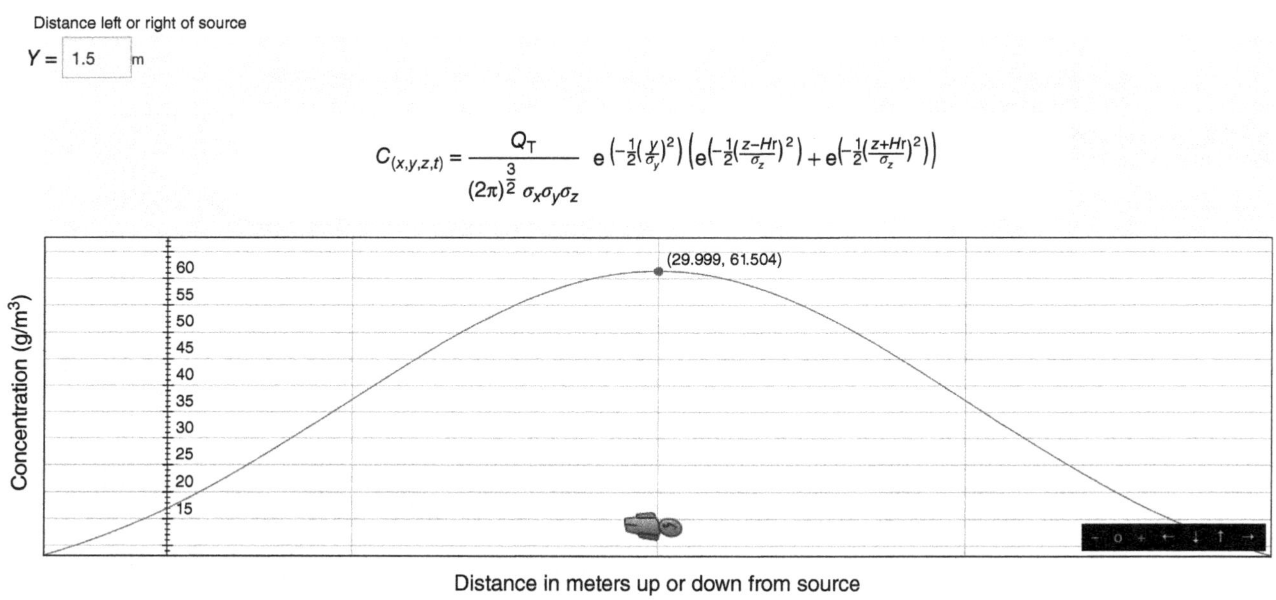

Figure 9.17 Concentration profile for the positive and negative *z* direction for the example problem (puff model). Source: Anders.

include the wind velocity flow (v) and dispersion (σ) in each direction. Thus, by systematically varying each of these, the modeler can tell how sensitive their model is to inaccurate or uncertain model parameters.

9.7 Limitations of Our Model

9.7.1 Chemistry

As noted earlier, one obvious parameter missing from the atmospheric modeling equations is chemical reactions. These can relatively easily included but are usually omitted because dilution of the pollution to insignificant concentrations usually occurs relatively quickly. This is due to the great dilution power of atmospheric mixing. Usually, we are concerned with minutes to hours of travel time, which are considerably short as compared to the degradation rates for pollutants. Extremely reactive pollutants such as chlorine gas may require that a degradation rate be included.

9.7.2 Dispersion and Mixing

Dispersion is difficult to accurately quantify in any environmental media, and this is especially true in atmospheric systems. The difficulty in estimating this model parameter is due to the inconsistent surface of the ground. We have included two ways of estimating dispersion, and each is specific to ground surface setting (i.e. urban, rural, etc.). To further complicate our efforts, it should be obvious that wheat fields of Midwestern United States yield different mixing scenarios than forested lands of the Southeast. Similarly, urban setting with low buildings yields different mixing events than cities with skyscrapers. Professionals use site-specific dispersion parameters for the area of interest and a numerical methods modeling approach.

9.7.3 Wind Velocity

A highly variable parameter in determining mixing and pollutant transport is the wind velocity. Wind velocity is generally much more variable than is water velocity in a river or aquifer. The only way to handle large variations in wind velocity is, again, to use a numerical methods approach to model the system.

9.8 Remediation

Although, in theory, technologies such as carbon absorption exist for the remediation of atmospheric gases, the enormous scale of such atmospheric pollutant outputs has prevented any such attempt at remediation on a large scale. The one tried-and-true approach to atmospheric remediation is to eliminate the source of pollution, and let nature take its course of dilution of the pollution in the prevailing wind. This may seem simplistic, but its efficacy was recently demonstrated during the electrical outage on the US east coast and in Canada in 2003. Scientists, meteorologists, and residents all reported cleaner air after the event when coal-fired plants and industry were closed down for hours or days. Of course, given our current dependence on coal, we cannot economically afford a long-term closure of these power-generating facilities. Still, it is interesting to consider the immediate impact of the removal of a large source of our air pollution. Again, the only cure for a polluted atmosphere is source reduction or elimination. An interesting side noted here is such elimination of pollution sources can be used as a technique to calibrate a model.

9.9 Models Used by Professionals

The simple analytical models presented here are perhaps the most inaccurate of all the models presented in this text because of real-world inconsistent atmospheric mixing scenarios at each location. There are a seemingly endless list of numerical models used in the real world that perform these calculations better, each specifically designed and verified for a specific set of site-specific atmospheric mixing conditions. A few of these are discussed in Chapter 10, and a more complete list can be found at https://en.wikipedia.org/wiki/List_of_atmospheric_dispersion_models.

Concepts

9.1 Name three point sources of pollution to the atmosphere in your area.

9.2 Describe three-step input scenarios of pollutants to the atmosphere.

9.3 Describe three-pulse input scenarios of pollutants to the atmosphere.

9.4 Study the dispersion tables for the urban and rural settings. How does an increase in wind velocity affect dispersion?

9.5 Why did we not include chemical degradation in our models?

9.6 Both of the atmospheric models used in this chapter are three-dimensional. Would you expect the pollutant concentration to increase or decrease if we used a one-dimensional model with the same input parameters?

9.7 Describe how one might use a cessation of inputs as a technique for model validation/calibration.

Suggested Reading for Class Discussions

Dunnivant, F.M., Environmental Success Stories, Chapter 6: Saving Our Atmosphere for Our Children, p 114. 2017, Columbia University Press. This chapter details our success in remediating SMOG, Acid Rain, and Ozone Depleting Chemicals.

Exercises

9.1 Using the background example data set in Fate, conduct a sensitivity analysis on dispersion in each direction by varying wind speed and cloud cover conditions. Which scenario gives the highest downwind concentration of pollutant?
Note: All plume and puff problems below can be solved using the Fate simulator at https://edusolns.com/fate/2016/welcome/home/.

Plume (step or continuous) Input Problems

9.1 A new coal-fired power plant was constructed near the banks of a large river where it is fairly windy. Even with the new technology, there is still some mercury in the emissions from the plant. Use Fate to construct an atmospheric dispersion graph assuming standard atmospheric pressure (approximately 1.0 atm) and the conditions listed in the following. Determine the concentration of mercury at a distance of 10.0 km downwind of the stack at a height of 1.0 m.
Height of initial release: 45 m
Stack exit velocity: 1.2 m/s
Wind speed: 3.33 m/s
Inside stack diameter: 1.5 m
Stack gas temperature: 450 °K
Air temperature: 298 °K
Source flow: 0.0025 g Hg/s
Atm. stability: B
Answer: Hg concentration equals 7.5×10^{-8}, very low.

9.2 An industrial smelting facility operating without an electrostatic precipitator on its effluent stack release lead at a rate of 0.33 g/s. Calculate the lead (Pb) concentration at 1.0 km downwind on the center line ($y = 0.0$ m) and at a height of 2.0 m. The stack is 50.0 m tall and 1.2 m wide with an exhaust rate of 0.80 m/s. The gas coming out of the stack is 400 °C, and the atmosphere is at 25 °C at 1.00 atm of pressure. The wind speed is 1.5 m/s, and the atmospheric stability is category B.
Answer: 2.19×10^{-7} g/m^3.

9.3 On a windy day, although working outside, you can occasionally detect the smell of rotten eggs coming from the pulp mill east (up wind) of your home. Since you know that hydrogen sulfide is highly toxic to birds and can kill at concentrations of several ppb, you decide to calculate what concentrations could be reaching your area from the mill about 6.0 km away. On the Internet, you find the concentration of H_2S emitted at the stack and other data you need to calculate the concentration downwind. Using data from the following table, calculate the concentration 6 km from the stack.
Height of initial release: 45 m
Stack exit velocity: 1.5 m/s
Inside stack diameter: 1.5 m
Atmospheric pressure: 1013 mbar
Stack gas temperature: 410 °K
Air temperature: 295 °K
Effective release height: 28 m
H_2S source flow: 2.18 g/s
Wind speed: 6.1 m/s
Atm. stability: C
Distance downwind: 6.0 km, $y = 0.0$ m, $z = 1.0$ m.
Answer: Concentration of H_2S 6.0 km downwind equals 3.57×10^{-7} g/m^3 or at 1.2 kg/m^3 of air, 0.30 ppb. The birds in your area are probably safe, but humans can just begin to smell the H_2S. Look up any human health ambient air quality standards that are applicable to your resulting concentration.

9.4 A train wreck has left a rusted gasoline tanker on its side and reeking of gasoline. The smell is emanating from a pool of gas on the ground next to the truck. The gasoline is evaporating at a rate of 10.0 g/s and is being dispersed by a 2-m/s wind during the middle of the day. Measurements are made at 2.0 m above the ground, 1.0 atm. of pressure, 25 °K, and the wind stability and radiation are described as type B. At 1.0 km away, what are the concentrations of gasoline? Create a plot of gasoline concentrations from 0.02 to 5 km away on the center line ($y = 0.0$ m) downwind from the spill. You will have to make a few modifications/assumptions in Fate to work this problem.
Answer: 5.0×10^{-6} g/m^3 or at 1.2 kg/m^3 of air, the concentration equals 4.17 ppb. Use Fate to calculate point in 0.5 km increments from 0 to 5 km downwind and convert your answers to units of ppb.

Puff (Pulse or Instantaneous) Pollutant Inputs

9.1 A chemical plant accidently releases a 25-kg plume of NO_2 gas into the atmosphere. The height of the release is ground level, the temperature of the gas and atmosphere is 300 °K, and the wind speed is 3.0 m/s (stability category B). What will be the concentration of gas at 100.0 m from the release site along the centerline line of the wind and at a height of 2.0 m? (N_2O reacts with atmospheric moisture to produce nitric acid but ignore this reaction at short distance of 100.0 m. Is it dangerous?)
Answer: 3.66 g/m^3 and using a unit conversion of 1.2 kg/m^3 of air yields a concentration of 3050 ppm which is very dangerous.

9.2 An industrial accident releases 13.6 kg of a slightly toxic gas 10 kg from a town. Determine the concentration when the plume reaches the town assuming the following conditions:

Center line: $y = 0.0$ m at a height of 2.0 m
$Q_m = 13.6$ kg
$u = 9.0$ m/s
Atm. stability = category D
Release height of the emission: 3.0 m.
Answer: 411 g/m^3.

9.3 A chemical manufacturer accidentally releases 5.0 kg in a plume from one of its tanks at ground level. The wind speed is 5 m/s with an atmospheric stability of category B. What is the ammonia concentration 1000.0 m from the source along the centerline ($y = 0.0$ m) and at a height of 2.0 m? *Answer:* 0.732 g/m^3.

9.4 A small tanker carrying chlorine gas crashes, releasing its contents into the surrounding area. Use Fate to create a plot showing the concentration in g/m^3 versus distance from 0.0 to 20 000 m directly downwind from the source. Using a conversion unit of 1.2 kg/m^3 for air, convert your concentration units to ppm (mg of pollutant per kg of air).

Mass of gas spilled: 900 g
Wind speed: 12 m/s
Height of release: 1.0 m
$y = 0.0$ m, $z = 2.0$ m
Atmospheric stability category: C.
Answer: Use Fate to calculate your point concentrations and Excel to create your plot.

Spreadsheet Exercise

Create a spreadsheet that performs the same calculations as Fate for both the step and pulse equation. Construct your spreadsheet so that it is interactive (so that you can change numeric values for parameters and the plot automatically updates itself).

Create your own problem:

Research the literature or your local town and identify an industry and a possible chemical release. Design a chemical release scenario and use Fate to investigate the downwind concentration as a function of distance from the source.

Bibliography

Bogard, W. (1989). *The Bhopal Tragedy: Language, Logic, and Politics in the Production of a Hazard*. San Francisco: Westview Press.

Briggs, G.A. (1972). Discussion: chimney plumes in neutral and stable surroundings. *Atmos. Environ.* **6**: 507–510.

Fjeld, R.A., Eisenberg, N.A., and Compton, D. (2006). *Quantitative Environmental Risk Analysis for Human Health*. New York: Wiley Interscience.

Gifford, F.A. (1976). Turbulent diffusion typing schemes: a review. *Nuclear Safety* **17** (1): 68–86.

Kumar, C.D. and Mukerjee, S.K. (1985). Methyl isocyanate: profile of a killer gas. In: *Bhopal: Industrial Genocide?* Hong Kong: Arena Press.

Kurtzman, D. (1987). *A Killing Wind: Inside Union Carbide and The Bhopal Catastrophe*. New York: McGraw-Hill Book Company.

Lapierre, D. and Moro, J. (2002). *Five Past Midnight in Bhopal*. New York: Warner Books/An AOL Time Warner Company.

McElroy, J.L. and Pooler, F. (1968). St. Louis Dispersion Study. U.S. Public Health Service, National Air Pollution Control Administration Report AP-53.

McGraw Hill (2003). *McGraw Hill Dictionary of Scientific & Technical Terms*, 6e. McGraw-Hill Companies, Inc.

Pasquill, F. (1961). The estimation of the dispersion of windborne material. *Meterol. Mag.* **90** (1063): 33–49.

Pasquill, F. (1976). *Atmospheric Dispersion Parameters in Gaussian Plume Modeling: Part II. Possible Requirements for Change in the Turner Workbook Values.* EPA-600/4-76-030b. Research Triangle Park, NC: U.S. Environmental Protection Agency.

Shrivastava, P. (1987). *Bhopal: Anatomy of a Crisis.* Cambridge, MA: Ballinger Publishing Company.

Turner, D.B. (1970). *Workbook of Atmospheric Dispersion Estimates.* Cincinnati, OH: U.S. Department of Health, Education, and Welfare.

Turner, D.B. (1994). *Workbook of Atmospheric Dispersion Estimates: An Introduction to Dispersion Modeling*, 2e. Ann Arbor, MI: Lewis Publishers.

10

Regulatory Environmental Modeling Practices and Software

Raymond C. Whittemore

Adjunct Professor of Civil and Environmental Engineering, NCASI, Tufts University, Medford, MA, USA

10.1 Introduction

There may always be an array of model choices available to environmental modelers for use in any single situation. In practice, however, modelers make this choice by convenience (availability of supporting data) and comfort (familiarity) by its repeated use over many differing scenarios. Rarely, a modeler would elect to work outside of their "comfort zone" by choosing a more complex tool unless necessary to deal with special circumstances such as legal challenges. The limited-choice scenario occurs because models are developed from our theoretical and empirical understanding of how nature guides and determines the fate of anthropogenic chemical emissions in its multimedia environment. There are a finite and limited number of understood ways to move them by river and ocean currents, transfer of phases by volatilization, immobilization by sorption and mass transfer, or reaction with other chemical constituents. Furthermore, alternate mass balances have no meaning in science, and there can only be one accurate mass balance so that models tend to be similar rather than different.

Many waste-load allocation models must be calibrated and validated with site-specific data by necessity to ensure its accuracy. Hence, a model might also be chosen based upon expected data sources or their absence, implying that a simpler, less accurate model may be selected based upon limited data. Implications on modeling utility in these cases are largely unknown but can be approximated by experienced modelers.

This chapter introduces an array of modern environmental models by media type. Inclusion does NOT imply any endorsement by the authors specifically or from EPA generally that maintains a clearinghouse for many models and historically have provided many public and private entities with the source codes and often the necessary training to use the models correctly. They are simply part of the model selection landscape modelers routinely face. A robust model is one that you grow to trust by experience or through peer review and acceptance. In many cases, a simpler model is sufficient if the prediction/forecast meets the modeling objectives.

10.2 Generic Model Types

Various names have been attributed to models based upon their use. Further, more detailed information on most of these models is available on Internet sites. Some generic types are briefly described here:

Regulatory Models are a class of models that pertain to a regulatory action or emission control activities (compliance, permitting, risk assessment, etc.). They exist in all segments of the environment – air, water, land, and for any human or ecological modeling end point.

A Receiving Water Model is a general term that refers to any waterbody receiving a wastewater discharge from a designated location or a widespread areal source including runoff and the atmosphere.

A Wasteload Allocation Model is a class of regulatory models that compute an allowable load based upon existing water quality and degradation potential. They have a technical and policy components based upon allowable discretion granted within existing federally approved water quality standards programs. They are founded on the concept of minimal degradation that is quintessential in the US water quality approach. This approach differs from ones based upon effluent standards and acknowledges that some contamination can be treated instream, allowing the water body to recover before they are polluted.

Watershed Models are those designed to apply to a large area delineated by defined areas accepting that area's water drainage and are usually a few square miles up to thousands of square miles in size. The Mississippi River drains the central United States

Pollutant Fate and Transport in Environmental Multimedia, First Edition. Frank M. Dunnivant and Elliot Anders.
© 2019 John Wiley & Sons, Inc. Published 2019 by John Wiley & Sons, Inc.
Companion website: www.wiley.com/go/Dunnivant/pollutant_fate&Transp_EnvironMultimedia

and encompasses dozens of state areas and several subbasins such as the Missouri and Red Rivers before entering the Gulf of Mexico below New Orleans, LA. There are 2100+ such delineated watershed areas in the conterminous US boundary. Also, there are many federal data products readily available to assist modelers to efficiently use these delineated land areas in their various missions.

Mixing Zone Models are those that purport to guide wastewater (point source) effluent mixing with the receiving water as guide to more effective designs that promote rapid integration with the ambient flow. Practically, however, their use is limited to water bodies to optimize effluent with ambient waters for design.

Point Source Models are those pertaining to defined wastewater effluents coming from a specific outfall pipe or diffuser. This designation also applies to an air stack source emission, and landfill leaks into the subsurface and the groundwater.

Non Point Source Models are applied to diffuse sources such as land runoff from agricultural, forested, urban streets, parking lots areas, and atmospheric deposition during precipitation events. They are often simply denoted as "NPS models."

Wastewater Treatment Plant Models are operational control models for biological wastewater treatment facilities such as those using the activated sludge tank (AST) or aerated stabilization basin (ASB) processes. They are sometimes used in a research forensic mode to learn more on the implications of remedial actions to correct a treatment plant upset.

Sediment Models are those that simulate sediment transport and detailed chemical fate processes occurring within their midst. These include basic sediment movement as well as biological and chemical processes in shallow and buried benthic materials.

Food Chain Models are a class of assessment tools used in human and ecological risk assessment to simulate chemical concentrations in food sources that store and, thereby, magnify exposure to those species that feed on and consume them. They are part of a modeling class known as ecological models that purport to represent other aspects of a species lifecycle or behavior and its preferred habitat. They require specialized knowledge of ecology and zoology that are beyond the scope of this book.

A model can fit multiple classes; therefore, it can be regulatory, receiving water, point source, nonpoint sources (NPSs), and used in wasteload allocation in a watershed. The terminology does not necessarily imply a widely different utility either. Students should be wary of the varied nomenclature usage in the literature.

10.3 Model Availability

The EPA Center for Exposure Assessment Modeling (CEAM) (EPA 2017a) and the related Support Center for Regulatory Atmospheric Modeling (SCRAM) (EPA 2017b) distribute simulation models and database software for several regulatory modeling needs. The former centers support models designed to quantify the movement and concentration of contaminants traveling through groundwater aquifers, surface water resources, and food chain media and as such are designated multimedia models. EPA's Council for Regulatory Environmental Modeling (CREM) provides a consistent and uniform policy for all agencies modeling activity (EPA 2017c).

The following sections describe some general information on model classes for each media type. The list of models included in each category is not complete by design, and those noted are only the major ones used currently by regulatory agencies in the United States. Students are encouraged to obtain more complete discussion in the documentation reports cited. Related user manuals provide more specific application detailed. Such information is beyond the scope of this introductory textbook, and interested students may want to track pending free EPA or commercial venues for more hands-on individualized training.

10.4 Atmospheric Quality Models

Models in this category are totally dependent on vital National Weather Service data for various meteorological conditions in the atmosphere within the troposphere close to the Earth surface where people live (often noted as the boundary layer and about 6 mi deep). This region is also characterized by a decrease in temperature with increasing height from the Earth's surface where anthropological emissions mix and accumulate and contaminate its overall quality. Earth's hydrologic, carbon, and nitrogen cycles inherently operate in this media phase.

Air dispersion models are used to estimate or predict ambient concentrations of gaseous and particulate constituents emanating from emission sources. The model outputs are commonly used as input into risk assessments or compared to regulatory standards (e.g. during permitting of a new combustion source or in prevention of significant deterioration [PSD] activities – both essential Clean Air Act requirements). EPA issued a revised guidance memorandum (EPA 1999a,b) to standardize the use of specific models for regulatory application. This revised guideline provided more details in these reporting or permit application processes.

The guideline noted further recommends air quality modeling techniques that should be applied to State Implementation Plan (SIP) revisions for existing sources and to new source reviews, including PSD. A SIP is simply the state's long-range plan to achieve compliance with primary pollutant exceedance (EPA 2017d), and a PSD permit (EPA 2017e) is required for any new or modification in compliance regions. Compliance regions are those meeting primary pollutant criterion. The guidance is intended for use by EPA Regional Offices in judging the adequacy of modeling analyses performed by EPA, state and local agencies, and by industry. Students should seek to understand the regulatory strategy for compliance with the human health National Ambient Air Quality Standards (NAAQS) adopted within the Clean Air Act in the 1960s as they are human health based.

Air quality modeling is frequently a two-phase process, utilizing models with two levels of complexity. The first level consists of general, relatively simple estimation techniques that provide conservative estimates of the air quality impact of a specific source or source category. These are screening techniques or screening models and not intended as the primary regulatory tool. The purpose of such techniques is to eliminate the need for more detailed modeling for those sources that clearly will not cause or contribute to ambient concentrations in excess of either the NAAQS (EPA 2017f) or the allowable PSD concentration increments. If the screening process indicates that the concentration contributed by the source exceeds the PSD increment or the increment remaining to just meet the NAAQS, then the second level of more sophisticated models should be applied.

The second level consists of analytical techniques (including refined models) that provide more detailed treatment of physical and chemical atmospheric processes, require more detailed and precise source and meteorological data, greater complexity in land topography and terrain, and provide more accurate concentration estimates. As a result, they provide a more refined and, at least theoretically, a more accurate estimate of source impact and the effectiveness of control strategies. These are referred to as refined models and require more training than with a screening tool that tends to rely upon simplifying assumptions and, as such, are generally less accurate or reliable, but still useful in many contexts.

Air models can apply to near- and far-field downwind areas or vent and stack emissions and/or accidental explosions or fires. The latter tend to be intermittent of shorter duration, while the former are more continuous in nature. EPA (2017g) describes all model types useful in the atmosphere as dispersion, photochemical, or receptor models. The three types noted reflect their use in the regulatory domain with important but differing objective as noted.

Dispersion models are usually used in permitting process to determine the primary air or hazardous pollutant concentrations at specific ground-level receptors surrounding emission sources either in the local area or miles away in neighboring states. Most air modeling done in the United States is with dispersion models using the Gaussian plume assumptions. Photochemical models are typically used in regulatory or policy assessments to simulate the impacts from all sources by estimating concentrations and deposition of both inert and chemically reactive pollutants over large spatial scale. Finally, receptor models are observational techniques that use chemical and physical characteristics of gases and particles measured at source and receptor to identify the presence of and to quantify source contributions to receptor concentration. Hence, they are used in a forensic mode in risk assessments.

The most commonly used models are AERMOD, ISC, CALPUFF, and a screening model, AERSCREEN. They are briefly described here.

AERMOD (EPA 2004, 2017h) and ISC (EPA 1995) are the major air regulatory models used, although AERMOD is the preferred source code for all but long-range transport less than 50 km. The American Meteorology Society developed AERMOD gradually over a decade-long period in the late-twentieth century. It is essentially a Gaussian plume model with advanced features that include differing emission configuration, atmospheric and weather conditions that affect airborne turbulence and mixing, and near- (urban buildings) and far-field terrain (mountains and valleys) features that affect long-term transport. This last feature is better represented in CALPUFF (Exponent Atmospheric Sciences 2017) and makes it preferred for distances greater than 50 km. The developers' headquarters is in the Menlo Park, CA area and offer training courses as does EPA SCRAM currently with AERMOD. A simplification of ISC3 is more commonly known as AERSCREEN (EPA 2016) and is used as a screening model to determine if more refined analyses is required.

In the case of a stationary source (e.g. a stack), the ISC or "Industrial Source Complex" Model, modified many times since its first inception in the 1970s, has been the more popular air dispersion model and has found the widest use in a number of different industrial settings. The major inputs to the model are mainly physical and/or geographical and include the terms that define the source (e.g. stack height, emission rate, concentration of gaseous and particulate phases, and size distribution of particulate) and the ambient meteorological conditions (e.g. local conditions, wind rose, precipitation, roughness of terrain, location of buildings). The model will analyze the weather and contaminant source terms and, given that the input is accurate, will estimate temporal (short- or long-term, depending on the version of the model) and spatial (over a receptor grid) concentrations of the various constituents within the Gaussian plume flow pattern.

10.5 Surface Water Models

Surface water models were created to address flow and quality issues as regulatory models and can be applied to conventional or toxic point source or diffuse NPS discharges. Surface waters exist in Earth's rivers, lakes (ponds), and swamp areas. Those rivers that drain into the oceans are affected by tidal forces and deemed estuaries. They represent by far the largest class and frequently used mathematical models in the public domain. EPA, USGS (USGS 2017), and the Army Corps of Engineers (COE 2017) have created and maintained such modeling tools. For several decades in the mid to late twentieth century, consulting engineers also developed their own proprietary models to meet client needs. The Army Corps of Engineers has had a long-term mission since the River and Harbors Act of 1899 (Wikipedia 2017a) to safeguard navigation in the Nation's rivers and harbors, and their hydrology models are widely used. More significantly, they are the oldest class of environmental models by virtue of their early twentieth century debut.

Surface water models trace their development to the 1920s and the Streeter–Phelps (USPHS 1925) analysis for the Ohio River and created the field of water quality modeling with the seminal work of O'Connor (1967), Thomann (1963), Velz (1971), and many others who they mentored throughout the twentieth century. In the wasteload allocation process, models may be used in concert with water quality or nutrient standards to ensure that discharges are protective of State and Federal instream levels and meet a variety of water quality standards (EPA 2017i). In a planning mode, these models help guide regional resource planning and development by issuing discharge permits wisely.

Historically, surface water models have also focused on water distribution and hydropower permitting (FERC 2017). In recent decades, toxic fate and excess nutrient problems have surfaced and remain as the remaining challenges for surface water protection in the twenty-first century. These likely will include sediment (benthic) contamination and bioaccumulation and biomultiplication of toxic compounds (EPA 2017j) that are primarily organic compounds known to be nonreactive and persistent in the environment. For example, a majority of the total mass of plastic material created in the twentieth century exists today in landfills or part of the litter that invades our oceans and landscapes worldwide (National Geographic 2017). The same conclusion probably applies to many insecticides and other chemicals invented by man.

The most frequently used models in the twentieth century include Qual2-E (and similar predecessors such as Qual1 and DOSAG). Qual-2K, Water Analysis System Program (WASP), Hydraulic Simulation Program Fortran (HSPF), and Streeter–Phelps in its expanded form gained prominence in the twentieth century and still used today. Their use was often mandated by State agency practices and sometimes specified in established regulatory rules. For many processes, water quality simulations are nearly identical due to their formulation/construction with similar assumptions. This is especially true for ultimate BOD degradation and atmospheric reaeration, although many other similarities exist between commonly used models in both the public and private domains. Model evolution is ongoing and never stagnant, but it is also common for modelers in some state agencies to continue to use modeling frameworks that are outdated or simply replaced by more efficient code (e.g. Windows format).

10.5.1 Qual-2E/Qual-2K

The Qual-2E water quality model evolved from a similar but simpler representation of water quality processes by the Texas Water Development Board and Water Resources Engineers in the 1970s (TWDB 1971). This model was heavily used by State and Federal Agencies throughout the twentieth century for water quality forecasting in river environments. Subsequent work by NCASI in collaboration with Tufts University and EPA led to a more robust version and named Qual-2E (EPA 1987). It simulated 1D steady-state river flow in a dendritic stream network, coupled C and N BOD kinetics, temperature, SOD, water column and benthic N, P cycling, and state-of-the-art algal productivity and respiration (see Figure 10.1). It received substantial regulatory attention and use throughout the United States, Asia, and Europe. The developers and EPA offered numerous training workshops that were widely attended in the United States and abroad. Additional information on the Qual-2E development is available in a NCASI Technical Bulletin (NCASI 1977).

Professor Steven Chapra (currently at Tufts University) and collaborators at the Washington Department of Ecology developed Qual-2K in the late twentieth century (Chapra et al. 2008). This model was built upon the successful Qual-2E model with further useful simulation capabilities. The hydraulic and water quality features were quite similar and would be expected to generate identical results with the same flow and loading conditions as Qual-2E. An extension of Qual-2K to lakes was also implemented by Chapra (Tufts 2017). Qual-2K

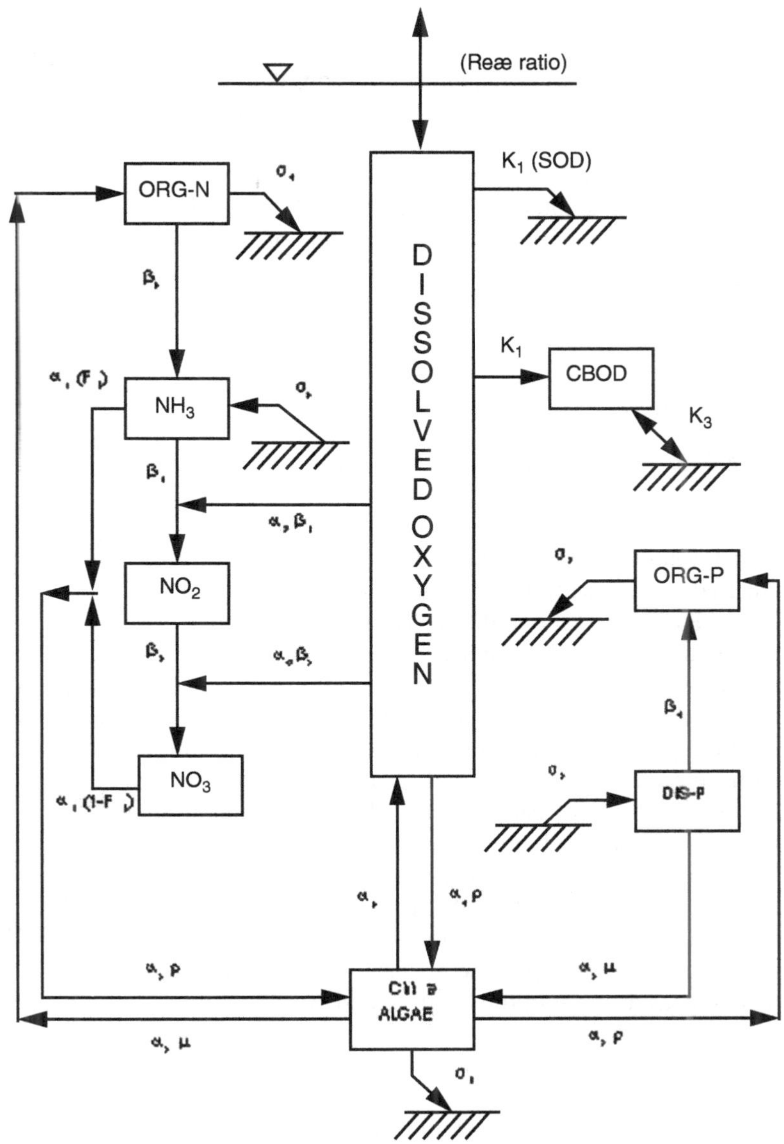

Figure 10.1 Original Qual – 2E processes. Source: adapted from EPA Training Workshop, circa 1985.

followed Qual-2E in the river model evolution but had some important differences. These differences included the following:

- Developed using macros on an EXCEL platform w/VISUAL BASIC programming
- Fully dynamic or steady-state simulation capability for river flow and wasteloads
- Included attached algae as a process (others identical to Qual-2E)
- Extensive use in nutrient modeling in rivers and lakes.

10.5.2 HSPF

HSPF was based upon the Stanford Watershed Model (ASCE 2003) of the 1960s that was the first NPS model used in agricultural watersheds with soil erosion and infiltration/runoff were determinants of river water quality. It used continuous simulation of these processes to achieve a complete water balance over periods of a few months (seasonal) to longer time periods (decades) for multiple end points. These simulations provided data with meteorological records to compute streamflow hydrographs and related pollutographs (concentration versus time graphs). HSPF simulates interception soil moisture, surface runoff, interflow, base flow, snow pack depth and water content, snowmelt, evapotranspiration, ground-water recharge, dissolved oxygen, biochemical oxygen demand (BOD), temperature, pesticides, conservatives, fecal coliform, sediment detachment and transport, sediment routing by particle size, channel routing, reservoir routing, constituent routing, pH, ammonia, nitrite-nitrate, organic nitrogen, orthophosphate, organic phosphorus, phytoplankton, and zooplankton.

The program can simulate one or many pervious or impervious unit areas discharging to one or many river reaches or reservoirs. Frequency-duration analysis can be done for any time series generated by the simulations. Also, any period from a few minutes to hundreds of years may be simulated; therefore, flexibility is provided. HSPF has been adapted for use within the EPA BASINS model toolkit (later this chapter), and one of the most frequently used models along with similar Soil and Water Assessment Tool (SWAT) to simulate NPS effects in watersheds.

10.5.3 SWAT

SWAT is a comprehensive, river basin model that requires a large number of input parameters and, similar to HSPF, is a nonpoint model (Arnold et al. 1995). It was developed at Texas A&M University and recently adapted to run within EPA BASINS Version 4.1 (EPA 2015). For both HSPF and SWAT, the shear number of parameters required complicates model parameterization and calibration, and several calibration techniques have been developed for SWAT, including manual calibration procedures and automated procedures, along with uncertainty analysis and can provide statistics for goodness-of-fit calculations. Hence, the model's use requires a good understanding of the governing hydrologic processes, sediment sources, crop yields, and nutrient balances. Successful use requires a significant learning curve and/or adequate prior training because many of the required parameters have little real physical meaning, as do most in the Qual-2E and WASP models.

10.5.4 WASP

WASP or Water Analysis System Program was originally developed by Di Toro et al. (1983) and EPA (1986a, b), and used extensively in the Great Lakes and Chesapeake Bay areas as part of EPA R&D efforts in the Center for Exposure Modeling (CEAM). Conceptually, the model segments water bodies into boxes (Figure 10.2) for lakes and performs mass balance water quality balances on its contents in 1D, 2D, or 3D, either in steady-state or fully dynamic modes. Table 10.1 provides general strategies for water body segmentation for water bodies. Hydraulic simulation is computed by linking an appropriate hydrodynamic submodel, usually DYNHYD for time varying tidal systems into the powerful mass balance routines. Table 10.2 displays some current modeling capabilities of the current WASP7 release. It is widely accepted in the United States as a comprehensive and flexibly integrated modeling system and used extensively in widely different hydraulic regimes ponds and lakes or rivers and estuaries. This effectively means that an agency uses a single model base for all its waterbody simulations – making its staffing and training more efficient.

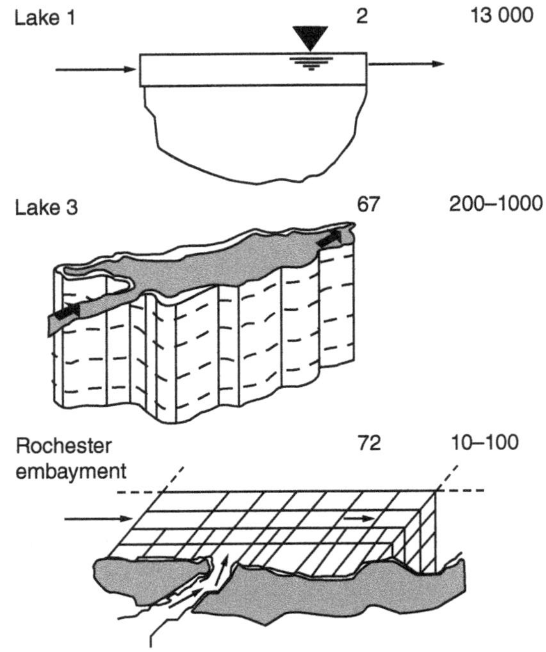

Figure 10.2 WASP model segmentation for a lake note numbers show boxes used in the examples depicted. Source: Adapted from EPA WASP COURSE.

Table 10.1 Generic river/WASP considerations (guidelines) for waterbody segmentation.

Spatial scale of the problem
Segment of a water body (e.g. reach, embayment)
Whole water body (e.g. main river, lake, estuary)
Dimensionality and spatial discretization of segments
Box, 1-, 2-, or 3- dimensional model
Horizontal: tens of meter to tens of kilometers
Vertical: whole water column to tens of centimeters
Physical components
Water column (epilimnion, hypolimnion)
Benthic sediments (surface, subsurface)
Model limitations
Maximum number of segments, 3000 (large lakes)

At its core, WASP is a box model that balances mass with several quality versions (Figure 10.3) that adds to its versatility. A state modeler, for example, could use the same or similar input files for conventional and toxicants alike with this modular structure format. Furthermore, WASP7 release has been added to BASINS4.1 allowing wider access to federal data products (watershed delineation, STORET, NPDES, NWS meteorology, land use, soil classification, etc.) conveniently integrated into the GIS user interface (Figure 10.4). As a result, Qual-2E

Table 10.2 WASP linked water quality models.

Streeter–Phelps type DO/BOD/SOD/algae
Eutrophication – N/P/Algae with possible sediment reactions (Eutro Qual 2E/2K – like)
Estuarine water bodies (DYNHYD5, EFDC)
Sediment diagenesis (nutrient)
HSPF (default NPS model)
Mercury-specific toxicant model (w/NPS land and air sources)
SWMM (storm water and urban sewer networks)
Simple toxicant (TOXI)
Unknown future environmental models and modeling tools

and Qual-2k are being used less than in prior decades and no longer supported by EPA personnel at the Regional Office. WASP outputs are large and require users to become familiar with a supplemental data management program, called WDM, for viewing and publication of results.

10.5.5 Estuarine Models – DEM/DYNHYD/EFDC

Estuary models began in earnest by the steady-state tidal mechanics with limited reaction processes attributed to O'Connor (1960). Estuary modeling was in its infancy when the Dynamic Estuary Model (DEM) was developed in California for use on the Sacramento-San Joaquin Delta and San Francisco Bay areas (EPA 1979). From its inception, complexities in tidal hydraulics were treated as a separate modeling module and called DYNHYD. Several versions have evolved from its original form (FWPCA 1970) and now at version 5 (DYNHYD5), but the hydraulic simulation is more versatile than the Qual models. Water quality simulation capabilities are

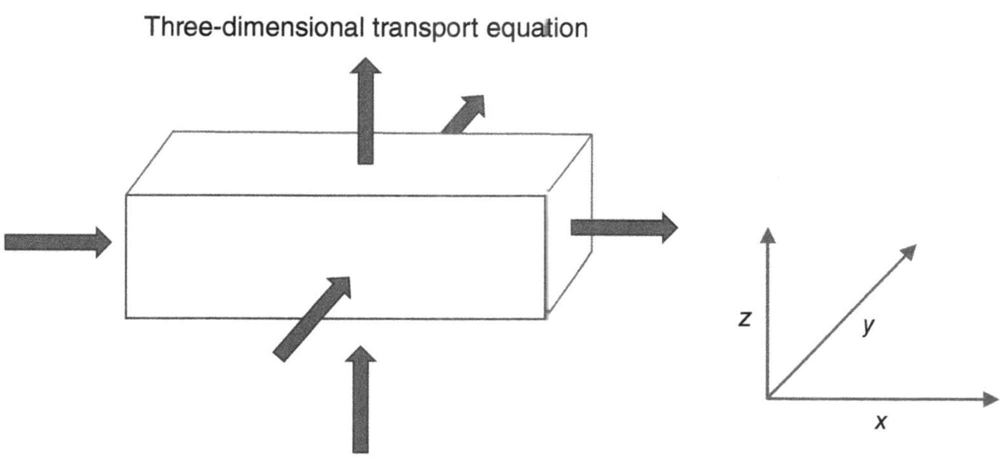

$$\frac{\partial C}{\partial t} = -\frac{\partial U_x C}{\partial x} + \frac{\partial}{\partial x}\left(E_x \frac{\partial C}{\partial x}\right) - \frac{\partial U_y C}{\partial y} + \frac{\partial}{\partial y}\left(E_y \frac{\partial C}{\partial y}\right) - \frac{\partial U_z C}{\partial z} + \frac{\partial}{\partial z}\left(E_z \frac{\partial C}{\partial z}\right) \pm \text{sources and sinks}$$

Figure 10.3 Model network with advective transport pathways. Source: From WASP8 documentation.

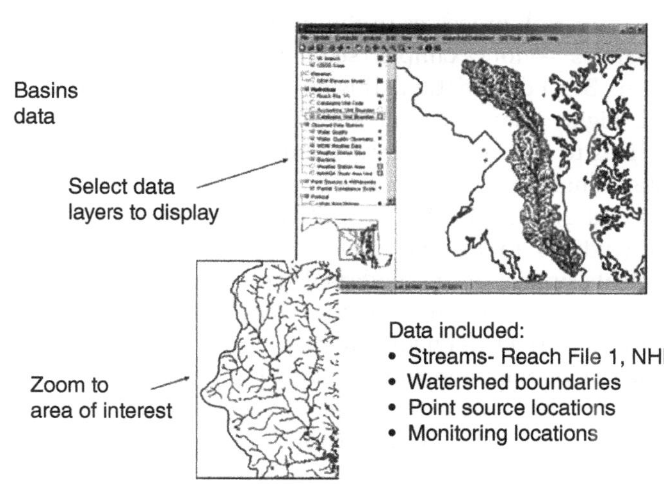

Figure 10.4 BASINS screenshot of GIS interface and typical use of data layers (in delineated watershed boundary). Source: Public domain, modified to black and white image and recreated.

equivalent to the state-of-the-art WASP or Qual-2E models. This model also has a significant learning curve and not intended for novice modelers. Students will find that it has been used in the Chesapeake Bay watershed, and the model code has evolved during this basin-specific use. DYNHYD in some form remains the model of choice in hydrodynamic simulation of estuarine hydraulics and the advective component of WASP mass balance simulation legacy.

10.5.6 EFDC

After early attempts in the Chesapeake Bay estuary, the WASP-like structure and mass balance routines emerged into fully dynamic representation of estuarine hydraulics and tidal cycles. Scientists at the nearby Virginia Institute of Marine Science (VIMS) continued to develop a full 3D model structure to simulate known variations in water quality, including vertical stratification due to salt intrusion from the ocean.

Dr. John Hamrick originally developed EFDC at VIMS and School of Marine Science of The College of William and Mary. The Commonwealth of Virginia passed special legislative research initiative to support this estuary of national commercial importance. EPA and NOAA (Sea Grant program) provided support for EFDC development. Recently, the EFDC code was added to the EPA BASINS modeling framework system (BASINS4.1), making simulation of estuary flow and water quality accessible with EFDC, DYNHYD, and WASP models (EPA 2015).

10.5.7 Streeter–Phelps – Expanded Formulation

The extension of the original Streeter–Phelps equation (model) to include an expanded number of source/sink processes was mathematically elementary and attributed to Manhattan College (Thomann 1974). Unit deficits provide for a useful way to evaluate the relative effects of source/sink processes in the river's dissolved oxygen dynamics – for example, sediment oxygen demand versus BOD degradation. This expanded equation form is a highly useful screening model to gain modeling insights to support monitoring and more refined (accurate) simulations. The Fate model now includes both this and the classical forms of Streeter–Phelps analyses for students to further learn modeling nuances and apply its simulations to modern day problems. The 1974 Thomann reference also includes the estuarine form of the basic Streeter–Phelps analyses. Expansion of this estuary equation to include expanded source/sink terms is beyond the scope of this book, but nevertheless useful for students to know it does exist and was used to address pollution in estuaries and bays.

10.6 Large-Scale Watershed Models

Watershed models are generally used to assess the impacts of multiple point and NPS contributions in a landscape setting. The source loadings are typically aggregated with information on climate, topography, geology, hydrology, and land use to simulate potential flow and water quality conditions at one or more locations of interest within the watershed system. Processes considered by most watershed models include the following:

(1) Rainfall abstractions (runoff, interception, infiltration and plant and tree evapotranspiration);
(2) Routing of water to and within a receiving water body;
(3) Sediment routing and yield; and the transport of chemical elements or factors from point and/or NPSs.

It is important to note that the "watershed" moniker refers to the application, not a fundamentally different waterbody and can include lakes, rivers, and estuarine components in the delineated watershed. The latter term is used to designate any particular watershed under study and includes land areas and topography of significance. In some instances, watershed models provide a geographic framework for other models to be applied to a landscape or combined with other spatially explicit information, such as soil types and land use. This data requirement makes watershed models unique and often more complex than most other groups of environmental models currently in use.

Early watershed models, first developed in the 1960s (Stanford Watershed Model) used empirical relationships to determine runoff and peak flow. Initially, sediment was the only component of NPS pollution studied. With the advent of sophisticated computer technology, however, many modern watershed models are now able to account for multiple and simultaneous processes including soil moisture and snow simulations, hydrologic routing, detachment, deposition and transport, sediment loadings, nutrient processes, oxidation–reduction, and many portions of the hydrologic cycle. The well-known Stanford Watershed Model and its subsequent HSPF form is the best example of these efforts. Most watershed models are divided into separate routines that simulate infiltration, evapotranspiration, percolation, and runoff routing.

The latest trend in watershed modeling has been to integrate the model algorithms with other research tools such as weather simulators, geographic information systems, and decision support systems. EPA BASINS (2015), USGS's MMS (USGS 1996), and GIBSI in Canada (Dupont and Villeneuve 1997) are the most developed by 2017, although BASINS remains the premiere model choice due to its practical versatility and growing user community.

In recent years, the array of chemical, biological, and ecological processes (see the AQUATOX model [https://www.epa.gov/ceam/aquatox]) represented in watershed models has improved, and some models now include handling of specialized scenarios such as bioaccumulation, lake effects, and impoundments such as stratification and nutrient upwelling, interbasin and subsurface routing, and unique land use impacts. Similarly, routing of pesticides, nutrients, and other chemicals in a water system has also been included in many watershed models. The application of this group of models within the regulatory community has focused on evaluating land use change, developing watershed improvement strategies, establishing best management practices (BMPs), and meeting water quality standards through calculations (and implementation) of TMDLs. A resource for some of the early models that were useful for watershed-scale analysis is EPA's Compendium of Tools for Watershed Assessment (EPA 1997a,b,c). See Table 10.3 below for a useful overview. EPA's new modeling paradigm developed for its CREM initiative fully justifies their continued development in the twenty-first century US regulatory practices.

ILWAS was developed to assess the acidification of lake and water systems (Gherini 1985). It was first applied in the Adirondack Region of New York State to study the impacts of acid rain in forested areas and rivaled the Stanford Watershed Model at this early time in watershed model development.

As previously stated, almost all watershed models are considered highly complex by design. This is primarily due to their inherent handling of multiple ecological processes over considerably large areas and in multiple dimensions. Fortunately, most watershed models utilize a graphical user interface (GUI) and a multiwindow environment. The GUI will typically allow the user to view the watershed area and the spatial information within the watershed at various scales. Multiple windows provide the user with the convenient opportunity to, for example, view input and output data at the same time and, in some cases, will demonstrate a graphical response to parameter adjustments over time. These factors generally allow for a user-friendly environment that will typically enhance the comfort level and learning curve that the user can achieve with the model. However, a model's ease-of-use can be, and quite often is, at the expense of input transparency, a bad feature that often hides relevant background data and assumptions. There are typically dozens and often hundreds of input options that users are expected to adjust knowledgeably. To gain the appropriate knowledge, one needs to study the model's user guide carefully and/or attend a specific training course. In order to use most watershed models properly, previous experience in water quality modeling and a background in geographic information systems is recommended if not required. Some students will see these GIS's intuitive by nature, and others may dislike the transparency it inherently removes.

In selecting a watershed model, user preference will primarily depend on the intended use of the program or objectives of the tasks to be completed because they are packaged with a considerable amount of data and function well within a GIS framework as noted with EPA BASINS modeling toolkit.

Table 10.3 Integrated watershed – scale modeling systems.

System	Organization	Domain	Models
BASINS	EPA	Watersheds	QUAL-2E, HSPF
			SWAT, AQUATOX
GIBSI	University of Quebec	Watersheds	ILWAS[a], QUAL-2E
MM.S	USGS	All waterbodies	Completely flexible
WASP	EPA	Rivers and estuaries	DYNHYD, TOXI
			EUTRO, EFDC

a) Integrated lake watershed acidification system – from Lake Adirondack acid rain studies.

Source: From EPA (1997b) and NSASI (2002).

10.6.1 BASINS/WARMF

EPA BASINS is a formidable and comprehensive framework for water quality and spatial data resources used in modeling. The BASINS model toolkit (now released as version 4.1) was created as a watershed scale-modeling framework that integrates useful federal data sources with several exposure models – SWAT, HSPF, QUAL-2E, WASP, EFDC, SWMM, and a host of other helpful algorithms. The integrated GIS Metadata and modeling platforms are available in BASINS (EPA 2015). Earlier versions document this evolutionary path (EPA 1999a). It now comes as (distributed by EPA) an Open-Source version of ESRI's ARCVIEW GIS software (ESRI 2017), saving the user from purchasing this necessary software.

BASINS with SWAT (Arnold et al. 1995) are highly recommended for data exploration and developing a landscape blueprint of an area considered for evaluation. These programs also have a built-in delineation tool that allows the user to define the area to be modeled. Both of these programs, in addition to MMS and AGNPS, are also particularly useful in evaluating multiple source impacts on the watershed system and for developing

Table 10.4 BASINS date layers environmental monitoring date in BASINS modeling system – from known federal sources (interior and others).

Water quality monitoring station summaries – STORET
Water quality observation data (user site-specific input)
Bacteria monitoring station summaries
Weather station summaries
USGA gaging stations
Permit compliance system (PCS) sites and loadings
Delineated watersheds in conterminous US, 48 states
River reach file (RF1)
NWS meteorological data
US soil data base
Road boundaries
Urban areas and county designations

management strategies. SWAT, similar to its related HSPF model, is especially valuable in the assessment of long-term impacts over a relatively large area. Similarly, the WARMF model is designed for river basin-scale studies, and a spatial information viewer was added to emulate the cartographic component found in most geographic information systems (Table 10.4).

The Watershed Analysis Risk Management Framework (WARMF) (Systech Water Resources 2017; EPRI 1998) is a generalized watershed modeling and analysis tool that can be applied to any watershed. It can be used for short- and long-term predictions, watershed management, and calculating Total Maximum Daily Load (TMDL). It development has been supported by the Electric Power Research Institute, the Duke Power Company, the US Bureau of Reclamation, and a number of state agencies. It has evolved considerably since its late 1990s development, used successfully on the Catawba River watershed in South Carolina. (SCDHEC 2017).

WARMF is specifically comprised of five modules: engineering, consensus, data, knowledge, and TMDL modules to mimic the varied aspects of total daily load development. The engineering module is used to simulate hydrology and water quality. The consensus module is used to evaluate management alternatives in a stakeholder group setting. The data module puts the model's input data into graphs and spreadsheets for easy viewing and editing. The knowledge module stores available information about a watershed. The TMDL module guides the user through a process to calculate TMDLs in the watershed. WARMF has a user-friendly GUI that displays a watershed map and allows data input and viewing simulation results by the point-and-click method. It further acts as a decision support system, providing detailed analysis in a form useful to watershed managers.

10.7 Subsurface or Groundwater Models

Groundwater models are often used in predicting the impact of solid waste disposal units (often called landfills) on human and wildlife exposure scenarios. These include contamination at prescribed down-gradient locations including receptor well sites and inflow to receiving waters. This use also includes migration of hazardous chemicals from an accidental release toward a valued resource such as a lake or river in addition to residential wells and homes. Typically, these models are useful in risk management scenarios associated with down-gradient exposure. The models provide a reasonable framework in which the exposure concentrations can be estimated based upon known and/or assumed management designs and operating scenarios.

Simulation capabilities in this class of models typically include the multiple processes of advection (movement), hydrodynamic dispersion, linear or nonlinear sorption, and chain-decay reactions that include details on daughter products. They are all time-variable, although the simulation timestep (user specified) can vary from minutes and days, to 1000 years, and depends upon the site-specific situation and toxicant properties. A distinction is often made between fate and transport in the unsaturated and saturated groundwater zones and finite versus infinite pollutant sources. Such models can be multidimensional (1D, 2D, or 3D) with the spatial domain selection based upon the modeling objective, appropriate margins of error, access to monitoring data sources, and user expertise and familiarity.

By the very nature of the groundwater subsurface environment (difficulty in sampling and the fact that groundwater movement is generally slow with years to decades being the appropriate time scale), appropriate models are difficult to effectively calibrate. These facts make the use of groundwater models speculative and often controversial when the time frames of regulatory decisions effectively preclude their effective calibration with site-specific data. Furthermore, there are presently few, readily used reliable integrated groundwater flow and quality models with a full suite of processes with compound-specific properties. Finite element subsurface models offer these options but are very difficult to implement and involve advanced mathematical numerical techniques. Some quality analyses are performed outside of the modeling system framework, but nevertheless can be credible.

Groundwater models exist to address pollution migration toward a valued resource such as a groundwater aquifer, surface water, or residential neighborhood where human health could be impacted. Such models are limited to the extent that water quality can be predicted, but some have nevertheless been attempted with uncertain results. It is difficult to easily sample these media to either calibrate and/or validate groundwater models. Although much is known about the properties of typical groundwater pollutants, such analyses are secondary to how fast the contaminant moves and disperses in the subsurface environment. The underlying science of such movement via Darcy's Law is sound, and limited sampling is required to determine hydraulic pressure is required. The dispersion components can be determined with laboratory studies with unique substrates. Therefore, given this reality of subsurface behavior, model constructs supported by the USGS and EPA are viable alternatives in practice. These are MODFLOW and BIOSCREEN. Other choices in the BIOPLUME family that may even have some quality capabilities are of limited utility at the present time.

Accordingly, models included in this overview are MODFLOW, BIOPLUME and BIOSCREEN, EPACMTP, and Hydrologic Evaluation of Landfill Performance (HELP). EPA's CEAM and related Ralph S. Kerr Research Laboratory in Ada, OK (EPA 2017k) support a few other legacy NPS models used in agriculture with implications to groundwater resources. Because of its familiarity within many states, MODFLOW is considered by some to be the premiere groundwater fate and transport model. Although EPACMTP and HELP have more specific uses related to hazardous waste regulatory management, MODFLOW is more generic in its application history. Similar to some of the receiving water quality models, groundwater models can have related versions and/or auxiliary tools to aid in postprocessing and simulation visualization. Readers should not confuse these add-ons as different model codes or versions.

10.7.1 MODFLOW

Since its original development in the early 1980s, the USGS has made four major releases of MODFLOW (Harbaugh and McDonald 1996) and is now considered to be the de facto standard code for aquifer flow simulation. There are several actively developed commercial and noncommercial GUIs for MODFLOW to aid in visualization and preprocessing and postprocessing of model generated files. It uses a finite-difference computer code to solve the groundwater flow equation (Darcy's Law) (Wikipedia 2017b) in up to its 3D formulation. It requires a significant learning curve and not suitable for novice modelers. Some training is available privately through consultants experienced in its many uses.

10.7.2 BIOPLUME

The BIOPLUME model family originated with the US Geological Survey, and its products are called Fate and Transport of Microbes and Chemicals (2DFATMIC [EPA 1997b,c] and 3DFATMIC [EPA 1998]) Models. They simulate subsurface flow, transport, and fate of contaminants that are undergoing chemical or biological transformation in unsaturated and saturated media. The processes simulated include advection, dispersion, sorption, first-order decay, and biodegradation using Monod kinetics. Both the models are applicable to transient conditions as well as steady groundwater flow conditions and use finite element analysis to solve the 3D time varying equations. As such, they are well beyond the scope of this course. In fact, most mathematics and engineering students never encounter finite element methods (FEMs) to solve partial differential equations, except potentially in graduate programs. The authors of this book have no expectation that most students using this textbook will successfully use either model. Both the 2D and 3D finite element models are included here for completeness and intellectual curiosity.

The FEM used in both 2DFATMIC and 3DFATMIC models is a numerical solution technique for partial differential equations that often appear in structural analysis, fluid mechanics, and heat transfer fields of science and engineering. They seek to reduce the more complex calculus representations to algebraic equations with finite boundary conditions. Commercial software such as MATHLAB is useful here (Mathlab 2017).

BIOSCREEN is a screening model and runs in MS EXCEL spreadsheet with helpful macros to enter required data sources. Functionally, it performs groundwater flow simulation and basic fate calculations using Fick's Law of Diffusion under 1D assumptions and simplified kinetics.

The software, programmed in the Microsoft Excel spreadsheet environment and based on the Domenico analytical solute transport model, has the ability to simulate advection, dispersion, adsorption, and aerobic decay as well as anaerobic reactions that have been shown to be the dominant biodegradation processes at many petroleum release sites. BIOSCREEN includes three different model types:

(1) Solute transport without decay,
(2) Solute transport with biodegradation modeled as a first-order decay process (simple, lumped-parameter approach),
(3) Solute transport with biodegradation modeled as an "instantaneous" biodegradation reaction (approach used by BIOPLUME models).

BIOSCREEN is designed to simulate biodegradation by both aerobic and anaerobic reactions. Groundwater Services, Inc., originally developed the model for the US Air Force Center for Environmental Technology Transfer Division at Brooks Air Force Base Houston, Texas.

A basic version known as BIOSCREEN and distributed by EPA simulates remediation through natural attenuation of dissolved hydrocarbons at petroleum fuel release sites. BIOSCREEN is also routinely used for simulating the movement of a wide variety of dissolved solutes in groundwater. As noted the software, programmed in the Microsoft ExcelTM spreadsheet environment, uses the Domenico analytical solute transport model for solute transport in three-dimensional porous media. The Domenico model has the ability to simulate advection, dispersion, adsorption, and aerobic decay as well as anaerobic reactions that have been shown to be the dominant biodegradation processes at many petroleum release sites (Domenico et al. 1996). BIOSCREEN Version 1.4 as distributed by the EPA, provides this analytical solution to the solute transport advection-dispersion equation with the following assumptions:

(1) The finite source dimension,
(2) The steady state source,
(3) Homogeneous aquifer properties,
(4) One-dimensional groundwater flow,
(5) First-order degradation rate,
(6) Contaminant concentration estimated at the centerline of the plume,
(7) Molecular diffusion based on concentration gradient is neglected, and
(8) No retardation (e.g. sorption) in transport process.

This analytical solution is an approximate solution to the solute transport equation that, under some circumstances, produces solutions that differ significantly from the exact solution to the solute transport equation. The analytical solution form is programmed into a Microsoft Excel spreadsheet. The analytical model is applied to estimate the plume length for dissolved organic contaminant in groundwater. The use of the analytical model requires contaminant spatial concentration data at a minimum of one source well and 1–2 downgradient wells. The groundwater data must show a reasonable plume pattern (i.e. contaminant concentration is highest in the source well and gradually decreasing in the downgradient monitoring wells). A model is calibrated by adjusting three model input parameters to fit groundwater concentration spatial pattern based on the spatial concentration distribution data. The model after calibration is then used to predict the horizontal plume length in groundwater. Prior to applying the spreadsheet model and interpreting the model results, full understanding of model assumptions is strongly advised.

A User's manual was developed by the California Water Resources Board in 1999 and revised in 2013 (Tong and Rong 2013) and offers insights on its use in common situations.

10.7.3 EPACMTP

EPACMTP (EPA 2003b) is a fate and transport model developed by EPA to simulate the release constituents from waste managed in land disposal units, and the subsequent impacts of these constituents to the subsurface environment. The acronym stands for Environmental Protection Agency Composite Model for Leachate Migration with Transformation Products. The model combines two modules to simulate one-dimensional downward flow and transport of constituents in the unsaturated zone beneath a waste disposal unit, as well as ground water flow and three-dimensional constituent transport in the underlying and deeper saturated zone. The model is designed to run in a probabilistic or deterministic mode and comes with built-in distributions of national and regional modeling parameters. The output of the model includes predicted concentrations of constituents arriving at a down gradient well under steady-state conditions or as a function of time.

10.7.4 HELP

The HELP (Schroeder 1994a b) model was developed by the US Army Corps of Engineers to facilitate rapid, economic estimation of the amounts of surface runoff, subsurface drainage, and leachate that may be expected to result from the operation of a wide variety of possible landfill designs. The program models the effect of hydrologic processes including precipitation, surface storage, runoff, infiltration, percolation, evapotranspiration, soil moisture storage, and lateral drainage using a quasi-two-dimensional approach. HELP does not consider contaminants explicitly, only the larger landfill facility design.

10.8 Modeling of Toxic Substances

Modeling of toxic compounds in surface waters is guided by EPA's 1971 strategy delineated in its published guidance (EPA 1971) that makes use of multiple criteria – compound-specific concentrations to protect human health and biological criteria to protect area-specific sensitive species such as fish and water-borne insects (anthropods and invertebrates) such as *Daphnia pulex*, a flea ubiquitous in North America and Europe.

A bioassay is a test to determine the concentration that is hazardous to living cells or tissue. Often times toxicity is a result of multiple unknown compounds; therefore, this required monitoring test becomes a regulatory tool, much like BOD for municipal and industrial wastewaters. The EPA guidance document provides strategies for toxicity identification evaluation (TIE) when a regulatory tool is needed for the discharger and based upon bioassays and tissue examination. Such solutions often focus on biological (not chemical specific) criterion. These bioassays are part of a useful monitoring process known as whole effluent toxicity (WET).

In prior decades, simple modeling tools were successfully used for those compounds that had reasonable half-lives in the aquatic phase. The Symtox model (LTI 1986; 1990) was one such model used and was based upon a mass balance that included advection, sorption, volatilization, and oxidation (Chapra 1997). Chapters 44 and 45 provides a useful derivation of the specifics for aquatic phase and suspended and benthic sorption routes. Regrettably, this specific model was archived by EPA managers as part of its model life cycle reviews under its CEAM process, although some state agencies may have retained the source code and use it to their advantage to better understand the behavior of its surface water resources. Many modern-day chemicals such as heavy metals and chlorinated organics do not readily degrade in reasonable time scales and often reside in benthic materials where their toxic effects act for decades. These are the challenges for the scientist of the future. It is further interesting to note that the major reasons States cited for nonattainment of water quality standards were bacteria, nutrient enrichment, and toxic compounds. Some even mentioned sediments in this context (EPA 2018b).

10.9 Human Health Risk Assessment

The range of risk assessment models is wide from spreadsheets to computer code. They attempt to aid in risk management in both arenas by organizing calculations using inputs from exposure assessment models, monitoring results, exposure scenarios, and dose-response data from the toxicology field. In humans, the end points are normally increased cancer risks, and in wildlife, one usually deals with large populations and species extinction (EPA 2003b). Modelers in this genre are almost always toxicologist to deal specifically with toxic effect end points. These modeling topics per se are beyond the scope of this book but very related to the toxic effects of chemicals in the environment and regulatory modeling, as are the remaining toxic fate models used in regulatory and research to address specific environmental issues. Chapter 9 explains how risk is managed and how a toxicologist may use mathematical models in these assessments.

10.10 Other Useful Regulatory Models

As noted in this chapter beginning, there are other miscellaneous uses for models that have not been discussed. These include wastewater treatment plants, mixing zone analysis, storm water models, and sediment models. They all have a specialized environmental niche and provide a useful role in design, operational optimization and troubleshooting, and/or research purposes. They will not be discussed further in this book. Students with such interests should refer to appropriate literature in government literature.

Suggestions For Class Discussion (*note:* students will require access to relevant Internet sites of EPA and/or State Agency Environmental sites).

Compare and contrast modeling practices in air, surface water, and groundwater media by your State Agency.

1. What are some of the resources available to modelers to aid in model building?
2. What do you think the following phrase means and what are some of the implications to modelers? "A model is more than just a collection of algorithms or equations."
3. What is meant by the terminology wasteload allocation? How does it differ by compound?
4. Review the current water quality standards process as outlined in the US EPA's Water Quality Standards Academy (epa.gov/) and discuss how discharge permits are structured and written for municipalities and industries for your local area.
5. Research the use of EPACMTP and list the laboratory tests commonly used in practice to characterize leachate materials.

Exercises

10.1 Contact modelers at your state Agency and obtain example model input and output files for some recently modeled river segments. Ask them to review their use of legacy models like Qual-2E in light of changes to WASP. This information should be public for any NPDES situation. They may even provide the student the data via executable files and get them installed with online resources provided on EPA websites.

10.2 Using older modeling textbooks develop a catalog of environmental models showing air, surface water, and groundwater entries. Your instructor will help you with sources of information.

10.3 What is a screening model? Give an example and describe its utility. Use State and federal web sites.

10.4 Discuss how refine and simplistic models may differ.

10.5 Describe the modeling process for air, surface water, and sub-surface environments. What are the regulatory uses in each case?

10.6 How do model accuracy, time to develop, and development cost relate? Use websites and TMDL or PSD modeling results to illustrate your findings.

10.7 Download and examine a few NPDES for major industries near your school.

10.8 How do they differ in general and monitoring requirements?

10.9 Compute and graph the DO/BOD for a 10 cfs stream with initial DO of 6 and 1 mg/l CBOD and receiving a1 MGD discharge of CBOD for the case with no SOD and for the 1 g/ft^2/d case. Assume that CBOD degrades at 0.10/day. Perform a sensitivity analysis for the CBOD decay rate and SOD rates with ±10% perturbations on each. What are the implications of such studies in regulatory modeling?

Advanced Exercises

1 Working as teams develop a working copy of WASP for some basic uses a 1-D River with BOD/DO, Lake CSTR assumption (instructor provide relevant case studies). In future years, the instructor may include the student efforts in case study examples of relevance.

2 Working in teams shows how HELP and EPACTMP work with basic landfill inputs in the most general case.

3 As a class meet with State modelers (if possible) and ask them to show you how and why they have used models. Alternately, ask a local industry to show you models used for their facility by consultants or a regulatory agency. Your success is related to the size of the facility (usually).

Bibliography

Arnold, J.G. (2012). SWAT, model use: calibration and validation. *ASABE* **55** (4): 1491–1508.

ASCE (2003). World Water Resources Congress, Philadelphia, PA: From the Stanford Watershed Model to BASINS: A Summary of Model Evolution.

Chapra, S.C. (1997). *Surface Water Quality Modeling*, 844. Long Grove, Illinois: Waveland Press.

Chapra, S.C., Pelletier, G.J., and Tao, H. (2008). QUAL2K: A Modeling Framework for Simulating River and Stream Water Quality, Version 2.11: Documentation and Users Manual.

Di Toro, D.M. (2001). *Sediment Flux Modeling*, 624. NY: Wiley-Interscience.

Di Toro, D.M., Fitzpatrick, J.J., and Thomann, R.V. (1983). *Water Quality Analysis Simulation Program (WASP) and Model Verification Program (MVP) – Documentation*. Westwood, NY, for U.S. EPA, Duluth, MN: Hydroscience, Inc., Contract No. 68-01-38.

Dupont, J. and Villeneuve, J.P. (1997). A watershed-based system for the integrated management of surface water quality: the GIBSI system. *Proceedings of the 4th International Symposium on Systems Analysis and Computing in Civil Engineering*.

EPA (1979). EPA 903/9-79-001, User's Manual for the Dynamic (Potomac) Estuary Model. *Technical Report 63*.

EPA (1986a). EPA-600/3-81-044, Documentation for Water Quality Simulation Program (WASP) and Model Verification Program. *US EPA-600/3-81-044, Environmental Research Laboratory, Office of Research and Development*, US Environmental Protection Agency, MN.

EPA (1986b). EPA/600/3-86/034, WASP3, A Hydrodynamic Water Quality Model – Model Theory, Users Manual, and Programmer's Guide, 374 pp.

EPA (1987). EPA/600/3-87/007 1987 Enhanced stream Water Quality Models Qual-2E and Qual-2E UNCAS, Documentation and User Manuals.

EPA (1991). *Technical Support Document for Water Quality-Based Toxics Control*. EPA/505/2- 90-001, 304. Washington: EPA Office of Water.

EPA (1992). EPA-454-92-019, Screening Procedures for Estimating the Air Quality Impact of Stationary Sources. www.epa.gov/ttn.

EPA (1995). EPA – 454/B-95 – 003b, Users Guide to the Industrial Source Complex Model (ISC3).

EPA (1997a). *Compendium of Tools for Watershed Assessment and TMDL Development*. EPA 841-B-97-006. Washington, DC: EPA Office of Water.

EPA (1997b). BIOPLUME II, Computer Model off Two-Dimensional Contaminant Transport Under the Influence of Oxygen-Limited Biodegradation in Ground water – User's Manual, Version 1.0 Publication No. EPA/600/8-88/093A

EPA (1997c). 2DFATMIC, Two-Dimensional Subsurface Flow, Fate, and Transport of Microbic Chemical Model – User's Manual, Version 1.0 (PDF) Publication No. EPA/600/R-97/052.

EPA (1999a). Appendix W to 40 CFR 51 – Guideline on Air Quality Models on 7/, 1999.

EPA (1999b). *Better Assessment Science Integrating Point and Nonpoint Sources.* EPA 823-F- 99-001. Washington, DC: EPA Office of Science and Technology http://www.epa.gov/OST/BASINS.

EPA (2003a). Background Technical Document, Office of Solid Waste (5305W) Washington, DC 20460 EPA 530-R-0 3-006, April 2003. https://www.e.pa.gov/smm/epas-composite-model-leachate-migration-transformation-products-epacmtp.

EPA (2003b). Generic Ecological Assessment Endpoints (GEAEs) for Ecological Risk Assessment, EPA/630/P-02/004F.

EPA (2004). AERMOD: Description of Model Formulation, EPA-454/R-03-004

EPA (2015). *BASINS 4.1 (Better Assessment Science Integrating point & Non-point Sources) Modeling Framework.* North Carolina: National Exposure Research Laboratory, RTP http://www.epa.gov/exposure-assessment-models/basins (accessed 7 December 2017).

EPA (2016). EPA-4054/B-16-004, AERSCREEN Screening Air Dispersion Model.

EPA (2017a). CEAM, Center for Exposure Assessment Modeling. https://www.epa.gov/exposure-assessment-models (accessed 7 December 2017).

EPA (2017b). SCRAM & Air Quality Models. https://www.epa.gov/scram/air-quality-models (accessed 7 December 2017).

EPA (2017c). CREM. https://wiki.epa.gov/watershed2/index.php/EPA_CREMCouncil_for_Regulatory_Environmental_Modeling (accessed 2017).

EPA (2017d). State Implementation Plan (SIP). https://www3.epa.gov/scram001/guidance_sip.html (accessed 2017).

EPA (2017e). Prevention of Significant Deterioration (PSD). https://www.epa.gov/nsr/prevention-significant-deterioration-basic-information (accessed 7 December 2017).

EPA (2017f). National Ambient Air Quality Standards (NAAQS). https://www.epa.gov/criteria-air-pollutants/naaqs-table (accessed 7 December 2017).

EPA (2017g). Air Quality Modeling. https://www3.epa.gov/scram001/aqmindex.htm (accessed 7 December 2017).

EPA (2017h). Air Quality Modeling AERMOD model. https://www3.epa.gov/scram001/7thconf/aermod/aermod_mfd.pdf (accessed 7 December 2017).

EPA (2017i). Sediment Contamination. https://www.epa.gov/environmental-topics/water-topics (accessed 7 December 2017).

EPA (2017j). Hydrologic Simulation Program Fortran (HSPF). https://www.epa.gov/exposure-assessment-models/hspf (accessed 7 December 2017).

EPA (2018a). National Report to Congress. https://www.epa.gov/national-aquatic-resources-surveys.

EPA (2018b). CALPUFF. https://www3.epa.gov/ttn/scram/dispersion_prefrec.htm#calpuff (accessed 7 December 2017).

EPRI Chen, C., Herr, J. and Ziemelis, L. (1998). Watershed Analysis Risk Management Framework – A Decision Support System for Watershed Approach and TMDL Calculation. *Documentation Report, Electric Power Research Institute, Palo Alto, CA., TR110709.*

Exponent Atmospheric Sciences (2017). CALPUFF. http://www.src.com (accessed 7 December 2017).

Federal Energy Licensing Commission (FERC) (2017). https://www.ferc.gov/industries/hydropower.asp (accessed 7 December 2017).

FWPCA, Feigner, K., Harris, H. (1970). Documentation Report for the Dynamic Estuary Model.

Gherini, S.A., Mok, L., Hudson, R.J.M. et al. (1985). The ILWAS model: formulation and application. *Water Air Soil Pollut.* **26** (4).

Harbaugh, A.W. and McDonald, M.G. (1996). User's Documentation for MODFLOW-96, an Update of the USGS Modular Finite-Difference Groundwater Model. *USGS Open File Report No. 91D484.*

Limno Tech, Inc. (LTI) (1986). Simplified Method Program for Toxics Users Manual, prepared for EPA, Washington, DC.

MATHLAB (2017). https://www.mathworks.com/products/pde.html (accessed 7 December 2018).

National Geographic (2017). Ocean Gyre. https://www.nationalgeographic.org/encyclopedia/ocean-gyre/ (accessed 7 December 2018).

NCASI (1977). Technical Bulletin No. 471: Computer Program Documentation for the Enhanced Stream Water Quality Model Qual 2E.

Newell, C., McLeod, R. and Gonzales, J. (1996). BIOSCREEN, Natural Attenuation Decision Support System, User's Manual Version 1.3. EPA/600/R-96-087. This document is available on the U.S. EPA web site. https://www.epa.gov/water-research/bioscreen-natural-attenuation-decision-support-system (accessed 7 December 2018).

O'Connor, D.J. (1960). Oxygen balance for an estuary. *JSED* **86** (3): 35–56.

O'Connor, D.J. (1967). The temporal and spatial distribution of DO in streams. *Water Resour. Res.* **3** (1): 89.

Schroeder, P.R. C.M. Lloyd, P.A. Zappi (1994a). Hydrologic Evaluation of Landfill Performance (HELP) Model: Volume 1, User's Guide for Version 3. EPA/600/R-94/.

Schroeder, P.R. C.M. Lloyd, P.A. Zappi (1994b). Hydrologic Evaluation of Landfill Performance (HELP): Volume II Users Guide, EPA/600/R-94/.

Streeter, W.H. and Phelps, E.B. (1925). A study of the pollution and natural purification of the Ohio River. Public Health Bull. 146, U.S. Public Health Service, Washington, DC.

Survey (USGS) (1996). The modular modeling system (MMS). USGS Open File Report 96-151.

Thomann, R.V. (1963). Mathematical model of dissolved oxygen. *J. Sanit. Eng. Div., Am. Soc. Civ. Eng.* **89** (SA5).

Thomann, R.V. (1974). *Systems Analysis in Water Quality Management*. US: McGraw-Hill Inc.

Tong, W. and Rong, Y. (2013). Underground Storage Tank Section, California Regional Water Quality Control Board – Los Angeles Region, Domenico Spreadsheet Analytical Manual revised 2013.

Tufts University (2017). http://www.qual2k.com (accessed 7 December 2017).

TWDB (1971). Simulation of Water Quality in Streams and Canals-Theory and Description of the QUAL-1 Mathematical Model Systems. With: Texas Water Development Board. F D Masch.

U.S. Army COE (2017). http://www.environmental-engineer.com/free_wq.html (accessed 7 December 2017).

US EPA (1998). BIOPLUME III, Natural Attenuation Decision Support System – User's Manual, Version 1.0." Publication No. EPA/600/R-98/010.

USGS (2017). https://www2.usgs.gov/water/ (accessed 7 December 2017).

Velz, C. (1971). *Applied Stream Sanitation*. NY: Wiley.

Wikipedia (2017a). Rivers & Harbors Act 1899. https://en.wikipedia.org/wiki/Rivers_and_Harbors_Act_of_1899 (accessed 7 December 2017).

Wikipedia (2017b). Darcy's Law. https://en.wikipedia.org/wiki/Darcy%27s_law (accessed 7 December 2017).

Part IV

Toxicology and Risk Assessment

All things are poisons, for there is nothing without poisonous qualities. It is only the dose which makes the thing a poison

Paracelsus 1493–1541

Pollutant Fate and Transport in Environmental Multimedia, First Edition. Frank M. Dunnivant and Elliot Anders.
© 2019 John Wiley & Sons, Inc. Published 2019 by John Wiley & Sons, Inc.
Companion website: www.wiley.com/go/Dunnivant/pollutant_fate&Transp_EnvironMultimedia

11

Toxicology, Risk Assessment, Cost–Benefit Analysis, and Life Cycle Assessment

11.1 Introduction

This chapter has considerably expanded since the 2016 edition. It will begin with a brief introduction to chemical toxicology, followed by the 2016 version of risk assessment. We will then tie risk assessment to the previous covered fate and transport modeling. Next new sections on life cycle assessment (LCA) and benefit–cost assessment are added. These diverse forms of assessment make the chapter more modern and complete.

11.2 Toxicology

Generally, toxicology is the study of chemical or physical effects when introduced to an organism or ecosystem. We will limit our discussions to environmental toxicology, where a chemical (pollutant) enters an organism, usually at relatively low concentrations, and look at the various modes of entry to the body, reactions in the body, and excretion of the chemical from the body.

The adsorption, distribution, biochemical transformation, and excretion of toxicants are a highly complicated area of study and are even more complicated by the chemical diversity of toxicants. The more classic modes of entry to an organism are dermal adsorption through the skin, inhalation into a more advanced organism through the lungs, oral ingestion into the gastrointestinal tract, and injection through the skin. This discussion only focuses on the first three since injection is mostly a medicinal mode of entry.

The skin of many higher-level organisms is an effective barrier to most chemicals, but a few chemicals, mostly liquids and gases, pass through the skin. Modes of entry through the skin include diffusion and active transport through ports in the skin. Once in the body, these chemicals are absorbed into the blood or lymph system and travel to the target biomolecule or organism and cause damage. Gasses and small solid particulates can pass into the lungs and present as a respiratory problems or get absorbed into the blood stream. Ingestion of chemicals in water and food is a common mode of entry. These three modes of entry are the main basis for risk calculations given in Section 11.3.

The dose of the chemical (pollutant) and duration of exposure are important concepts in determining risk. The dose or concentration of a chemical is an obvious variable in determining exposure and can vary from low parts per trillion concentrations to the presence of a pure chemical (no dilution). The duration of exposure of a chemical is commonly divided into (i) acute exposure (usually just one exposure for a short time), (ii) subchronic exposure (multiple exposures of a day to month duration), and chronic exposure (long-term exposures or even a life-long exposure). Again, these three categories are used in risk assessment calculation in Section 11.3.

Various modes of toxic effects are possible given the wide range of chemicals nature provides us with and given the large chemical inventory humans have synthesized and released into the natural, home, and industrial environments.

Organisms have developed a wide range of detoxification approaches. These include simple and rapid excretion by the kidneys, a series of complex biotransformations such as hydrolysis, reduction, oxidation, and conjugations (the addition of functional groups, i.e. methylation, sulfonation, and phosphorylation) performed mostly in the liver, and deposition and storage in more permanent parts of the body (Pb in teeth and bones; DDT and PCBs in fatty tissue).

Toxic effects of chemicals entering the body include *nonspecific actions* where a chemical depresses the action of a biological process and *specific actions*, usually at low toxic concentrations, where the toxicant modifies or inhibits a biological process by binding at a specific site or molecule. Example of specific actions includes oxidative phosphorylation (electron transfer and ATP

Pollutant Fate and Transport in Environmental Multimedia, First Edition. Frank M. Dunnivant and Elliot Anders.
© 2019 John Wiley & Sons, Inc. Published 2019 by John Wiley & Sons, Inc.
Companion website: www.wiley.com/go/Dunnivant/pollutant_fate&Transp_EnvironMultimedia

formation inhibition), acetylcholinesterase inhibition in nerve synapses, irritants that cause an inflammatory action in tissues, central nervous system agents that inhibit cellular signaling, and respiratory blockers that interfere with respiration. One of the more chronic effects of toxicants is, of course, cancer in a variety of organisms and biological systems. In the next section, we will explore ways to predict the dose at which each of these toxic effects can occur.

Several text and reference books are available that extensively document environmental and chemical toxicology. The standard chemical toxicology reference is by Klaassen (2013). Chemical-specific examples of environmental pollutants are given in Gilbert (2004).

11.3 Risk Assessment

The EPA's mandated mission by Congress is to "protect human health and the environment." In the early days of EPA, it was felt that the agency regulated by decree – they defined a situation or practice as "bad" and ordered that it be stopped. Since the 1970s, the EPA's approach has been slowly transformed through the civil court process so that they now have to more effectively substantiate why a practice is "bad" before they can regulate it. This prior litigation is probably the origin of risk assessment, which seeks to mathematically quantify how bad a situation is, and which was and is viewed by some environmentalists as an attempt by industry to undermine the power of EPA. Unfortunately, this dependence on a documented risk only serves half the mission of the EPA: the protection of human health. We cannot so easily and numerically quantify the effects or place a monetary value on the environment.

The overall goal of this textbook was to develop ways to determine whether a polluted site warranted remediation or clean up and ultimately to understand how the health effects of such a hazardous waste site or action are evaluated. We started by reviewing the chemistry of pollution and next presented basic fate and transport modeling procedures for predicting pollutant concentration as a function of chemical and physical conditions in the system. In this chapter, we will finally use all of the previous chapters to calculate the concentration of pollutant reaching a human (the receptor in our models). Based on this information, combined with the results of chemical exposure experiments involving animals, we will attempt to estimate the risk to humans. As you will see, our approaches are simple, and when you consider all of the possible sources of error or uncertainty, from the beginning of the process (pollutant source and mass characterization) through fate and transport modeling (with uncertainties in chemical degradation and physical dispersion), and now to risk assessment (where we extrapolate animal study data to human risk), we should recognize that our risk estimates may be orders of magnitude in error. Thus, risk assessment should be viewed as only one tool in our approach to environmental management. At the end of this Section 11.3.7, we will attempt to summarize and evaluate our overall approach.

11.3.1 The Concept of Risk

What is an acceptable risk? This is a major, and largely unanswerable, question. In order to attempt to answer this question, we will first look at commonly accepted risks that each of us exposes ourselves to every day. Table 11.1 shows the annual risk associated with a variety of actions. These are listed in decreasing degree of risk and show that smokers routinely accept an increased mortality risk of 3600 per million; that is for every one million smokers, 3600 or 0.36% of them will die each

Table 11.1 Some commonplace risks.

Action	Annual risk (deaths per 1 000 000 persons at risk)
Cigarette smoking (1 pk/d)	3600
All cancers	2800
Mountaineering	600
Motor vehicle accident (total)	240
Police killed in line of duty	220
Air pollution	200
Home accidents	110
Frequent flying (professors)	50
Alcohol (light drinker)	20
Sea-level radiation	20
Eating 4 Tbsp peanut buffer/d (liver cancer from aflatoxin)	8
Electrocution	5.3
One-in-a-million chance of death per year	1
One-in-a-million chance of death per life	0.0014
Drinking water with EPA limit of chloroform (cancer over a 70-year period)	0.6
Drinking water with EPA limit of chloroform (cancer)	0.002

Source: From Wilson and Crouch (1987).

year. As you can see, smoking is more risky than being a police officer or driving a car, yet millions of citizens willingly purchase cigarettes every day. Because this is a self-posed risk, it is perceived as more acceptable to most citizens than a risk that is imposed on them.

Another example of a popular perception of unknown risk can be seen in the question of drinking water treatment. Many citizens are becoming aware that the chlorination of our drinking water can result in the formation of cancer-causing agents, such as trihalomethanes (THMs). Some residents using treated drinking water purchase relatively costly water filters or boil their drinking water prior to use to remove the THMs. But when you look at the associated risk of drinking chlorinated water (as we will do in this chapter), you can clearly see that drinking chlorinated water is far safer than driving a car, working in the home, or even eating four tablespoons of peanut butter (refer to Table 11.1). But again, these are common practices we routinely accept – often, the main difference in terms of our risk perception is the distinction between a deliberate decision and an imposed situation.

At the time of the writing of this textbook, the news media and the Center for Disease Control and Prevention (CDC) have been focusing on the problem of obesity in the US population. This undoubtedly is linked to some of the leading causes of death in the United States, as seen in Table 11.2. The top three or four leading causes of death can be related to body weight. Such a consideration of the leading causes of death is one way of assessing risk.

Another way of looking at risk is to estimate what action will increase a person's chance of dying by one in a million. Table 11.3 lists a variety of activities that will increase your annual chances of death by one in a million. For example, it has been estimated that only smoking 1.4 cigarettes per year, eating 100 charcoal-broiled steaks per year, or living in New York for a year will increase your risk of death, by the magic number of one in a million. These are risks that most of us are willing to assume, and all risk assessment calculations now consider a risk of one in a million to be acceptable, whatever the cause. Thus, if the risk associated with living near a chemical factory, drinking polluted water, or eating fruit containing a pesticide is less than one in a million, we (through our government) consider this activity, and the associated risk, acceptable. So now we have a point of reference for our risk assessment calculations. We will accept a risk of one in a million.

In Chapters 5 through 9, we studied ways to estimate the concentration of a pollutant at various locations and times in river, lake, groundwater, and atmospheric systems. The modeling calculations estimated the concentration of pollutant that a receptor (human) consuming the water and air would be exposed to. First, we will use these concentrations to calculate the actual mass of a pollutant that a human would be exposed to under a set of "standard" conditions. The risk factors (Table 11.4) used in such dose calculations depend on certain standard conditions. For example, most calculations for adults use a body weight of 70 kg, while 15 kg is used for a child. Other suggested risk factors are the average water consumption at home and at the workplace and the average volume of air that we breathe. The use of recommended values makes comparisons of polluted waste sites and cancer clusters more feasible. In addition, EPA has compiled data from a variety of sources that can be used in risk assessment calculations. All of the data can be found at www.epa.gov/iris/, and data for

Table 11.2 Leading causes of death in the United States for the year 2000.

Rank	Cause of death (all races, both sexes, and all ages)	Number	Percent of total deaths
	All causes	2 403 351	100.0
1	Diseases of the heart	710 760	29.6
2	Malignant neoplasms	553 091	23.0
3	Cerebrovascular diseases	167 661	7.0
4	Chronic lower respiratory diseases	122 009	5.1
5	Accidents (unintentional injuries)	97 900	4.1
6	Diabetes mellitus	69 301	2.9
7	Influenza and pneumonia	65 313	2.7
8	Alzheimer's disease	49 558	2.1
9	Nephritis, nephritic syndrome and nephrosis	37 251	1.5
10	Septicemia	31 224	1.3
	All other causes	499 283	20.8

Source: www.cdc.gov.

Table 11.3 Activities that increase annual mortality risk by one in a million.

Activity	Type of risk
Smoking 1.4 cigarettes	Cancer, heart disease
Living 2 months with a cigarette smoker	Cancer, heart disease
Eating 40 tablespoons of peanut butter	Liver cancer from aflatoxin
Eating 100 charcoal-broiled steaks	Cancer from benzo(*a*)pyrene
Flying 6000 mi by jet	Cancer by cosmic radiation
Living two summer months in Denver (at a mile elevation rather than at sea level)	Cancer by cosmic radiation
Traveling 300 mi by car	Accident
Flying 1000 mi by jet	Accident
Traveling 10 mi by bicycle	Accident
Traveling 6 min by canoe	Accident
Spending 1 h in a coal mine	Black lung disease
Living 2 d in New York or Boston	Air pollution

Source: Wilson (1979).

specific pollutants are given in Tables 11.5–11.8. There is an extensive amount of data for each chemical in the IRIS database. For example, the printout for benzene is more than 20 pages long.

Now we will look at the process of estimating risks to human health. First, we will show equations used to calculate the dose rate, also referred to as the Chronic Daily Intake (CDI), and later use these values to estimate rates of carcinogenic and noncarcinogenic effects. Recall the general approach of environmental risk assessment outlined in Chapter 1: (i) identify a potentially hazardous situation or polluted site, (ii) identify the source of the pollutant and inputs to the systems, (iii) estimate the concentration of pollutant reaching the receptor (human), use fate and transport modeling, and (iv) evaluate health risk, the subject of this chapter. Finally, a remediation decision is made based on the results of all four steps and upon agreement between the liable party (the polluter), federal, state, and local governments, and the public.

11.3.2 Dose Rates from Various Sources

There are a number of exposure routes by which humans can receive and uptake pollutants. These include not only drinking polluted water and breathing polluted air but also ingestion during swimming activities, adsorption through the skin upon exposure to any contaminated material, direct intake of pollutants on soil, sediment, and dust, dermal adsorption from soil, sediment, and dust, and intake from foods. In the following pages, we will present calculations detailed by the EPA (1989a) that allow the estimate of pollutant intake from each of these sources. The overall process is referred to as a CDI and is expressed as the mass of pollutant taken up per unit body weight and unit time. We will address each common exposure route separately. In the next section, we will use equations from Chapters 5 to 9 to estimate the receptor (human) concentration exposure and use the CDI equations below to determine health risks.

11.3.2.1 Ingestion of Pollutants from Drinking Water

Pollutant CDIs from drinking polluted water are calculated using

$$\text{Intake from drinking (mg/kg d)} = \frac{CW * IR * EF * ED}{BW * AT} \tag{11.1}$$

where CW is the pollutant concentration in the drinking water (mg/l), IR is the ingestion rate of water (l/d) (2.00 l/d for an adult at the 90th percentile) (1.40 l/d for an average adult), EF is the exposure frequency (d/yr) (usually 365 d/yr), ED is the exposure duration (number of years) (70.0 years for a conventional lifetime) (30.0 years, upper 90th percentile at one residence) (9.00 years, 50th percentile median time at one residence), BW is the body weight (kg) (70 kg is used as an average), and AT is the average time period of exposure (days) (for a pathway-specific noncarcinogen, the value is usually ED × 365 d/yr) (for a carcinogen, use 70 years × 365 d/yr).

Specific values of each parameter can be adjusted as needed. For example, a child would drink less water each day, the receptor may be away from home most of the day or year, or the body weight of a specific receptor or group of receptors may differ from the average.

11.3.2.2 Ingestion of Water While Swimming

Exposure from ingestion of water during swimming can be calculated by

$$\text{Intake from water (mg/kg d)} = \frac{CW * CR * ET * EF * ED}{BW * AT} \tag{11.2}$$

where CW is the pollutant concentration in the water (mg/l), CR is the contact rate (0.050 l/h, EPA 1989b), ET is the exposure time (hours/swimming event), EF is the exposure frequency (swimming events/year, national average is 7 d/yr), ED is the exposure duration (70 years

Table 11.4 EPA-recommended exposure factors for risk assessment.

Exposure source	Exposure pathway	Daily intake	Exposure frequency (d/yr)	Exposure duration (yr)	Body weight (kg)
Residential	Ingestion of potable water	2.00 l (adult)	350	30	70 (adult)
		1.00 l (child)			15 (child)
	Ingestion of soil and dust	100. mg (adult)	350	6	70 (adult)
		2000. mg (child)		24	15 (child)
	Inhalation of contaminants	20. m^3 (adult)	350	30	70 (adult)
		12 m^3 (child)			15 (child)
Industrial and commercial	Ingestion of potable water	1.00 l	250	25	70 (adult)
	Ingestion of soil and dust	50. mg	250	25	70 (adult)
	Inhalation of contaminants	20. m^3 (workday)	250	25	70 (adult)
Agricultural	Consumption of produce	42 g (fruit)	350	30	70 (adult)
		80. g (vegetable)			15 (child)
Recreational	Consumption of fish	54 g	350	30	70 (adult)
					15 (child)

Source: EPA (1991).

for lifetime; 30 years for national upper-bound time (90th percentile) at one residence; nine years for national median time (50th percentile) at one residence, BW is the body weight (70 kg for average adult; for age-specific values, see EPA (1989b)), and AT is the averaging time (period over which the exposure is averaged, in days; for noncarcinogenic effects use ED × 365 days; for carcinogenic effects use 70 years (lifetime) × 365 d/yr.

11.3.2.3 Dermal Contact with Pollutants in Water While Swimming

The CDI of pollutants from dermal contact with pollutants in water can be calculated by

$$\text{Absorbed dose (mg/kg d)} = \frac{CW * SA * PC * ET * EF * ED * CF}{BW * AT} \quad (11.3)$$

where CW is the pollutant concentration in water (mg/l), SA is the skin surface area available for contact (cm^3)

50th percentile total body surface area (m^2)		
Age (yrs)	**Male**	**Female**
3 < 6	0.728	0.711
6 < 9	0.931	0.919
9 < 12	1.16	1.16
12 < 15	1.49	1.48
15 < 18	1.75	1.60
Adult	1.94	1.69

50th percentile body part-specific surface area for males			
Age (years)	**Arms**	**Hands**	**Legs**
3 < 4	0.096	0.040	0.18
6 < 7	0.11	0.041	0.24
9 < 10	0.13	0.057	0.31
Adult	0.23	0.082	0.55

and PC is the chemical-specific dermal permeability constant (cm/h). These must be obtained from the open literature or an average value can be used (8.4×10^{-4} cm/h), ET is the exposure time (h/d) (the national average for swimming is 2.6 h/d), EF is the exposure frequency (d/yr) (the national average for swimming is 7 d/yr), ED is the exposure duration (years) (70 years for lifetime; 30 years for national upper-bound time (90th percentile) at one residence; 9 years for national median time (50th percentile) at one residence), CF is the volumetric conversion factor for water (1.00 l/1000 cm^3), BW is the body weight (kg) (70.0 kg for an average adult; for age-specific values, see EPA (1989b)), and AT is the average time (period over which the exposure is averaged, days) (for noncarcinogenic effects use ED × 365 d/yr; for carcinogenic effects, use 70 years × 365 d/yr).

11.3.2.4 Ingestion of Pollutants in Soil

Ingestion of polluted soil can result from the ingestion of unwashed root crops and from children playing in dirt. The CDI of pollutants from the ingestion of polluted soil

Table 11.5 Oral RfDs for selected pollutants.

Compound	CASRN	Experimental doses (mg/kg d, unless other wise noted)	UF	MF	RfD (mg/kd d)	Critical effect
Acetone	67-64-1	NOAEL: 900 LOAEL: 1700 BMDL: not determined	1 000	1	0.9	Nephropathy
Acrylamide	79-06-1	NOAEL: 0.2 LOAEL: 1	1 000	1	2E−4	Nerve damage
Alachlor	15972-60-8	NOAEL: 1 LOAEL: 3	100	1	1E−2	Hemosiderosis, hemolytic anemia
Aldicarb	116-06-3	NOAEL: 0.01 LOAEL: 0.025	10	1	1E−3	AChe inhibition
Aldrin	309-00-2	NOAEL: none LOAEL: 0.5 ppm diet 0.025	1 000	1	3E−5	Liver toxicity
Antimony	7440-36-0	NOEL: none LOAEL: 0.35	1 000	1	4E−4	Longevity, blood glucose, and cholesterol
Barium and compounds	7440-39-3	NOAEL: 65 LOAEL: 115	3	1	7E−2	Increased kidney weight
Benzene	71-43-2	BMDL = 1.2	300	1	4.0E−3	Decreased lymphocyte count
Beryllium and compounds	7440-41-7	BMD10: 0.46	300	1	2E−3	Small intestinal lesions
Bromoform	75-25-2	NOEL: 25	1 000	1	2E−2	Hepatic lesions
Cadmium	7440-43-9	NOAEL (water): 0.005	10	1	5E−4	Proteinuria
Carbon tetrachloride	56-23-5	NOAEL: 1 LOAEL: 10	1 000	1	7E−4	Liver lesions
Chloroform	67-66-3	NOAEL: none LOAEL: 15	1 000	1	0.01	Fatty cyst formation
Chlordane (technical)	12789-03-6	NOAEL: 0.15 LOAEL: 0.75	300	1	5E−4	Hepatic necrosis
Chromium(VI)	18540-29-9	NOAEL: 25 (mg/l)	300	3	3E−3	None reported
2,4-Dichloro-phenoxyacetic acid (2,4-D)	94-75-7	NOAEL: 1.0 LOAEL: 5.0	100	1	1E−2	Hematologic, hepatic, and renal toxicity
Dichlorvos	62-73-7	NOAEL: 0.05 LOAEL: 0.1	100	1	5E−4	Plasma and RBC ChE inhibition
Dieldrin	60-57-1	NOAEL: 0.1 (ppm)	100	1	5E−5	Liver lesions
Endrin	72-20-8	NOEL: 0.025 LOAEL: 0.05	100	1	3E−4	Lesions in liver occasional convulsions
Ethyl acetate	141-78-6	NOEL: 900 LOAEL: 3600	1 000	1	9E−1	Mortality body weight loss

Table 11.5 (Continued)

Compound	CASRN	Experimental doses (mg/kg d, unless other wise noted)	UF	MF	RfD (mg/kd d)	Critical effect
Ethylbenzene	100-41-4	NOEL: 136 LOAEL: 408	1 000	1	1E−1	Liver and kidney toxicity
Formaldehyde	50-00-0	NOAEL: 15 LOAEL: 82	100	1	2E−1	Reduced weight gain histopathology
Heptachlor	76-44-8	NOEL: 3 (ppm diet) 0.15 LEL: 5 (ppm diet) 0.25	300	1	5E−4	Liver weight increases
Hexachloro-benzene	118-74-1	NOAEL: 1.6 (ppm diet) 0.08 LOAEL: 8.0 (ppm diet) 0.29	100	1	8E−4	Liver effects
Hexachloro-ethane	67-72-1	NOAEL: 1 LOAEL: 15	1 000	1	1E−3	Atrophy
Isophorone	78-59-1	NOEL: 150 LEL: None	1 000	1	2E−1	No observed effects
Methyl parathion	298-00-0	NOEL: 0.5 (ppm) 0.025 LEL: 5.0 (ppm) 0.25	100	1	2.5E−4	ChE inhibition
Methyl mercury (MeHg)	22967-92-6	BMDL05: 0.857 1.472 (μg/kg d)	10	1	1E−4	Developmental neuropsychological impairment
Paraquat	1910-42-5	NOEL: 0.45 LEL: 0.93	100	1	4.5E−3	Chronic pneumonitis
Pentachloro-benzene	608-93-5	NOAEL: none LOAEL: 8.3	10 000	1	8E−4	Liver and kidney toxicity
Pentachloro-phenol	87-86-5	NOAEL: 3 LOAEL: 10	100	1	3E−2	Liver and kidney pathology
Tetraethyl lead	78-00-2	NOAEL: none LOAEL: 1.7 (μg/kg/d) converted to 0.0012 mg/kg/d	10 000	1	1E−7	Histopathology of liver and thymus
Toluene	108-88-3	NOAEL: 312 mg/kg converted to 223 mg/kg/d LOAEL: 625 mg/kg converted to 446 mg/kg/d	1 000	1	2E−1	Changes in liver and kidney weights
Trifluralin	1582-09-8	NOEL: (30 ppm) 0.75 LEL: (150 ppm) 3.75 mg/kg/d	100	1	7.5E−3	Increased liver weights
Xylenes	1330-20-7	NOAEL: 250 LOAEL: 500	1 000	1	0.2	Decreased body weight, increased mortality

Source: http://www.epa.gov/iris/.

Table 11.6 Inhalation RfCs for selected pollutants.

Compound	CASRN	Experimental doses* (mg/m^3 unless otherwise noted)	UF	MF	RfD (mg/m^3 unless otherwise noted)	Critical effect
Acetonitrile	75-05-8	NOAEL: 336	100	10	6E−2	Mortality
Ammonia	7664-41-7	NOAEL: 6.4 (9.2 ppm)	30	1	1E−1	Pulmonary function or changes
Aniline	62-53-3	NOAEL: 19 (5 ppm) LOAEL: none	3000	1	1E−3	
Benzene	71-43-2	BMCL = 8.2	300	1	3E−2	Decreased lymphocyte count
Beryllium and compounds	7440-41-7	LOAEL(HEC): 0.55 $\mu g/m^3$	10	1	2E−2 $\mu g/m^3$	Beryllium sensitization
Chlordane (technical)	12789-03-6	NOAEL 1.0	1000	1	7E−4	Hepatic effects
Chromium(VI)	18540-29-9	NOAEL: none LOAEL: 2E-3	90	1	8E−6	Nasal septum atrophy
Dichlorvos	62-73-7	NOAEL: 0.05 LOAEL: 0.48	100	1	5E−4	Decreased brain colinesterase activity
Ethylbenzene	100-41-4	NOAEL: 434 (100 ppm) LOAEL: 4340 (1000 ppm)	300	1	1E+0	Developmental toxicity
Hydrogen sulfide	7783-06-4	NOAEL: 0.64 LOAEL: 41.7 (30 ppm)	300	1	2E−3	Nasal lesions
Mercury, elemental	7439-97-6	NOAEL: None LOAEL: 0.025 (converted to LOAEL [ADJ] of 0.009)	30	1	3E−4	Hand tremor memory disturbance
Methyl *tert*-butyl ether (MTBE)	1634-04-4	NOAEL: 1453 (403 ppm) LOAEL: 10899 (3023 ppm)	100	1	3E+0	Increased absolute and relative liver and kidney weights
Toluene	108-88-3	NOAEL: None LOAEL: 332 (88 ppm)	300	1	4E−1	Neurological effects
Xylenes	1330-20-7	NOAEL(HEC): 39 LOAEL(HEC):	300	1	0.1	Impaired motor coordination

Source: www.epa.gov/iris/.

Table 11.7 Carcinogenicity assessment: carcinogenic risk from oral exposure for selected pollutants.

Compound	CASRN	Classification	Oral slope factor per (mg/kg d)	Drinking water unit risk per (μg/l)
Acrylamide	79-06-1	B2	4.5	1.3E−4
Aldrin	309-00-2	B2	1.7E+1	4.9E−4
Aniline	62-53-3	B2	5.7E−3	1.6E−7
Azobenzene	103-33-3	B2	1.1E−1	3.1E−6
Benzene	71-43-2	A	1.5E−2 to 5.5E−2	4.4E−4 to 1.6E−3
Benzo[*a*]pyrene	50-32-8	B2	7.3E+0	2.1E−4
Bromoform	75-25-2	B2	7.9E−3	2.3E−7
Carbon tetrachloride	56-23-5	B2	1.3E−1	3.7E−6
Chloroform	67-66-3	B2	6.1E−3	—
Chlordane (technical)	12789-03-6	B2	3.5E−1	1E−5
3,3′-Dichloro-benzidine	91-94-1	B2	4.5E−1	1.3E−5
Dichlorvos	62-73-7	B2	2.9E−1	8.3E−6
Dieldrin	60-57-1	B2	1.6E+1	4.6E−4
2,4-/2,6-Dinitro-toluene mixture	NA	B2	6.8E−1	1.9E−5
Heptachlor	76-44-8	B2	4.5E+0	1.3E−4
Hexachloro-benzene	118-74-1	B2	1.6	4.6E−5
Hexachloro-ethane	67-72-1	C	1.4E−2	4.0E−7
Isophorone	78-59-1	C	9.5E−4	2.7E−8
Pentachloro-phenol	87-86-5	B2	1.2E−1	3E−6
Trifluralin	1582-09-8	C	7.7E−3	2.2E−7

Note the water unit risk = risk per μg/l = slope factor * 1/70 kg * 2.00 l/d × 10^{-3}.
Source: www.epa.gov/iris/.

Table 11.8 Carcinogenicity assessment: carcinogenic risk from inhalation exposure for selected pollutants.

Compound	CASRN	Classification	Slope factor[a] per (mg/kg d)	Inhalation unit risk per (μg/m³)
Acrylamide	79-06-1	B2	45	1.3E−3
Aldrin	309-00-2	B2	17	4.9E−3
Benzene	71-43-2	A	2.9E−2	A range of 2.2E−6 to 7.8E−6 is the increase in the lifetime risk of an individual who is exposed for a lifetime to 1 μg/m³ benzene in air
Beryllium and compounds	7440-41-7	B1	8.4	2.4E−3
Bromoform	75-25-2	B2	3.9E−3	1.1E−6
Cadmium	7440-43-9	B1	6.1	1.8E−3
Chloroform	67-66-3	B2	8.1E−2	2.3E−5
Chlordane (technical)	12789-03-6	B2	1.3	1E−4
Chromium(VI)	18540-29-9	A	41	1.2E−2
Dieldrin	60-57-1	B2	16	4.6E−3
Formaldehyde	50-00-0	B1	4.5E−2	1.3E−5
Heptachlor	76-44-8	B2	4.5	1.3E−3
Hexachloro-ethane	67-72-1	C	1.4E−2	4.0E−6
Nickel refinery dust	No CASRN	A	8.4E−1	2.4E−4

Note the air unit risk = risk per μg/m³ = slope factor * 1/70 kg * 20 m³/d × 10^{-3}.
a) Louvar and Louvar (1998).
Source: www.epa.gov/iris/.

can be calculated from

$$\text{Intake (mg/kg d)} = \frac{\text{CS} * \text{IR} * \text{CF} * \text{FI} * \text{EF} * \text{ED}}{\text{BW} * \text{AT}} \quad (11.4)$$

where CS is the pollutant concentration in the soil (mg/kg), IR is the ingestion rate (mg soil/d; suggested values (EPA 1989c): 200 mg/d children 1–6 years old, 100 mg/d >6 years old), CF is a conversion factor (10^{-6} kg/mg), FI is the fraction of soil ingested from the polluted site (versus soil ingested from a nonpolluted site or playground; this value varies based on population activity patterns), EF is the exposure frequency (365 d/yr), ED is the exposure duration (years) (70 years for a lifetime exposure; 30 years for the national upper-bound time (90th percentile) at one residence; 9 years for the national median time (50th percentile) at one residence), BW is the body weight (70 kg for an average adult; 16 kg for children aged 1–6 years (50th percentile)), and AT is the averaging time (period over which the exposure is averaged, in days) (for noncarcinogenic effects, this is equal to ED × 365 d/yr; for carcinogenic effects, use 70 years × 365 d/yr).

11.3.2.5 Intake from Dermal Contact with Pollutants in Soil

Humans can also obtain a pollutant dose from working with polluted soil, via absorption through the skin. This dose can be calculated by

$$\text{Adsorbed dose (mg/kg d)} = \frac{\text{CS} * \text{CF} * \text{SA} * \text{AF} * \text{ABS} * \text{EF} * \text{ED}}{\text{BW} * \text{AT}} \quad (11.5)$$

where CS is the pollutant concentration in the soil (mg/kg), CF is a conversion factor (10^{-6} kg/mg), and SA is the skin surface area available for contact (cm^2/event).

Suggested values for SA are the following:

50th percentile total body surface area (m^2)		
Age (years)	Male	Female
3<6	0.728	0.711
6<9	0.931	0.919
9<12	1.16	1.16
12<15	1.49	1.48
15<18	1.75	1.60
Adult	1.94	1.69

50th percentile body part-specific areas for males (m^2)			
Age (years)	Arms	Hands	Legs
3<4	0.096	0.040	0.18
6<7	0.11	0.041	0.24
9<10	0.13	0.057	0.31
Adult	0.23	0.082	0.55

AF is the soil-to-skin adherence factor (mg/cm^2) (1.45 mg/cm^2 for commercial potting soil for hands; 2.77 mg/cm^2 for kaolin clay for hands), ABS is the absorption factor (unitless) that accounts for desorption of the pollutant from the soil matrix and absorption of the pollutant across the skin (literature data for this parameter are limited), EF is the exposure frequency (events/year), ED is the exposure duration (for a lifetime use 70 years; use 30 years for the national upper-bound (90th percentile) at one residence; use 9 years for the national medium time (50th percentile) at one residence), BW is the average body weight (use 70 kg for an average adult), and AT is the averaging time (period over which the exposure is averaged, days) (for noncarcinogenic effects use ED × 365 d/yr; for carcinogenic effects use 70 years × 365 d/yr).

11.3.2.6 Inhalation of Airborne (Vapor Phase) Pollutants

Chapter 9 discussed ways of estimating the downwind pollutant concentration of a pollutant from a chemical spill. Other atmospheric exposure routes include vapors from household products, gasoline fumes from automobile filling operations, and volatilization of pollutants from household water. Another common inhalation route is breathing vapors while showering with polluted water. Estimates of CDI from vapors can be calculated by

$$\text{Intake (mg/kg d)} = \frac{\text{CA} * \text{IR} * \text{ET} * \text{EF} * \text{ED}}{\text{BW} * \text{AT}} \quad (11.6)$$

where CA is the pollutant concentration in the air (mg/m^3), and IR is the inhalation rate (m^3/d).

Suggested values are

Adult (upper-bound): 30 m^3/d
Adult (average): 20 m^3/d
For hourly rates see EPA (1989b)
Use 0.6 m^3/h for showering events

ET is the exposure time (days) (use 12 minutes converted to days for upper-bound shower duration [90th percentile]; use 7 minutes converted to days for average shower duration [50th percentile]), EF is exposure frequency (times/d), ED is the exposure duration (years) (use 70 years for a lifetime exposure; use 30 years at one residence (national upper-bound time (90th percentile); use 9 years at one residence (national median time (50th percentile)), BW is the body weight (kg) (use 70 kg for an average adult; age-specific values are given in EPA (1989b)) and AT is the averaging time (time period over which the exposure is averaged, days) (for noncarcinogenic effects upper-bound shower duration use ED × 365 d/yr; for carcinogenic effects, use 70 years × 375 d/yr).

11.3.2.7 Ingestion of Contaminated Fish and Shellfish

We not only have to be concerned with the intake of pollutants from water but also with foods that come from polluted water sources. Aquatic species often accumulate pollutants into their muscle and organ tissue, which we use for food. Thus, significant doses of pollutants can result from eating these polluted organisms. A recently discovered example is the presence of mercury in several species, especially those at the top of the food chain. The dose of a pollutant can be calculated by

$$\text{Intake (mg/kg d)} = \frac{\text{CF} * \text{IR} * \text{FI} * \text{EF} * \text{ED}}{\text{BW} * \text{AT}} \tag{11.7}$$

where CF is the pollutant concentration in the fish or shellfish (mg/kg), IR is the ingestion rate (kg/meal) (suggested values: 0.284 kg/meal) (90th percentile for fin fish; Pao et al. 1982); 0.113 kg/meal (50th percentile for fin fish; Pao et al. 1982); 132 g/d (95th percentile daily intakes averaged over three days for consumers of fin fish; Pao et al. 1982); 38 g/d (50th percentile daily intake, averaged over three days for consumers of fin fish; Pao et al. 1982); 6.5 g/d (daily intake averaged over a year; EPA 1989b); other specific values for age, sex, race, region, and other fish species are available in EPA (1989b), FI is the fraction of the daily fish intake from the polluted source (a case-specific value), EF is the exposure frequency (meals/year) (general value: 48 d/yr; average per capita for fish and shellfish, EPA 1989d), ED is the exposure duration (for lifetime exposure use 70 years; for 30 years at one residence (national upper-bound time, 90th percentile), EPA 1989b); for 9 years at one residence (national median time (50th percentile), EPA 1989b), BW is the body weight (kg) (70 kg for an average adult, EPA 1989b; for age-specific values, see EPA 1989b), and AT is the period of exposure (for noncarcinogenic effects, use ED × 365 d/yr; for carcinogenic effects, use 70 years × 365 d/yr).

11.3.2.8 Ingestion of Contaminated Fruits and Vegetables

Some plants grown on polluted soil will take up pollutants through their root systems, and these pollutants may concentrate in the plants' fruits and vegetables. We can calculate the CDI from eating contaminated produce by

$$\text{Intake (mg/kg d)} = \frac{\text{CF} * \text{IR} * \text{FI} * \text{EF} * \text{ED}}{\text{BW} * \text{AT}} \tag{11.8}$$

where CF is the pollutant concentration in the fish (mg/kg d), IR is the ingestion rate (kg/meal) (food-specific values can be found in Pao et al. 1982), FI is the fraction of food ingested from the contaminated source (as opposed to food from a noncontaminated source; unitless), EF is the exposure frequency (meals/year) (region-specific values must be used), ED is the exposure duration (years), for lifetime exposure use 70 years; use 30 years for living at one residence (national upper-bound time, 90th percentile, EPA 1989b); use 9 years for living at one residence (national median time 50th percentile: EPA 1989b), BW is the body weight for an average adult use 70 kg; for age-specific values see EPA (1989b), and AT is the averaging time (period over which the exposure is averaged, days) (for noncarcinogenic effects, use ED × 365 d/yr; for carcinogenic effects, use 70 years × 365 d/yr).

11.3.2.9 Ingestion of Contaminated Meat, Eggs, and Dairy Products

Contaminated food can result from a variety of sources. A more general equation for calculating CDIs from food is

$$\text{Intake}\left(\frac{\text{mg}}{\text{kg d}}\right) = \frac{\text{CF} * \text{IR} * \text{FI} * \text{EF} * \text{ED}}{\text{BW} * \text{AT}} \tag{11.9}$$

where CF is the pollutant concentration in the food (mg/kg), and IR is the ingestion rate (kg/meal).

Suggested values:

0.28 kg/meal for beef (95th percentile; Pao et al. 1982)
0.112 kg/meal for beef (50th percentile; Pao et al. 1982)
For specific values for other meats see Pao et al. (1982)
0.150 kg/meal for eggs (95th percentile; Pao et al. 1982)
0.064 kg/meal for eggs (50th percentile; Pao et al., 1982)
For specific values for milk, cheese, and other dairy products see Pao et al. (1982).

FI is the fraction of food ingested from the contaminated source (as opposed to food from a noncontaminated source; unitless), EF is the exposure frequency (meals/year) (region-specific values must be used), ED is the exposure duration (years) (for lifetime exposure use 70 years; use 30 years for living at one residence for the national upper-bound time) (90th percentile, EPA 1989b); use 9 years for living at one residence for the national median time (50th percentile: EPA 1989b), BW is the body weight (for an average adult, use 70 kg; for age-specific values, see EPA 1989b), and AT is the averaging time (period over which the exposure is averaged, days) (for noncarcinogenic effects, use ED × 365 d/yr; for carcinogenic effects, use 70 years × 365 d/yr).

11.3.3 Health Risk Calculations for Carcinogens

Perhaps the most disputed, but very important, aspect of determining the effects of a pollutant on human health is our attempt to estimate the increase in cancer risk in a population exposed to the pollutant at a specific concentration. We start with the standard approach of developing a dose–response curve, such as the one shown in Figure 11.1, by exposing an animal to increasing concentrations of pollutant. After a given time period – days, months, or even a lifetime – the animal is sacrificed and examined for signs of cancer development. In rare cases, human data are available, not from experiments, of course, but from legal, illegal, or in many cases accidental exposures to pollutants. The obvious problem with animal experiments is determining how nonhuman responses in exposure studies relate to human responses to the same pollutant. In most cases, we do not know but rather approximate the relationship. To complicate the situation further, dose–response experiments are conducted at relatively high dose rates (CDIs) so that we can easily see a percentage level response (cancer rate). Clearly, a percentage response rate in humans is unacceptable given our general accepted level of one in a million. Thus, we must severely extrapolate the experimental animal data not only from species to species but also from intense exposures down to very small doses and response rates.

We begin our extrapolation process by assuming that the dose–response curve is linear at low pollutant concentrations (the extreme lower left-hand section of Figure 11.1). Note that for cancer risk assessment, we assume that there is no lower threshold where the pollutant concentration does not have an effect. As a result of our linearization assumption, we will have a straight-line with the slope of y over x, or risk per dose. The units of risk are the fraction of the population **developing** cancer (not dying from cancer). Risk is a unitless fraction, while the dose has units of mg/kg d. Thus, the slope of the plot has units of $(\text{mg/kg d})^{-1}$.

This slope of the plot, referred to as the slope factor or potency factor, is the basis of our risk assessment calculations. There are a number of models used to estimate slope factors, most of which produce conservative estimates, conservative in the sense that they maximize the likelihood of detecting cancer development risk. Usually, the 95th percentile value is given. In addition, a weight-of-evidence classification is given to each slope factor. A summary of the classification system is given in Table 11.9 and is based on the conclusiveness (certainty) of the data (i.e. human test data versus animal test data, correlation of results, etc.).

The slope factor is used, along with the CDI calculations, to estimate the risk of developing cancer, usually over a 70-year exposure period. We will work several example problems to illustrate the use of the slope factor.

Example Problem

Most municipalities in the United States treat their water with some form of chlorine. Since almost all waters have some natural organic matter present, a by-product of the chlorine treatment is the formation of chlorinated organics referred to as trichloromethanes (THMs). Calculate the cancer risk of drinking water containing THMs (in the form of chloroform, $CHCl_3$). Assume that an adult drinks 2.00 l/d for 70 years with the chloroform at the THM drinking water limit (the governmentally regulated pollution concentration limit) of 0.10 mg/l.

Solution

To solve this problem, we would first use Section 11.3.2.1 to calculate the CDI.

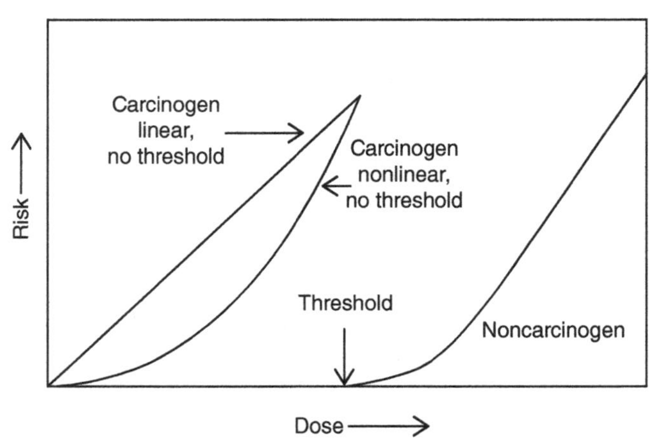

Figure 11.1 The general relationship between risk and dose of carcinogen and noncarcinogen pollutants.

Table 11.9 EPA weight of evidence classification system for carcinogenicity.

Group	Description
A	Human carcinogen
B1 or B2	Probable human carcinogen B1 indicates that limited human data are available B2 indicates sufficient evidence in animals and inadequate or no evidence in humans.
C	Possible human carcinogen
D	Not classifiable as to human carcinogenicity
E	Evidence of noncarcinogenicity for humans

From Table 11.7, the slope factor for chloroform is 6.1×10^{-3} (mg/kg d)$^{-1}$ for the oral route, thus the risk of cancer is

$$\begin{aligned} \text{Risk} &= \text{CDI}^*\text{slope factor} \\ &= 2.86 \times 10^{-3}\ \text{mg/kg d}^*6.1 \times 10^{-3}(\text{mg/kg d})^{-1} \\ &= 1.74 \times 10^{-5}\ \text{or 17.4 per one million people} \end{aligned}$$

Recall that the slope factors given by the US EPA represent the upper-bound (95th percentile) of developing cancer, not dying from cancer. This means that over a 70-year period, we can estimate that 17.4 more cases of cancer per million people will develop in the population drinking the water.

Suppose the city supplying the water has a population of 3.5 million. How many actual cases of cancer per year will result from drinking the water over a 70-year period?

$$\begin{aligned} &3\,500\,000\ \frac{17.4\ \text{cancers patients}}{1\,000\,000\ \text{people}}\ \frac{1}{70\ \text{years}} \\ &= 0.87\ \text{cancers per year} \end{aligned}$$

The average annual cancer death rate in the United States is 193 per 100 000 people. How does the calculated cancer development rate compare in our city of citizens that drink the chlorinated water?

$$\begin{aligned} &3\,500\,000\ \frac{193\ \text{cancers/yr}}{100\,000\ \text{people}} \\ &= 6755\ \text{cancer deaths per year} \end{aligned}$$

Given that 6755 people living in the city will die from cancer, it is unlikely that we will statistically notice the additional 0.87 cases of cancer (not cancer deaths). This last type of calculation, comparing the added risk of cancer development to the existing rate of cancer deaths, should be used in risk assessment for all carcinogens.

For a thorough risk assessment, we must also consider what the rate of death would be if we did not disinfect our drinking water and sewage waste, by chlorination or any other treatment. Numbers are difficult to estimate accurately, but for undeveloped countries where waterborne disease is common, we find that approximately 4 million people die each year of diarrheal diseases that infect approximately 2 billion people per year (UNEP 1993). This translates into a death rate of 2000 per million citizens. This clearly illustrates the benefit of disinfecting our drinking water and sewage waste (17.4 cases of cancer per million people as opposed to 2000 deaths per million people from not treating the water).

Slope factors are also used to estimate when an unacceptable risk will develop. In our case, the risk is in terms of the number of cancer cases and, as noted earlier, it is becoming commonly accepted to accept any risk below one in a million. When the risk equals or becomes greater than one in a million, governments usually impose pollutant exposure limits, referred to as drinking water equivalent levels (DWEL), at the concentration that causes this highest acceptable risk. We will next use the slope factor to illustrate how to estimate this pollutant concentration.

Example Problem

A well-known, cancer-causing chemical is benzene. Table 11.7 quotes a slope factor range from 1.5×10^{-2} to 5.5×10^{-2} (mg/kg d)$^{-1}$ for benzene.

Calculate the DWEL that would result in a one-in-a-million risk of developing cancer over a 70-year time span.

Solution

$$\begin{aligned} \text{CDI} &= \frac{\text{Risk}}{\text{Slope factor}} = \frac{10^{-6}}{5.5\ \times 10^{-2}\ (\text{mg/kg d})} \\ &= 1.82 \times 10^{-5}\ (\text{mg/kg d}) \end{aligned}$$

For a 70.0-kg person drinking 2.00 l/d, the DWEL would be

$$\text{CDI} = 1.82 \times 10^{-5}\ \text{mg/kg d} = \frac{X\ \text{mg/l} \times 2.00\ \text{l/d}}{70\ \text{kg}}$$

$$X\ \text{mg/l} = 6.36 \times 10^{-4}\ \text{mg/l or}\ 0.636\ \mu\text{g/l benzene}$$

Thus, benzene concentrations greater than 0.636 μg/l should be avoided to keep our cancer risk below one in a million. Similar approaches can be used for any of the CDI estimation equations in Section 11.3. We will next work an example of cancer risk assessment for a worker in an agricultural setting.

Example Problem

A farmworker is exposed to Trifluralin, a commonly used pesticide, while working in the fields. Calculate the risk of the worker exposed to 0.01 mg/m^3 who works 8 h/d, 7 days a week, for 4.0 months a year, and for 25 years. Assume a body weight of 70.0 kg and an inhalation volume of 20.m^3 per 8 hours. The slope factor for Trifluralin is 8.4×10^{-1} (mg/kg d)$^{-1}$.

Solution

$$\text{Intake by inhalation (mg/kg d)} = \frac{CA * IR * ET * EF * ED}{BW * AT}$$

$$= \frac{(0.010\ \text{mg/m}^3)\ (20\ \text{m}^3/\text{d})\ (29\ \text{d/mo})\ (4\ \text{mo/yr})\ (25\ \text{years})}{(70\ \text{kg})\ (25\ \text{years})\ (365\ \text{d/yr})}$$

$$= 9.08 \times 10^{-4}\ \text{mg/kg d}$$

$$\text{Risk} = \text{CDI} \times \text{Slope factor}$$

$$= 9.08 \times 10^{-4}\ \text{mg/kg d}^* 8.4 \times 10^{-1} (\text{mg/kg d})^{-1}$$

$$= 7.6 \times 10^{-4} \text{or 763 cancer cases per million}$$

$$= 7.6 \times 10^{-4} \times 10^{6} = 763 \text{ cancer cases per million}$$

Obviously, this is a significant risk and adequate protection of the workers, such as a breathing apparatus, should be required.

There is one final note concerning the risk of developing cancer from exposure to a pollutant. Cancer risks are assumed to be cumulative, which means the total cancer risk is the sum of all individual cancer risks. For example, if you have a cancer risk of 0.000 04 from pollutant exposure route one, a cancer risk of 0.000 08 from pollutant route two, and 0.000 005 from exposure route three, then your total cancer risk would be the sum of these, or 0.000 125 (a cancer risk of 125 cases per million people).

11.3.4 Health Risk Calculations for Noncarcinogens

In considering the carcinogenic effects of chemicals, there is no threshold dose of a chemical to which an animal species can be exposed safely. In risk calculations for noncarcinogens, though it is accepted that there is a threshold chemical concentration below which the chemical is not harmful. These values are calculated by taking a population of animals, exposing them to increasing doses of the chemicals, and observing health effects. A generic plot of such an experiment is shown in Figure 11.2, where the response is shown on the y-axis and the chemical concentration is on the x-axis. Note the presence of the threshold, which is essentially just an estimate. There are several important points on the plot that need to be mentioned. The first is the lowest-observed-effect level (LOEL; not shown in Figure 11.2) that is the lowest dose administered to the animal that results in a response (death or health effect). The next point is the no-observed-effect level (NOEL; not shown in Figure 11.2) or the highest dose that does not produce a health effect or response. The LOEL and NOEL can be refined further by noting whether the health effect is an adverse effect. This changes the names to the no-observed-adverse-effect level (NOAEL) and the lowest-observed-adverse-effect level (LOAEL). These latter two levels are the ones commonly reported by researchers. A list of these doses and effects for selected pollutants are summarized in Tables 11.5 and 11.6.

The final dose level given in Figure 11.2 is the reference dose (RfD). The RfD listed by EPA is considered the maximum acceptable daily intake (ADI) of the pollutant. Concentrations or masses below this dose are considered acceptable and should not cause adverse health effects in humans. But how do we arrive at the RfD? The RfD is obtained by dividing the NOAEL by several safety factors. The safety factors are also referred to as uncertainly

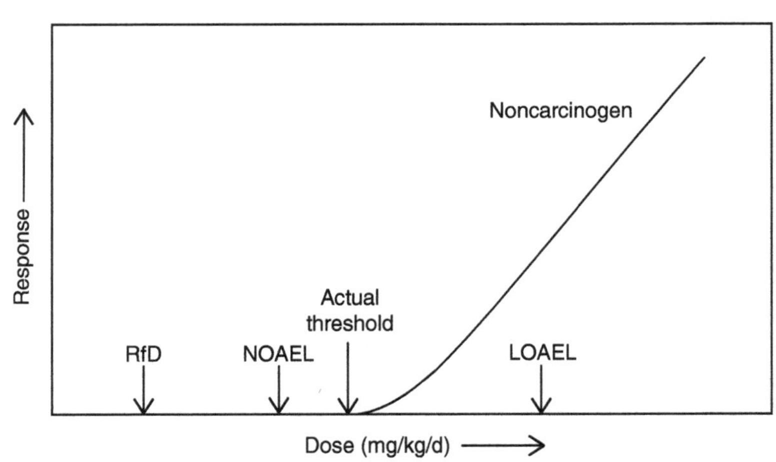

Figure 11.2 The general relationship between response and dose for a noncarcinogen pollutant illustrating the concept of threshold level.

factors because they reflect the uncertainties associated with extrapolating the results between individuals in a population and from animal testing to humans. The first uncertainly factor divides the NOAEL by a factor of 10 to account for sensitivity differences in the exposed human population. This accounts for variations between individuals, including pregnant women, infants, the elderly, and normal, healthy adults. A second adjustment (the magnitude depends on the situation) occurs if the NOAEL results from animal testing data that are extrapolated to humans. A final factor of 10 is used when no data are available on human responses to the pollutant. Values of these latter two adjustment factors range from 1 to 10. Thus, the RfD will usually be only a fraction of the NOAEL. Note the difference between the NOAEL and RfD given in Tables 11.5 and 11.6.

In many ways, the predicted response from noncarcinogenic risk assessment is less quantitative than the series of calculations involved in cancer risk assessment. For example, we cannot exactly tell the magnitude of the observed effect but only whether one should occur. To do this, we use a hazard quotient

$$\text{Hazard quotient} = \frac{\text{Average daily dose during the exposure period (mg/kg d)}}{\text{RfD}} \tag{11.10}$$

Hazard quotient values of 1.00 or greater indicate that an adverse health effect or risk of toxicity should occur. Values below 1.00 are considered safe. Another major difference between cancer and noncarcinogenic risk assessment is that, for noncarcinogens, we use the specific exposure time, not the average exposure over a 70-year period. As in cancer risk assessment, risks are additive.

Example Problem

A contaminated drinking water contains the following concentrations of gasoline products: benzene at 0.010 mg/l, tetraethyl lead at 0.015 mg/l, toluene at 0.050 mg/l, and xylene at 0.050 mg/l. Calculate the cumulative hazard quotient for a 70.0 kg person who drinks the water at a rate of 2.00 l/d. The RfD values in mg/kg d are 4.0×10^{-3} for benzene, 1.00×10^{-7} for tetraethyl lead, 2×10^{-1} for toluene, and 0.2 for xylene.

Solution

First, calculate the average daily doses (ADD) for each pollutant

$$\text{ADD}_{\text{Benzene}} = \frac{(0.010\ \text{mg/l})\,(2.00\ \text{l/d})}{70.0\ \text{kg}} = 2.86 \times 10^{-4}\ \text{mg/kg d}$$

$$\text{ADD}_{\text{Tetraethyl lead}} = \frac{(0.000\,015\ \text{mg/l})\,(2.00\ \text{l/d})}{70.0\ \text{kg}} = 4.29 \times 10^{-7}\ \text{mg/kg d}$$

$$\text{ADD}_{\text{Toluene}} = \frac{(0.050\ \text{mg/l})\,(2.00\ \text{l/d})}{70.0\ \text{kg}} = 1.43 \times 10^{-3}\ \text{mg/kg d}$$

$$\text{ADD}_{\text{Xylene}} = \frac{(0.050\ \text{mg/l})\,(2.00\ \text{l/d})}{70.0\ \text{kg}} = 1.43 \times 10^{-3}\ \text{mg/kg d}$$

Next, calculate the Hazard Quotient (HQ) for each pollutant

$$\text{HQ}_{\text{Benzene}} = \frac{2.86 \times 10^{-4}\ \text{mg/kg d}}{4.0 \times 10^{-3}\ \text{mg/kg d}} = 0.0715$$

$$\text{HQ}_{\text{Tetraethyl lead}} = \frac{4.29 \times 10^{-7}\ \text{mg/kg d}}{1.0 \times 10^{-7}\ \text{mg/kg d}} = 4.4$$

$$\text{HQ}_{\text{Toluene}} = \frac{1.43 \times 10^{-3}\ \text{mg/kg d}}{0.2\ \text{mg/kg d}} = 0.007\,15$$

$$\text{HQ}_{\text{Xylene}} = \frac{1.43 \times 10^{-3}\ \text{mg/kg d}}{0.2\ \text{mg/kg d}} = 0.007\,15$$

The sum of the Hazard Quotients is 4.3 units, where tetraethyl lead is responsible for essentially all of the risk.

11.3.5 Bioconcentration Calculations

In Chapter 2 and 3, we discussed bioconcentration and looked at DDT data from a food chain on Long Island, NY (Table 2.3). The data from Table 2.5 is reproduced in Table 11.10, where we used it to calculate bioconcentration factors (BCFs) for each animal species. Each BCF is calculated by dividing the DDT concentration in each animal by the water concentration. As you can clearly see, DDT is increasingly bioconcentrated as we move up the food chain, from 800 in the first tropic level to 528000 in birds that eat fish.

BCFs, given in Table 11.11, can be used to estimate cancer risk for a person eating contaminated fish, as is illustrated in the following problem.

Example Problem

EPA estimates that an average person might eat 54 g of fish per day. Calculate the cancer risk of a 70.0 kg person who eats 54 g of fish per day, 350 days each year for 30 years. Assume that the water the fish was taken from has a chlordane concentration of 0.0100 ppb (μg/l). The oral slope factor for chlordane is 3.5×10^{-1}.

Table 11.10 Bioconcentration factor for DDT in a Long Island Food Web (USA).

Organism	DDT residues (ppm)	BCF (l/kg)
Water	0.000 05	—
Plankton	0.04	800
Silverside minnows	0.23	4 600
Sheepshead minnows	0.94	19 200
Pickerel (predatory)	1.33	26 600
Needlefish (predatory)	2.07	41 400
Heron (feeds on small aquatic animals)	3.57	71 400
Herring gull (scavenger)	6.00	120 000
Osprey egg	13.8	276 000
Merganser (fish-eating duck)	22.8	456 000
Cormorant (feeds on large fish)	26.4	528 000

Source: Woodwell et al. (1967).

Solution

First, we must calculate the concentration in the fish.

$$\begin{aligned}&\text{Chlordane concentration in fish}\\&= \text{pollutant concentration in fish}^{*}\ \text{BCF}\\&= 0.000\,010\,0\ \text{mg/l}^{*}\ 14\,000\ \text{l/kg}\\&= 0.14\ \text{mg Chlordane/kg fish}\end{aligned}$$

Using Eq. (11.7) we can then calculate the CDI

$$\text{Intake (mg/kg d)} = \frac{\text{CF} * \text{IR} * \text{FI} * \text{EF} * \text{ED}}{\text{BW} * \text{AT}}$$

$$= \frac{(0.14\ \text{mg/kg})\ (0.054\ \text{kg/d})\ (1.00\ 350\ \text{d/yr})\ (30\ \text{years})}{(70\ \text{kg})\ \ (30\ \text{years})\ \ (365\ \text{d/yr})}$$

$$= 1.04 \times 10^{-4} \text{mg/kg d}$$

$$\begin{aligned}\text{Risk} &= \text{CDI}^{*}\text{Slope factor}\\&= 1.04 \times 10^{-4}\ \text{mg/kg d}^{*}3.5 \times 10^{-1} (\text{mg/kg d})^{-1}\\&= 3.62 \times 10^{-5} \text{ or 36 cancer cases per million}\end{aligned}$$

people eating the fish

Thus, 36 cancer cases should develop among one million people consuming the contaminated fish.

Table 11.11 Bioconcentration factors (BFCs) for selected pollutants.

Compound	BCF for fish (l/kg)
Acenaphthene	242
Acrylonitrile	48
Aldrin	28
Antimony	1
Arsenic	44
Benzene	5.2
Benzidine	87.5
Beryllium	19
Bis(2-chloroethyl)ether	6.9
Bis(chloroethyl)ether	0.63
Cadmium	81
Carbon tetrachloride	19
Chlordane	14 000
Chlorobenzene	10
Chloroform	3.75
Chromium(III)	16
Chromium(VI)	16
Copper	200
DDE	51 000
DDT	54 000
1,2-Dichlorobenzene	56
1,3-Dichlorobenzene	56
1,4-Dichlorobenzene	56
3,3′-Dichlorobenzidine	312
1,2-Dichloroethane	1.2
1,1-Dichloroethylene	5.6
1,2-Dichloroethylene	1.6
1,2-Dichloroethylene (*cis*)	1.6
1,2-Dichloroethylene (*trans*)	1.6
Dichloromethane	5
2,4-Dichlorophenol	41
1,3-Dichloropropene	1.9
Dieldrin	4 760
Diethyl phthalate	117
2,3-Dinitrotoluene	3.8
2,4-Dinitrotoluene	3.8
2,5-Dinitrotoluene	3.8
2,6-Dinitrotoluene	3.8
3,4-Dinitrotoluene	3.8
N,N-Diphenylamine	30
1,2-Diphenylhydrazine	25
Ethylbenzene	37.5
Fluoranthene	1 150
Fluorine	1 300

Table 11.11 (Continued)

Compound	BCF for fish (l/kg)
Heptachlor	15 700
Heptachlor epoxide	14 400
Hexachlorobenzene	8 690
Hexachlorobutadiene	2.8
Hexachlorocyclopentadiene	4.3
α-Hexachlorocyclohexane (HCCH)	130
β-HCCH	130
γ-HCCH (lindane)	130
Hexachloroethane	87
Kepone	8 400
Lead	49
Mercury (alkyl)	3 750
Mercury (inorganic)	5 500
Methyl parathion	45
Nickel	47
Pentachlorobenzene	2 125
Pentachlorophenol	770
Phenanthrene	2 630
Phenol	1.4
Polychlorinated biphenyls (PCBs)	100 000
Selenium	16
Silver	3 080
1,2,4,5-Tetrachlorobenzene	1 125
2,3,7,8-TCDD (dioxin)	5 000
1,1′,2,2′-Tetrachloroethane	42
Tetrachloroethylene	31
2,3,5,6-Tetrachlorophenol	240
Toluene	10.7
Toxaphene	13 100
1,2,4-Trichlorobenzene	2 800
1,1,1-Trichloroethane	5.6
1,1,2-Trichloroethane	5
Trichloroethylene	10.6
2,4,5-Trichlorophenol	110
2,4,6-Trichlorophenol	150
Tris(2,3-dibromopropyl) phosphate	2.7
Vinyl chloride	1.17
Zinc	47

Source: US EPA (1986).

11.3.6 Putting It All Together: Margin of Error (Uncertainty) of the Entire Estimation Process

We are finally at the culmination of a long and complicated discussion of source characterization, chemistry, fate and transport modeling, and now risk assessment. But how good are our risk assessment values? We have discussed a number of precautions in every chapter. The major ones include the following:

(1) Every number we generate in our modeling efforts depends on an accurate characterization of the waste site or polluted system with respect to pollutant mass dumped or contained in the system. These values are usually based on company records, employee memory, or varying amounts of sampling and analysis, and these may not be the most reliable sources. Thus, there may be considerable error with our source estimate.
(2) In Chapters 2 and 3, we discussed the chemical aspects that are important in successful fate and transport modeling. As we clearly saw, many degradation rate constants depend on pH, E_H, the number of viable microorganisms, the concentration of degrading chemical (such as natural organic matter), and the temperature. Sorption phenomena can be important in lake, river, and especially in groundwater systems. Finally, chemical speciation will also influence toxicity and therefore risk. Chemical data, produced by complex experimentation and analysis, are often used in our fate and transport modeling equations, and the accuracy of our modeling efforts is closely linked to the accuracy of this chemical data.
(3) A major factor that is difficult to estimate in fate and transport modeling, but one that greatly influences pollutant concentration, is mixing or dispersion in the system. As we saw for groundwater systems, estimates of the dispersion can vary by four orders of magnitude. Errors in such estimates will again significantly impact pollutant concentration, and therefore risk.
(4) One fact that has become distinctly clear in collecting risk assessment data for this textbook is the lack of human-specific toxicity data on many pollutants. We must therefore attempt to account for this paucity of data by including uncertainty factors in the RfD, RfC, and slope factors.

So, are our efforts adequate, and do they provide an adequate margin of safety given the uncertainty in our source, fate and transport, and toxicity data? In general, we feel that the answer to this question is yes and that our accuracy will improve as more toxicity data become available. At every point in our assumptions, we tend to make *conservative estimates* of parameters that will affect pollutant concentrations and risk. For example, it is always best to assume *maximum* pollutant input to a system, *minimal* chemical and biological degradation, and *minimal* dispersion, since this will result in the calculation of the *maximum pollutant concentration*. To this end, we also include relatively large uncertainty factors in risk assessment. And perhaps the most conservative of all of our approaches is declaring that the highest acceptable risk is one-in-a-million, which in most cases is beyond our abilities to detect given our day-to-day life practices.

We should also keep in mind that risk assessment calculations targeted toward human populations, such as those considered here, should only be one tool in our approach to determining acceptable risk and especially determining whether or not a site should be remediated. For example, although a toxic metal mining operation in the middle of proverbial nowhere may pose no risk to any human, the mission of EPA also includes protecting the environment. Both factors must be given weight in our decision-making processes. Because no formal risk assessment calculations address environmental protection, it must often be built into the dialogue in other ways.

To summarize, the errors and uncertainties are very large, but by making conservative assumptions, we can provide meaningful information to assist in making policies to protect the human public health.

11.3.7 Connecting Fate and Transport Predictions with Risk Assessment

Equations in Chapters 5 through 9 can be used to estimate the pollutant concentrations to the receptor (humans) from water and air sources. These pollutant concentrations can then be used in risk assessment calculations to determine if a health risk is present.

For example take the following problem statement:

An industrial processing plant is near a river synthesizes aniline for large-scale use in polyurethanes. A leakage has been discovered in a tank that holds excess nitrobenzene. The leakage appears to have been happening for a considerable time as a step input. Calculate the concentration of nitrobenzene in the river 100 km downstream given the following river parameters:

Average depth: 85 ft
Average width: 300 ft
Input of nitrobenzene: 0.015 kg/s
Average water velocity: 3.7 m/s
Channel slope: 0.001
Half-life of nitrobenzene due to photolysis degradation: 133 days.

Using the River: Step equation and Fate®, the following results were obtained:

Inputs	
Stream depth (d)	25.908 m
Stream width (w)	91.44 m
Mass input (W)	0.015 kg/s
Water velocity (v)	3.7 m/s
Stream flow rate (Q)	8765 m^3/s
Channel slope (s)	0.001
Gravity constant (g)	9.81 m/s^2
Half life $t_{1/2}$	0.364 year
Distance (m)	100 km
Output	
Longitudinal dispersion (E)	96.40 m^2/s
First-order rate constant (k)	1.902 $year^{-1}$
First-order rate constant (k)	6.03E−08 s^{-1}
Concentration (μg/l; ppb)	1.71

Now, use this result to assess health risk from drinking this water with the following parameters:

A 77-kg person drinks the water 100 km downstream from the plant at a rate of 1.5 l/d. Calculate the hazard quotient for this noncarcinogenic consumption.

Results:

Mass of person	77 kg
RfD	0.002 mg/kg d
Rate of consumption	1.5 l/d
Average daily dose	3.33E−05 mg/kg d
Hazard quotient	0.0166
Risk	<1 for hazard quotient

Similar exposure-risk scenarios can be formulated for each of the fate and transport equations and risk calculations.

11.4 Life Cycle Assessment (LCA)

The concept of LCA can likely be traced back to a similar concept under Resource Conservation and Recovery Act (RCRA) discussed in the US Environmental Law chapter (12) where we will introduce the "cradle to grave" concept. Curran (2015) defines LCA as "a holistic, cradle-to-grave environmental approach which provides a comprehensive view of the environmental aspects of a product or process throughout its life cycle. A properly conducted LCA identifies and quantifies the potential impacts of an industrial system (aiming to assess products, processes, and activities)." This definition and approach to environmental assessment is a culmination of decades of assessment evolution. Again, as summarized by Curran (2015) prior to the environmental revolution of the 1960s and 1970s, we used end-of-the pipe technologies to treat pollution after it had been created and just prior to release into the natural environment. The mid-1980s saw a shift to waste minimization and pollution reduction, followed by the beginning of the green chemistry developments in the early 1990s with pollution prevention and cleaner production industrial processes. LCA came into existence c. mid-1990s and in the following century evolved into sustainable development and life cycle sustainability assessment, with the goal of selecting the best technology or process prior to the initiation of any industrial construction. What should have been common sense decades ago is now reality and common practice.

As one can imagine from the definition above, LCA is a highly involved and time-intensive process, and assessment range from minor to major processes. For example, take a question everyone has been asked, "Do you prefer paper or plastic shopping bags?" There are several assessments addressing this issue in the technical literature, each with its own set of starting assumptions (type of paper, plastic, cloth, or other material) so we will use summary data here from Appropedia. To start, paper bags are made from a renewable resource, trees, plastic bags are made from a nonrenewable resource, fossil fuels, and cloth bags are made from the growing of highly energy, water, and chemical intensive cotton. Plastic bags consume less water and energy in their production, and consume less energy in their recycling as compared to paper. As a result, this website claims "The evidence from multiple scientific studies have demonstrated that plastic disposable bags are less harmful to the environment than disposable paper bags." But the final decision depends on the intended outcome goal of the assessment. This studies mostly based "environmental cost" on energy, water, and global warming potential consumption. Depending on where you live, aesthetic costs may be the desired outcome of a LCA. Many countries have outlawed plastic bags due to their lack of effective disposal and waste collection, where discarded bags are seen hanging from every tree. Coastal communities, even the environmental friendly Seattle, Washington (USA), have disregarded technical LCA results in favor of not potentially contributing to the plastic garbage patch in the Pacific Ocean, and prefer the more LCA and monetarily costly paper bags. A current common theme seen at the checkouts in many stores are manufactured cloth or cloth-plastic bags, and is the clear LCA loser choice in all LCA bag studies given the extensive cost of growing cotton, using fossil-fuel–derived plastic, and manual labor involved in manufacturing the bag. Of course, home-made, postconsumer product making of the bag is a more environmental friendly story. So, depending of the goals and interpretation of your LCA, your conclusions and environmental implementation efforts or choices may differ. These LCAs of plastic need to now look at the burden placed on marine wildlife and the ever-increasing collection of plastic in the Pacific Ocean.

Next, look at a much more technical product light bulbs. Versions of a light bulb have been around since the early 1800s and have evolved greatly in recent years, from the conventional incandescent light bulb, to the long tube fluorescent lamp, to the compact fluorescent lamp (CFL), to and the light-emitting diode lamp. Which is "better?" Again, we will avoid the technical literature and look at less technical ones such as OSRAM (2009). Their executive summary of their LCA states.

"OSRAM Opto Semiconductors" study on the LCA compared the environmental performance of three different types of lamps: a conventional light bulb (GLS), a CFL and a light-emitting diode lamp (LED lamp). In order to evaluate lamps and how they actually deal with energy and resources, it is not enough just to consider energy consumption while they are in use. The aim of this LCA is therefore to analyze the environmental impact of an LED lamp over its entire life and to compare it to a CFL and a GLS. The relevant material and energy supplies, including all input and outputs, were determined in detail for all the LED lamp's components and production processes. Apart from a detailed analysis of each individual production stage, for LED chips and lamp housings, for example, these also include all necessary transports. Apart from the primary energy and resource consumption, the impact on the environment corresponding to selected environmental impact categories, i.e. acidification, eutrophication, photochemical ozone depletion, and human toxicity was evaluated.

A comparison of the life cycle of three lamp types was accomplished: a 40 W GLS, an 8 W CFL Dulux Superstar, and an 8 W Parathom LED lamp.

Main findings:

- Less than 2% of the total energy demand is needed for production of the LED lamp. The manufacturing phase is insignificant in comparison to the use phase for all three lamps as it uses less than 2% of the total energy demand. This study has dismissed any concern that production of LEDs particularly might be very energy-intensive. Merely about 0.4 kWh are needed for production of an LED (OSRAM Golden Dragon Plus), about 9.9 kWh for the production of the Parathom LED lamp including six LEDs.

- LED lamps are competitive to CFL today. In contrast to the primary energy consumption of incandescent lamps of around 3302 kWh, CFL and LED lamps use less than 670 kWh of primary energy during their entire life. Thus 80% of energy can be saved by using CFL or LED lamps. The bottom line is that LED lamps are more efficient than conventional incandescent lamps and also ahead in terms of environmental friendliness. Even today, LED lamps show nearly identical impact on the environment compared to CFL.
- Future improvements of LED lamps will further cut down energy demand. As the efficiency of LEDs continues to increase, LED lamps will be capable of saving more energy and achieving even better LCA results in future. This LCA proves that LED lamps are among the most environmentally friendly lighting products."

Note that this study was conducted in 2009 and, as predicted in the summary above, considerable improvements have been made in LED technology and manufacturing. As many other sources state, current LED products are 95% more energy efficient than standard incandescent light bulbs and LCAs can be used in many other product assessments to choose the most environmental friendly technology or product.

LCA can be conducted on products as simple as the most environmentally friendly container to use (i.e. glass, metal, paper, or plastic) or can be conducted on an entire chemical process or product manufacturing. The utility and popularity of LCA has led to the development of an entire industry of consultants to conduct these assessments and a variety of computer tools to conduct LCAs. LCA leads to our next assessment topic benefit–cost analysis, and while LCAs seem to be redundant, remember that LCA consider more inputs in their analysis, while Benefit–Cost Analysis primarily focuses on the cost of a product or process versus profit made from the product.

11.5 Benefit–Cost Analysis

Readers will be more familiar with the term "cost–benefit," but the reverse term will be used here since almost any additional cost to a product or process will increase the benefit. Benefit–cost analyses predate LCA and only evaluate the comparison cost of two or more options. Wikipedia defines benefit–cost analysis as "a systematic approach to estimate the strengths and weaknesses of alternatives (for example, in transactions, activities, functional business requirements or projects investments); it is used to determine options that provide the best approach to achieve benefits while preserving savings. The CBA is also defined as a systematic process for calculating and comparing benefits and costs of a decision, policy (with particular regard to governmental policy or (in general) project." (https://en.wikipedia.org/wiki/Cost%E2%80%93benefit_analysis). While benefit–cost analysis can be applied to virtually every decision, they have found extensive use in the environmental arena.

A topic more relevant to energy and climate change is a "cost effectiveness analysis of energy sources" where "levelized cost of energy" or LCOE are used. Two excellent explanations and examples can be found at US Energy Information Administration (https://www.eia.gov/tools/faqs/faq.php?id=19&t=3) and Lazard's levelized cost of energy analysis (https://www.lazard.com/perspective/levelized-cost-of-energy-2017). The latter (Lazard's LCOE) *"shows a continued decline in the cost of generating electricity from alternative energy technologies, especially utility-scale solar and wind,"* and shows promise in addressing climate change issues with renewable energy sources.

11.6 Summary

A variety of assessments tools are now being used in the environmental area to justify choices. While scientists can make relatively accurate and effective calculations with respect to chemical toxicology, life cycles, health risk, and benefit costs, there is still a lack of expertise and competency in policy-makers using these well-established and documented tools.

Concepts

11.1 List the modes of toxicant inputs to the human body.

11.2 Research and explain the common biochemical mechanisms of toxicants.

11.3 Put a time scale of the acute and chronic toxicity.

11.4 Explain the difference between noncarcinogens and carcinogens.

11.5 Using Tables 11.1–11.3, justify the acceptability of the one-in-a-million approach used by EPA with regard to risk assessment.

11.6 Draw a typical dose–response curve for a carcinogenic pollutant.

11.7 Draw a typical dose–response curve for a non-carcinogenic pollutant and label appropriate dose levels.

11.8 Define RfD, NOAEL, threshold, and LOAEL.

11.9 Research a LCA on the Internet and summary it. State the comparison products or processes, summary the actual costs, and explain which product or process is more environmentally friendly.

11.10 Research a benefit–cost analysis or levelized cost of energy on the Internet and summary it. State the comparison products or processes, summary the actual costs, and explain which product or process is more environmentally friendly.

Exercises

11.1 Mercury is released into the atmosphere by burning coal. When it rains, mercury enters the ocean, where it is bioconcentrated in marine life forms. Calculate the risk of cancer for a 65 kg person who eats 45 g of tuna, 325 days per year for 35 years. Assume that mercury is present in the water at 2.5 ppb. Mercury's oral slope factor is 3×10^{-4}, and it has a BCF of 5500 l/kg.
Answer: Use Eq. (11.7).

11.2 It has been found that exposure to high levels of aniline results in cancerous effects on the lungs, as well as upper respiratory tract irritation and congestion. Aniline has been classified as carcinogenic and is very toxic to humans. It may be produced during the burning of plastics or from burning tobacco. Assume that an adult drinks 2.00 l of water per day for 70 years with an aniline concentration of 0.60 mg/l. Calculate the CDI and then calculate the risk. The slope factor is 5.7×10^{-3} (mg/kg d)$^{-1}$. Next, suppose that the city supplies water to 5 million people. How many actual cases of cancer are predicted to result from drinking the water over a 70-year period?
Answer: Use Eq. (11.1).

11.3 Studies indicate that the long-term intake of bromoform in drinking water can cause cancer in animals and humans. The slope factor for bromoform is 7.9×10^{-3} (mg/kg d)$^{-1}$. Calculate the DWEL that would result in a one-in-a-million risk of developing cancer over a 70-year time span.
Answer: Use Eq. (11.1).

11.4 A worker works 8 hours a day, 5 days a week, 10 months a year, and is exposed to 0.01 mg/m^3 of nickel refinery dust for 21 years. Assume that his body weight is 70.0 kg and that he inhales a volume of 20 m^3/d. The slope factor of nickel refinery dust is 8.4×10^{-1} (mg/kg d)$^{-1}$. What is the risk of developing cancer?
Answer: Use Eq.(11.6).

11.5 Contaminated drinking water near a farm contains the following concentrations of pesticide products: Aldrin at 0.0025 mg/l, Dieldrin at 0.0003 mg/l, and Endrin at 0.0007 mg/l. Calculate the cumulative hazard quotient for a local 70.0 kg person that drinks 2.0 l/d of the water. The RfD values in mg/kg d at 3×10^{-5} for Aldrin, 5×10^{-5} for Dieldrin, and 3×10^{-4} for Endrin.
Answer: Use Eq. (11.1).

11.6 Calculate the risk of a 70.0 kg person who eats fish 300 days a year for 50 years. The water that the fish was taken from has a 3,3′-dichlorbenzidine concentration of 1.1 ppb. Assume that the person eats 54 g of fish per day. The oral slope factor for the pollutant is 4.5×10^{-1}.
Answer: Use Eq. (11.7).

11.7 In the past, before labor laws and intensive study on the health effects of pesticides were completed, fumigation planes were guided by people on the ground with brightly colored flags. Pesticides were sprayed directly onto the flaggers. One common pesticide applied in this manner was chlordane. What would the CDI be for a person who lived near a fumigated field for 30 years, if the pollutant concentration was 0.02 mg/m^3 for 12 minutes once every week?
Answer: Use Eq. (11.6).

11.8 Calculate the risk of developing cancer for an average adult that inhales $0.003\,\text{mg/m}^3$ of gaseous carbon tetrachloride per day from a nearby leaking dry cleaning facility. The person is only exposed immediately after they return home from work for a total of 23 minutes each day, for 5 days a week. The person lives by the facility for 9 years. Assume an inhalation rate of $0.6\,\text{m}^3/\text{h}$. The slope factor for chloroform is 1.3×10^{-1}.
Answer: Use Eq. (11.6).

11.9 Determine the average daily doses (ADD) for the inhalation of elemental mercury for an average adult living only a few miles away from the coal-fired plant. Each day an average person breathes in $20\,\text{m}^3$ of air. Every day the person inhales an average of $0.04\,\text{mg/m}^3$ of mercury. What is the hazard quotient?
Answer: Use Eqs. (11.6) and (11.10).

11.10 Polychlorinated biphenyls (PCBs) are no longer produced in the United States, but they are still found in relatively high concentrations in animals around the world. Many people enjoy consuming fish on a regular basis. What is the cancer risk for a 70-kg person who eats 6 oz of swordfish steak on Mondays and an 8 oz tuna steak on Thursdays? This pattern of fish consumption is constant for 25 years. (There are 28.35 g in an ounce.) The BCF for PCBs is 100 000 l/kg. The water the swordfish was taken from has a PCB concentration of 5.0 ppb, while the tuna was raised in water containing 3.5 ppb. The slope factor for PCBs is 2.0 (www.epa.gov/iris/).
Answer: Use Eq. (11.7).

11.11 Azobenzene is a commonly used herbicide. An average adult weighing 70 kg lives a few miles from a spraying operation that creates an average risk of 840 cancers per million. What is the CDI for this adult? The slope factor is 0.11.
Answer: Use Eq. (11.6).

11.12 A painter has painted houses for 45 years. There are many chemicals in paint that, upon volatilization, pose a risk to human health. Two chemicals found in paint are benzene (a carcinogen) and toluene (a noncarcinogen). The painter weighs 70 kg, breathes $22\,\text{m}^3/\text{d}$ of air, and works 5 days a week. Calculate the painter's lifetime intake of each of the chemicals assuming an average air concentration of $0.010\,\text{mg/m}^3$. Using this calculated intake, calculate the cancer risk associated with benzene (slope factor $= 2.9\times10^{-2}\,\text{mg/kg}\,\text{d}^{-1}$) and the hazard quotient of toluene (RfD $= 4.00\times10^{-1}\,\text{mg/m}^3$).
Answer: Use Eq. (11.6).

11.13 You are a researcher who is constantly exposed to appendages and organs preserved in formaldehyde. You inhale an average dose of 0.02 mg/kg d. The RfD for inhaled formaldehyde is 2.0×10^{-1} mg/kg d. Calculate the hazard quotient for the exposure. Are you at risk of developing a problem?
Answer: Use Eq. (11.6).

11.14 MTBE, an additive in gasoline, has recently been in the media as a drinking water contaminant due to leaking gasoline storage tanks. Calculate the cancer risk from drinking water that has an MTBE concentration of 40.0 ppb (considered safe according to the EPA website). Assume that a 70 kg adult drinks 2.0 l/d for 70 years. Finally, calculate the DWEL that would result in a one-in-a-million risk of developing cancer over a 70 year time span.
Answer: Use Eq. (11.1).

11.15 Drinking water contains the following concentrations of contaminants: Aldrin at 0.0400 mg/l, toluene at 0.060 mg/l, benzene at 0.010 mg/l, and trifluralin at 0.05 mg/l. Calculate the cumulative hazard quotient for a 70.0 kg person who drinks the water at a rate of 2.00 l/d. The RfD values in mg/kg d are 7.5×10^{-3} for trifluralin, 4.0×10^{-3} for benzene, 2×10^{-1} for toluene, and 3×10^{-5} for Aldrin.
Answer: Use Eq. (11.1).

11.16 Dichlorvos is a contact insecticide that is used as a household and public health insecticide. As the active ingredient of Nuvan 500 EC, Dichlorvos was used extensively in salmon farming to control the salmon louse *Lepeophtheirus salmonis*. The principal, direct routes of entry for Dichlorvos into waters include industrial effluents and accidental discharges (e.g. from pesticide manufacturing plants, formulation plants, and marketing outlets), use in salmon fisheries, disposal of unused insecticide, and the cleaning of application and mixing equipment. Dichlorvos may also indirectly enter the aquatic environment via spray drift during application and in land runoff. Dichlorvos has a potential for bioaccumulation into marine life, such as in

the salmon consumed by humans. Calculate the Dichlorvos-associated cancer risk of a 70.0 kg person who eats 54 g of fish for 350 days out of the year for 30 years. Assume that the water the fish was taken from has a dichlorvos concentration of 2.00 ppb. The oral slope factor for dichlorvos is $2.9 \times 10^{-1}(\text{mg/kg d})^{-1}$.
Answer: Use Eq. (11.7).

11.17 A secretary reviewing records at a local hospital notices that an unusually high number of patients have had liver lesions in the past few years and alerts the health officials. Finding no significant correlation between these patients, investigators decide to test the drinking water. They find high concentrations of carbon tetrachloride, a compound known to cause liver lesions, far above the LOAEL, and link the pollutant to discharges from an upstream industrial site. If the measured concentration of CCl_4 is 55 ppb, what is the average CDI of those people drinking the water? How does this concentration compare to the maximum acceptable contaminant level (MCL) of CCl_4 (0.005 mg/l)? Assume an intake of 2.00 l water per day, an exposure frequency of 350 d/yr, exposure duration of 5 years, a body weight of 70 kg, and an average time of exposure of 5 years (at 365 d/yr). Finally, from the calculated daily intake (CDI) you just determined, find the risk of people drinking the contaminated water if the slope factor is 0.13 kg d/mg. Judging by this estimated risk, how many cases, on average, would you expect to see each year in a population of 150 000 people?
Answer: Use Eq. (11.1).

11.18 Suppose that the polluting industry in Problem 21 also discharged other pollutants, such as 1,2,4,5-tetrachlorobenzene. If this compound was discharged into a lake, it could enter the food chain by being adsorbed by or sorbing to microorganisms in the water. If the concentration of tetrachlorobenzene is the lake is 0.30 ppm, what is the concentration in local fish if the BCF is 1125 l/kg? Suppose that local fisherman in the area consume a third of the EPA average fish consumption (54 g). What is their intake of tetrachlorobenzene in mg/kg d?
Answer: Use Eq. (11.7).

11.19 A worker is exposed to chromium(VI) every day as a by-product from welding vapor. Assuming that he/she works 8 h/d, 5 days a week, for 10 years, find the risk of the welder exposed to 0.0001 mg/m^3. Assume an inhalation rate of 20 m^3/d.
Answer: Use Eq. (11.6).

11.20 A contaminated water source contains many hazardous materials. The water contains 0.025 mg/l acrylamide, 0.010 mg/l chloroform, 0.050 mg/l ethyl acetate, and 0.030 mg/l trifluralin. Find the cumulative hazard quotient for a 70 kg person who drinks 2.00 l/d. RfD values are listed in Table 11.5.
Answer: Use Eq. (11.1).

11.21 Some shellfish have recently been discovered to contain Dieldrin, a pesticide. If a person eats 54 g of shellfish in one sitting, find the cancer risk of a 70 kg person that eats shellfish 200 days out of the year for 40 years. Assume that half the fish eaten is contaminated and that half the fish eaten in not contaminated. The water that the shellfish was taken from contains 0.100 ppb Dieldrin.
Answer: Use Eq. (11.7).

11.22 A twofold excess risk of lung cancer was observed in cadmium smelter workers. The study group consisted of 602 white males, with an average body weight of 70 kg, who have been employed during the years 1940–1969. The researchers were able to ascertain that the increase in lung cancer was probably not due to the presence of arsenic or to smoking. The men are assumed to have worked eight-hour days, 365 days a year, in order to assess a maximum exposure value. The average concentration of cadmium dust in the air was 5 mg/m^3. The inhalation rate for an average adult is 20 m^3/d. Use this data to confirm or dispute the stated twofold excess risk of lung cancer, assuming that the risk of cancer was 2500 per million persons at risk. The slope factor for cadmium is 6.1 $(\text{mg/kg d})^{-1}$. How many actual cases of cancer are predicted to occur among this worker group?
Answer: Use Eq. (11.6).

11.23 Sixty-four production workers exposed to ammonia in a soda ash facility agree to participate in a study to assess the cumulative hazard for workers at this site. The mean age of the workers was 38.9 years, with an average weight

of 70 kg, and a mean duration of exposure of 12.2 years. The mean ammonia exposure was 6.4 mg/m^3. The RfD is 1×10^{-1} mg/m^3. Calculate the hazard quotient.
Answer: Use Eq. (11.6).

Cap Stone Exercise

11.1 Research a pollution scenario, as presented in Section 11.3.7, and conduct a rate and transport calculation and risk assessment calculation.

Bibliography

Appropedia (2018). http://www.appropedia.org/Paper_versus_plastic_bags (accessed 28 June 2018).

Klaassen, C.D. (2013). *Casarett & Doull's Toxicology: The Basic Science of Poisons*, 8e. New York: McGraw-Hill Education: Medical.

Cheremisinoff, N. (2001). *Handbook of Pollution Prevention Practices*. New York: Marcel Dekker.

Curran, M.A. (2015). *Life Cycle Assessment Student Handbook*, Chapter 1, 1–18. New York: Wiley; Scrivener Publishing.

Environmental Protection Agency (EPA) (1989a). *Risk Assessment Guidance for Superfund, volume 1, Human Health Evaluation Manual (Part A), Interim Final*. Office of Emergency and Remedial Response EPA/540/1-89/002.

Environmental Protection Agency (EPA) (1989b). *Exposure Factors Handbook*. Office of Health and Environmental Assessment EPA/600/8-89/043.

Environmental Protection Agency (EPA) (1989c). *Interim Final Guidance for Soil Ingestion Rates*. Office of Solid Waste and Emergency Response (OSWER Directive 9850.4).

Environmental Protection Agency (EPA) (1989d). *Guidance Manual for Assessing Human Health Risks form Chemically Contaminated Fish and Shellfish*. Office of Marine and Estuarine Protection EPA/503/8-89/002.

Environmental Protection Agency (EPA) (2018a). https://www.epa.gov/p2 (accessed 9 July 2018).

Environmental Protection Agency EPA (2018b). https://www.epa.gov/iris (accessed 10 August 2018).

Gilbert, S.G. (2004). *A Small Dose of Toxicology*. New York: CRC Press.

ISO 9000. https://en.wikipedia.org/wiki/ISO_9000 (accessed 9 July 2018).

Louvar, J.F. and Louvar, B.D. (1998). *Health and Environmental Risk Analysis: Fundamentals with Applications*. New Jersey: Prentice Hall.

OSRAM (2009). http://seeds4green.net/sites/default/files/OSRAM_LED_LCA_Summary_November_2009.pdf (accessed 28 June 2018).

Pao, E.M., Fleming, K.H., Gueuther, P.M., and Mickle, S.J. (1982). *Food Commonly Eaten by Individuals: Amount per Day and per Eating Occasion*. U.S. Department of Agriculture.

Royal Society (2018). https://en.wikipedia.org/wiki/Biochemical_oxygen_demand (accessed 10 July 2018).

UNEP (1993). *Environmental Data Report, 1993-1994*, United Nations Environment Programme. Oxford: Blackwell.

US EPA (1986). *Exhibit A-1 in Superfund Public Health Evaluation Manual*. Washington, DC: Office of Emergency and Remedial Response.

US EPA (1991). *Human Health Evaluation Manual, Supplemental Guidance: Standard Default Exposure Factors, OSWER directive 9285.6-03*. Washington, DC: Office of Emergency and Remedial Response.

Wikipedia (2018). https://en.wikipedia.org/wiki/Sewage_treatment (accessed 23 July 2018).

Wilson, R. (1979). Analyzing the daily risks of life. *Technol. Rev.* **81** (4): 41–46.

Wilson, R. and Crouch, E.A.C. (1987). Risk assessment and comparisons: an introduction. *Science* **17**: 267–270.

Woodwell, G.M., Wurster, C.F., and Isaacson, P.A. (1967). DDT residues in an east coast estuary: a case of biological concentration of a persistent insecticide. *Science* **156**: 821–824.

Part V

Environmental Legislation in the United States

A law is valuable not because it is law, but because there is right in it.

Henry Ward Beecher

Pollutant Fate and Transport in Environmental Multimedia, First Edition. Frank M. Dunnivant and Elliot Anders.
© 2019 John Wiley & Sons, Inc. Published 2019 by John Wiley & Sons, Inc.
Companion website: www.wiley.com/go/Dunnivant/pollutant_fate&Transp_EnvironMultimedia

12

US Environmental Laws

Frank Dunnivant, Lance DeMuth, Savanna Ferguson, Rose Kormanyos, Loren Sackett, and Jill Schulte

Department of Chemistry, Whitman College, Walla Walla, WA 99362, USA

12.1 Environmental Movements in the United States

A myriad of social and historical factors may have contributed to the environmental movement and to the eventual formation of the Environmental Protection Agency. Some argue that environmentalism has been an American value since the United States' earliest history. Some historical events date to England and the Royal Society Commission on Sewage Disposal (Wikipedia 2018). The pilgrims and colonialists monitored depletion of their natural resources – "as early as 1626 the members of the Plymouth Colony passed ordinances regulating the cutting and sale of timber on colony lands" (Switzer 1994, p. 3). Following is a timeline outlining the important eras and turning points of the modern environmental movement.

Timeline of the US environmental movement:

1800–1900: The nascent environmental debate is split into two opposing sides: the conservationists and the preservationists. The conservationists "believed sustainable exploitation of resources was possible," while the preservationists "sought to preserve wilderness areas from all but recreational and educational use" (Switzer 1994, p. 7). They are both right for a sustainable future society.

In literature, Henry David Thoreau, Ralph Waldo Emerson, and John Muir appeal to peoples' natural longing for the good life, by portraying nature as the ultimate utopia, offering freedom, escape, and fulfillment for every person. Writing at the close of the nineteenth century, Muir uses poetic language to portray the destruction of nature as the ultimate evil. He hopes people will identify with the need for wildness. Many are affected by his essays, including then-President Theodore Roosevelt.

1900–1920: Progressive Era marked by the mobilization of government and private organizations such as the 1919 National Parks Association and the 1905 National Audubon Society.

1901–1909: Under President Theodore Roosevelt, conservationism gains dominance. Roosevelt sets aside millions of acres for national forests (Barton 2002, p. 102).

May 1908: White households a Conference on Resource Management, which concludes with the creation of a National Conservation Commission to monitor and take an inventory of natural resources.

1912: National Conservation Congress focuses its session on "the conservation of human life" (Switzer 1994, p. 8), foreshadowing a new debate, that between human welfare and the welfare of the environment.

1930s: Out of the dust bowl and the beginning of the Great Depression, environmental consciousness emerged as part of a new, holistic, and communal way of thinking about life and the world. People became more concerned with "maintaining the whole community of life in stable equilibrium with its habitat" (Barton 2002, p. 70), and began to include the protection of the environment as one of their new priorities.

1935–1945: World War II and the New Deal shape the next development in environmental philosophy, as people move from the cities to the suburbs and "upwardly mobile white collar workers left crowded cities for localities with clean air, gardens, and grass," while "at the same time some rural folk watched with dismay as their small towns became urbanized" (Landy et al. 1994, p. 22). This dramatic environmental change opens the public eye to concrete examples of deforestation, development, and natural resource issues.

Meanwhile, the "population was becoming younger, more financially secure, and better educated" (Landy et al. 1994, p. 22). The population's youth coupled with the economic prosperity of the time, allowed "affluence, leisure, mobility, and a greater understanding of

Pollutant Fate and Transport in Environmental Multimedia, First Edition. Frank M. Dunnivant and Elliot Anders.
© 2019 John Wiley & Sons, Inc. Published 2019 by John Wiley & Sons, Inc.
Companion website: www.wiley.com/go/Dunnivant/pollutant_fate&Transp_EnvironMultimedia

physical and biological science [to combine] to create a new awareness of, and interest in, the natural world" (Landy et al. 1994, p. 22). The luxury of environmental awareness, which would have been mocked during the Great Depression, found itself embraced in this new society, a society which would learn from the New Deal that "government could be used to achieve social goals," a realization that the youth of the 1960s eagerly took to heart (Landy et al. 1994, p. 22).

1962: Rachel Carson's *Silent Spring* begins to raise awareness of the gravity of environmental issues and the danger that our society might essentially be welcoming future generations into silent, desolate, and bleak "uncontrolled environmental decay" (Quarles 1976, p. 11). The work "attracted immediate attention and wound up causing a revolution in public opinion" (Lewis 1985). *Silent Spring* was responsible for the controversial ultimate worldwide banning of DDT for insect control and malaria control in areas in Asia, Africa, and South America.

1969: The moon landing offers a patriotic and poetic metaphor for environmentalism – "when astronauts turned their cameras homeward, capturing the image of a delicate blue planet, the world looked upon itself with fresh understanding," although "their photographs showed clouds of pollution hovering over North America (EPA 1995, Dec. 1; Quarles 1976, p. 11).

1965–1970: The momentum of the anti-Vietnam-War protest clears the way for the environmental movement. The movement toward environmental awareness "has been one of American's most effective protests. It gained power from the rebellion of youth, which dominated the civil rights and antiwar movements of the 1960s" (Quarles 1976, p. xiv). Popularly, the environmental movement was carried on the shoulders of America's youth and activists. The author of an editorial in the *New York Times* in 1969 observed, "call it conservation, the environment, ecological balance, or what you will, it is a cause more permanent, more far-reaching, than any issue of our era – Vietnam and Black Power included" (Lewis 1985, p. 2).

1970: The first Earth Day enjoys mass participation. Its widespread popularity influences Congress to pass legislation protecting the environment, giving the movement added momentum. Earth Day also results in "environmentalism" becoming a household word. Not until after Earth Day did surveyed members of the public rank environmentalism among the most important problems facing the nation—in fact, it ranked second in a May 1970 Gallup poll (Switzer 1994, p. 15).

1975: Edward Abbey's *The Monkeywrench Gang* portrays radical environmentalists as heroes sabotaging the attempt to destroy nature, a moving but unrealistic approach to environmentalism.

1978: At Love Canal, where the Hooker Chemical Company had buried toxic material in a landfill that was later covered over and developed with housing, the pollution begins to seep into groundwater and basements, poisoning residents of the community built on the land. The event had the dramatic elements of a great story ... irresponsible corporations, indifferent bureaucrats, and arrogant scientists.
The Love Canal crisis receives extensive media coverage, causing people nationwide to sympathize with the affected residents (Taylor 1995, p. 38). During the 1960s and 1970s, television networks had begun to realize the tremendous profitability of the news, given that "the news is substantially cheaper to produce than entertainment programs, and widely viewed" (Landy et al. 1994, p. 23). And environmental stories held particular appeal: "Oil covered birds, belching smokestacks, rusting storage drums, and inspection crews in 'moon suits' are all visually compelling. Environmental stories have the particular advantage that often the crew can set up at its leisure, obtain the desired angles, and have plenty of vivid footage to show at the six and the eleven PM new cases" (Landy et al. 1994). While the media surely could not plant concern for an issue into the mind of an apathetic individual, the dramatics of the television media likely catalyzed the environmental movement and improved the force it would carry.

These are just a few of the events that helped shape and force environmental policy development in the United States, and as we will see in this chapter, they helped form comprehensive and effective laws for the protection of human health and to some extend to protect the environment.

12.2 The History of the Environmental Protection Agency (US EPA)

As we saw in the last section, the idealism, activism, and momentum of the environmental movement coalesced in the first celebration of Earth Day on 22 April 1970. Earth Day "lives in popular memory to this day as a joyous and life-affirming moment in American history" (Lewis 1985, p. 3). Protestors delivered strong messages to the community, the president, and the government as "oil-coated ducks were dumped on the doorstep of the Department of the Interior ... A student disguised as the Grim Reaper stalked a General Electric Company stockholder's meeting ... Demonstrators dragged a net filled with dead fish down Fifth Avenue, and shouted to passers-by 'this could be you!'" (Lewis 1985, p. 3). The government heard their message loud and clear, and as early as 9 July, Nixon sent his Reorganization Plan No. 3, which would become the EPA, to Congress. The intention of the agency was to "establish and enforce environmental protection standards, conduct environmental research, provide assistance to others combating environmental pollution, and assist the Council on Environmental Quality (CEQ) in developing and recommending to the president new policies for environmental protection (Lewis 1985, p. 4).

Combining offices and responsibilities from the Department of the Interior, the Department of Agriculture, the Department of Health, Education, and Welfare, the Department of Atomic Energy Commission, the Federal Radiation Council, and the CEQ, the EPA began strongly, setting the following mission statement (https://www.epa.gov/aboutepa, EPA 2018a):

> The mission of the US Environmental Protection Agency is to protect human health and to safeguard the natural environment – air, water, and land – upon which life depends.

EPA's purpose is to ensure that

- All Americans are protected from significant risks to human health and the environment where they live, learn, and work.
- National efforts to reduce environmental risk are based on the best available scientific information.
- Federal laws protecting human health and the environment are enforced fairly and effectively.
- Environmental protection is an integral consideration in the US policies concerning natural resources, human health, economic growth, energy, transportation, agriculture, industry, and international trade, and these factors are similarly considered in establishing environmental policy.
- All parts of society – communities, individuals, businesses, state and local governments, and tribal governments – have access to accurate information sufficient to effectively participate in managing human health and environmental risks.
- Environmental protection contributes to making our communities and ecosystems diverse, sustainable, and economically productive.
- The United States plays a leadership role in working with other nations to protect the global environment.

Unfortunately, the director of the EPA is appointed by the president, yet they are to be confirmed by the ruling political party, and the ruling political party determines environmental policy, not current science. Looking at the environmental movement, from the 1960s to present, the United States has seen radical swings in environmental policy, generally based on deregulation by conservative's administrations and regulation via science by liberal administrations. This periodic shifting between politics and science has caused instability in chemical and industrial regulation, and a shifting between protection of human health and the environment. However, in recent decades, industry seems to have taken a more steady programmatic approach than politicians have because industry leaders recognize that if they are compliant during conservative administration they will be held accountable for their actions when liberal administration return to power.

12.3 Major US Environmental Laws

As we saw in the previous section, there were many factors that led to legislative environmental action by the Federal government. Twelve major environmental laws were passed from the 1960s to 1990 and amended later. These are listed in Table 12.1 by year of passage and

Table 12.1 Summary and timeline of the major environmental laws.

1947	Federal Insecticide, Fungicide, and Rodenticide Act (FIFRA)
1955	Clean Air Act (CAA)
1965	Solid Waste Disposal Act
1969	National Environmental Policy Act (NEPA)
1970	Clean Air Act (CAA)
1970	Occupational Safety and Health Act (OSHA)
1972	Federal Insecticide, Fungicide, and Rodenticide Act (FIFRA)
1972	Federal Water Pollution Control Act (FWPCA)
1974	Safe Drinking Water Act (SDWA)
1976	Toxic Substances Control Act (TSCA)
1976	Resource Conservation and Recovery Act (RCRA)
1976	Solid Waste Disposal Act (SWDA)
1977	Clean Air Act Amendments
1977	Clean Water Act Amendments
1980	Comprehensive Environmental Response, Compensation, and Liability Act (Superfund)
1984	Hazardous and Solid Waste Amendments (HSWA)
1986	Safe Drinking Water Act Amendments
1986	Superfund Amendments and Reauthorization Act (SARA)
1987	Clean Water Act Amendments
1990	Oil Pollution Act
1990	Pollution Prevention Act
1990	Clean Air Act Amendments

include names that should be familiar to anyone with environmental interests. But how does an environmental law come about? The following gives a relative simply path for the formation of a law. Generally, a member of Congress proposes a bill, and then the bill is discussed, debated, and if necessary revised. If both houses of Congress approve the bill, then the president has the option to approve or veto it. If approved, the new law is called an act, and the text passed by Congress is known as a public statute. Once an act is passed and signed by the president, the House of Representatives standardizes the text which it publishes in the US Federal Register (https://www.epa.gov/laws-regulations, EPA 2018b).

Yet the laws themselves do not, and indeed cannot, contain all the details required for their implementation. Laws give a Federal agency the power to regulate, and then a specified agency must then develop regulations that set specific rules about what is legal and what is not legal. For example, the Clean Air Act (CAA) gave EPA the power to regulate sulfur dioxide, but EPA scientists had to determine what level (emission concentration) was safe. This latter process, the defining of what is safe and how to monitor it, is a monumental task, and in many cases, it takes years to actually enact a regulation after the law has been passed.

Many interested parties provide information to lawmakers during the early and later stages of a law and regulation. Industry certainly has vested economic interests in any environmental regulation and can contribute significantly to the formation or blockage of a law and the setting of standards of a regulation, and in recent years, even have helped some conservative administrations write the regulations. This industry involvement is often a source of controversy. The most famous recent case is the Energy Task Force formed by Vice-President Dick Cheney, for which the list of members was kept from the public. Some claim that this represents an attempt to hide industry influence on the task force, and this has resulted in a major legal fight for Freedom of Information seekers. Likewise, environmental organizations, who see themselves as politically defending the health of the public and the environment, attempt to influence the formation, structure, and passage of laws and regulations. Some of the most influential environmental organizations are shown in Table 12.2.

Table 12.2 Major US environmental organizations.

Organization	Founded
Sierra Club	1892
National Audubon Society	1905
National Parks and Conservation Association	1919
Izaak Walton League	1922
Wilderness Society	1935
National Wildlife Federation	1936
Environmental Defense Fund	1967
Friends of the Earth	1970
National Resources Defense Council	1970
Greenpeace (worldwide)	1971

Laws in the United States can essentially be formed by two avenues: either as statutes and regulations formed by Congress and the president, or by case law, under which industry, citizens, or groups file suit in order to question or attempt to clarify a law. In case law, a plaintiff who feels a private or civil wrong or injury has been committed may claim a "tort" has been committed and ask the court to provide a remedy in the form of an action for damages. The alternative to a tort is a "nuisance claim," for cases of alleged unreasonable or unlawful use by an industry or person of their property or injury to the rights of another person or group. Examples of judicial clarification include court cases from both sides (industry and environmental organizations). Industry can sue stating that a regulation by EPA is outside the intent of the law passed by Congress, and citizens or groups can sue EPA or an industry stating that they are not following established regulations. Both types of action, statute and case law, have been instrumental in the formation of our environmental regulations. Industry and environmental advocates often get involved by arguing that the Agency does not have the authority to regulate in the proposed manner, their science is flawed, or that the Agency is not acting in a timely manner. A noteworthy example of science flaws date back to Rachel Carson's claim that DDT causes cancer, but at that time *no* such data were available to validate the claim. She endured personal attacks on her credibility as a scientist until vindication by President's Kennedy's Commission a few years after her death (http://www.rachelcarson.org/SilentSpring.aspx, 2018). Many other examples occurred in the annals of environmental legislation.

In the remainder of this section, we will look at brief summaries of major environmental laws, with some interpretations as needed. Summaries for some of the relatively simpler laws are very short, while others deserve more discussion to attain a true appreciation of their intent. We will start in chronological order for the laws shown in Table 12.1.

12.3.1 The Federal Insecticide, Fungicide, and Rodenticide Act (FIFRA)

The passing of the FIFRA in 1947 was the first attempt to regulate the manufacture and use of pesticides across the nation. Although the enforcement efforts, and thus the act itself, were weak, the act was a starting point for a major movement. The act was amended in 1972; in 1988, to regulate all phases of pesticide sale, use, handling, and disposal; and in 1996, under the Food Quality Protection Act (FQPA), to establish tolerances for pesticide residues in food. The heart of FIFRA is the pesticide registration program. Before a pesticide can be manufactured, distributed, or imported, it must have approval from the EPA (the EPA assumed this responsibility upon its creation). In determining whether to register a pesticide, EPA considers its economic, social, and environmental costs and benefits.

FIFRA also defined the terms "pesticide" and "pest." A pesticide is any substance, or mixture of substances, intended for a pesticidal purpose (for preventing, destroying, repelling, or mitigating any pest or for use as a regulator, defoliant, or desiccant). A pest is defined as an insect, rodent, nematode, fungus, weed, terrestrial and aquatic plant, virus, bacteria, or any other living organism that the US EPA designates as a pest.

Some of the most significant aspects of FIFRA include

- Pesticides and their containers must be labeled with ingredients, instructions for use, EPA registration number, and all necessary warnings or restrictions.
- Pesticides are divided into two categories:

General-use pesticides may be applied by anyone; no permit is required.

Restricted-use pesticides may only be applied by (i) private applicators in an agricultural setting on their own land, (ii) commercial applicators who apply pesticides to other people's lands for a fee, and (iii) experimental-use applicators (manufacturers or researchers).

States were given the right to regulate the sale and use of pesticides within their borders, so long as the regulations are not less stringent that those specified in the FIFRA regulations.

The 1972 amendments stated that a balanced approach to considering the benefits and risk of a pesticide should be used in licensing. It also prohibited the use of cancer-causing agents, but defined an acceptable risk as less than a one-in-a-million chance of causing cancer. (This one-in-a-million concept was developed and discussed in Chapter 10.)

The 1988 amendments sped up the process of registering a pesticide and shifted the costs of cancelled pesticides from the EPA (and the taxpayer) to the manufacturer (costs such as removal from retail store shelves and disposal).

The 1988 amendment called for a nine-year toxicology testing period of the 700–900 active ingredients found in pesticides approved prior to 1984. (The same negligible risk of one cancer in a million was applied in these analyzes.)

The 1996 amendments required that there must be a reasonable certainty of no harm from aggregate exposure (exposure to more than one pesticide at one time). These amendments also required that existing tolerances (acceptable levels of exposure) be reassessed under new standards, and required EPA to make an explicit determination of acceptable residue levels in foods for infants and children.

FIFRA established penalties for misuse of a pesticide. These include (i) a $5000 per-offense fine for commercial applicators, wholesalers, dealers, retailers, or distributors who violate FIFRA or their permit (violators are also subject to a $25 000 fine and one-year imprisonment for knowingly violating a permit for FIFRA), (ii) producers of pesticides who knowingly violate FIFRA are subject to criminal penalties for up to $50 000 and one-year imprisonment, and (iii) private applicators are subject to a $1000 fine after a written warning. Due to the complexity of environmental laws and overlapping laws, violators of FIFRA can also be subject to penalties under the FQPA; the Federal Food, Drug, and Cosmetic Act; the Resource Conservation and Recovery Act (RCRA); and the Occupational Safety and Health (OSH) Act.

Virgil Freed, an Agricultural Extension scientist at Oregon State University was instrumental in advocating that such pesticides should be applied safely to minimize unintended consequences of pesticide application and use. Dr. Freed coined the term "chemodynamics" to deal with fate and transport of environmental chemicals. Dr. Freed had a great influence on Louis Thibodeaux, then at the University of Arkansas who extended these fate science precepts to the larger earth boundaries as presented in his 1976 and 1996 books on *Environmental Chemodynamics: Movement of chemicals in air, water, and soil* (Thibodeaux 1996). Hence, a new field was created.

12.3.2 The Air Quality Act, The Clean Air Act (CAA), and Amendments

The CAA is probably the most amended environmental act in US history. While we would like to think an act is amended because of increased knowledge or tightening of pollutant emissions, this has not necessarily been the case with the CAA. Although, in general, air quality has improved since the introduction of clean air legislation in 1955, some critics of the process feel we have regressed in our efforts by deregulating what we are having trouble controlling or what we find too expensive to control.

Two examples of these problematic pollutants are acid rain and carbon dioxide. But strides have been made in improving the air quality in most cities.

Given the high number of amendments to the CAA, and in order to give a more clear explanation of the Act, we will present only the general goals of the Act and amendments.

In the early years, the goal of the Act was to regulate air emissions from area, stationary, and mobile sources. This was attempted by setting National Ambient Air Quality Standards (NAAQS) that set (specified) the maximum amount of pollutant allowed in the air anywhere in the nation. The responsibility to accomplish or meet the NAAQS was given to the individual states, but if a state did not meet the NAAQS, the EPA would take control. Individual states developed State Implementation Plans (SIPs) applicable to their particular industrial pollutant sources. There are two basic types of sources defined by EPA: stationary sources such as power plants and factories, and mobile sources such as cars, trucks, and other motor vehicles. Geographical areas meeting primary and secondary standards are referred to as attainment areas, while those not meeting the standards in Table 12.3 are referred to as nonattainment areas.

The amendments in the 1970s, especially the 1977 amendment, set new goals (dates) for achieving the NAAQS, since many areas of the nation had failed to meet the earlier deadlines. The 1977 amendments and previous NAAQS focused on a set of common air pollutants referred to as Criteria Air Pollutants. These include ozone, volatile organic compounds (VOCs), nitrogen dioxide, carbon monoxide, particulate matter, sulfur dioxide, and lead. Health effects, major sources, and primary and secondary standards for these pollutants (established in 1977) are shown in Table 12.3. VOCs, nitrogen dioxide, and ozone are contributors to the formation of smog, the brown haze that exists over large cities such as Los Angeles and Atlanta. Nitrogen dioxide and sulfur dioxide are precursors to acid rain and are especially important over the Midwest,

Table 12.3 Criteria air pollutants, effects, sources, and primary and secondary standards.

Pollutant	Effects	Sources	Primary standard	Secondary standard
Ozone	Respiratory tract problems (difficulty breathing, reduced lung function, possible premature aging of lung tissue, and asthma), eye irritation, nasal congestion, reduced resistance to infection, and damage to trees and crops	Chemical reaction of pollutants, volatile organic compounds, and nitrogen oxides	0.12 ppm (160 μg/m^3) for 1 h	Same as primary standard
Particulate matter	Eye and throat irritation, bronchitis, lung damage, impaired visibility, and cancer	Burning of wood, diesel, and other fuels; industrial plants; agriculture; and unpaved roads	260 μg/m^3 for 24 h	150 μg/m^3 for 24 h
Carbon monoxide	Impairment of bloods ability to carry oxygen; damage to cardiovascular, nervous, and pulmonary systems	Burning of gasoline, natural gas, coal, and oil	9 ppm (10 mg/m^3 for 8 h); 35 ppm (40 mg/m^3 for 1 h)	Same as primary standard
Sulfur dioxide	Respiratory tract problems (diminished lung capacity and permanent damage to lung tissue); primary component of "killer fogs"; precursor to acid rain which damages trees, vegetation, and aquatic life	Burning of coal and oil, especially high-sulfur coal from the Eastern US; industrial processes	0.14 ppm (365 μg/m^3) for 24 h	None
Nitrogen dioxide	Respiratory illnesses and lung damage; damage to immune system; precursor to acid rain	Burning of gasoline, natural gas, coal, and oil	0.05 ppm (100 μg/m^3) arithmetic mean	Same as primary standard
Nonmethane hydrocarbons (NMHC or VOCs)	Cancer; precursor to acid rain	Incomplete combustion of hydrocarbons (gasoline); industrial processes	0.24 ppm (160 μg/m^3) for 3 h	Same as primary standard
Lead	Brain damage and retardation, especially in children	Leaded gasoline, paint, smelters, manufacture of lead storage batteries	1.5 μg/m^3 arithmetic mean	Same as primary standard

Source: Bryner (1995), Lave and Omnenn (1981), www.epa.gov/oar/oaqps/peg_caa/pegcaa02.html.

where the majority of the US coal-fired power plants are located. Much of the acid precipitation (rain, snow, and dry precipitation) then falls over the northeastern United States and Canada.

The 1990 amendments established a permit system so that each state issues a permit to polluters, stating which pollutants are being released, how much pollutant may be released, and what kinds of steps the source's operator is taking to reduce pollution, including plans to monitor (measure) the pollution. Operators must pay a fee for the permits and the money from the fee helps pay for state air pollution control activities. The 1990 amendments also gave new enforcement powers to the EPA, allowing the EPA to fine violators for exceeding the limits of their permits. As with most CAA amendments, these established new deadlines for the EPA and states to meet and reduce air pollution. "Market-based approaches" were also introduced in the 1990 amendments, enabling owners of permits to trade pollution allowances. For example, emissions of the precursors to acid rain from coal-fired power plants are openly traded on the market.

While ozone is considered a pollutant at ground level, its concentration in the stratosphere is essential for life on Earth, as it absorbs (blocks) the entry of harmful high-energy radiation (ultraviolet radiation). Certain pollutants have been shown to destroy this essential stratospheric ozone, and the 1990 CAA set a schedule for ending the production and use of these chemicals. This schedule is shown in Table 12.4. To give one example of ozone-destroying chemicals, chlorofluorocarbons (CFCs), common propellants in aerosol cans in the twentieth century, are not readily degraded and thus are able to migrate upward in the atmosphere. Once in the stratosphere, they destroy ozone through a series of complex reactions. The removal of ozone allows more harmful radiation to enter the lower atmosphere, potentially increasing the incidence of skin cancer and cataracts, and harming the basis of our food chain (photosynthesis-based organisms).

The Clear Skies Initiative was signed by President George W. Bush. This Initiative has received intense criticism from environmentalists and has even been dubbed the "Clean Air Sham" and "Clean Lies." The long-term effect of this Act is yet to be determined, but the Bush Administration claims that the Initiative will slash three major pollutants by 70% over the next 15 years. Environmentalists' predictions range widely and include among the worst cases a 50% increase in sulfur dioxide, a 190% increase in mercury, and a 36% increase in nitrogen dioxide.

Two major pollutants have yet to be addressed by the CAA amendments, not due to lack of science or health risk but due to a lack of leadership in Washington. These pollutants are mercury and carbon dioxide. The major source of mercury to the atmosphere is from the burning of coal across the world, which accounts for 52.7 million grams of mercury per year (Seignecr et al. 2004). Waste incineration produces another 32.2 million grams of mercury per year, while chlor-alkali facilities, in manufacturing chlorine bleach, used to generate another 6.8 million grams per year (Seignecr et al. 2004). Chlor-alkali facilities in the United States have all converted to a new membrane technology that does not use liquid mercury as an electrode. Technologies exist today that would eliminate the vast majority of these emissions, but implementation of such technologies worldwide are often fought by industry in the name of economics. Still, mercury was one of the major hazards facing us in recent times but actions by the Obama administration forced most coal-fired plants to install relatively inexpensive removal technologies for mercury from their effluent gasses. This administration action has since been reversed by the Trump administration, but most coal-fired plants have already installed the necessary pollution abatement equipment.

Most scientists find it amazing that carbon dioxide has not been regulated. We regulate nitrogen dioxide, a minor contaminant leaving the exhaust of our

Table 12.4 Ozone-destroying chemicals and proposed dates for ending production.

Chemical name	Use	When US production ends
CFCs (chlorofluorocarbons)	Solvents, aerosol sprays, and foaming agents in plastic manufacture	1 January 1996
Halons	Fire extinguishers	1 January 1994
Carbon tetrachloride	Solvents, chemical manufacture,	1 January 1996
Methyl chloroform (1,1,1-trichloroethene)	Widely used solvent: in the work place, in consumer solvents, used in auto repair and maintenance products	1 January 1996
HCFCs (hydro CFCs)	CFC substitutes, chemicals slightly different from CFCs	1 January 2003 or 2030

Source: www.epa.org.

automobiles, while the major component, carbon dioxide, is now considered a pollutant by the EPA, and has been upheld by the US Supreme Court. Carbon dioxide, of course, is the major culprit in global warming, a topic that is beyond the scope of this textbook.

12.3.3 The National Environmental Policy Act

The National Environmental Policy Act (NEPA) is the Magna Carta of environmental law. Yet there are few today, even among environmentalists, who know what the law contains or understand the potential it once had, and may still have, to shape the US federal policy toward the environment. The statute has been misunderstood ever since it was signed into law by President Nixon on 1 January 1970. Yet it is fundamental to environmental protection in the United State and exemplifies fair and equitable for all its citizens for generations that follow and might be a model for a sustainable future.

There are four main purposes to the Act (NEPA 1969): (i) "to declare a national policy which will encourage productive and enjoyable harmony between man and his environment," (ii) "to promote efforts which will prevent or eliminate damage to the environment and biosphere and stimulate the health and welfare of man," (iii) "to enrich the understanding of the ecological systems and natural resources important to the Nation," and (iv) "to establish a Council on Environmental Quality." Thus, NEPA seeks to provide guidance to federal agencies not through budgeting or strict regulations, but through an ethical code (Lindstrom and Smith 2001, p. 8). Historically, environmental laws have focused on specific environmental concerns; this is not the case with NEPA. *It is a broad, umbrella law that seeks to incorporate ecological thinking and systems theory into the political process itself*. The Act was an attempt by Congress to institute a law that would prevent future environmental destruction rather than perpetuate the crisis mentality so common in environmental legislation, which focuses on reclamation. NEPA, as envisioned by its authors, had the potential to be the most powerful and most important environmental law ever enacted in the United States. If implemented correctly, it is capable of improving the quality of both human and nonhuman life. Especially in the context of the political process, where progress typically is approached incrementally with constant compromise, the NEPA is a monumental piece of legislation. However, this also makes it difficult to incorporate the necessary science behind the act into effective policy.

Public misunderstanding of the law's intent was evident from its inception. The *New York Times* and *Washington Post* articles regarding NEPA upon its passage referred to it as Nixon's new anti-pollution law rather than what its authors intended it to be: a revolutionary, ecological approach to governmental interaction with the environment (Caldwell 2001). It has subsequently been degraded so thoroughly by judicial misinterpretation and executive neglect that it has degenerated into one of the least effective and certainly one of the least well known of all environmental laws. We will now look at some of the features of the Act in detail.

Section 101 of Title I outlines a powerful and all-encompassing but well-defined policy for federal interaction with the environment. There are six points in this section, declaring that it is the intention of the Federal Government "to improve and coordinate" Federal action and resources in order that the Nation (NEPA, SEC. 101 [42USC $ 4331, b]): (1) fulfill its responsibility as "trustee of the environment" for future generations, (2) assure for all Americans a healthy environment, including aesthetic and cultural components, (3) discover the maximum number of "beneficial uses of the environment" without degrading the environment, or causing unintended or undesirable outcomes, (4) preserve national heritage and maintain an environment that is diverse and offers a variety of "individual choice", (5) "achieve a balance between population and resource use that will permit high standards of living and a wide sharing of life's amenities", and (6) enhance renewable resources and provide the highest level of recycling for non-renewable resources.

The second major section, Section 102 of Title II, of the Act has come to be known as the action forcing mechanism. The first requirement set out in Section 102 is that all government agencies use a systematic and interdisciplinary methodology in decision making in order to insure that the environment and social factors are considered in agency proposals. It is further required that federal agencies develop a system to be used in decision-making, with the aid of the Council on Environmental Quality (established in Title II of the Act), which will guarantee that unquantifiable environmental factors are given consideration alongside technical and economic factors. The third requirement, (C), has five parts and has been manifested in the ubiquitous Environmental Impact Statement (EIS), the single greatest legacy of NEPA: "In every recommendation or report on proposals for legislation and other major Federal actions significantly affecting the quality of the human environment, a detailed statement by the responsible office on (1) the environmental impact of the proposed action, (2) any adverse environmental effect that cannot be avoided should the proposal be implemented, (3) alternatives to the proposed action, (4) the relationship between local short-term used of man's environment and the maintenance and enhancement of long-term productivity, and (5) any irreversible and irretrievable commitments of resources which would be involved in the proposed action should it be implemented" (NEPA, SEC. 102 [42USC $ 4332] (C)).

Furthermore, prior to making any detailed statement, the responsible Federal official shall consult with and obtain comments of any Federal agency that has jurisdiction by law or special expertise with respect to any environmental impact involved. Copies of such statements and the comments and views of the appropriate Federal, state, and local agencies, which are authorized to develop and enforce environmental standards, shall be made available to the president, the Council on Environmental Quality and to the public (NEPA, SEC. 102 [42USC $ 4332], (C)).

The EIS was meant to force agencies to consider, and take steps to minimize, their effect on the environment and be punished if they did not. The EIS is not only requisite for the ongoing activities of regular federal agencies but also applies to any major actions made by independent agencies that are licensed by the government.

Another important section of the Act is Title II, Section 202, which achieves the fourth purpose of NEPA by establishing the CEQ. The Council is to have three members who are appointed by the president with the advice and consent of the senate. One of these members serves as chairman. The major portion of the section outlines the rigorous guidelines for the selection of a chairman who shall be "exceptionally well qualified to analyze and interpret environmental trends and information of all kinds." It was the intention of the authors that the CEQ members be environmental experts, free of political ties and constraints; individuals who would not be prone to administration favoritism. The specific duties of the Council include (i) assisting and advising the president with the annual Environmental Quality Report (discussed below), (ii) gathering information on environmental trends and referring it to the president in a timely manner, (iii) reviewing the policies and programs of the Federal government with respect to NEPA, (iv) developing policies for the president that promote environmental quality, (v) researching and analyzing ecological systems and the environment, (vi) documenting changes in the environment and collecting data that aids in understanding the causes of such changes, (vii) reporting annually to the president on the condition of the environment, and (viii) creating and supplying to the president studies, reports, and recommendations for policy and legislation that the president may request.

Section 102, again of Title II, establishes that the president shall submit an annual Environmental Quality Report to Congress on the environmental condition of the United States. In this report, the president must not only relate the environmental health of all of the varying biomes and urban communities of the United States, but he must report current and foreseeable trends in environmental quality and use, and predict how these trends will affect the nation.

In total, the Supreme Court has ruled 12 times against the full enforcement of NEPA and never ruled for it (Lindstrom and Smith 2001, p. 120). The Supreme Court's downsizing of NEPA has led to a weakened enforcement of the Act by the lower courts, which had previously been NEPA's greatest chance for success.

Diminishment of the Act in the courts has led to a similar degradation of NEPA among government agencies. Even the creation of the EPA and subsequent environmental laws have weakened NEPA by granting exemptions from the EIS. These exemptions are frequently given to actions that are condoned by the EPA, an agency that is prone to administration preferencing. All of the presidents since NEPA's enactment have sought to abolish the CEQ or lower its budget. Even the EIS, the only remaining stronghold of NEPA, has been undermined by the lack of support for the Act among high-ranking government officials. Apparent compliance with the EIS often serves to improve the appearance of an agency's consideration of environmental effect while, in reality, affecting the decision-making process not at all. After working for the Department of Energy, one of the authors of this text (Dunnivant) can strongly vouch for the truth of this statement.

To date, NEPA has resulted in over 1000 lawsuits (Lindstrom and Smith 2001, p. 100). Due to the judicial downsizing of the Act, environmental organizations have turned their efforts toward projects that may be applicable to other, better-enforced statutes. Indeed, even the impact statements that are prepared today are frequently researched and compiled not by federal officials, but by special environmental research firms that are hired to conduct EIS and Environmental Assessment (EA) studies. While individuals within such organizations may be far better qualified to research the environmental effects of Federal projects, they are also legal mercenaries, inclined to create reports that will ensure the approval of their client's proposal. Even if the EIS determines that the proposal will do irreparable harm to the environment, and even if the EIS contains the comments of thousands of outraged citizens and agency officials, the action may still proceed. Having fulfilled the procedural requirements of the Act, the agency will have satisfied the level of NEPA compliance required by the courts. To restore NEPA to its original potential, there must be a complete reinterpretation and enforcement of the Act. NEPA in its simplicity is nonetheless radical, and despite the growing wave of environmentalism, a law like NEPA would never pass Congress today.

12.3.4 The Solid Waste Disposal Act, Resource Conservation and Recovery Act (RCRA), and Amendments

Disposal of solid wastes produced by human activities has posed problems since humans gave up the nomadic lifestyle. As humans settled in larger and larger groups in isolated areas, wastes reached sufficient concentrations such that they were no longer naturally removed and waste management plans had to be created. Continued development of sessile and ever larger communities exacerbated the problem. Congress began to address an impending crisis after World War II, as the US faced dealing with the solid and hazardous waste generated during the war effort and as cities continued to deal with growing solid waste burdens. As the problem grew during the 1950s and 1960s, it led to the passage of the Solid Waste Disposal Act of 1965. The Act has subsequently been amended several times and restated in the RCRA of 1976 and the Hazardous and Solid Waste of Amendments of 1984. Note that the original Act was passed prior to the creation of the EPA in 1970. However, amendments to this act have set up an Office of Solid Waste within the EPA, which is responsible for establishing a regulatory program for waste that includes methods for identifying solid and hazardous wastes, determining the degree of hazard, and establishing recycling programs. The amendments also created the "cradle to grave" concept of legal responsibility for the waste and required the Office of Solid Waste to dictate how a waste is tracked and monitored between the source and its ultimate destination, whether a recycling plant or disposal site.

We will address each of these aspects of the Act and its successors and amendments further, but first, how did the RCRA define solid and hazardous wastes (both of which it addresses)? "Solid waste means any garbage, refuse, sludge from a waste treatment plant, water treatment plant, or air pollution control facility and other discarded material, including solid, liquid, semi-solid, or contained gaseous material from industrial, commercial, mining, and agricultural operations" (Section 1004 of RCRA). It does not include discharges from permits issued under the Federal Water Pollution Control Act (FWPCA) or nuclear waste as defined by the Atomic Energy Act of 1954.

Hazardous waste, in turn, "means a solid waste or combination of solid wastes, which because of its quantity, concentration, or physical, chemical, or infectious characteristics may

(a) cause or significantly contribute to an increase in mortality or an increase in serious irreversible, or incapacitating reversible, illness; or
(b) pose a substantial present or potential hazard to human health or the environment when improperly treated, stored, transported, or disposed of, or other managed…" (Section 1004 of RCRA).

Although the RCRA thus relatively clearly defined solid and hazardous wastes, it did not specify the exact standards for their disposal. As we saw earlier in Section 12.1, these types of decisions are left up to the EPA to research and set appropriate standards.

The EPA established three ways that a solid waste can be classified as hazardous: (i) if it is specifically listed in EPA regulations as such (for example, waste halogenated solvents or pesticides), (ii) if it is tested by exact procedures and meets one of the four characteristics defined by EPA, including ignitability, corrosiveness, reactivity, or toxicity, and (iii) if it is declared to be hazardous by the generator (the manufacturer or source of the waste). Procedures were developed by the EPA for conducting each of the tests mentioned in (ii) above. Toxicity testing was the most complicated and involved subjecting the waste to two standardized procedures, the Extraction Procedure Toxicity Test (EP Toxicity Test) and the Toxicity Characteristic Leaching Procedure (TCLP).

A procedure for tracking wastes from source to disposal was also developed, hinging on a manifest system. A manifest form is shown in Figure 12.1. The manifest form is initiated at the source of the waste, and the form must be present with the waste as it travels to recycling, disposal, or destruction facilities. If an inspection is conducted by the EPA at the site of origin, during transport, temporary storage, or recycling, or at the disposal or destruction site, the manifest must be present or fines and imprisonment can result. The purpose of this manifest tracking system is to record who is responsible for the waste at each step of the process, as well as the volumes of waste generated and disposed, in order to provide a means of establishing liability if a future problem occurs. This obviously leads to the concept of "cradle to grave" liability, which states in simple terms that anyone dealing with the waste can be held accountable for future cleanup costs or lawsuits. Thus, it is in the best interest of the creator of the waste to hire reputable firms at each step of the handling, recycling, distribution, and disposal process.

RCRA also established detailed requirements for the construction and operation of solid and hazardous waste landfills, recycling facilities, and destruction facilities, including transportation and labeling of containers involved in transport. Specific design requirements of interest for landfills include (i) double liners in the bottom of the landfill, (ii) leachate collection systems, (iii) groundwater monitoring around the site, and (iv) leak detection systems. Incinerators were required to have

Please print or type *(Form designed for use on elite (12 - pitch) typewriter)* *Form Approved. OMB No. 2050 - 0039 Expires 9 - 30 - 91*

UNIFORM HAZARDOUS WASTE MANIFEST

1 Generator's US EPA ID No. | Manifest Document No. | 2. Page 1 of | Information in the shaded areas is not required by Federal law

GENERATOR

3. Generator's Name and Mailing Address

A. State Manifest Document Number

B. State Generator's ID

4. Generator's Phone ( )

5. Transporter 1 Company Name | 6. US EPA ID Number | C. State Transporter's ID | D. Transporter's Phone

7. Transporter 2 Company Name | 8. US EPA ID Number | E. State Transporter's ID | F. Transporter's Phone

9. Designated Facility Name and Site Address | 10. US EPA ID Number | G. State Facility's ID | H. Facility's Phone

11. US DOT Description *(Including Proper Shipping Name, Hazard Class, and ID Number)*	12. Containers No.	12. Containers Type	13. Total Quantity	14. Unit Wt/Vol	I. Waste No.
a.					
b.					
c.					
d.					

J. Additional Descriptions for Materials Listed Above

K. Handling Codes for Wastes Listed Above

15. Special Handling Instructions and Additional Information

16. **GENERATOR'S CERTIFICATION:** I hereby declare that the contents of this consignment are fully and accurately described above by proper shipping name and are classified, packed, marked, and labeled, and are in all respects in proper condition for transport by highway according to applicable international and national government regulations.
If I am a large quantity generator, I certify that I have a program in place to reduce the volume and toxicity of waste generated to the degree I have determined to be economically practicable and that I have selected the practicable method of treatment, storage, or disposal currently available to me which minimizes the present and future threat to human health and the environment; **OR**, if I am a small quantity generator, I have made a good faith effort to minimize my waste generation and select the best waste management method that is available to me and that I can afford.

Printed/Typed Name | Signature | *Month Day Year*

TRANSPORTER

17. Transporter 1 Acknowledgement of Receipt of Materials

Printed/Typed Name | Signature | *Month Day Year*

18. Transporter 2 Acknowledgement of Receipt of Materials

Printed/Typed Name | Signature | *Month Day Year*

FACILITY

19. Discrepancy Indication Space

20. Facility Owner or Operator: Certification of receipt of hazardous materials covered by this manifest except as noted in item 19.

Printed/Typed Name | Signature | *Month Day Year*

EPA Form 8700 - 22 (Rev. 9 - 88) Previous editions are obsolete.

Figure 12.1 The Hazardous Waste Manifest Form required by EPA to ship hazardous waste. Source: EPA.

removal efficiencies of 99.99% for principle organic hazardous constituents and 99.9999% for acutely hazardous wastes (such as dioxin). Landfills and incinerators were licensed for specific types of waste, and some specific pollutants were banned from these facilities (a common example being polychlorinated biphenyls). Later amendments banned the disposal of bulk liquid hazardous waste in landfills or impoundments.

One other important program set up under RCRA required the oversight and management of storage tanks, specifically underground storage tanks (USTs). One of the most common uses of USTs in the 1970s (when RCRA was passed), and today, is at gasoline filling stations. These tanks were originally made of metal that eventually rusted when buried in wet soil. Thus, numerous tanks did leak gasoline into the surrounding soil and groundwater. Cleanup and prevention of such leaks required a major effort by the EPA, but an end is in sight, and the program has been extremely successful in eliminating or minimizing releases from gasoline storage tanks. Operators were required to monitor for leaks through testing and through inventory assessments. When a leak was detected, the tank had to be dug up and replaced, and thousands of tanks have been replaced since the UST program began.

One very important distinction should be made. RCRA was responsible for controlling the generation, transport, and disposal or destruction of new and future solid and hazardous waste, and initially to clean up dangerous waste sites through the "cradle to grave" liability actions. Yet RCRA did very little to address the issue of waste already disposed of or released to the environment. Such waste, in abandoned or illegal waste sites, had to be addressed by a later act, the Comprehensive Environmental Response, Compensation, and Liability Act (CERCLA) of 1980, which will be discussed in Section 12.3.8.

12.3.5 Occupational Safety and Health Act (OSH Act)

While the OSH Act is not specifically considered an environmental legislative act, it is the Act most responsible for safety in the workplace, where humans are often at risk of being exposed to pollutants. Thus, we will include a brief description of the Act here. Prior to 1970, studies strongly indicated to Congress that excessive economic costs were being imposed on workers, in the form of lost productivity, wage loss, medical expenses and disability compensation payments, due to personal injuries and illnesses arising from poor work conditions. While all of the injuries were not due to chemical or pollutant exposure, many were, and these are the focus of our concern. The OSH Act created the National Institute for Occupational Safety and Health (NIOSH) as the research institution for the Occupational Safety and Health Administration (OSHA), with the purpose of researching and setting standards for workplace health and safety. These standards include chemical exposure limits and restrictions on conditions that can harm workers. For example, time-weighted exposure limits now exist for many chemicals, specifying the acceptable level (concentration) of each chemical that a worker can be exposed to over a specified period of time (usually based on an eight-hour work day). It is widely recognized that the creation and enforcement of these standards have been instrumental in reducing illness and cancer in industrial workers. Right-to-Know provisions after the Bhopal incident aftermath have strengthened this Act in the United States.

12.3.6 The Federal Water Pollution Control Act, The Clean Water Restoration Act, The Safe Drinking Water Act, and Amendments

Today, we are concerned with both the availability (or scarcity) of water and the quality of the water. These concerns are best summarized by Charles C. Johnson, Former Assistant Surgeon General of the United States who said (https://archive.epa.gov/epa/aboutepa/epa-history-water-challenge-environment-primer-epas-statutory-authority.html, 2018):

> Our water resources, more perhaps than any other, illustrate the interaction of all parts of the environment and particularly, the recycling process that characterizes every resource of the ecosystem…Everything that man himself injects into the biosphere—chemical, biological or physical—can ultimately find its way into the Earth's water. And these contaminants must be removed, by nature or by man, before that water is again potable.

This statement also encompasses the need for maintaining the quality of our dwindling water resources. But this is not a recent development. Water treatment to improve the taste and odor were recorded as early as 4000 BCE. Ancient Sanskrit and Greek writings indicated the use of water treatment methods such as filtering dirty water through charcoal, exposing the water to sunlight, boiling by placing a hot metal rod in the water, and straining. Chemical treatment, such as adding alum to remove suspended particles, was used as by Egyptians as early as 1500 BCE (EPA 2000) Indications of poor quality such as visible cloudiness (turbidity), taste, and odor were the driving forces behind early water treatment, since bacteria, the source of disease, had not been discovered.

Large-scale filtration was introduced in the 1700s as an effective way to remove particles from water, with sand filtration being regularly used in Europe in the 1800s. The epidemiologist Dr. John Snow proved in 1855 that cholera was a waterborne disease by linking an outbreak in London to a single public well contaminated by sewage (EPA 2000). In the 1880s, Louis Pasteur developed the "germ theory." During the late 1800s and early 1900s, water treatment focused more on disease-causing microbes or pathogens in public water supplies. Research at the time showed that turbidity was not only an aesthetic problem but that particles also harbored pathogens. In 1908, chlorine was first used as a primary disinfectant in the Jersey City (New Jersey) drinking water plant.

The government first stepped in 1914, when the US Public Health Service set standards for the bacteriological quality of drinking water. These standards were subsequently revised in 1925, 1946, and 1962. In 1962, 28 substances were regulated in drinking water. The first major legislation was the River and Harbor Act of 1886, which was basically an act to protect waterways from substances or material that would interfere with shipping. The FWPCA of 1948 was enacted to "enhance the quality and value of our water resources and to establish a national policy for the prevention, control, and abatement of water pollution" (https://www.epa.gov/laws-regulations, EPA 2018bb). The Water Pollution Control Act of 1956 strengthened enforcement provisions of the 1948 Act, and the Federal role in regulating states was further expanded in 1965 with the enactment of interstate water quality standards. The Clean Water Restoration Act of 1966 imposed fines on a polluter who failed to submit a required report. Additional acts and amendments on water quality include the FWPCA Amendments of 1972, the Safe Drinking Water Act of 1974, Clean Water Act Amendments of 1977, and the Safe Drinking Water Act Amendments of 1986 and 1996. As you can see, there has been considerable legislative activity with respect to water, which emphases the importance of water to human life and stable civilization.

So what have these Acts actually done to improve water quality? They defined sources of pollutants in two ways. Point sources are broadly defined to include any "discernible, confined, and discrete conveyance" including any pipe, ditch, channel, tunnel, conduit, well, or container from which pollutants are or may be discharged. Point sources are usually subdivided into industrial and municipal sources. Nonpoint sources include a range of inputs to surface waters, including runoff from agricultural fields, feedlots, paved streets and parking areas, mining sites, forestry operations, and atmospheric deposition. Pollutants or contaminants of concern specified in these Acts include (i) conventional pollutants such as human wastes, food from sink disposals, laundry and bath waters, and waters containing fecal coliforms, and oil and grease, (ii) toxic pollutants such as organics (pesticides, solvents, polychlorinated biphenyls, and dioxins) and metals, and (iii) nonconventional pollutants such as nutrients containing nitrogen and phosphorus. The Acts also established a list of water quality criteria that originally included 115 "priority pollutants" and set maximum concentration levels (MCLs) for particular pollutants or contaminants in drinking water. Thus, these Acts gave EPA the power to (i) identify pollutants or contaminants and (ii) determine the maximum contaminant level.

One of the most important elements of these Acts established the National Pollutant Discharge Elimination System (NPDES), which stated that all individual, government, and industrial point sources must have a permit to discharge waste into any waterway. The permit states what pollutant can be emitted and at what level (mass loading) the pollutant can be released. Violations of the permit can result in fines and imprisonment.

Government funding to build sewage and water treatment plants for municipalities was provided under the FWPCA of 1948, the Clean Water Act Amendments of 1956, and additional funding in 1961 and 1971. The construction and operation of these plants were instrumental in cleaning up sewage and associated bacteria from our waterways.

A chronological account of the evolution of recent water quality legislation follows:

(1) The original Safe Drinking Water Act of 1974
    (a) federally mandated the regulation of contaminants that pose a health risk to the public based on frequency of the chemical and contaminant nature,
    (b) established a maximum contaminant level goal (MCLG), which is the limit below which there is no known or expected threat to public health,
    (c) designated that EPA would specify a maximum contaminant level (MCL) that is as close to the MCLG as deemed feasible through technology and cost effectiveness, and
    (d) provided that, in the event that no MCL can be determined, the EPA set a required treatment technique that specifies a way to treat the water to remove contaminants.
(2) In 1979, the presence and concentrations of 6 synthetic organic chemicals, 10 inorganic chemicals, turbidity, total coliform bacteria, radium-226, radium-228, and total trihalomethanes were added to the control of EPA. Also added was the provision that in the event of a violation, the established system (State or Federal government) must notify the public.

(3) The Amendments of 1986
 (a) established regulations for 83 contaminants,
 (b) required disinfection of all public water supplies,
 (c) specified filtration requirements for nearly all water systems that draw water from surface sources,
 (d) developed additional requirements to protect groundwater supplies (Wellhead Protection programs and Sole Source Aquifer Program),
 (e) established monitoring requirements on nonregulated contaminants every five years so the EPA could decide if these should be regulated in the future,
 (f) implemented a new ban on lead-based solder, pipe, and flux materials in water distribution systems, and
 (g) specified the "best available technology" for the major drinking water contaminant groups (pathogens, organic and inorganic contaminants, and disinfectant by-products).

(4) The Amendments of 1996
 (a) required consumer confidence reports,
 (b) required source water assessments,
 (c) specified state capacity development strategies,
 (d) gave water plant certification revisions,
 (e) required public notification improvements,
 (f) provided new publicly accessible drinking water contaminant databases,
 (g) required annual compliance reporting,
 (h) required health-care provider outreach and education, and
 (i) developed the Drinking Water State Revolving Fund (DWSRF) for treatment facility upgrades.

To summarize, the Clean Water Acts (i) gave EPA the authority to implement pollution control programs such as the setting of water quality standards, (ii) made it unlawful for any person to discharge any pollutant from a point source into navigable waters without a permit, (iii) funded the construction of water and sewage treatment plants, and (iv) recognized the need for planning to address the critical problems posed by nonpoint source pollution.

12.3.7 The Toxic Substances Control Act

The Toxic Substances Control Act was signed by President Carter in 1977 and gave EPA the authority to track and regulate more than 75 000 industrial chemicals produced or imported into the United States. TSCA authorized the EPA to obtain information on all existing and newly created chemical substances, including chemicals that were determined to cause a risk to public health or the environment. TSCA was considerably different from earlier laws; previously, the EPA only had the authority to intervene after damage became evident. TSCA gave the EPA the authority to regulate before a chemical was commercially manufactured.

In controlling and testing chemicals, TSCA gave the EPA certain powers. These include requiring a "pre-manufacture notice," so that chemical manufactures had to apply to EPA to commercially produce a substance. Prior to the manufacture of a chemical, the producer or importer had to conduct testing as defined by the EPA to ensure the safety of the chemical. Based on the results of the testing, the EPA had the right to ban the chemical from being manufactured or imported if it imposed an unreasonable risk to human health or the environment. The EPA was also given authority to act in cases of imminent health hazards, and to seize a chemical substance, mixture, or article containing a specified chemical. TSCA also required the EPA to create the Toxic Substances Release Inventory (TSRL), which contains information on the safety and toxicity of specific chemicals and how much of each compound can safely be released to the environment.

TSCA was divided into four titles:

Title 1 – Control of Toxic Substances – included provisions for chemical substances and mixtures; manufacturing and processing notices; regulating hazardous chemicals, substances and mixtures; managing imminent hazards; and reporting and retaining information.

Title II – Asbestos Hazard Emergency Response (enacted by Congress in 1986) – authorized EPA to amend its TSCA regulations to impose more requirements on asbestos abatement in schools. It also required that EPA determine the extent of the danger to human health posed by asbestos in public and commercial buildings.

Title III – Indoor Radon Abatement – added in 1988 to respond to the human health threat posed by exposure to radon. It required EPA to publish an updated citizen's guide to radon health risk and to perform studies on the radon levels in schools and federal buildings.

Title IV – Lead Exposure Reduction – added in 1992 in an attempt to reduce environmental lead contamination and prevent adverse health effects as a result of lead exposure, particularly to children, from lead-based paint hazards and lead contamination in paint and toys.

But TSCA had not been updated since its passage in 1977, yet the chemical industry has exponentially grown. The decades-long debate to update TSCA was finally realized in the Frank R. Lautenbery Chemical Safety for the twenty-first century Act. The main components of this Act were

- mandatory duty was placed on the EPA to evaluate existing chemical with clear and enforceable deadlines,
- that chemicals be assessed against a risk-based safety standard,
- that any unreasonable risks identified in the risk evaluation must be eliminated, and
- gave EPA expanded authority to more quickly require development of chemical information when needed. Similar laws that surpass the updates to TSCA have been successfully implemented in Europe. These will be discussed in the next chapter.

12.3.8 The Comprehensive Environmental Response, Compensation, and Liability Act

The RCRA and its amendments established a much-needed and extensive program to manage newly created hazardous waste, and also set out rules for its proper disposal. However, RCRA did little to nothing for the hundreds to thousands of abandoned waste sites from previous poor and irresponsible disposal practices. Examples of these include Love Canal, Times Beach, and the Valley of the Drums, as well as the numerous contaminated sites resulting from the improper disposal of hazardous wastes generated during the Cold War by the US government. The need to clean up these sites led Congress to pass the CERCLA of 1980, which was amended in 1986 by the Superfund Amendments and Reauthorization Act (SARA). Collectively, the resulting legislation is referred to as "Superfund." Again, the primary purpose of Superfund was to clean up unsafe and in many cases abandoned hazardous waste sites.

We will start by looking at the differences between CERCLA and SARA. CERCLA established a \$1.6 billion fund, derived mostly from feedstock taxes on the chemical industry, to clean up the waste sites over a five-year period. We now see, over 20 years later, that this was a naive and optimistic effort, since there are hundreds of sites yet to be cleaned up. CERCLA also established prohibitions and requirements concerning closed and abandoned hazardous waste sites, provided for liability of persons responsible for the releases of hazardous waste at these sites, and established a trust fund to provide for cleanup when no responsible party could be identified. CERCLA also authorized two types of responses: short-term removals for sites requiring prompt response and long-term remedial actions to reduce dangers that are serious but are not immediately life-threatening.

SARA made several important changes and additions to the program. SARA (i) stressed the importance of permanent remedies, (ii) required Superfund actions to consider the standards and requirements found in other state and federal environmental laws and regulations, (iii) provided new enforcement authorities and settlement tools, (iv) increased state involvement in Superfund, (v) increased the focus on human health, (vi) revised the Hazardous Ranking System (HRS) to ensure that it accurately assessed the relative degree of risk to human health and to the environment, (vii) encouraged citizen participation in deciding how best to deal with a site, (viii) created the Emergency Planning and Community Right to Know Act (EPCRA) to give the public access to information concerning a site, and (ix) increased the size of the cleanup fund from \$1.5 to \$8.5 billion.

So how does the Superfund process work? This is outlined in Figure 12.2. First, a site is identified in any of a number of ways and evaluated using criteria specified in the Hazard Ranking System (HRS), including proximity to a large population, nature of the contaminants, and potential exposure pathways. The result is an HRS score related to the relative risk, and if the score exceeds a threshold value, the site is placed on the National Priorities List (NPL). The site then undergoes a Remedial Investigation (RI) that describes the history of the contamination, the current concentration of pollutant in the air, water, soil, and any waste, and a detailed risk assessment. Next a Feasibility Study (FS) is conducted to evaluate remediation (cleanup) options for the site. The FS contains a variety of remediation options, from no action (in which the site is eliminated from public access but not actually remediated) to extensive remediation. A risk assessment is conducted for each remediation approach. Nine criteria are used to evaluate each remediation option: (i) short-term effectiveness, (ii) long-term effectiveness, (iii) implementability, (iv) reduction in toxicity, mobility, and volume, (v) cost-effectiveness, (vi) compliance with established standards, (vii) human health protection, (viii) state concurrence, and (ix) local acceptance. In the end, a Record of Decision (ROD) is agreed upon and issued by EPA, specifying the selected remedy and time scale for the remediation.

One major aspect of Superfund is that it created a method of determining liability, and thus who should pay for the remediation efforts. The Superfund process establishes Potentially Responsible Parties (PRPs) who can be held accountable for these costs, which follows the "cradle-to-grave" concept created under RCRA. Any party who has generated the waste, owned a facility associated with the waste, stored the waste, transported the waste, or disposed of the waste can be held accountable for remediation costs. It is well known that government projects are usually most costly than a private approach to the same problem, so EPA will allow a PRP to conduct the remediation effort, but under the strict oversight of EPA or the state where the site is located.

At the time of this research, Superfund has completed work at 40 sites across the country, for a total of 886

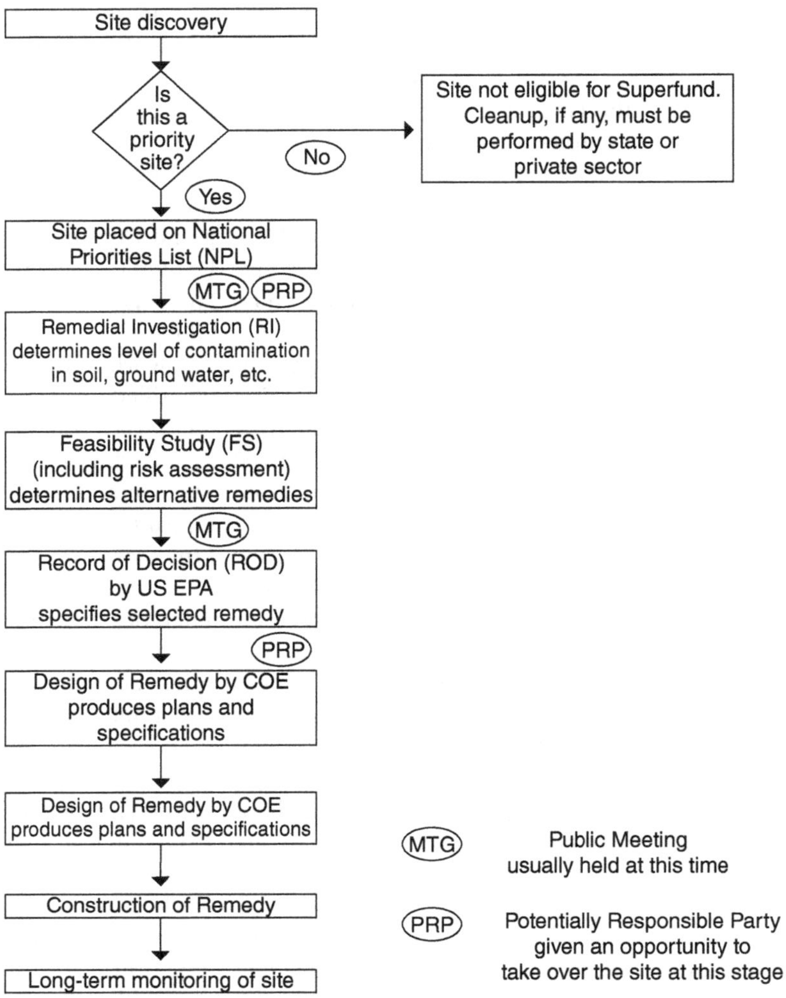

Figure 12.2 An outline of the Superfund process. Source: EPA.

cleanups or 59% of the sites listed on the NPL. It has conducted 699 ongoing cleanups at 436 projects, but the NPL increases in number by about ten sites per year. Controls are now in place at 82% of the NPL sites to prevent unacceptable human exposures.

An informative map and search engine of SuperFund sites near you is available at https://www.epa.gov/superfund/search-superfund-sites-where-you-live (2018).

12.3.9 The Oil Pollution Act

On March 24, 1989, the *Exxon Valdez* struck the Blight Reef in Prince William Sound in Alaska, releasing 11 million gallons of crude oil. This accident sparked public and legislative action to attempt to prevent future accidents and establish strict liability for such accidents. The Oil Prevention Act of 1990 was Congress's response. This Act (i) strengthened the EPA's ability to prevent and respond to future catastrophic oil spills, (ii) established a trust fund, financed by a tax on oil, to clean up future spills when the responsible party is incapable or unwilling to do so, (iii) required oil storage facilities and shipping vessels to submit to the Federal government plans detailing how they will respond to large oil discharges, and (iv) required the development of Area Contingency Plans to prepare and plan for oil spill response on a regional scale. Other major provisions of the Act include the following:

- the establishment of legally responsible parties for the vessel or facility from which oil is discharged and determination of liability for specified damages resulting from the discharged oil and removal costs,
- establishment of new liability determination procedures by
  (a) increasing the liability for tank vessels larger than 3000 tons to $1200 per gross ton or $10 million, whichever is greater,
  (b) making responsible parties at onshore facilities and deep water ports liable for up to $350 million per spill, and

(c) making holders of leases or permits for offshore facilities liable for up to $75 million per spill, plus removal costs.

The Oil Pollution Act of 1990 seemed to work well until the Deepwater Horizon disaster (also known as the BP disaster) on 20 August 2010, where an estimated 4.9 million barrels (210 million US gal; 780 000 m^3) were released in the Gulf of Mexico. One of the most massive cleanup operations ensued, and while success was declared, no effort could completely remediate such a large oil spill. Numerous law suits followed and as of February 2013, the cost to BP was $42.2 billion. After a series of court actions, an additional agreement due to gross negligence and reckless conduct of $18.7 billion was paid by BP, the largest corporate settlement in the US history. The difference in legal action and in monetary settlements between the Exxon oil spill and BP should be noted.

12.3.10 The Pollution Prevention Act

Much as the NEPA was designed to change the way we think in approaching projects with environmental impacts, the Pollution Prevention Act (PPA) of 1990 was designed to promote a change in attitude among polluters and reduce pollutant volumes through voluntary source reduction and recycling. The PPA focused on industry and government in laying out a plan to reduce the amount of pollution through cost-effective changes in production, operation, and raw material use. The act was designed to (i) reduce the amount of hazardous substance, pollutant, or contaminant entering any waste stream or released into the environment prior to recycling, treatment, and disposal and (ii) reduce the hazards to public health and the environment related to the release of these chemicals.

Note the focus on reduction of pollution prior to treatment or disposal. This distinct focus is in contrast to the more common focus on compliance and proper treatment and disposal, which often failed to reduce the amount of pollution generated. The PPA focused on source reduction, increasing the efficiency of use of energy, water, or other natural resources, and protecting our resource base through conservation. To achieve these goals, the PPA fosters pollution prevention through the collection and dissemination of data and technical assistance. These reductions are tracked by requiring the owner or operator of a facility that uses chemicals or produces pollution to file a toxic source reduction and recycling report. The regulations do not impose additional regulator obligations on the chemical industry other than increased reporting of their efforts to reduce the creation of pollution. Thus, the act is essentially designed to promote attitude change. These activities are noted as P2 in some industrial and regulatory groups, and its content are the focus of some engineering programs in chemical engineering as an elective. Cheremisinoff (2001) published a handbook of pollution practices that elaborates on what industry is doing in this arena and includes multiple chemicals and processes. EPA's website includes others (EPA 2018).

12.3.11 The Endangered Species Act of 1966 and Amendments

The Endangered Species Act (ESA) and its Amendments provide a program for the conservation of threatened and endangered plants and animals and the habitats where they live. Species of interest include birds, insects, fish, reptiles, mammals, crustaceans, flowers, grasses, and trees. The US Fish and Wildlife Service and the Department of the Interior maintain the Threatened and Endangered Species System (TESS) that includes 632 endangered species (326 plants) and 190 threatened species (78 plants). These include well-known species such as birds of prey, the bald eagle, and the grizzly bear.

While this Act is not directly involved with EPA, the EPA has historically been required to consult with the Fish and Wildlife Service and the Department of Interior when reviewing a pesticide or a chemical, in order to evaluate its adverse exposure effects on endangered species as well as the environmental fate of the chemical. Under FIFRA, the EPA could issue emergency suspension of certain pesticides to cancel or restrict the chemical's use if an endangered species will be adversely affected. This all changed in August 2004, when the Bush Administration developed new rules for EPA to approve pesticides without consulting the federal wildlife agencies about the potential harm to endangered species (Hileman 2004). Thus, while the ESA is an important environmental law, it no longer affects chemical regulations. This diminishes the Act's holistic, habitat-based approach to management of endangered species.

12.3.12 Marine Protection, Research, and Sanctuaries Act (MPRSA) of 1972

This Act is commonly known as the Ocean Dumping Act. The purpose of this law is to regulate the dumping of all types of materials into ocean water and to prevent or limit the dumping of any material that would adversely affect human health or welfare, or the marine environment. The Act specifically prohibits (i) transportation of material from the United States for the purpose of ocean dumping, (ii) transportation of material from anywhere for the purpose of ocean dumping by the US agencies of US flagged vessels, (iii) the dumping of material transported from outside the United States into the US territorial zone (12 nautical miles). Penalties from breaking laws established under the Act include (i) a civil penalty of up to $50 000 against persons in violation of MPRSA or a permit, (ii) a fine of up to $125 000 for the illegal dumping of medical waste, (iii) criminal penalties including a fine and up to five years in jail for knowingly violating MPRSA, its regulation, or a MPRSA permit. In addition, each day of violation is considered as a separate offense. The MPRSA is enforced by the EPA working with the US Coast Guard to monitor and conduct surveillance operations on shipping vessels.

12.4 EPA's Record

So how has the EPA done in controlling or reducing environmental pollution since its creation in 1970? There are mixed views on the answer to this question, but most of the comments are positive. Environmental groups such as the Sierra Club contend that EPA has not acted soon enough and industry still questions EPA's authority or its science basis. Although a bit dated now, two feature articles that evaluate the EPA's performance are "Reform or Reaction: EPA at a Crossroads," in *Environmental Science & Technology* (Andrews 1995), and a series of short articles in *Chemical and Engineering News* (Ember 1995a, b; Kirschner 1995; Hileman 1995; Rawls 1995; Lepkowski 1995).

There are basically three important groups that judge EPA: the regulated industries, the public affected by the pollution, and the scientists that monitor and study the effects of the pollution. Almost every member of these groups notes the following undisputable accomplishments by EPA (Hileman 1995):

- Sewage, oil, grease, and visible industrial pollutants have largely disappeared from rivers, streams, and lakes in the United States.
- The number of fishable and swimmable rivers has doubled since the passage and initial enforcement of the Clean Water Act. But, thanks to effective biological treatment, these segments that are listed as impaired for decreased dissolved oxygen have virtually disappeared in the last 10 years, and fishable and drinkable river waters in sight across the nation. EPA is required to submit an inventory to Congress every two years that simply counts the number of river segments that fail to meet applicable water quality standards. Unlike reports from the 1980s and 1990s, the top causes of water quality impairment today are pathogens, sediment contamination, and nutrients. The assessment also reveals probable contamination sources such as agricultural activities, atmospheric deposition, and hydrologic modification. Earlier summaries in prior decades did list organic wastes as probable sources. Students are also encouraged to read a report authored by the EPA Alumni Association provides relevant summaries of EPA success stories across all media and respective legislation in the past 50 years (http://www.epaalumni.org/hcp/Overview.pdf, 2018).
- Urban air quality in most cities has dramatically improved, but more work needs to be done in this area.
- Our efforts toward controlling ozone-depleting chemicals are unprecedented.
- The enforcement of the Toxic Release Inventory and the Community-Right-To-Know provision of the Superfund Amendment and Reauthorization Act of 1986 have greatly improved the safety and knowledge, and decreased the toxin exposure, of the general public.
- Coastal cities no longer simply dump garbage and sewage into the oceans.
- Domestic and hazardous waste landfills are now lined to inhibit leaching into surrounding soil and groundwater.
- Raptors, or birds of prey, have made a dramatic comeback since the banning of DDT and other chlorinated pesticides.

These are major accomplishments by any measure, and most of the world is still catching up to our environmental living standards, while a few countries are admittedly slightly ahead (most of Europe). To paraphrase Boris Yeltsin, if Russia had the environmental problems of the United States, it would have no problems. The United States is currently attempting to address the environmental disasters that resulted from our side of the Cold War, but Russia has not even begun to address this problem. Most of our Cold War pollution was in rural, very isolated portions of the western United States. Russia lived (and still lives) in the middle of their Cold War pollution.

Many environmental leaders and organizations have also made negative statements about the EPA. EPA has done a relatively good job at addressing visible, higher pollutants, but done a considerably poorer job at addressing "invisible toxics" that are known hazardous, persistent in the environment, found at very low concentrations (often at the part per trillion level), and bioaccumulate (with the obvious exceptions of PCBs and DDT). A new and highly aggressive area of research is the ubiquitous presence of many endocrine-disrupting compounds, including human-made estrogens, plasticizers, and pharmaceuticals that have been found in many, if not all, of our waterways. Another major criticism is that EPA has done well at regulating large Fortune 500-type industries, while largely ignoring the hundreds of thousands of small local businesses and municipalities that contribute significantly to stream and air pollution. It is difficult for EPA to prosecute small public operations, as they are perceived at enforcing CAA and NPDES regulations on entities that do not have the resources to challenge these violations in the courts with highly paid consultants to present their case and necessary science.

Many also criticize the handling of Superfund sites and the slow and costly cleanup of these sites. Much of the problem with cleaning up these sites is, of course, associated with assigning legal blame for who is responsible for the abandoned polluted site. However, considerable legal maneuvering is also involved in determining *how* the site should be cleaned and *to what level* the site should be cleaned. The Superfund problem is an issue that will take decades to settle. One last concern is the shift in environmental policy from equal "protection of the environment and human health" to a human risk-based approach to regulation and cleanup that has the potential to ignore or lessen the emphasis on protection of the environment where risk cannot be quantified. Still, we must give EPA credit; the large issues have been addressed, and the public and the EPA are moving on to the smaller, but still significant, issues facing us today.

So what does industry think of the EPA and its regulations? Most industry leaders truly want a strong, consistent, and effective regulatory agency such as the EPA (Baum 1995), not only on the national level but also on the global level. A consistent set of rules and controls on pollution, across the world, would provide a level playing field on which to base industry. The export of jobs to developing countries is not only due to the low wages that employers can pay workers in these countries but also due to lax or nonexistent environmental regulations. Consistent regulations also help ensure that if an industry uses a chemical today and disposes of it legally, the industry will not be responsible for the chemical in the future. It is important to note that various industrial certification programs (such as ISO 9000) require this demonstrated support and compliance with applicable regulatory efforts, but do not prevent the party from protesting and suing EPA when its legal rights have been infringed upon (ISO 9000, available at numerous sources on the Internet).

The pollution that occurs in the United States today is mostly due to regulated sources, unlike the "midnight dumping" that occurred prior to the strong regulation by the EPA and other Federal agencies. The 12 major environmental laws discussed in the previous section are largely responsible for this. But we still have violations. A recent study by the EPA and the Chemical Manufacturers Association (CMA) found that *noncriminal violations* largely occur due to a failure to understand the regulations or human error, and not intentional releases of chemicals. The most common violations include (McCoy 1999): (i) facilities being unaware of the applicability of specific regulations, (ii) human error in judgment or responsibility, (iii) failure to follow procedures, (iv) faulty equipment design or installation, (v) problems with compliance by contractors, and (vi) various communication difficulties. But, as the old saying goes, "ignorance of the law (or not operating your equipment [factory] correctly) is no excuse, and the violators are usually fined or closed down."

Still, clearly, pollution must exist in a regulated environment. Not all pollution can be eliminated, given our population density and the relatively limited carrying capacity of the land we live on and the air we breathe. In today's world, if you breathe air, drink water, walk on soil, or live in a modern dwelling, you will incorporate some form of pollution into your body. It's simply a fact of modern life; our goal in enacting legislation is to limit this intake to an acceptable, below threshold level.

Before we move on to the major US international environmental laws, one point should be made. Recent presidential candidates have expressed the need for a cabinet-level environmental post but have conveniently forgot this interest when elected to office. One would think that the quality of the air we breathe, the water we drink, and the soil we live on would rank as important as the economy and defense. But, sadly, no direct voice to the president has taken shape. The motivation for change often comes from disasters; this is true at local intersections in need of traffic lights or stop signs, and it is also true at the national level, where change is slow even as thousands of Americans die each year from pollution-related illnesses (asthma, cancer, etc.). Unfortunately, additional and probably significant environmental degradation will have to occur before our elected officials will take the dramatic actions ultimately necessary to truly protect the environment in the long-term.

Recently, there have been numerous positive interpretations of our environmental progress since the 1960s and 1970s. These success stories are a must read for conservative and liberals alike and be can found in

- Richard Nixon and the Rise of American Environmentalism, by Meir Rinde, *Distillations Magazine*, Spring 2017, pp. 16–29.
- EPA Alumni Association, http://www.epaalumni.org/hcp/.

- *Environmental Success Stories: Solving Major Ecological Problems & Confronting Climate Change*, by Frank M. Dunnivant, Columbia University Press, 2017.
- EPA Alumni Association, Half Century of Progress http://www.epaalumni.org/hcp/.

12.5 Environmental Permitting and Compliance

Public and industrial facilities are required to have basic operating permits from state and/or EPA Regional Offices, which is the meat of complying with the various environmental acts. For example, these include a CAA permit if the facility has gaseous emissions from a power boiler or process vent that can be from a combustion source or a known hazardous compound from a process. If the facility operates a landfill to dispose of biological sludges from wastewater treatment units or otherwise waste materials, it would usually require a permit to govern its construction and ongoing operation – necessary liners and monitoring wells. The Clean Water Act requires a NPDES permit if the wastewater is discharged into waters of the United States, even if the waste is treated in a municipal treatment plant (POTW): pretreatment standards and conditions may apply. Carbonaceous and nitrogenous BOD, temperature, and specific compound restrictions may apply.

Wastewater toxicity is addressed by bioassays of sensitive organisms like invertebrates on a site-specific basis. These are useful when a toxicant cannot be identified specifically as the cause for the toxicity.

The main point is there are numerous permitting systems and compliance reports that are used to enforce all of the US environmental laws.

12.6 International Agreements/Treaties Involving the United States

Two acts that we have already discussed have international implications: the ESA, which protects migratory species such as birds and ocean species, and the Marine Protection, Research, and Sanctuaries Act, which regulates ocean dumping (although focusing on the US waters). Many other explicitly international environmental agreements have been made during the past century.

12.6.1 US–Canada Environmental Agreements

One of the first international agreements that the United States made was with Canada. The 1909 Boundary Water Treaty that states the rights and restrictions of each nation in using boundary waters (LeMarquand 1993, p. 67). It declares that each country's government maintains the exclusive jurisdiction and control over the use and diversion of all waters on its side of the boundary line. The treaty also addresses the issue of pollution, citing that "boundary waters and waters flowing across the boundary shall not be polluted on either to the injury of health or property on the other" (LeMarquand 1993, p. 67). Most significantly, the Boundary Waters Treaty created the International Joint Commission (IJC), which is composed of half Canadian and half US advisory boards that prevent and resolves disputes between the United States and Canada under the 1909 Boundary Waters Treaty.

In the 1960s and 1970s, as the Great Lakes shoreline became the site of more and more industrial and residential development, the IJC found itself unable to handle the large volume of complaints related to pollution and contamination of the Great Lakes, especially dangerously high levels of phosphorous. In response, the governments of Canada and the United States created the Great Lakes Water Quality Agreement of 1972, which was later amended in 1978 and 1987. The stated goal of these agreements was to restore and maintain the chemical, physical, and biological integrity of the waters of the Great Lakes Basin Ecosystem, though they were largely aimed at removing toxic substances from the waters. The agreements also created the Great Lakes Water Quality Board, which is responsible for advising the IJC on all environmental issues involving the Great Lakes.

In the past 15 years, a number of national and bi-national programs and follow-up agreements have emerged to enable Environment Canada (EC) and the US EPA to work toward elimination of toxic substances from the lakes, as required by the 1987 amendment. These additions were necessary because researchers still found dangerously high levels of PCBs, mercury, toxaphene, and other pollutants in the lakes. Among the programs is the 1997 Great Lakes Binational Toxics Strategy (GLBTS), which established a four-step process for the EC and EPA to work toward elimination of toxic substances in the Great Lakes. The GLBTS also defined Level I Substances (of the highest concern), including benzo(*a*)pyrene, chlordane, DDT, hexachlorobenzene, mercury and mercury compounds, PCBs, PCDD, and toxaphene.

Though water is arguably the most complex aspect of Canada–US environmental relations, air quality and acid rain are also crucial concerns for each country. Canada struggled to convince the United States to establish emission regulations until the 1990s. The US Department of Energy in 1976 began a research program on

Long-Range Transport of Airborne Pollutants and two years later joined the Canada–US Research Consultation Group on transport of these pollutants. In 1980, both governments signed a Memorandum of Intent to cooperate in fighting acid rain (Doern and Conway 1994, p. 150). Yet this agreement collapsed two years later when the Canadian government confronted the Reagan administration with a study attributing the pollution of the eastern Canadian lakes to sulfur and nitrogen oxide emissions from Midwestern US industry. The United States declared that more research was necessary before regulations could be established (Harris 2001, pp. 9–10).

Throughout the 1980s, the United States rejected repeated Canadian requests to limit emissions harmful to Canadian environmental quality. Meanwhile, in 1984, Canada and nine European countries vowed to reduce their sulfur dioxide emissions by 30% in the next ten years. The United States refused to join what became known as the "Thirty Percent Club." In 1986, the US Court of Appeal overturned a lower court ruling that would have compelled seven states to reduce acid rain-causing emissions, while a year later an agreement was formed among six Canadian provinces to cut sulfate emissions in half. It is interesting to note that the United States did sign an agreement with Mexico in 1987 concerning air quality and transboundary pollution; however, this followed the publication of a report that stated that Mexico's pollution emissions were causing environmental damage in the United States. The Bush (senior) administration finally gave into pressure from environmentalists and the Canadian government and passed the new CAA in 1990 that included a provision to reduce sulfur dioxide emissions and acid rain-causing pollution (Doern and Esty 2002, pp. 150–151).

Another the US–Canada agreement addressed ground-level ozone and the Antarctic ozone hole. The 2000 Protocol amended the 1991 Air Quality Agreement to include provisions for the reduction of ground-level ozone, following studies by Health Canada showing that ground-level ozone is a major contributor to premature pollution, and that emissions from the US Midwest and Eastern seaboard were largely responsible for high ozone concentrations. The Protocol designates Pollution Emission Management Areas (PEMAs), sets vehicle emission standards, and requires that both countries cut nitrogen oxide emissions in their respective PEMAs.

12.6.2 Multinational Agreements

Obviously, pollutant transport is not limited to states, regions, countries, or even continents. Only by joining forces and working toward common environmental goals can countries achieve broad environmental goals. This is especially true for issues of atmospheric pollution, which affects a globally shared resource. More and more international treaties like the ones discussed are needed. Citizens of all countries must force government leaders to enter well-planned international agreements.

12.6.2.1 The Montreal Protocol to Limit Ozone-Depleting Chemicals

In 1985, a team of British researchers discovered unusually low levels of ozone above Halley's Bay in Antarctica, representing a "hole" in the stratosphere that 9 million square miles with little to no ozone protection (Makhijani and Gumey 1995, pp. 30–31). Continuing studies have shown that his hole appears every spring over Antarctica; there are reports of thinning ozone in other parts of the world as well. This information, along with the framework already established from the Vienna Convention, led to the Montreal Protocol on Substances that Deplete the Ozone Layer (http://sedac.ciesin.columbia.edu/entri/). This Protocol established measures by which controlled substances, such as CFCs, would be phased out in developed nations. Those countries that signed the Protocol were under obligation to follow the guidelines to help reduce destruction of the ozone layer. Twenty-five countries signed the agreement on 16 September 1987 (Newton 1995, p. 102). By 1999, all countries that signed the Protocol had agreed to cut CFC production by 50% (Newton 1995, p. 16).

The United States, the world's largest producer and consumer of ozone-depleting chemicals, acted very aggressively. The Amendments to the CAA of 1990 allowed government agencies to ban the use of many ozone-depleting substances. As a result, the EPA ruled that after 16 February 1993, certain products using CFCs would be banned. The EPA limited the sale of certain types of material only to essential products, such as medical sprays, after 17 January 1994, and it planned to completely ban the production and importation of methyl bromide. However, recently EPA has extended its phase-out of certain agricultural ozone-depleting chemicals (*Environmental Science and Technology* 2004a). Unfortunately, recent measurements of CFCs in the atmosphere indicate that bad actors in China are still producing using banned CFCs (www.upi.com/Science_News/2018/07/18/China-to-beef-up-CFC-inspections-as-UN-investigates-illegal-emissions/3041531928505/).

12.6.2.2 The Kyoto Protocol for Limitation of Greenhouse Gas Emissions

With the success of the Montreal Protocol, environmentally conscience leaders soon sought a similar agreement for controlling global warming gases, especially carbon dioxide. Extensive scientific evidence has led to near-consensus on the fact that humans are rapidly contributing to global warming by their unrestricted use of fossil fuels. Important data sets include the CO_2 monitoring data from Mauna Loa and long-term (tens of thousands of years) data from the Vostok ice core monitoring project. Even the US Department of Defense recognizes climate change as a threat to national security (*Environmental Science and Technology* 2004b). Among the numerous voices calling for action, perhaps one of the most convincing groups is the usually conservative insurance companies, who realize the future potential for economic disaster. Predictions of not acting soon include increased ocean levels, shifting ocean currents, warmer atmospheric temperatures (especially at the poles), more dramatic and destructive violent weather, further loss of species diversity, changing habitats, and increased development and spread of disease vectors.

From 1 to 11 December 1997, leaders from 160 nations met in Kyoto, Japan, to consider and develop an agreement with binding limitations on greenhouse gases. This agreement came to be known as the Kyoto Protocol and calls for developed nations to limit their greenhouse gas emissions, relative to 1990 levels. From the period between 2008 and 2012, countries were to decrease their greenhouse emissions to below 1990 levels: 7% for the United States, 8% for the European Union, and 6% for Japan. Specific points of the Kyoto Protocol are (EPA 1999)

- a five-year period for the initial phase reductions (for an average of 9% below 1990 level),
- reductions in six major greenhouse gases, including carbon dioxide, methane, nitrous oxide, and three synthetic ozone-depleting CFCs,
- and using activities that absorb carbon, such as planting trees, to offset emissions.

But what are the implications for the US economy? The Energy Information Administration of the US Department of Energy (www.doe.gov) estimates that US compliance with the Kyoto Protocol would require (i) a reduction in CO_2 emissions requiring the use of 18–77% less coal and 2–13% less petroleum, (ii) an increase of between 2% and 16% in natural gas use and an increase in renewable energy sources between 2% and 16%, and (iii) a market-based means of reducing energy use by increasing the price of energy from 17% to 23%.

While the Clinton administration agreed to the terms of the Kyoto Protocol, the subsequent Bush administration rejected these terms, citing economic concerns due to a slowing of the economy. This ignores potential economic costs of not acting to avoid or minimize the impacts of climate change on the inundation of coastal areas, on a potential increase in violent weather, on water scarcity in some areas, and on agriculture. For ratification, the Protocol requires that industrialized countries producing 55% of the world's greenhouse gases sign the treaty. Even without the cooperation of the United States, the treaty is likely to come into force with the recent commitment of Russia to sign on (www.newscientist.com, 30 October 2004 issue). This may put pressure on the United States and Australia which could be restrictive enough to really change the course of global warming – a daunting challenge at best.

More information can be obtained from the Intergovernmental Panel on Climate Change (IPCC) that was established by the World Meteorological Organization and the United Nations Environment Programme. Their website and publications contain the definitive source of information on global warming.

12.6.2.3 The Basel Convention on Hazardous Waste

The most publicized treaty regarding hazardous waste came out of the Basel Convention on the Control of Transboundary Movement of Hazardous Wastes and Their Disposal, held in Basel, Switzerland. The Convention is a response by the international community to the problems caused by the annual worldwide production of hundreds of millions of tons of wastes. The convention was adopted in 1989 and came into effect in May 1992. This global environmental treaty strictly regulates the transboundary movement of hazardous wastes and requires that its parties ensure that such wastes are managed and disposed of in an environmentally sound manner. So far, the total number of parties (countries) who have ratified the treaty is up to 155: 35 from Africa, 40 from Asia and the Pacific, 27 from Western Europe and others, 22 from Central and Eastern Europe, and 30 from Latin America and the Caribbean. Countries that have yet to ratify the treaty are Afghanistan, Haiti, and the United States (the largest generator of hazardous wastes). The global acceptance of this treaty without US ratification clearly calls into question the quality of US environmental leadership.

12.6.2.4 Stockholm Convention on Persistent Organic Pollutants

The most recent treaty that the United States has entered into resulted from the 2004 Stockholm Convention on Persistent Organic Pollutants (POPs). Under this agreement, participating countries will stop the production of aldrin, chlordane, DDT, dieldrin, endrin, heptachlor, morex, toxaphene, and furans. Some exemptions are possible, including the use of DDT to control malarial mosquitoes in some countries. Still, the agreement prioritizes ways to reduce or eliminate these uses.

12.6.2.5 Paris Climate Accord – Paris Agreement

This is the positive, but nonbinding result of a UN-sponsored meeting held in December 2015, where

196 nations and state parties agree to limit global temperature rise from human causes. Again, 196 nations agreed to this policy, including the United States. From the United States prospective, President Trump has withdrawn from the Accord, but US cities, states, and many industries have agreed to meet their goals in spite of this leadership failure.

12.7 Summary

The United States and the world have made great strides in improving human health and the environment since the global uprising of the environmental movements of the 1960s and 1970s. While there is still significant work to be accomplished, most citizens, local, state, and national governments, and industries are committed to positive action and may build upon prior success.

In recent years, there have been several positive decisions that indicate that pollution does not pay, but polluters do pay dearly for their misdeeds. Examples include the courts multibillion dollar additional fines to Kerr-McGee and BP's record of $46 billion plus cleanup and fines associated with the Deepwater Horizon Gulf disaster. Recently, the Sierra Club has called for Climate Reparations just as the tobacco, asbestos, and other industries knowingly polluted and killed humans. The wheels of justice may finally be turning toward long overdue environmental retributions.

Many students contributed to this chapter including members of my spring 2004 section of Chemistry of the Natural World and my spring 2004 section of Environmental Chemistry. The noted authors of this chapter wrote, in part, the following sections: the Environmental Movements in the US section was taken from work of Lance DeMuth, Loren Sackett, Rose Kormanyos, the History of EPA administrators was written by Rose Kormanyos, the section on NEPA was prepared by Savanna Ferguson, the section on International Treaties was taken from the work of Jill Schulte. I thank you all for your contributions.

Exercises

12.1 Research an industry in your area and try and obtain an RCRA hazardous waste manifesto.

12.2 Research the permitting system on two of the major environmental laws (and amendments) given in Table 12.1. What permits are required and how are inspections conducted?

Disclaimer

The authors are not responsible for any misuse of material extracted from this chapter that is intended as a general overview of regulatory laws and regulations and not intended as *guidance* that only an attorney licensed to practice environmental law could provide. Regulations are a serious component of residual management with necessary safeguards to protect the public.

Discussions of environmental laws can become burdened by technicalities and legal interpretations that are decided by judges not scientists and engineers. Thus, we have decided to take a different approach to this chapter. The contents of this chapter are the result of presentations and papers written by many students in Environmental Chemistry during the spring of 2004 and Chemistry of the Natural World during the spring of 2003 and 2004, and re-edited in 2018. The students listed as authors are those whose papers were incorporated in one form or another in this chapter. I have added my own interpretations and literature research to the students' work. We feel this approach, in the words of students, provides a more interesting presentation of this relatively dry, but important material.

Bibliography

Andrews, R.L. (1995). Reform or reaction: EPA at a crossroads. *Environ. Sci. Technol.* **29** (11): 505A–510A.

Barton, G. (ed.) (2002). *American Environmentalism*, 70. Cengage Gale.

Baum, R.M. (1995). EPA's Silver Anniversary. Editor's Page, *Chemical and Engineering News*, October 30, p. 5.

Bryner, G.C. (1995). *Blue Skies, Green Politics*. Washington (DC): Congressional Quarterly Inc.

Caldwell, L.K. (2001), Forward in). *The National Environmental Policy Act: Judicial Misconstruction, Legislative Indifference, and Executive Neglect*, Number 17, the Environmental History Series (ed. M.J. Lindstrom and A.A. Smith). Dan Flores, College Station: Texan A&M University Press.

Cheremisinoff, N.P. (2001). *Handbook of Pollution Prevention Practices*. CRC Press.

Doern, B.G. and Conway, T. (1994). *The greening of Canada*. Toronto: University of Toronto Press.

Doern, B.G. and Esty, D.C. (2002). *Greening the Americas: NAFTA's Lessons for Hemispheric Trade*. Cambridge, MA: MIT Press.

Ember, L.R. (1995a). EPA at 25. *Chemical and Engineering News* (30 October), pp. 16–17.

Ember, L.R. (1995b). EPA administrators deem agencies first 25 years bumpy but successful. *Chemical and Engineering News* (30 October), pp. 18–23.

Environmental Science and Technology (2004a). Ozone pollutant phaseout delayed (June 15), pp. 217–218A.

Environmental Science and Technology (2004b). Perspective: Abrupt climate change threatens national security (May 15), p. 179A.

EPA (1999). Fact Sheet on the Kyoto Protocol, October 1999, www.epa.gov.

EPA (2000). The History of Drinking Water Treatment, Office of Water, EPA-816-F-00-006, February.

EPA (2003). Plain english guide to clean air act.www.epa .gov/oar/oaqps/peg_caa/pegcaa02.html (accessed 6 February 2003).

EPA (1995). A Look at EPA Accomplishments: 25 years of protecting public health and the environment. EPA Press Release—DEC. 1, 1995. http://epa.history/topics/epa/ 25b.htm (accessed 5 February 2003).

EPA (2018a). https://www.epa.gov/aboutepa (accessed 20 August 2018).

EPA (2018b). https://www.epa.gov/laws-regulations (accessed 20 August 2018).

Harris, P.G. (2001). *The Environment, International Relations, and United States Foreign Policy*. Washington, DC: Georgetown University Press.

Hileman, B. (2004). Pesticides Registration Streamlined: EPA will no longer consult with wildlife agencies on pesticides. *Chemical and Engineering News* (9 August), p. 9.

Hileman, B. (1995). Environmental leaders give EPA mixed reviews on its performance. *Chemical and Engineering News* (30 October), pp. 30–37.

Kirschner, E.M. (1995). Industry see maturation, contradiction in EPA's quarter-century history. *Chemical and Engineering News* (30 October), pp. 24–29.

LaGrega, M.D., Buckingham, P.L., and Evans, J.C., and the Environmental Resources Management Group (1994). *Hazardous Waste Management*. New York: McGraw-Hill, Inc.

Landy, M.K. et al. (1994). *The Environmental Protection Agency: Asking the Wrong Questions from Nixon to Clinton*. New York: Oxford.

Lave, L.B. and Omnenn, G.S. (1981). *Clearing the Air: Reforming the Clean Air Act*. Washington, DC: The Brookings Institution.

LeMarquand, D. (1993). The International Joint Commission and changing boundary relations. *Nat. Res. J.* **Winter**: 59–91.

Lepkowski, W. (1995). Government seeks new balance in environmental protection. *Chemical and Engineering News* (30 October), pp. 44–49.

Lewis, J. (1985). The birth of the EPA. *EPA J.* http://www .epa.gov/history/topics/eap/15c.htm (accessed 5 February 2003).

Lindstrom, M.J. and Smith, Z.A. (2001). *The National Environmental Policy Act: Judicial Misconstruction, Legislative Indifference, and Executive Neglect*, Number 17, the Environmental History Series (ed. D. Flores). College Station: Texan A&M University Press.

Makhijani, A. and Gurney, K.R. (1995). *Mending the Ozone Hole: Science, Technology, and Policy*. Cambridge, MA: The MIT press.

McCoy, M. (1999). Why companies go astray. *Chemical and Engineering News* (19 July), p. 9.

National Environmental Policy Act of 1969, Public Law 91-190, 42 U.S.C. 4321-4327, January 1, 1970, as amended by Public Law 94-52, July 3, 1975, Public Law 94-83, August 9, 1975, and Public Law 97-258, 4(b), September 13, 1982 SEC. 2 [42 U.S.C. 4321].

Newton, D. (1995). *The Ozone Dilemma*. Santa Barbara, CA: Instructional Horizons, Inc.

Quarles, J.R. (1976). *Cleaning Up America: An Insider's View of the Environmental Protection Agency*. Boston: Houghton Mifflin Co.

Rawls, R.L. (1995). Environmental scientists fault EPA for its shifting, short-term research focus. *Chemical and Engineering News* (30 October), pp. 38–43.

Seignecr, C., Vijayaraghavan, K., Lohman, K. et al. (2004). Global source attribution for mercury deposition in the United States. *Environ. Sci. Technol.* **38** (2): 555–569.

Sierra (2004). The bush record. (September–October), pp. 3–88.

Switzer, J.V. (1994). *Environmental Politics: Domestic and Global Dimensions*. New York: St. Martin's.

Taylor, B. (ed.) (1995). *Ecological Resistance Movements: The Global Emergence of Radical and Popular Environmentalist*. State University of New York Press.

Thibodeaux, L.J. (1996). *Environmental Chemodynamics: Movement of Chemicals in Air, Water, and Soil*, 2e. Wiley-Interscience.

Wallace, H. and Wallace, W. (1996). *Policy Making in the European Union*. New York: Oxford University Press, Inc.

Wikipedia (2018). https://en.wikipedia.org/wiki/Sewage_ treatment (accessed 20 August 2018).

13

Environmental Policy in the European Union

Steven Woolston and Aisha Kimbrough

Whitman College, Walla Walla, WA 99362, USA

13.1 Introduction to the European Union

The European Union (EU) contains roughly twenty member states such as Austria, Croatia, France, Greece, Italy, Spain, etc., and it was established to create a sense of economic unity and peace among European countries. The formation of the EU started in 1945 with an emphasized mission of peace, and later, in 1960, economic growth became a motivating factor. It was post-World War II and Europeans wanted a common market and a grander community. Europe became further unified in the 1970s due to the fall of the Berlin Wall and the downfall of Communism. In November 1993, the EU was officially formed, and in the 2000s, the Euro was established as the currency of most European countries. Additionally, the EU created a Banking Union that aided several European counties through its alleviation of economic downturn in the early 2000s.

The unification of European countries demonstrated the social, economic, and political benefits to the formation of the EU, which would not have been possible if not for the Founding Fathers of the EU: Konrad Adenauer, Joseph Bech, Johan Beyen, Winston Churchill, Alcide De Gasperi, Walter Hallstein, Sicco Mansholt, Jean Monnet, Robert Schuman, Paul-Henri Spaak, and Altiero Spinelli. The dream of unification had finally been achieved due to the formation of the EU, which also has played a substantial role in the development of environmental policies and laws – some of which this chapter discusses in further detail. An important distinction to consider is the difference between the European Union and the European Council. The European Council outlines the general political direction and priorities of the EU, and its members consist of the heads of state or government of countries that comprise the EU. It was established in 1974, and it essentially acts as the formal legislation focused side of the EU, which is why so many of the laws mentioned in this chapter are drawn from the European Council (European Union 2018).

13.2 The Environment and the European Union

The unification of the European countries led to economic, political, and social successes. Behind these achievements is a set of core standards that the EU wishes to maintain:

1. Preserving, protecting, and improving the quality of the environment.
2. Protecting human health.
3. Ensuring a prudent and rational utilization of natural resources.
4. Promoting measures at the international level to deal with regional or worldwide environmental problems.

Such standards clearly demonstrate that the EU firmly believes that preserving the environment goes hand-in-hand with economic success and growth. The EU has spearheaded progress toward sound environmental responsibility and growth through collaboration and has ultimately served as one of the best examples of environmental responsibility and policy implementation.

13.3 The Early Stages of the EU's Environmental Efforts

The early treaties of the EU and the European Economic Community demonstrate the shifting of institutional power toward the strong relationship between the member states and the EU. Since the EU governs both internally and internationally, establishing a legal foundation is utterly essential. In the 1960s and 1970s, the establishment of environmental policy within the EU

Pollutant Fate and Transport in Environmental Multimedia, First Edition. Frank M. Dunnivant and Elliot Anders.
© 2019 John Wiley & Sons, Inc. Published 2019 by John Wiley & Sons, Inc.
Companion website: www.wiley.com/go/Dunnivant/pollutant_fate&Transp_EnvironMultimedia

was heavily influenced by the internal interest and focus on unification; however, although the members of the EU were passionate about their recent amalgamation, they lacked sound legislation. This is apparent in the early insubstantial regulation that was scrutinized by critics, as they deemed it a reactionary form of environmental action. Additionally, a passion for environmental action was sparked in the early life of the EU following the establishment of the Stockholm Declaration on the Human Environment in 1972, the Environment and Consumer Protection Service, and the first Environmental Action Program (EAP) in 1973. These acts were established to further hold the EU accountable for environmental policy and push environmental concerns onto their agenda. Although the first EAP lacked a stable framework, it succeeded in providing a sense of direction for the EU and its environmental action. More refined EAPs were later established, and the third EAP in the 1980s provided stability in policy-making; it emphasized prevention and a strong approach toward emissions reduction. During the early stages of EU environmental action, the legal foundation was not yet stabilized, but soon after, the EU finally adopted a stronger legal basis. In 1986, the Single European Act was created and brought a sound legal foundation for environmental legislation; therefore, on an internal level, this was a progressive step toward a stronger initiative on environmental action.

13.4 Existing Environmental Legislation

Legislation regarding the environment is split between Waste Management, Water Legislation, Air Quality, and Environmental Disasters. Several programs have been passed since 1972 that are known as EAPs. The fifth EAP is focused on sustainable development, seeing that it addresses government, enterprise, public, and industry when considering new policy, and that it includes *Voluntary* programs from the public and *Market* incentives from industry. Furthermore, it uses a horizontal approach to environmental regulation, which means that it considers different factors that contribute to environmental issues and conservation: industry, energy, tourism, transportation, agriculture, etc. This is the first step in recognizing that not one culprit for climate change – rather that the blame is very much shared but that these factors are significant contributors to the problem.

There are several different forms of environmental legislation, but the laws that pertain to environmental protection primarily focus on waste management, water management, air quality, and environmental disasters.

13.5 Waste Management Legislation

On its own, waste management legislation addresses existing laws that pertain to the ways that the EU handles its waste – whether it is maintaining it, reducing it, or expelling it. Regarding the reduction of waste, a couple of main strategies that are used include minimizing waste through product design and the encouragement of recycling and reuse of waste (EC 2006). There are a few important laws that address criteria that must be met by garbage before it enters a landfill:

The waste has been treated (prevention of leachate generation – which poses a major threat to surface water and ground water (Hegazi et al. 2013).

Any hazardous waste must be assigned to a hazardous waste landfill.

Landfills for nonhazardous waste must be used for both municipal waste and nonhazardous waste.

Any landfills that do not operate under these instructions are not permitted to operate until the criteria are completely met, and they are required to update the status of their landfill every three years (EC 2006).

For hazardous waste, the disposal process is similar to that of nonhazardous waste, but it is removed and treated more carefully to reduce potential contamination of other waste. It must be properly disposed, recovered, and managed. Waste can be qualified as hazardous waste based on a set of criteria established by European Union Law, or the Directive, which states "waste which is capable by chemical reaction of producing gas at such a temperature and pressure and at such a speed as to cause damage to the surroundings" (EC 2008a, b). Furthermore, not all landfills are eligible to handle hazardous waste; they must have a permit to safely dispose of it. This permit renders said that plant is eligible for periodic inspections by Member-State Authorities, and the transportation and production of the hazardous waste must be monitored and recorded for the member states authorities' use and records.

The main form of expelling waste is by incineration, which is not the most preferred method of waste management. Incineration of waste is an issue due to the air pollution that it causes – therefore, the EU member states regulate current waste incineration plants. The regulation strives to reduce soil and water or air pollution that stems from incineration by only using incineration when it is considered energy efficient, when the waste is not recyclable, and when/if the incineration plants are permitted to operate by the member-state authorities (EC 2008a, b). The incineration plants must also have public records on information regarding composition, generation, and hazardous characteristics of waste, as well as reduce any harmful residue that they produce.

Seeing that technology is substantial to help reduce emissions and harmful residue, a separate directive requires incineration facilities to have the advanced technology necessary to keep very precise measurements of said factors. Additionally, the plant is required to take monthly emissions tests, and if they fail an emissions test, then the plant will not be permitted to continue operations until it proves that their methods will adhere to the emissions requirement set by the member states.

13.6 Water Legislation

With regard to Water Legislation, three of the most important directives that have been implemented focus on the following topics: fresh water for drinking and bathing, and wastewater. Additionally, the EU has participated in conventions that focus on protecting marine life through environment preservation. The water management policy that the EU abides by is rather broad in the way that it attempts to hit multiple issues simultaneously. It focuses on maintaining a fresh water supply that is able to sustain human consumption and economic demands while considering aquatic environment protection efforts and limiting natural disasters such as droughts. This piece of legislation is very intricate seeing as it hits three water-related issues as once. The EU holds its member states to very high standards regarding the maintaining of this broad water policy (EC 2000).

A drinking water directive was established in 1980, and it enforces strict quality and control standards that are to be followed by all EU member states. The kicker is that not all water is the same: some water has different substances than other water sources; therefore, the properties of the water determine the standards of which the drinking water directive is defined. The drinking water standard is determined by nonharmful amounts of certain substances that exist in the drinking water; this standard is determined by the EU, but each member-state monitors its own drinking water and is responsible for its water meeting the minimum standards set by the EU (EC 2000).

Some of the substances that are previously mentioned are explicitly addressed in an EU Directive established in 2001:

> "For substances occurring naturally, or produced through natural processes, such as cadmium, mercury and polyaromatic hydrocarbons (PAHs), complete phase-out of emissions, discharges and losses from all potential sources is impossible. When the relevant individual directives are drawn up, this situation must be properly taken into account and measures should aim at the cessation of emissions, discharges and losses into water of those priority hazardous substances which derive from human activities".
>
> (EC 2001)

The surface water directive is very similar to the criteria for drinking water in case of the event that surface water should be used for drinking water. Once again, the same concept of the substances present in water determine the standards for which the surface water legislation can operate – the standards are set by the EU, and they are maintained and monitored by each individual member-state (EC 2000).

Wastewater treatment: Collection, treatment, and discharge of industrial urban waste are all monitored by the EU; therefore, the directive for wastewater treatment can apply to all member states: There are no specific wastewater treatment directives per member state – there is a directive that encompasses all member states, and it is determined through the analysis of industrial collection, treatment, and discharge of wastewater. Similar to the drinking water directive, individual member states are responsible to maintaining and monitoring the waste-water quality. The EU is responsible for making the information public every other year (EC 2000). There are specific rules for nitrogen content and temperature of wastewater and including a substitute method for testing various temperature-climatic zones.

13.7 Air Quality Legislation

The EU has been at the forefront of reducing air pollution worldwide through the creation and implementation of air quality policy. They have established a list of goals of which they strive to achieve with regard to air quality, while also striving to adhere to the standards set by the World Health Organization. Harmful air pollutants that are recognized and limited include sulfur dioxide, nitrogen dioxide and oxides of nitrogen, particulate matter, and lead in ambient air. One of the more significant contributors to air pollution is large combustion plants; therefore, the EU has established limitations on harmful air pollutants emitted from these plants. The EU member states' general responsibilities for sustaining appropriate air quality include the following:

(A) Assessment of ambient air quality.
(B) Approval of measurement systems (methods, equipment, networks, and laboratories).
(C) Ensuring the accuracy of measurements.
(D) Analysis of assessment methods.
(E) Coordination on their territory if community-wide quality assurance programs are being organized by the Commission.
(F) Cooperation with the other Member States and the Commission (EC 2008a, b).

One of the most important commitments made by the EU in 1997 was to the Kyoto protocol. This agreement that was signed by 38 developed countries aimed to cut greenhouse gas emissions by 8% between 2008 and 2012 relative to the 1990 greenhouse gas levels. The agreement was not a complete success, but not necessarily a failure either: "Although overall global emissions rose, the 38 countries collectively reduced their output by 2 $GtCO_2$ per year from 2008 to 2012 compared with 1990 levels" (Le Page 2016). The United States and Canada committed to the agreement, although their emissions were not on par with the projected requirements for annual declining greenhouse gas emissions. If Canada and the United States were not included in the agreement, then the 36 other developed countries would have adequately achieved the goals set by the Kyoto Protocol.

13.8 Environmental Disasters

Legislation on environmental disasters mostly pertains to industrial accidents that greatly harm the environment, rather than natural disasters such as floods, hurricanes, earthquakes, etc. The deep-water horizon oil spill in 2010 is a very well-known example of an industrial accident that substantially harmed the environment, spilling 210 million gallons of oil into the Gulf of Mexico. Former environmental disaster legislation has included the Transboundary Impacts on Industrial Accidents, which requires all member states to cooperate to help resolve the industrial accident that can range from nuclear accidents to dam failures and anything in between. There are currently 41 parties and 27 signatures on the agreement, and it was formerly registered on 19 April 2000 (Helsinki 1992). Furthermore, the EU explicitly explains that environmental disasters fall on the communities in which they occur and that communities are required to help repair damage caused by industrial disasters.

> "Since the objective of this Directive, namely to establish a common framework for the prevention and remedying of environmental damage at a reasonable cost to society, cannot be sufficiently achieved by the Member States and can therefore be better achieved at Community level by reason of the scale of this Directive and its implications in respect of other Community legislation...Community may adopt measures in accordance with the principle of subsidiarity as set out in Article 5 of the Treaty. In accordance with the principle of proportionality, as set out in that Article, this Directive does not go beyond what is necessary in order to achieve that objective".
>
> (EC 2004)

Bibliography

EUR-Lex - 31995L0046 – EN (2000). EUR-Lex Access to European Union Law. OPOCE, 23 October 2000. eur-lex.europa.eu/legal-content/EN/TXT/?qid=1535754708273&uri=CELEX:32000L0060 (accessed 4 September 2018).

EUR-Lex - 31995L0046 – EN (2001). EUR-Lex Access to European Union Law. OPOCE, 20 November 2001. eur-lex.europa.eu/legal-content/EN/TXT/?qid=1534885489251&uri=CELEX:32001D2455 (accessed 4 September 2018).

EUR-Lex - 31995L0046 – EN (2004). EUR-Lex Access to European Union Law. OPOCE, 21 April 2004. eur-lex.europa.eu/legal-content/EN/TXT/?qid=1535755275212&uri=CELEX:32004L0035 (accessed 4 September 2018).

EUR-Lex - 31995L0046 – EN (2006). EUR-Lex Access to European Union Law. OPOCE, 6 September 2006. eur-lex.europa.eu/legal-content/EN/TXT/?qid=1535755941470&uri=CELEX:32006L0066 (accessed 4 September 2018).

EUR-Lex - 31995L0046 – EN (2008a). EUR-Lex Access to European Union Law. OPOCE, 19 November 2008. eur-lex.europa.eu/legal-content/EN/TXT/?qid=1535754291064&uri=CELEX:32008L0098 (accessed 4 September 2018).

EUR-Lex - 31995L0046 – EN (2008b). EUR-Lex Access to European Union Law. OPOCE, 21 May 2008. eur-lex.europa.eu/legal-content/EN/TXT/?qid=1535755143615&uri=CELEX:32008L0050 (accessed 4 September 2018).

European Union (2018). European Council – European Union – European Commission. europa.eu/european-union/about-eu/institutions-bodies/european-council_en (accessed 24 May 2018).

Hegazi, H.A., Meguid, A.M., and Raghab, S.M. (2013). Treatment of leachate from municipal solid waste landfill. *HBRC J.* **9** (2): 187–192. https://doi.org/10.1016/j.hbrcj.2013.05.007.

Helsinki (1992). UN, United Nations, UN Treaties, Treaties. *United Nations*, United Nations, 17 March 1992. treaties.un.org/pages/ViewDetails.aspx?src=TREATY&mtdsg_no=XXVII-6&chapter=27&clang=_en (accessed 4 September 2018).

Le Page, M. (2016). Was Kyoto climate deal a success? Figures reveal mixed results. *New Scientist*, 14 June 2016. www.newscientist.com/article/2093579-was-kyoto-climate-deal-a-success-figures-reveal-mixed-results/ (accessed 4 September 2018).

14

Environmental Laws in China

Zeyu Liu (刘泽宇)[1] and Yi Xu (徐逸)[2]

[1] Chemicals and Health Branch, UN Environment, Geneva 1203, Switzerland
[2] Civil and Environmental Engineering, Stanford University, Y2E2 Building, Stanford, CA 94305, USA

14.1 Environmental Law and Policy in the People's Republic of China

The previous two chapters focus on environmental movements and legislations in the United States and the European Union. In this chapter, we shift our gear and take a close look at an increasing important country – the People's Republic of China (China). China has been attracting a lot of attention in the twenty-first century due to its increasing influence in the political, economic, and ecological arena. China's expanding power and impact are certainly going to persist in the future. China's evolving economic model and political regime have determined the evolution of its environmental governance. Although China has already passed its infancy stage of environmental governance, China's law and policy are far from mature and comprehensive. Although this chapter is not a detailed history of China's environmental governance, it will start its discussion by taking a historical perspective on the evolution of law and policy in China. This chapter compares the success and failure of law and policy in the developing China to those of developed countries.

14.2 Brief Introduction to China

In the past century, China has drawn attention in the world after she implemented an economic reform. This reform was described as "Socialism with Chinese characteristics" and was initiated in December 1978 by reformists within the Communist Party of China (CPC), led by Xiaoping Deng. This reform introduced market principles into China's economic system and enabled the private sector to grow remarkably, accounting for as much as 70% of China's gross domestic product by 2005 (Engardio 2005). From 1978 to 2013, the economy in China increased by 9.5% a year (Scissors 2009). Large-scale government planning programs alongside market characteristics have greatly decreased poverty, whereas incomes and income inequality have increased. The rapid economic growth was accompanied by the correspondingly rising political power in the world as well. China operates in a framework of a socialist republic run by a single party, the CPC, headed by the General Secretary. The Communist Party, the Central People's Government (State Council), and their provincial and local representations together exercise state power within China. Economic reform during the 1980s and the devolution of much central government decision-making, along with strong interest of local officials in benefiting themselves, made it increasingly difficult for the central government to maintain its authority (PRC 1996). Despite a change in economic and political regime, environmental issues persist throughout Chinese modern history. The introduction of market system certainly increased the complexity of the environmental issues in China, but environmental issues have actually been long observed before the reform. It was frequently argued that the western capitalism is the culprit for pollution and environmental degradation. In fact, several large-scale ecological disasters in China occurred during Mao's communism era. It is particularly interesting to study environmental issues in contemporary China because China's "dual-track economy," one that combines government planning and market forces, is very unique.

14.3 Economy and the Environment

Before the economic reform, several environmental issues already raised people's concern in China. By the mid-1800s, large parts of northern China were deserts; land suffered from deforestation and poor agricultural practices; overfishing damaged river systems; and small factories started to pollute country's

Pollutant Fate and Transport in Environmental Multimedia, First Edition. Frank M. Dunnivant and Elliot Anders.
© 2019 John Wiley & Sons, Inc. Published 2019 by John Wiley & Sons, Inc.
Companion website: www.wiley.com/go/Dunnivant/pollutant_fate&Transp_EnvironMultimedia

water resources (Smil 1993; Rock 2002a; Economy 2004). China's increasing population and her frequent turbulent history also took a heavy toll on the environment (Smil 1993). After the economic reform, China's economy grew rapidly and robustly. New government policies encouraged the privatization of agriculture, the urbanization of rural area, the development of thousands of small rural factories, and a flush of foreign investment (Carter and Mol 2007; Fengshi 2009). However, China's ecological landscape paid a high price for this rapid economic growth. Water pollution, air pollution, and soil degradation posted danger to fragile ecosystem and human health. Fortunately, these pollution issues are receiving increasing attention from the Chinese government, given that corresponding policies and plans are still under way (Cai and Wang 2010).

14.4 History of Environmental Law and Policy

The first environmental law in China was passed by the Standing Committee of the National People's Congress (NPC), the legislative committee, in 1979. During the past 40 years, environmental law is one of the fastest evolving law in China. This evolution *was urged by* results from urgency of stopping ecological degradation and is also a manifestation of China's transition to a country that is governed by the rule of law.

In 1972, Chinese representatives attended the first United Nations Conference on the Human Environment in Stockholm, Sweden. During the conference, Chinese representatives argued that environmental pollution is an unavoidable result of capitalist societies. In fact, ecological damage and environmental pollution already existed in China. In order to prove the superiority of a socialist system, high-level officers within the Chinese government started to consider taking effort to avoid environmental damage that had been occurring in the western hemisphere (Cai and Wang 2010).

In 1974, the Chinese central government established the Leading Group for Environmental Protection as well as its equivalences in provinces. Then, in 1979, the first environmental law, the Environmental Protection Law of the People's Republic of China, was passed by the fifth Standing Committee of the NPC during the eleventh conference. The law combined experiences from the preceding years with *some effective parts of other countries' environmental laws*. In particular, four aspects of this law have since played pivotal roles in the later environmental protection process in China (Cai and Wang 2010). First, it established environmental protection institutions, defined their goals and tasks, and set rules for incentives and punishments. It gave rise to a formal definition of environmental protection as the usage of natural resources, as well as the prevention and control of environmental pollutions. Second, it defined environmental liability and sets up the Pollution Charging System. Third, it put forward the mandate of environmental impact assessment (EIA) practices. Fourth, it stipulated the "Three Synchronies" system, which required that the design, construction, and operation of a project have to be accompanied by the same three processes of corresponding pollution treatment facilities.

Following the establishment of the first environmental law, China's environmental protection process slowly grew from the 1980s to the 1990s. The environmental policy system has been becoming more mature; systematic routine procedure has been forming; and certain progress has been achieved in pollution prevention and control. In 1983, the Chinese government announced that environmental protection would become a state policy. Then, in 1988, the Leading Group for Environmental Protection evolved into the National Environmental Protection Agency (Editorial Board Environmental Protection in China 2000).

However, by the mid-1990s, it was apparent that the not so strongly established relative comprehensive environmental regulatory framework established in the late 1970s had failed to prevent the overall degradation of China's environmental quality. According to Shi and Zhang (2006), there are a number of reasons why the performance of China's early environmental regulation was unimpressive.

First, China developed its environmental regulation in the 1970s, with little experience and virtually no institutional capacity. This initial environmental regulation was also a model of those of Western countries, but China's international isolation in the 1970s and early 1980s made the learning process difficult and deficient (Jahiel 1998).

Second, China lacked a strong environmental regulatory system, with large and effective monitoring and enforcement capacity, to oversee the initial regime of industrial environmental pollution control (Wang et al. 2002). Due to Environmental Protection Bureaux's (EPB) lack of authority, capacity, and resources to monitor and enforce industrial environmental compliance, the regime became ineffective in handling and addressing industrial pollution. This is especially true at the town level because as at 2004, there were no environmental monitoring stations and environmental supervision and inspection agencies at the town level (Zang 2002; Shi and Zhang 2006).

Third, the regime was principally designed to target large sources of emissions (SOEs) within a centrally planned economy via direct command-and-control interventions. The most important basis of the regulation was the concentration-based pollution discharge standards, and in consequence, all the regulatory programs were biased in favor of end-of-pipe pollution abatement, which proved technologically unreliable and economically costly in China (Wang et al. 2004). Moreover, the system was ineffective and costly in dealing with the large quantity of emerging small-scale industrial entities, especially in rural areas (Zang 2002).

Finally, industry experienced constant and rapid change in the 1990s, both in scale (quantitatively) and in structure (qualitatively). The industrial transformation formed a moving target, and it proved difficult to develop a corresponding environmental regulatory system (Zang and Chen 2003).

During the turn of the century, increasing attention to environmental matters from citizens caused the Chinese government to display an increased level of concern toward environmental issues. For example, from 2001 to 2005, Chinese environmental authorities received more than 2.53 million letters and 430 000 visits by 597 000 petitioners seeking environmental redress (Wang 2013). In response, in his 2007 annual address, Wen Jiabao, the Premier of the People's Republic of China, made 48 references to "environment," "pollution," and "environmental protection," and stricter environmental regulations were subsequently implemented. Subsidies for some polluting industries were cancelled, whereas other polluting industries were shut down. However, many internal environmental targets were missed (Kahn and Yardley 2007). In an effort to improve the efficiency of regulatory enforcement and to incorporate environmental protection in the national policy development, on 27 March 2008, the State Environmental Protection Administration (SEPA, founded in 1982 as the National Environmental Protection Agency [NEPA]) was upgraded to the Ministry of Environmental Protection (MEP). During the 2010s, several pieces of major environmental law were modified and are discussed below. A more comprehensive list of existing environmental law and policy is presented in Section 14.5.

14.4.1 Survey of Pollution Sources

On 12 October 2006, China launched its first nation-wide pollution resource survey, led by Zeng Peiyan, Vice-Premier, and in October 2007, the State Council issued specific regulations for the survey. Recently a public advertisement about the survey was promoted via China's central television channels with quite a high broadcasting density. On 9 February 2010, the survey results were presented via a public formal media press release from the MEP which, according to the writer's personal experience, is also the first time such press media communication channels have been used for environmental protection, especially for pollution discharge. The public media interpretation is that the survey results would become essential statistics for decision-making on important environmental topics, such as environmental tax, soil contamination control, etc.

14.4.2 Soil and Groundwater Legislation

The Chinese Government is establishing/revising soil and groundwater standards, technical guidelines for investigation and prevention laws. The MEP set up an expert team to draft the soil standards and a technical guideline on EIA reporting on soil. The team revised the soil quality standard with updates to more than 100 compounds including all the industrial organic chemicals and pesticides. They also collected information about standards from the international community, and different categories of soil may be considered for agricultural, residential, commercial, and industrial purposes in terms of background value, impacted value, and contamination value. In 2009, the MEP published for public consumption a series of regulations and standards on soil and groundwater contamination investigation, assessment, monitoring, and remediation. It is expected that the revised soil quality standards will be finalized using an accelerated approach.

14.4.3 New Chemical Legislation

A revised *Provisions on the Environmental Management of New Chemical Substances* was adopted by the MEP on 30 December 2009, to take effect on 15 October 2010. Compared with the 2003 Provisions, the new law is expected to contain more lenient requirements, larger geographical scope, such as Free Trade Zones, and a broader concept of applicable substances, such as articles, imported or domestic. In addition, risk assessment would be regarded as a necessary base for review and approval, and more in line with the REACH mechanism.

14.4.4 Waste Legislation

The *Solid Waste Pollution Prevention Law*, initially issued in 1995, was adopted on 29 December 2004, with effect from 1 April 2005. Making polluters responsible and encouraging recycling are the key principles driving this amendment. The amended law provides requirements for solid waste pollution prevention liabilities when an enterprise is to be separated, merged or acquisitioned, dissolved, or goes bankrupt. Before termination, an enterprise is required to deal properly with its generated solid wastes and related solid wastes locations, and facilities.

14.4.5 Water Legislation

On 28 February 2008, the *Law of the People's Republic of China on the Prevention and Control of Water Pollution* was promulgated by the Central Government, taking effect on 1 June 2008 to substitute the existing law approved by the NPC on 11 May 1984. The new law features many changes in addressing issues such as including water environmental quality into the performance evaluation of the local governments; implementing severe legal punishment for unapproved water pollution discharges; and the enforcement of mass loading control for major water pollutants. The most significant modification relates to the section on legal liability/penalty, which has been updated to include much more detailed penalty rules and harsher punishment.

14.5 Existing Environmental Law and Policy

Law, policy, and regulation can be complicated and tedious for nonprofessionals. In this section, current environmental regulatory framework and regulatory authorities to briefly present the main structures and main actors.

14.5.1 Legislation Authorities

Law concerning the environment includes the following:

- Laws promulgated by the NPC and its Standing Committee.
- Administrative regulations formulated by the State Council.
- Department rules formulated by ministries and commissions under the State Council.
- Local regulations formulated by each province and municipality.

14.5.2 The Regulatory Authorities

Multiple administrative authorities jointly supervise and manage environmental matters including:

- Environmental protection authorities at county level and above and their authorized environmental supervision institutions.
- The housing and urban–rural construction authorities, land authorities, agriculture authorities, public security departments, transportation departments, territorial resources departments, forestry departments, oceanic departments, and water departments, which undertake environmental protection supervision and administrative responsibilities.

14.5.2.1 Ministry of Environmental Protection (MEP)

Main Activities These include the following:

- Establishing a sound basic system for environmental protection.
- Taking precautionary measures against pollution at the source and controlling environmental pollution and environmental damages.
- Supervising and administering prevention and control of environmental pollution.
- Guiding, coordinating, and overseeing ecological conservation efforts.

http://english.mep.gov.cn

14.5.2.2 Ministry of Water Resources

Main Activities These include the following:

- Formulating water resources development strategies, plans, and policies, providing draft legislation and promulgating water administrative rules and regulations, and making integrated river basin management plans and flood control plans for major rivers and lakes.
- Ensuring water resources are rationally developed and used.

www.mwr.gov.cn/english

14.5.2.3 National Development and Reform Committee (NDRC)

Main Activities These include the following:

- Formulating and implementing strategies of national economic and social development, annual plans, and medium and long-term development plans.
- Compiling plans for major functioning zones, coordinating the implementation of these plans, and carrying out monitoring and assessment.
- Promoting the strategy of sustainable development and undertaking comprehensive coordination of energy saving and emission reduction.
- Organizing the formulation of key strategies, plans, and policies in addressing climate change.

http://en.ndrc.gov.cn

14.6 Challenges and the Future of Environmental Governance

Chinese state apparatus remains dominantly important in environmental protection and reform. Both the nature of the contemporary Chinese social order and the character of the environment as a public good will safeguard the crucial position of the state in environmental protection and reform for some time (Carter and Mol 2007). Currently, articulations of environmental interests have become the sole responsibility of the EPB, which is a prominent department at various governmental levels. However, the most common complaints from Chinese and foreign environmental analysts focus precisely on this system of (local) EPBs. It is evident that the local EPBs depend so much on both the higher level environmental authorities and on local governments and as such they have been criticized for their poor environmental capacity (in both qualitative and quantitative terms) and, more generally, for the lack (and distortion) of environmental information (Ma and Ortolano 2000). Yet, the system of environmental state organizations in China is showing remarkable changes since the turn of the millennium, of which four deserve mentioning (Mol 2009).

First, the content of China's environmental policy has been modernized and updated. Over the last decade, China has formulated new environmental laws, which include the Cleaner Production Promotion Law (of 2002), the Environmental Information Disclosure Decree (of 2008), the Law on Promoting Circular Economy (of 2009), the new environmental policy instruments, such as voluntary agreements (Eichhorts and Bongrardt 2009) and emission trading (Tao and Mah 2009). All these laws point at a harmonization of China's environmental policy with that of Organization for Economic Cooperation Development (OECD) countries. However, operating system and mode of implementation of these modern environmental strategies still maintain some of their Chinese characteristics: low fines, an emphasis on promotion rather than regulation, and a complicated division of responsibilities between horizontal and vertical lines (Mol 2009). Additionally, some typical Chinese environmental policies, such as the three simultaneous principles and the National Environmental Model City program are still intact despite the strong international influence and cooperation (MacBean 2007; Economy 2006).

The second development is that China is moving away from a rigid, hierarchical, command-and-control system to a decentralized and more flexible environmental policy-making and implementation system. Local and provincial Environmental Protection Bureaus and local governments are more (actively participating) prominent in developing environmental priorities, strategies, financial models, and institutional arrangements. This is as a result of past failures on the state level in national environmental policy, though it is seen also as a part of a wider tendencies of less centralized control in Chinese society. The tendency is one toward larger influence and decision-making power by the local authorities and diminishing control by Beijing, by the central state structures, and by the CPC (for example, decentralization in energy policy). Proponents of decentralization have argued that decentralized and more flexible environmental policies may be more effective as they can be better adapted to the local physical and socioeconomic situation; however, citizen's perceptions on the (environmental) performance of local urban governments are not very positive (Saich 2007). In China, as elsewhere, decentralization does not automatically result in better protection of the environment, as local authorities typically give preference (priority) to economic growth and investments over the progressive development of environmental policies and stringent enforcement of environmental regulation and standards (Carter and Mol 2007). Therefore, decentralized policies need to be paralleled by performance measurement, (environmental) auditing, and accountability mechanisms to make sure that local authorities give priority to the environmental concerns of citizens (and not to their own interest). In contemporary China, the Quantitative Examination System for Comprehensive Urban Environmental Control and an emerging environmental auditing system have been developed in this regard. They ensure some form of national control, where environmental policy-making and implementation becomes decentralized.

The third major development in China's system of environmental governance which is also instrumental to the overall performance of all the decision-makers is the strengthening of the rule of (environmental) law. This can be identified as a modernization in environmental policies, closely tied to the emergence of a market economy (Carter and Mol 2007). The system of environmental laws has led to the setting of environmental quality standards and emission discharge levels. There are signs that the rule of law is taken more seriously in the field of environments, and this has been triggered by the opening up of China to the global economy and polity. The strengthening of the rule of environmental law goes together with more resources for enforcement, a more formal enforcement style of EPBs, and stronger (financial) punishments of companies. For instance, courts (legislatures) are getting more involved in enforcement in addition to legal procedures initiated by citizens and environmental NGOs, such as the well-known Centre for Legal Assistance to Pollution Victims (CLAPV) (Wang 2006).

14.7 Can China Take on the Leading Role in the Global Environmental Governance?

As of October 2018, the United States under the Trump administration seems to be taking an exit from the global coalition, especially on the war against climate change, China is expected to take on the leading role to bring together countries and regions amidst the pressing circumstances, especially at the juncture when International Panel on Climate Change published the new report warning the average temperature is 1.5 °C above the preindustrial level.

In the recent years, we have seen that China has been collaborating with UN Environment to contribute to international environmental protection. UN Environment and Chinese Ministry of Environmental Agency have published the study *China's Green Long March*, which shows that only a green economy can sustain China's economic growth and mitigate climate change. Besides the effort from the state level, Chinese environmental NGOs have also been acting actively in supporting environmental movements. Groups such as Beijing Energy Network and Global Environmental Institute have already contributed in making environmental progress.

However, as China is shifting from the economy from carbon-intensive to environment-friendly, the international community still needs to be patient with this rising power. China needs to invest domestically in educating the general public to become environmentally conscious and act to offset carbon footprints. Only when internal efforts on environmental regulation are achieved, China will be a powerful global leader leading and collaborating with the international community.

Bibliography

Cai, D. and Wang, C. (2010). *China's Journey Toward the Rule of Law: Legal Reform, 1978–2008*. Brill Publishers.

Carter, N.T. and Mol, A.P.J. (2007). *Environment Governance in China*. London: Routledge.

Economy, E. (2004). *The River Runs Black: the Environmental challenge to China's Future*. Ithaca, NY: Cornell University Press.

Economy, E. (2006). Environmental governance: the emerging economic dimension. *Environ. Politics* **15** (2): 171–189.

(2000). *Editorial Board Environmental Protection in China*. Beijing: China's Environmental Science Press.

Eichhorts, U. and Bongrardt, D. (2009). Towards cooperative policy approaches in China: drivers for voluntary agreements on industrial energy efficiency in Nanjing. *Energ. Policy* **37**: 1855–1865.

Engardio, P. (2005). China is a private-sector economy. *Bloomberg Businessweek*.

Environmental Governance in China (2018). https://en.wikipedia.org/wiki/Environmental_governance_in_China#cite_note-58 (accessed 16 October 2018).

Fengshi, W. (2009). Environmental politics in China: an issue area in review. *JCPS* **14** (4): 383–406.

He, Q. (1996). Pitfalls of Modernization 现代化的陷阱, published in Peoples' Republic of China, 1996, never translated.

Jahiel, A. (1998). The organization of environmental protection in China. *China Quart.* **156**: 757–787.

Jing, C., Richard, G., and Mun, S.H. (2009). China's 11th five year plan and the environment: reducing SO_2 emissions. *Rev. Environ. Econ. Policy* **3** (2): 231–250.

Kahn, J. and Yardley, J. (2007). As China roars, pollution reaches deadly extremes. *The New York Times* (retrieved 28 July 2013).

Ma, X. and Ortolano, L. (2000). *Environmental Regulation in China. Institution, Enforcement and Compliance*. Lanham, M.D: Rowman and Littlefield.

MacBean, A. (2007). China's environment: Problems and policies. *World Econ.* **30** (2): 292–307.

Mol, A.P.J. (2009). Urban environmental governance innovations in China. *Curr. Opin. Environ. Sustain.* **1**: 96–100.

Mu, Z., Bu, S., and Xue, B. (2018). Environmental legislation in China: achievements, challenges and trends. www.mdpi.com/2071-1050/6/12/8967/pdf (accessed 16 October 2018).

Rock, M. (2002a). *Pollution Control in East Asia*. Washington, DC: Resources for the Future.

Saich, T. (2007). Citizens' perceptions of governance in rural and urban China. *JCPS* **12** (1): 1–28.

Scissors, D. (May–June 2009). Deng undone: the costs of halting market reform in China. *Foreign Aff.* **88** (3).

Shi, H. and Zhang, L. (2006). China's environmental governance of rapid industrialization. *Environ. Politics* **15** (2): 271–292.

Smil, V. (1993). *China's Environmental Crises*. Armonk, NY: M.E. Sharpe.

Tao, J. and Mah, D.N. (2009). Between markets and states: dilemmas of environmental governance in China's sulphur dioxide emission trading system. *Environ. Plann. C: Gov. Policy* **27**: 175–188.

Wang, A. (2006). The role of law in environmental protection in China: recent development. *Vt J. Envtl. L.* **8**: 195–223.

Wang, A. (2013). Environmental protection in China: the role of law. *China Dialogue* (retrieved 12 February 2013).

Wang, H., Bi, J., Wheeler, D. et al. (2002). *Environmental Performance Rating and Disclosure: China's Green*

Watch Program, Program Research working paper 2889. Washington, DC: World Bank.

Wang, J., Yang, J., Ge, C. et al. (2004). Controlling Sulphur dioxide in China: will emission trading work? *Environment* **46** (5): 28–39.

Zang, L. (2002). Ecologizing industrialization in Chinese small towns. PhD thesis. Wageningen University.

Zang, T. and Chen, J. (2003). Industrial environmental management in China. *Environ. Chang.* **45**: 129–140.

Part VI

World Class Pollutants

If we are going to live so intimately with these chemicals—eating and drinking them, taking them into the very marrow of our bones—we better know something about their nature and their power

from Silent Spring, Rachael Carson

Pollutant Fate and Transport in Environmental Multimedia, First Edition. Frank M. Dunnivant and Elliot Anders.
© 2019 John Wiley & Sons, Inc. Published 2019 by John Wiley & Sons, Inc.
Companion website: www.wiley.com/go/Dunnivant/pollutant_fate&Transp_EnvironMultimedia

15

World Class Pollutants

Frank Dunnivant and Emily Welborn

Department of Chemistry, Whitman College, Walla Walla, WA 99362, USA

This chapter focuses on several "world class pollutants," that is pollutants that are spread around with world and that can be found in every environmental compartment (air, surface water, groundwater, oceans, soil, biota, etc.). Although there are several chemicals that could be presented here, we will focus on four classic pollutants: mercury, lead, polychlorinated biphenyls (PCBs), DDT, plastics, endocrine disruptors, and carbon dioxide. We will close this chapter with a different type of ubiquitous pollutant, endocrine disruptors, which may turn out to be one of the most serious anthropogenic threats to human and animal health.

15.1 Mercury

15.1.1 Sources

15.1.1.1 Mineral

Cinnabar (HgS) is by far the most important source of mercury, typically consisting of 0.6–0.7% of mined mercury ore (Gribble 1988). The most important mine in the world is in Almaden, Spain, and has been in operation since Roman times (Berry and Mason 1959). In 1985, the major producers of mercury were the former USSR (2415 Mg), Spain (1725 Mg), China (1190 Mg), Algeria (862 Mg), the USA (100 Mg), Turkey (>690 Mg), Finland (100 Mg), and Yugoslavia (90 Mg).

15.1.1.2 Other Anthropogenic Sources

While mining ore is a major source of mercury for industrial use, other sources of mercury to the environment include (i) the combustion of coal, (ii) oil production combustion, (iii) cement production, (iv) lead production, (v) zinc production, (vi) pig iron and steel production, (vii) caustic soda production (chloralkaline plants), (viii) gold production, and (ix) waste disposal (such as batteries in landfills) (Pacyna and Pacyna 2002). The burning of coal is one of the major sources of mercury to the environment. Coal produced in the United States contains mercury at concentrations between 0.07 (Uinta coal) and 0.24 ppm (northern Appalachian coal) (USGA Fact Sheet FS-095-01 2001). With this mercury concentration, any other material would be classified as hazardous waste. Yet, in burning coal for energy, we emit all or most of its mercury to the atmosphere, from which it is deposited into our air, soil, and water. Mercury has even been found in atmospheric deposition falling on the Arctic and Antarctic (Rouhi 2002). Major annual anthropogenic sources of Hg emissions to North America (southern Canada, United States, and northern Mexico) are the (i) electrical utilities (52.7 Mg/yr; US contribution 41.5 Mg/yr), (ii) waste incineration (32.2 Mg/yr; US contribution 28.8 Mg/yr), (iii) residential, commercial, and industrial coal burning (12.8 Mg/yr; US contribution 12.8 Mg/yr), (iv) mining (6.7 Mg/yr; US contribution 6.4 Mg/yr), (v) chlor-alkali facilities (6.75 Mg/yr; US contribution 6.7 Mg/yr), (vi) mobile sources (24.8 Mg/yr; US contribution 24.8 Mg/yr), and (vii) miscellaneous other sources (64.1 Mg/yr; US contribution 30.9 Mg/yr). These add up to 200.1 Mg of mercury per year released to the environment (Seigneur et al. 2004). The USA contributes 76% of this. In recent years, considerable improvements have been made to coal-fired power plants operated in the United States (summarized in Dunnivant 2017, Chapter 4).

15.1.1.3 Biosynthetic

Mercury from all of these sources is generally deposited in its relatively less toxic, inorganic form. Yet, after mercury pollution has been deposited in soils and sediments, microorganisms in anaerobic environments can transform the inorganic mercury to organic forms, specifically methyl and dimethyl mercury. These forms are very toxic and are a major concern especially when they bioaccumulate, as they do in fish; this is discussed later in more detail.

Pollutant Fate and Transport in Environmental Multimedia, First Edition. Frank M. Dunnivant and Elliot Anders.
© 2019 John Wiley & Sons, Inc. Published 2019 by John Wiley & Sons, Inc.
Companion website: www.wiley.com/go/Dunnivant/pollutant_fate&Transp_EnvironMultimedia

15.1.2 Production/Use

Cinnabar (HgS) is processed by roasting the ore in the presence of oxygen where the sulfide is oxidized to sulfurous acid, and the freed metal (Hg) is volatized and recovered through condensation. The major uses of processed mercury ore are in the manufacture of drugs and chemicals and as a cathode in chlorine and soda ash production. It is also used in the electrical industry for batteries, rectifiers, automatic switches, and mercury vapor lamps and in the instrument industry in the construction of thermometers and barometers. Some insecticides and fungicides for agriculture contain mercury, and mercury is still used in dental fillings (dental amalgam contains 40–50% mercury, 25% silver, and 25–35% copper, zinc, and tin; www.epa.gov). At one time, mercury was used in the United States for extraction and recovery of gold and silver from mine tailings, as it still is today in some parts of the world (Gribble 1988).

15.1.3 Fate and Environmental Distribution

Mercury enters the environment via two primary pathways: into the atmosphere from the smelting of ores, fossil combustion sources (primarily coal), and chlor-alkaline bleach production facilities; and into waterways via atmospheric deposition and industrial pollution. Regardless of the original source, mercury is initially deposited into the water in its inorganic, ionic form (Hg^{2+}). Inorganic mercury is toxic and can be taken in through the intake of water or food. A more toxic form of mercury is the methylated form (CH_3HgCH_3 and CH_3Hg^-), which is produced by bacteria in anaerobic environments, such as those found in some groundwaters and especially in river and lake sediments. Both forms of mercury, organic and inorganic, bioaccumulate in the food chain and can easily reach health-threatening levels in fish.

Recently, governmental agencies and the news media have focused on mercury contamination in our foods due to its bioaccumulation tendency. Several fish consumption warnings for pregnant women and children have been issued (Renner 2004; Crenson 2002; Wright 2005). Fish, especially ocean fish, are a major source of mercury exposure in the human diet. The higher the fish on the food chain, the more mercury it will contain, as illustrated in a *New York Times* article on seafood consumption. The article reports mercury levels for animals at increasing trophic levels: the concentration increased from 0.023 ppm in clams, to 0.035 in salmon, to 0.042 in scallops, to 0.047 in shrimp, to 0.092 in flounder, to 0.117 in crab, to 0.121 in cod, to 0.150 in Pollock, and to 0.206 in tuna (Saar 1999). Hence, environmentalists have two reasons to avoid tuna: not only to protect dolphins, as has long been advertised, but also to avoid mercury intake. Still, not all tuna is a problem, since mercury levels vary significantly depending on the type of tuna processed into food. "White tuna" is one of the safest forms of tuna with respect to mercury levels.

Efforts are underway, although slowly, to reduce mercury emissions. Nine chlor-alkaline plants that manufacture sodium hydroxide, chlorine, and chlorine bleach in the United States have gradually replaced mercury in their systems. This change over will account for approximately 65 tons of mercury that is "lost" in the system annually (Chemical and Engineering News 2004a,b,c). One of the major impediments to reducing mercury from coal burning is, not surprisingly, the coal and energy lobbyists, as well as the current Bush administration, who fear the economic costs to industries required to update their equipment and processes (Chemical and Engineering News 2005a,b). These impediments have slowed installation of mercury recovery equipment, even though existing technologies have the potential to virtually eliminate mercury from emissions passing through smokestacks (Betts 2003). Although the US federal government has slowed the reduction of mercury emissions, other countries are calling for enforceable reductions in emissions (Chemical and Engineering News 2005a,b). Also, individual states in the United States are starting to require emission control efforts for coal burning facilities under their jurisdiction (Chemical and Engineering News 2004a,b,c; Christen 2004).

15.1.4 Health Effects

Although it will come as a surprise to many college-age students, liquid mercury was a common medicine in the distant past, when people would even drink small quantities of the liquid metal to treat a variety of illness. Fortunately, this elemental form of mercury is not significantly absorbed across the intestinal lining. This is why doctors are not alarmed when a child breaks a mercury-filled thermometer. Still, there is a movement to replace mercury thermometers for medical use, in order to decrease the amount of mercury pollution created. Mercury vapor, on the other hand, is very toxic because it can pass the lung-blood vessel barrier and proceed directly to the brain. Ionic mercury in water or food is easily absorbed into the blood and is toxic. Methylmercury, the most toxic form of mercury, can be directly absorbed through the skin, but this is mostly a concern for chemists working in the laboratory.

Sufficiently high exposure to any form of mercury can damage the gastrointestinal tract, the nervous system, and the kidneys. Symptoms of exposure to inorganic mercury include skin rashes and dermatitis, mood swings, memory loss, mental disturbances, and muscle weakness. The most common source of methylmercury to the average citizen is from eating fish that have accumulated methylmercury from the water. Infants can also be exposed via their mothers, who bioaccumulate mercury especially in their breast milk. Fetuses, infants, and children exposed to methylmercury have impaired neurological development (www.epa.gov). Methylmercury exposure also impacts cognitive thinking, memory, attention, language, and fine motor and visual spatial skills (www.epa.gov).

15.2 Lead

15.2.1 Sources

15.2.1.1 Minerals

There are a number of minerals that contain lead. These include galena (PbS; the most economically important lead mineral in the Earth's crust), boulangerite ($Pb_5Sb_4S_{11}$; found in vein deposits associated with galena, stibnite, sphalerite, pyrite, quartz, siderite, and other lead sulfosalts), cerussite ($PbCO_3$; a common mineral in oxidized zones of ore deposits containing galena), anglesite ($PbSO_4$; a secondary mineral found in the oxidized zone of ore deposits containing galena), and pyromorphite ($Pb_5(PO_4)_3Cl$), minetite ($Pb_5(AsO_4)_3Cl$), and vanadinite ($Pb_5(VO_4)_3Cl$) (three minor secondary minerals found in the oxidized zone of galena deposits) (Berry and Mason 1959, pp. 307, 345, 417, 430, and 455).

15.2.1.2 Synthetic Sources

The four most common sources of synthetic lead are lead-lead oxide batteries, lead in paints (typically present in the past, at a level of 0.5% or 5000 ppm), the pesticide lead arsenate ($Pb_3(AsO_4)_2$), and most importantly, tetra-ethyl lead (as a gasoline additive).

15.2.2 Production/Use

To reduce lead from its principle ores (galena, anglesite, and cerussite) to its base metal (Pb), the ores are roasted (in a simple process also referred to as calcining or smelting). Lead has a variety of uses in our industrialized world and is common in maintenance-free batteries, metal sheeting, piping, cable-covers, ammunition, and foil and in alloys such as pewter, solder, babbitt-metal, bronzes, and anti-friction metals (Lead 2018). Lead has been used as pigments, especially in paints (as $Pb(OH)_2(CO_3)_2$, replaced in modern paints with TiO_2), glassmaking (as PbO_2), and in the rubber industry. One of the most common uses of lead was in an anti-knock compound (tetra-ethyl lead) in leaded gasoline. Production figures for 1985 show that 4.11 megatons (Mt) of lead were smelted, with the dominant producers being the United States (1.0 Mt), the former USSR (0.50 Mt), West Germany (0.36 Mt), the United Kingdom (0.31 Mt), Canada (0.24 Mt), Australia (0.22 Mt), and France (0.22 Mt) (Gibble 1988, pp. 180–183).

15.2.3 Fate

Lead is one of the most common forms of heavy metal pollution throughout the world, due primarily to its use as a gasoline additive. In the environment, lead is predominately present, and absorbed in organisms, as a free cation (Pb^{2+}), and is classified as a toxic heavy metal. Lead is only slightly soluble in water and is mostly found adsorbed to minerals in soil and sediments. Lead from smelting operations and leaded gasoline was dispersed into the atmosphere, so respiration is another possible intake pathway. Lead has not been found to be readily adsorbed by plants but does enter the food chain through deposits on food crops. Lead is not known to bioaccumulate in fish nor is it methylated like mercury.

15.2.4 Environmental Distribution

As previously noted, lead is one of the most commonly found heavy metals although it is at part per million and part per billion levels. The US EPA requires action when the level of lead in drinking(?) water reaches 15 ppb but prefers a concentration of "zero," recognizing that almost any level of lead can be harmful to infants. Inhalation and ingestion of lead-contaminated paint chips are a common source of lead to infants and have received much public attention in recent years. In the past, however, widespread combustion of leaded gasoline was the major source of lead to all living organisms.

Lead has been known to be toxic for thousands of years, and the addition of lead to gasoline in the early 1900s came as a shock to most scientists, who certainly realized that if you put a toxic compound like tetra-ethyl lead in your gasoline tank, it would of course be spewed out into the world as you drive your car. However, the gasoline industry chose the cheaper anti-knocking agent regardless, and we faced decades of lead exposure even though other anti-knocking agents for our combustion engines were readily available (see Kittman 2000, a must-read for environmentalists, listed as the suggested reading for Chapter 9, available for viewing at http://www.globalleadnet.org/advocacy/initiatives/nation.cfm or for download at http://www.globalleadnet.org/pdf/TheSecretHistoryofLead.pdf).

Human intake and blood levels of Pb have been closely correlated with the twentieth century leaded gas use, and an excellent plot that demonstrates this tread has been published by Lippmann (1990). Fortunately, this means that biological lead levels dropped after leaded gasoline was replaced by unleaded, during the 1970s. Lead blood levels fell from 16 μg/dl in 1976, at the height of our leaded gasoline consumption, to less than 10 μg/dl in 1980 and have continued to fall in countries using unleaded gasoline (Lippmann 1990). Today, our estimated intake of Pb from all sources is approximately 50 μg/d in the United States. In the United States, unleaded gasoline was introduced in the 1970s and slowly took over the market as all cars were equipped with catalytic converters. The US EPA banned the use of leaded gasoline in highway vehicles in December of 1995; however, it is still used in airplane fuel. These efforts resulted in a decrease of lead in the air over the United States by 94% between 1980 and 1999 and a 78% decrease in the levels of lead in human blood between 1976 and 1991. These improvements are clearly indicated by a survey of lead concentrations by EPA shown in Figure 15.1 and is a direct result of the reduction in lead shown in Figure 15.2a,b.

Many developing countries still use leaded gasoline. By the early 1990s, however, Brazil and Canada had

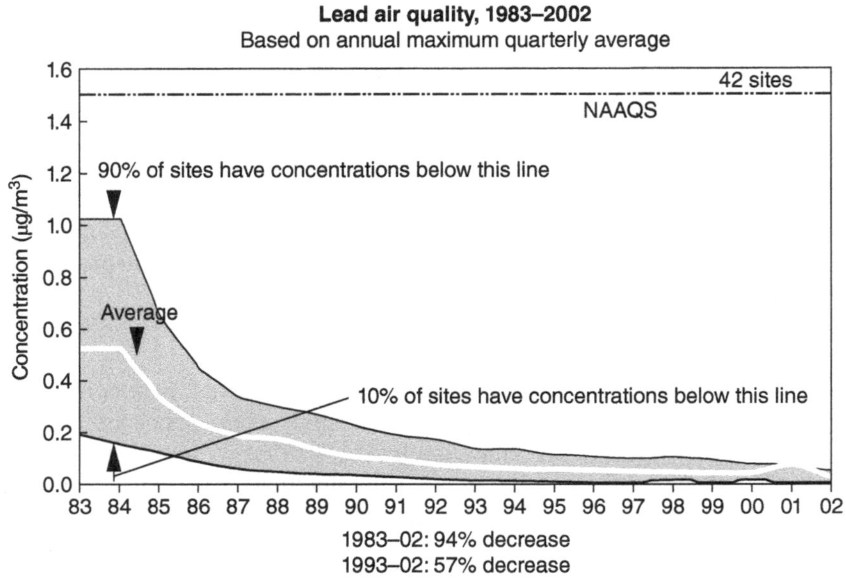

Figure 15.1 Lead air quality for a variety of sites in the United States from 1983 to 2002. Source: US EPA at www.epa.gov.

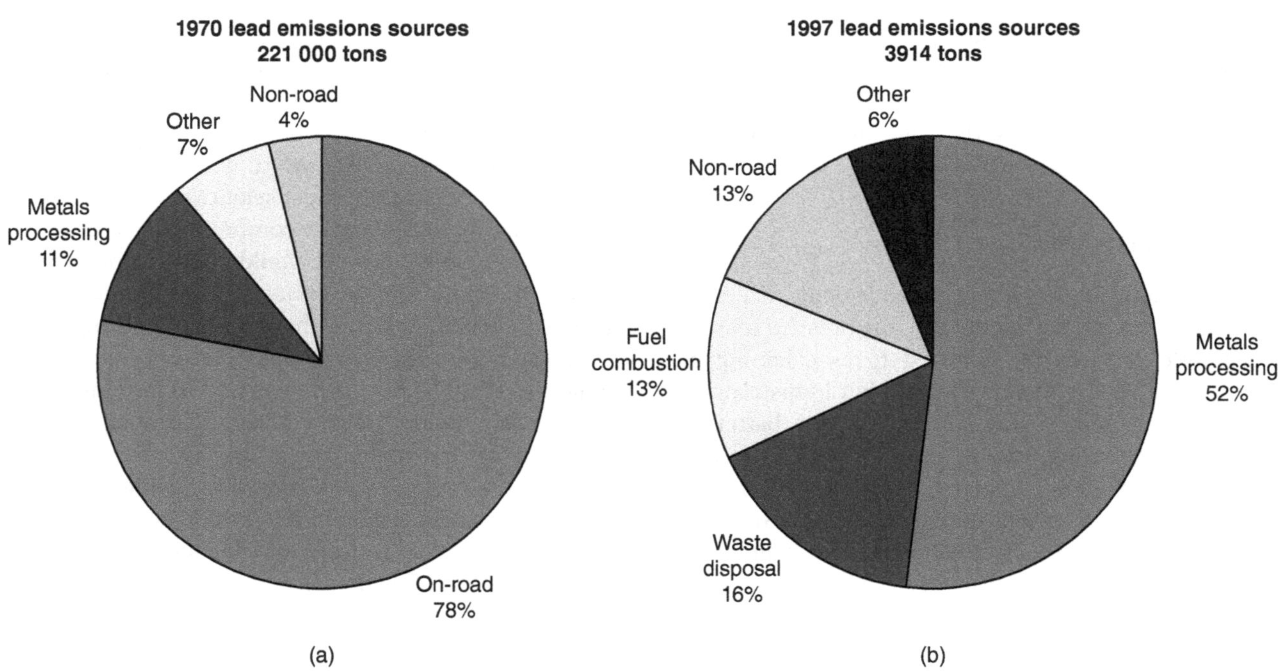

Figure 15.2 Lead emissions in the United States for years 1970 (a) and 1997 (b). Source: US EPA at www.epa.gov.

phased out lead in gasoline, and many other countries (Argentina, Iran, Israel, Mexico, Taiwan, Thailand, and most EU countries) significantly reduced the lead concentration in gasoline. Unfortunately, lead exposure and emissions on the global scale are on the increase, due primarily to the continued use of leaded gasoline in Africa, Asia, and South America. A complete history of our removal of lead from our human environment, for the most part, can be found in Dunnivant (2017).

15.2.5 Health Effects

Humans: Lead has been known to be toxic to humans as far back as 3800 BCE, when the Greeks noticed that drinking acidic beverages from lead containers could result in illness. Once lead is absorbed into the body, it enters the bloodstream and is transported to all parts of the body. Some is deposited into our bones, where it replaces calcium because Ca^{2+} and Pb^{2+} have similar ionic radii. This can result in brittle bones after years of accumulation. Lead also inhibits the function of enzymes involved in the synthesis of heme, the porphyrin-binding complex in hemoglobin that serves as the binding site of O_2. Lead affects nerve cells by decreasing the nerve conduction velocity, even at relatively low blood levels. Lead also causes damage to the kidneys, liver, brain, and nerves and can result in seizures, mental retardation, behavioral disorders, memory problems, and mood changes. It also causes high blood pressure and increases heart disease and anemia. Lead toxicity primarily occurs due to lead's ability to bind to critical proteins that are also nitrogen and sulfur ligands, and thus to interfere with their function. Lead can be removed from the body by intravenous injection of metal chelators that compete for the binding of Pb with these proteins. Chelated Pb is then excreted from the body by the kidneys.

A common source of lead toxicity to wildlife is the deposition of fishing weights and lead shot from munitions in river and lake sediments. Waterfowl, sorting through these sediments for food, ingest the lead particles, which are then dissolved in their acidic stomachs and spread throughout their bodies in high doses. The United States has replaced lead shot with other metals.

15.3 PCBs

15.3.1 Sources

Industrial Synthetic: PCBs are produced by the chlorination of biphenyl with chlorine gas. There are 10 possible sites for chlorination on the biphenyl ring, and more than one site can be chlorinated. This leads to 209 possible PCB compounds, referred to as congeners. One such structure is shown in Figure 15.3

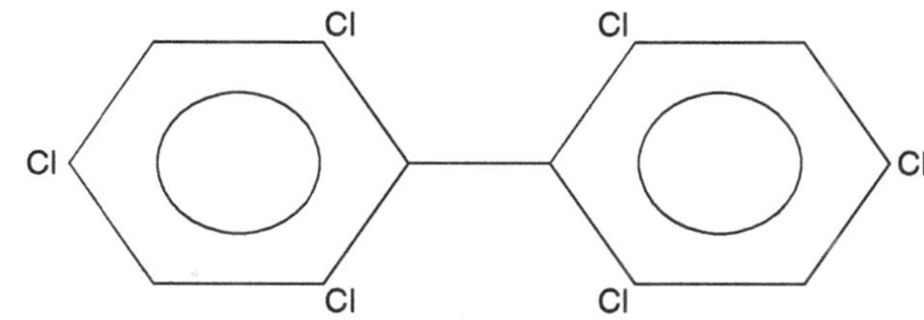

Figure 15.3 Structure of 2,2′,4,4′,6,6′-hexachlorobiphenyl, a PCB.

for 2,2′,4,4′,6,6′-hexachlorobiphenyl. For simplicity purposes, the governing body of chemical nomenclature, IUPAC, has developed a system for numbering the PCB congeners from 1 to 209.

15.3.2 Production/Use

The industrial synthesis of PCBs is a bit crude and does not result in pure congeners but in complex mixtures referred to in the United States as Aroclors, the trade name patented by the US Swann Chemical Company in 1929. The US Swann Chemical Company was later bought in 1935 by Monsanto Chemical Company, which became the main manufacture of PCBs in the US Trade names for industrial-grade PCBs include Clophen (Bayer, Germany), Phenoclor (Caffaro, Italy), Pyralene (Kanegafuchi Chemical Company, Japan), Kanechlor (Prodelec, France), Fenchlor (Chemko, former Czechoslovokia), and Delor (from the former USSR), but there are many others.

The common Aroclor mixtures include Aroclors 1221, 1232, 1242, 1248, 1254, 1260, 1262, 1268, and 1270. The basic rules of Aroclor nomenclature are that the first two numbers in the name refer to the 12-carbon basic structure of the biphenyl structure. The last two numbers refer to the average degree of chlorination of the biphenyl rings. Note that any given Aroclor mixture will contain tens of different PCB molecular structures. The degree of chlorination is a result of the reaction time with chlorine, which is determined by the desired chemical and physical properties. As a side note, it is interesting to note the shift in nomenclature when PCBs started receiving a "bad name." At this time, a "new" Aroclor was released that was supposed to have less detrimental environmental properties. The new Aroclor was numbered 1016. The selection of "10" was unclear, but in keeping with the established numbering system, the "16" was clearly meant to imply a lower degree of chlorination. In fact, Aroclor 1016 was essentially the same and Aroclor 1242.

During the early years of their production, PCBs were thought of as the perfect chemical – they were highly unreactive and did not degrade. However, this means that PCBs are also very persistent in the environment, as we will discuss in the next section. The unreactive nature of PCBs lends them to many functions, including use as dielectric fluids in capacitors and transformers, as heat transfer fluids, as hydraulic fluids, and in lubricating and cutting oils. PCBs were also employed as additives in pesticides, paints, copying paper, carbonless copy paper, adhesives, sealants, and in plastic formulations.

Little was known about the dangers of PCBs prior to the 1980s. For example, at the height of the environmental movement in the mid-1970s, PCBs were presented in undergraduate microbiology classes as an ideal fluid for preparing microscope slides, due to the fact that they do not cause refraction of light. Students were neither told the name of the fluid nor advised to take precautions in handling it, and therefore came into extensive contact with the fluid. Of course, in the 1970s, there was no such things as material safety data sheet (MSDS) or the Citizen's Right to Know Act. Still, this example demonstrates the importance of always having and reading the label of chemicals that you work with.

One of the most common and unfortunate uses of PCBs was as dielectric fluid, used as a fluid insulator in electrical systems. In order for a chemical to be a good dielectric fluid, it should be an excellent electrical insulator (a nonconductor) and be heat resistant and not thermally degrade, since most electrical equipments become very hot. Aroclor mixtures were ideal for these uses, and for a time were legally placed in all transformers and capacitors. But transformers wear out, on occasion explode, and sometimes leak, spreading PCBs wherever transformers were used or disposed of.

The Toxic Substances Control Act of 1976 gave EPA the authority to regulate all aspects of the manufacture, distribution, use, and disposal of chemicals, including PCBs. The US Congress banned the manufacture, processing, and distribution in commerce or use of PCBs after 1 January 1978. Congress further specified that after 1 January 1979, no person could process or distribute in commerce any PCBs unless granted permission by the EPA. But the US Congress only regulates the United States, and some chemical companies simply shifted their operations to countries where PCBs were, and in some cases still are, legal.

15.3.3 Fate and Environmental Distribution

PCBs are hydrophobic compounds, and therefore, they prefer to sorb to virtually any surface, especially soil and sediment media, rather than remain free in water. Thus, dissolved water concentrations will be very low in contaminated environments compared to soils and sediments, especially those containing organic matter. This distribution of PCBs between phases occurs by a partitioning process, as discussed in Chapters 2 and 3. Even higher concentrations of PCBs will be found in the lipid deposits of biota, enabling PCBs to readily bioaccumulate in animals.

In general, PCBs are considered to be nonbiodegradable, although some laboratory studies do show considerable degradation in enriched, highly specialized bacterial cultures. Some in the chemical industry like to use these studies to claim that the PCBs are not a persistent problem, since they have been shown to degrade. But if this were true, and the laboratory rates and conditions were applicable to the natural world, PCBs would have been degraded decades in the past. The clearly documented presence of PCBs in soils, sediments, and biota around the world show that PCBs do in fact persist in natural environments.

PCBs are truly one of our ubiquitous pollutants, in that they can be detected in every environmental sample around the world, given sufficiently low detection limits. In general, PCBs levels around the world are decreasing but slowly given the refractory nature of PCBs. A few examples of PCB concentrations in animals are marine organisms (0.003–212 μg/g), fish (0.1–190 μg/g), birds (0.1–14 000 μg/g), and humans (0.001–75.5 μg/g). Concentrations for environmental compartments range from 0.004–36 ng/m^3 for ambient air, from 5 to 10 800 ng/m^3 for indoor air, from 9.8 to 58 000 ng/m^3 for stack gas, from 0.01 to 10×10^6 ng/m^3 for occupational sites, from 0.004 to 4200 ng/l for water, from 0.1 to 250 ng/l for rain, from 0.02 to 17 800 μg/g for soil, and from 0.000 08 to 61 000 μg/g for sediment.

15.3.4 Health Effects

PCBs are listed by EPA as a probable human carcinogen, and studies indicate that toxicity is directly related to the chlorine substitution pattern. The lack of ortho-chlorine (chlorines in the 2, 2′, 6, or 6′ position) gives the PCB a similar conformation as chlorinated dioxins (2,3,7,8-tetrachlorodioxin is one of the most toxic chemicals known to humans). Toxicology for PCBs is also very species-specific, with some mammals showing high susceptibility to birth defects and cancer related to exposure.

15.4 DDT

15.4.1 Sources

15.4.1.1 Synthetic

DDT was first synthesized in 1873 by Othmar Zeidler, who was working at the University of Strasbourg. The structure for DDT and its degradation produces, DDD and DDE, are shown in Figure 15.4.

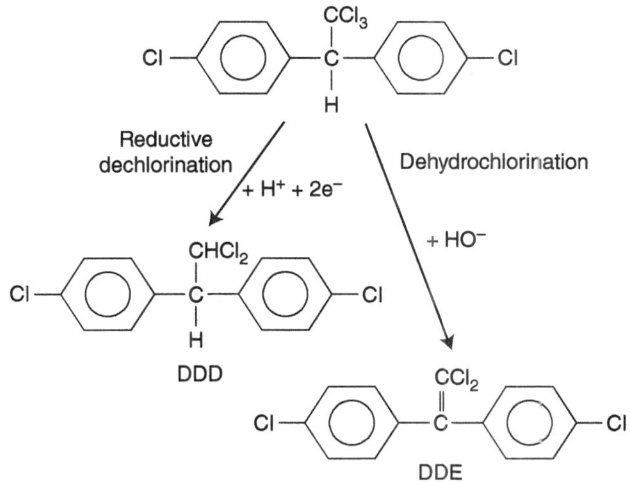

Figure 15.4 Structure of DDT and its major degradation products.

15.4.2 Production/Use

DDT is one of many chlorinated pesticides and persistent organic pollutants (POPs) that have been in the news for decades. Recently, efforts have focused on an international treaty to ban their production and use (Chemical and Engineering News 2004a,b,c; Burke 2004; Hogue 2004). POPs of immediate interest include aldrin, chlordane, DDT, dieldrin, dioxins, eldrin, furans, heptachlor, hexachlorobenzene, mirex, PCBs, and toxaphene. In 2001, the Stockholm Convention on POPs entered into effect stating that treaty partners will begin evaluating additional chemicals for possible control starting in May 2005. Table 15.1 shows the cumulative world production and use of each chemical.

DDT was (and is) primarily used as an insecticide for mosquito control. While it is not used in most developed countries, it is commonly used in areas affected by malaria, although its effectiveness as an insecticide has diminished over time due to adaptation and evolution of the mosquito. Many people make the statement that "DDT has saved more lives than it has hurt." This is definitely true if you are only concerned with human life. DDT has been a very effective tool during wartime and in mosquito infested countries in fighting human disease. But it has also been a disaster for other species, especially raptors, or birds of prey, such as the American bald eagle.

15.4.3 Fate

All the POPs, including DDT, are hydrophobic chlorinated hydrocarbons and exhibit very low solubility in water (sub-ppm levels). Their behavior in the environment is similar to that of PCBs. Chlorinated hydrocarbons have high soil/water partition coefficients and highly partition to organic matter in soil and sediment, increasing the pollutant concentration on the soil and decreasing the concentration in the water. POPs are well known to bioaccumulate and are found in lipid regions in plants and organisms.

15.4.4 Environmental Distribution

DDT, and many other POPs, have been found in every environmental compartment in the world, including remote areas such as high alpine lakes (Catalan et al.

Table 15.1 Production and use of 12 POPs.

POP	Production start date	Cumulative world production (tons)	Designed use
DDT	1942	2.8–3 million	Domestic and agricultural insecticide against mosquitoes
Aldrin	1950	240 000	Insecticide to control soil pests (termites) on corn and potatoes; fumigant
Dieldrin	1950	240 000	Insecticide used on fruit, soil, and seeds
Endrin			Rodenticide and insecticide used on cotton, rice, and corn
Chlordane	1947	70 000	Insecticide used in fire ant control and on a variety of crops
Heptachlor	1952	No data	Insecticide in seed grain and crops; termiticide; used for fire ant control
Hexachlorobenzene	1945	1–2 million	Used as a fungicide; also a byproduct in pesticide manufacture
Mirex	1959	No data	Insecticide and fire retardant
Toxaphene	1948	1.33 million	Insecticide, especially against ticks, mites, maggots; used on cotton
PCBs	1929	1–2 million	Used as weatherproofers, hydraulic fluids in transformers, liquid insulators, dielectrics, and to prolong residual activity of plastics
Dioxins			By-product of combustion – especially plastics – and of chlorine products manufacture and paper production
Furans			By-product, often bonded to dioxins

Source: Christen (1999). Reproduced with permission of American Chemical Society.

2004; Vives et al. 2004) and in the Antarctic food web (Chiuchiolo et al. 2004).

15.4.5 Health Effects

DDT and the other POPs are considered to be endocrine disruptors, interfering with the hormonal systems of the human body (and other animals). An increase in reproductive abnormalities in humans and wildlife has occurred over the last 20–40 years, and many scientists are looking closely at chlorinated hydrocarbons as the cause of these observations. Studies indicate a decrease in fertility in men and a significant increase in still births, neonatal deaths, and congenital defects.

DDT adversely affects animals by causing reproductive and developmental failure, possible immune system effects, and widespread deaths of birds. Long-term exposure to DDT has been tied to neurological, hepatic, renal, and immunologic effects in animals. DDT prevents androgen from binding to its receptor, thereby blocking androgen from guiding normal sexual development in male rats. The best known DDT effect is the interference of the estrogen system in birds of prey that results in the thinning of eggshells and premature breaking of the eggs during nesting. DDT almost single-handedly decimated the populations of bald eagles, hawks, and owls in the US countryside.

15.5 Endocrine Disruptors

Compounds that disrupt endocrine systems in animals have been widespread for decades but were generally thought to be active only in animals, not humans, at the low concentrations commonly found in the environment. Today, however, scientists have discovered that many widely used chemicals can act as endocrine disruptors in the human body, in much the same way that DDT is an endocrine disruptor for bald eagles and other raptors. The story of endocrine disruptors is long and involved, and is still unclear, but scientists are making daily headway into the cases and effects of endocrine disruptors. The following section is a summary of what we currently know or suspect.

15.5.1 Sources

15.5.1.1 Synthetic

Endocrine disruptors include a variety of POPs, including DDT. In recent years, numerous chemicals used in plastic production have been found to be or suspected to be endocrine disruptors, with the most publicly known as BPA, a component of polycarbonate products. BPA has received considerable media coverage and is the subject of legislation in more progressive states. A modern summary of endocrine disruptors can be found in Dunnivant (2017; end of Chapter 4).

15.5.2 Uses and Points of Contact

The group of commonly used endocrine disruptors is highly varied, and they are used in a wide variety of modern consumer products. Basically, anything resembling a type of plastic contains potential endocrine disruptors.

Perhaps, the most direct source of endocrine disruptors to people is from foods, which are exposed to EDCs either by to direct application or as a result of leaching from packaging materials (Casajuana and Lacorte 2003). Many food products are contained in materials, including plastic bottles and lined aluminum cans, that contain or consist of endocrine-disrupting compounds (EDCs) that can leach into the food (Casajuana and Lacorte 2003).

However, there are many sources of EDCs in our environment other than food, including medical supplies, children's plastic toys, and the food and waste produced by livestock treated with hormones. Estrogen hormones are excreted by both sexes and all species of farm animals in both urine and feces, creating the potential for environmental contamination (Hanselman et al. 2003). Bovine waste in the United States is a much larger source of hormones, even, than human waste, due to the widespread use of hormones and large scale of cattle production (Raman et al. 2004). However, the use of livestock waste in agricultural fields for fertilizer does dilute the estrogen chemicals, whereas human wastes from wastewater treatment plants are dispersed directly into the environment, acting as point sources of EDCs (Thacker 2004). The persistence of EDCs also makes them difficult to eliminate in the wastewater treatment process.

Polybrominated diphenyl ethers (PBDEs), one specific type of endocrine disruptors, are used as flame retardants and are used on indoor objects such as furniture, carpeting, mattresses, televisions, and certain plastic products (Stapleton et al. 2005). They are also used in certain manufactured products, such as polymers, resins, electronic devices, building materials, textiles, and polyurethane foam padding (Oros et al. 2005), and on many consumer products, such as electrical and office equipment, including computers, televisions, copiers, and printers, and in house products such as upholstery, carpeting, wall coverings, and ceiling materials. Because of their effectiveness in suppressing fires, they have been applied to many items in large amounts, to prevent property damage. They may also be spread into the environment by combustion, either in routine garbage incineration or in major events such as the destruction of the World Trade Center in 2001 (Litten et al. 2003).

The health effects of PDBEs include reproductive and developmental toxicity and cancer, as discussed in the following (Schecter et al. 2004).

Much public attention has been given to the possible dangers of polycarbonate water bottles. Polycarbonate contains the EDC bisphenol-A, which interferes with estrogen uptake by binding to estrogen receptors (Whittelsey 2003). There have been many warnings against washing polycarbonate bottles with harsh detergents and against the use of polycarbonate bottles that have been harshly used or are old because they are more likely to release bisphenol-A into their contents. One such product is Nalgene bottles. They are made of polycarbonate and are thought to release bisphenol-A under these conditions (*Non-Toxic Times* 2004). Unfortunately, the specific effects of small bisphenol-A to the human body are not well enough understood to know whether the levels leached from water bottles could be harmful.

15.5.3 Fate and Environmental Distribution

Endocrine disruptors, although they include a wide variety of compounds, are generally hydrophobic, with behavior in the environment similar to that of PCBs and DDT. EDCs are also known to bioaccumulate through food chains.

A study of the bioaccumulation of two perfluorinated compounds and known endocrine disruptors – perfluorooctanesulfonate (PFOS) and perfluorooctanoate – in an Eastern Arctic marine food web showed that there exists a positive linear relationship between PFOS concentrations in organisms based on their wet weight and trophic level, indicating that a higher trophic level corresponds to more PFOS (Tomy et al. 2004).

Endocrine disruptors are found in nearly every natural and human environment. PBDEs in particular have been measured in harbor seals, fish, and a local municipal wastewater treatment plant effluent in San Francisco. Additionally, a study in 2002 measured PBDEs in water, surface sediments, and bivalves in the San Francisco Estuary. Bivalves populations are known to reflect water conditions in an ecosystem and to demonstrate the level of contaminant bioavailability in the water column. They are also an important food source for various fish. When compared to levels of PBDEs in European water sources, the concentrations in the San Francisco Estuary were much higher but similar to other water sources in the United States, even though California has some of the strictest state regulations regarding the use of PBDEs.

A study comparing PBDE concentrations in supermarket salmon from eight different farming areas in North America and Europe found that the highest concentration of PBDEs in salmon came from farmed salmon in Europe, followed by farmed salmon in North America, and then wild salmon from the Pacific Ocean. However, all the different fish samples showed the presence of PBDE. The highest PBDE levels, in Chinook salmon, have been attributed to the Chinook perhaps being at a higher trophic level than other salmon, resulting in more bioaccumulation of these compounds (Hites et al. 2004).

Some studies have revealed possible solutions to the presence of EDCs in various environmental compartments. As previously mentioned, there are high amounts of estrogen compounds in sewage from sources such as human and animal waste. A few previously suggested solutions to this problem have been ozonation, UV radiation, membrane filtration, and activated carbon adsorption. However, all these suggestions would add significant costs to the sewage treatment process. An article by Andersen et al. (2003) states that an "activated sludge system for nitrification and denitrophication including sludge recirculation can appreciably eliminate natural and synthetic estrogens." For example, some data reveal that E1 and EE2 (two forms of estrogen) were reduced by 50% and 70%, respectively, after the first denitrification tank. Natural estrogens can then be reduced further, biologically, in the second denitrification tank, achieving an overall removal of more than 98%. A better understanding of the fate of estrogens is necessary in order to develop more efficient and less expensive methods to eliminate estrogens from wastewater (Andersen et al. 2003).

Another suggested method of removing EDCs from the environment is by using plants to "mobilize and translocate DDT" and other POPs from soil environments. With more weathering of a soil, it is more difficult to remove POPs, which tend to bioaccumulate due to their persistence in the environment. Plants may ingest these contaminants through soil water, where they are translocated to the upper part of the plant and may be metabolized or mineralized. This process, called phytoremediation, uses vegetation to treat polluted soils, sediments, groundwater, or surface. One study tested five plants for their ability to phytoremediate: zucchini, tall fescue, alfalfa, rye grass, and pumpkin. Pumpkin had the highest value of pollutants removed from the soil, suggesting that it is the most valuable plant for phytoremediation (Lunney et al. 2004).

15.5.4 Health Effects

EDCs interfere with the many functions of the endocrine system. These functions include the regulation of cellular proliferation and differentiation, growth, development, reproduction, senescence, behavior, regulating mood, tissue function, and metabolism (Berne and Levy 1990, p. 779). Endocrine cells or glands send messages through the body in the form of hormones or other molecular signals, which are carried through the bloodstream and act on other targeted cells that contain receptors for specific hormones. The hormones come in contact with many cells but only interact with those containing the correct receptor (Encyclopedia Britannica 2002, p. 288). The response is either to stimulate or repress the transcription of proteins that are necessary to perform certain bodily functions (Campbell and Reece 2002, p. 960). This pathway controls the performance of many bodily functions.

EDCs cause harm by either mimicking hormones or blocking their receptors, and they have been a source of concern for humans and wildlife for many years because of the serious effects they have on the endocrine system. Some possible effects of EDCs on the body are increased likelihood of testicular, prostate, and breast cancer, decreased sperm count and quality, and other reproductive disorders (Lee et al. 2003). In the recent years, the concern about EDCs present in the environment has increased as more people have become aware of their widespread use and potentially harmful effects. In response, scientists around the world have performed many studies to determine the levels of EDCs in various consumer products, identify their specific effects, and evaluate the potential dangers associated with their ubiquity in the environment.

In addition to DDT, other EDCs present in natural waters mimic steroidal estrogen hormones and can have strong effects on the reproductive abilities of aquatic wildlife such as fish, turtles, and frogs (Hanselman et al. 2003). In 2003, a group of scientists from Europe performed a study examining the extent of the effects of EDCs on fish from aquatic ecosystems in Europe. Estrogen in fish stimulates the synthesis of the protein vitellogenin, used in the development of oocytes. Therefore, the presence of estrogen-active endocrine disrupters in the fish's environment also stimulates the production of vitellogenin, which results in feminization of males. Although not every fish responds to contact with contaminated water, an overall increase in vitellogenin levels in a population can indicate the presence of endocrine disrupters in their environment (Pickering and Sumpter 2003).

In the Potomac River near Sharpsburg, Maryland, feminization of male fish has been observed to produce eggs inside their sex organs. The discovery has been linked to endocrine disrupters inhibiting normal sexual development due to inhibition of hormonal signals. Of course, the Potomac also supplies drinking water for the Washington, DC area (Associated Press 2004). Studies such as this have led to wider concern about the effects of EDCs and their widespread presence in the environment.

PBDEs, a group of endocrine disruptors discussed previously, have been detected in human serum, adipose tissue, and breast milk. Their levels are 17 times higher in people living in the United States than in those living in Europe (Stapleton et al. 2005). The presence of PBDEs in mice has resulted in neurotoxicity (Oros et al. 2005), and the metabolism of PBDEs is thought to be similar to that of PCBs because of their similar structures. Therefore, PBDEs accumulate in organisms and may affect them similarly to PCBs (Tuerck et al. 2005). PBDEs alter the thyroid hormone homeostasis and weakly prevent the function of estrogen receptors. More specifically, in neonatal mice, they "disrupt spontaneous behavior, impair learning and memory, and induce other neurotoxic effects ..." (Hites et al. 2004).

So what do we do with all this information about EDCs in our bodies and the environment? The various studies mentioned previously reveal many different sources of EDCs: wastewater effluent, food, plastic, furniture, medical supplies, livestock waste, etc. The levels (concentrations) of EDCs in the human body seem to be on the same magnitude as the concentrations of our own hormones. Additionally, many of these chemicals bioaccumulate through the food chain; therefore, animals that eat higher on the food chain will contain even higher levels of EDCs. Ongoing animal testing seeks to discover the effects of EDCs on humans, although, as always, interspecies extrapolation will no doubt prove very difficult.

EDCs may seem completely unavoidable in modern society, due their widespread presence, persistence, and tendency to bioaccumulate. Clearly, continued research into their effects, their alternatives, and their environmental fates is necessary to provide the knowledge necessary to evaluate the costs and benefits of their omnipresence in our lives.

15.6 Plastics

Since c. 1970s, plastics have become an intricate part of our lives, from automotive components to clothing fabrics, to food storage containers, to medical devices, and to the ever-present plastic shopping bags. Given that their density is often less than that of water, they float, and when they enter our waterways, they eventually end up in the nearest oceans. Today, the northern Atlantic and Pacific Oceans, Southern Pacific, Indian Ocean, as well as the Caribbean Gulf have accumulated these waste produces into ocean garbage patches. Most notably, the Pacific Ocean patch is now estimated to be three times of the US. State of Texas. These sad environmental situations are individually discussed at https://en.wikipedia.org/wiki/Garbage_patch. In addition, every beach tested has found the presence of microplastics, ground up microscopic pieces of plastics washing in with the tides. These plastics are now being found in marine wildlife (https://en.wikipedia.org/wiki/Plastic_pollution).

Three opinion pieces in *Environmental Science & Technology*, by Burton (2017), Hale (2018), and Kramm et al. (2018), serve as an interesting class discussion on the topic.

15.7 Carbon Dioxide and Climate Change

Decades ago, even a century ago, the hypothesis of human-caused climate change was appropriately debated. But for several decades now, scientists have made a fact-based argument for this case. Today, over 99% of scientists researching climate change are in perfect agreement: humans are causing climate change. One of the largest and most technical group of people, the Intergovernmental Panel on Climate Change, have methodically studied the hypothesis of human-caused climate change and agree that humans and fossil fuels are the culprit. The Intergovernmental Panel on Climate Change reports are readily available on the Internet and summarized on Wikipedia, as well as other places. The stunning, and disturbing summary, by Fischetti (2014), goes as follows:

The causes of climate change are concluded as follows:

- "CO_2 emissions are by far the largest cause of global warming and ocean acidification, and they are rising.
- Methane emissions are the second largest cause of warming, and they are rising.
- Since 1950, human activities have led to virtually all temperature rise.
- Natural forces have caused virtually none of the temperature rise.
- The largest human sources of CO_2 emissions are burning fossil fuels, making cement, and burning off gas ("flaring") from oil and gas production."

The impacts are as follows:

- "Sea level is rising at an increasing pace.
- Glaciers are melting, ice sheets are thinning, and Arctic sea ice is disappearing.
- Permafrost is thawing.
- In North America, snow pack is decreasing.
- The number of cold days and nights are decreasing.
- The number of hot days and nights are increasing.
- Heat waves will occur more often and last longer.
- Heavy rainstorms and snowstorms will become more intense and frequent.
- Overall, precipitation will rise in high latitudes and the equatorial Pacific. In midlatitudes, dry areas will get drier, wet areas will get wetter.
- Species are vanishing at an alarming and ever-increasing rate.
- Most plants, small mammals, and ocean organisms cannot adapt fast enough to keep up with changes.
- Global temperature rise greater than 2 °C will compromise food supplies globally.
- Human health problems will get worse.
- Risks to poorer people are greater than for others, in all countries."

What we must do:

- "To avoid severe damage to natural and human systems, the world should keep global warming to less than 2 °C above preindustrial levels.
- Without more mitigation than is being done today, the temperature is more likely than not to rise by 4 °C by 2100.
- Significant reductions in greenhouse gas emissions by 2050 can significantly reduce warming by 2100.
- Keeping greenhouse gases in the atmosphere below the equivalent of 450 parts per million of CO_2 can keep warming below 2 °C.
- Levels are likely to stay below 450 ppm if human emissions are reduced 40–70% by 2050 compared with 2010 levels.
- Allowing levels to reach 530 ppm by 2100 gives the planet slightly better than 50–50 odds of staying below 2 °C; that would require reducing emissions to 25% from 55% by 2050 versus 2010.
- To hit a target of 430–530 ppm by 2100, the world must invest several hundred billion dollars a year in low-carbon electricity sources and energy efficiency.
- It is highly unlikely the world will stay below 450 ppm without the widespread use of carbon capture and storage technologies."

These statements are clear, confident, and conclusive. This work was endorsed by another internationally recognized society, the Norwegian Nobel Committee, when they awarded the Nobel Peace Prize to the IPCC in 2007.

Today, we only have a political argument that remains, funded by fossil fuel industry and paid-in-full politicians. Dunnivant (the first author of this text; 2017) has summarized the topic in detail and is a must read for anyone interested in saving the human race on Planet Earth.

Bibliography

Andersen, H., Siegrist, H., Halling-Sorensen, B., and Ternes, T.A. (2003). Fate of estrogens in a municipal sewage treatment plant. *Environ. Sci. Technol.* **37** (18): 4021–4026.

Associated Press (2004). Male fish growing eggs found in potomac (21 December).

(2004). On the trail of water bottle toxins…are hikers and others quenching outdoor thirst with H_2-UH-OH?. *The Non-Toxic Times* (April).

Baird, C. (1995). *Environmental Chemistry*. New York: W.H. Freeman and Company.

Berne, R.M. and Levy, M.N. (1990). *Principles of Physiology*. St. Louis: The C.V. Mosby Company.

Berry, L.G. and Mason, B. (1959). *Mineralogy: Concepts, Descriptions, and Determinations*. San Francisco: W.H. Freeman and Company.

Betts, K. (2003). Dramatically improved mercury removal. *Environ. Sci. Technol.* **37** (15): 283–284A.

Brossa, L., Pocurull, E., Borrull, F., and Marcé, R.M. (2002). A rapid method for determining phenolic endocrine disrupters in water samples. *Chromatographia* **56** (9/10): 573–576.

Burke, M. (2004). POPs treaty takes flight. *Environ. Sci. Technol.* 157A.

Burton, G.A. Jr., (2017). Stressor exposures determine risk: so, why do fellow scientists continue to focus on superficial microplastics risk? *Environ. Sci. Technol.* **51**: 13515–13516.

Campbell, N.A. and Reece, J.B. (2002). *Biology*. San Francisco: Benjamin Cummings.

Carson, R. (1962). *Silent Spring*. Boston, Cambridge: The Riverside Press.

Casajuana, N. and Lacorte, S. (2003). Presence and release of phthalic esters and other endocrine disrupting compounds in drinking water. *Chromatographia* **57** (9/10): 649–655.

Catalan, J., Ventura, M., Vives, I., and Grimalt, J.O. (2004). The roles of food and water in the bioaccumulation of high mountain lake fish. *Environ. Sci. Technol.* **38** (16): 4269–4275.

Chemical and Engineering News (2004a). Senators criticize mercury controls at chlorine plants (31 May), p. 18.

Chemical and Engineering News (2004b). Massachusetts cuts mercury at power plants (31 May), p. 18.

Chemical and Engineering News (2004c). Persistent pollutant pact enters into force (24 May), p. 24.

Chemical and Engineering News (2005a). EPA hit for mercury air standard (14 February), p. 24.

Chemical and Engineering News (2005b). EPA's mercury analysis biases, GAO says (14 March), p. 29.

Chiuchiolo, A.L., Dickhut, R.M., Cochran, M.A., and Ducklow, H.W. (2004). Persistent organic pollutants at the base of the Antarctic marine food web. *Environ. Sci. Technol.* **38** (13): 3551–3557.

Christen, K. (1999). U.N. negotiations on POPs snag on malaria. *Environ. Sci. Technol.* **33** (21): 444–445A.

Christen, K. (2004). States going it alone on mercury controls. *Environ. Sci. Technol.* **38** (9): 157A.

Crenson, S.L. (2002). Study records elevated mercury. *Associated Press* (20 October Sunday).

Dunlap, T.R. (1981). *DDT: scientists, citizens, and public policy*. Princeton: Princeton University Press.

Dunnivant, F.M. (2017). *Environmental Success Stories: Solving Major Ecological Problems & Confronting Climate Change*. New York: Columbia University Press.

Erickson, M.D. (1997). *Analytical Chemistry of PCBs*, 2e, 1–98. New York: Lewis Publishers.

Fischetti, M. (2014). 29 Bullets that all about climate challenge. *Scientific American; Blogs* (3 November 2014). http://environmentalfuture.org/29-bullets-tell-all-about-climate-challenge/ (accessed 25 November 2014).

Gribble, C.D. (1988). *Rutley's Elements of Mineralogy*, 27e. London: Unwin Hyman.

Guenther, K., Heinke, V., Thiele, B. et al. (2002). Endocrine disrupting nonylphenols are ubiquitous in food. *Environ. Sci. Technol.* **36** (8): 1676–1680.

Hale, R.C. (2018). Letter to the editor: are the risks from microplastics truly trivial? *Environ. Sci. Technol.* **52**: 931–931.

Hanselman, T.A., Graetz, D.A., and Wilkie, A.C. (2003). Manure-borne estrogens as potential environmental contaminants: a review. *Environ. Sci. Technol.* **37** (24): 5471–5478.

Hileman, B. (2002). FDA suggests replacing DEHP in plastics. *Chemistry and Engineering News* (16 September).

Hileman, B. (2003). Clash of views on bisphenol a: finding gives rise to controversy over interpretation of scientific literature. *Chemistry and Engineering News* (5 May).

Hileman, B. (2004). Government regulation, phthalates in toys: EU bans some phthalates in toys and child care products. *Chemistry and Engineering News* (4 October).

Hites, R.A., Foran, J.A., Schwager, S.J. et al. (2004). Global assessment of polybrominated diphenyl ethers in farmed and wild salmon. *Environ. Sci. Technol.* **38** (19): 4945–4949.

Hogue, C. (2004). U.S. vote at treaty meetings threatened. *Chem. Eng. News* **82** (13): 22–23.

Inoue, K., Kato, K., Yoshimura, Y. et al. (2000). Determination of bisphenol A in human serum by high-performance liquid chromatography with multi-electrode electrochemical detection. *J. Chromatogr.* **749**: 17–23.

Kittman, J.L. (2000). *The Secret History of Lead: Special Report*. The Nation. http://www.thenation.com/article/secret-history-lead (accessed 3 June 2014).

Kramm, J., Völker, C., and Wagner, M. (2018). Superficial or substantial: why care about microplastics in the anthropocene? *Environ. Sci. Technol.* **52**: 3336–3337.

Lead (2018). Lead. https://en.wikipedia.org/wiki/Lead (accessed 8 January 2019).

Lee, L.S., Strock, T.J., Sarmah, A.K., and Rao, P.S.C. (2003). Sorption and dissipation of testosterone, estrogens, and their primary transformation products in soils and sediment. *Environ. Sci. Technol.* **37** (18): 4098–4105.

Lippmann, M. (1990). Lead and human health: background and recent findings. *Environ. Res.* **51**: 1–24.

Litten, S., McChesney, D.J., Hamilton, M.C., and Fowler, B. (2003). Destruction of the world trade center and PCBs, PBDEs, PCDD/Fs, PBDD/Fs, and chlorinated biphenylenes in water, sediment, and sewage sludge. *Environ. Sci. Technol.* **37** (24): 5502–5510.

Lunney, A.I., Zeeb, B.A., and Reimer, K.J. (2004). Uptake of weathered DDT in vascular plants: potential for phytoremediation. *Environ. Sci. Technol.* **38** (22): 6147–6154.

North, K.D. (2004). Tracking polybrominated diphenyl ether releases in a wastewater treatment plant effluent, Palo Alto, California. *Environ. Sci. Technol.* **38** (17): 4484–4488.

Oros, D.R., Hoover, D., Rodigari, F. et al. (2005). Levels and distribution of polybrominated diphenyl ethers in water, surface sediments, and bivalves from the san francisco estuary. *Environ. Sci. Technol.* **39** (1): 33–41.

Pacyna, E.G. and Pacyna, J.M. (2002). Global emission of mercury from anthropogenic sources in 1995. *Water Air Soil Pollut.* **137**: 149–165.

Personal communication with Paul H. Yancey, Biology Department, Whitman College, (2004).

Pickering, A.D. and Sumpter, J.P. (2003). Comprehending endocrine disrupters in aquatic environments. *Environ. Sci. Technol.* **37** (17).

Raman, A.R., Williams, E.L., Layton, A.C. et al. (2004). Estrogen count of dairy and swine wastes. *Environ. Sci. Technol.* **38** (13): 3567–3573.

Renner, R. (2004). Mercury woes appear to grow. *Environ. Sci. Technol.* **38** (8): 144A.

Rouhi, A.M. (2002). Mercury showers. *Chemical and Engineering News* (15 April).

Saar, R.A. (1999). New efforts to uncover the dangers of mercury, health and fitness section. *New York Times* (Tuesday 2 November), p. D7.

Schecter, A., Papke, O., Tung, K.C. et al. (2004). Polybrominated diphenyl ethers contamination of United States food. *Environ. Sci. Technol.* **38** (20): 5306–5311.

Seigneur, C., Vijayaraghavan, K., Lohman, K. et al. (2004). Global source attribution for mercury deposition in the United States. *Environ. Sci. Technol.* **38** (2): 555–569.

Sherwood, L., Klandorf, H., and Yancey, P.H. (2005). *Animal Physiology: From Genes to Organisms*. Australia: Brooks/Cole.

Spiro, T.G. and Stigliani, W.M. (2003). *Chemistry of the Environment*, 2e. Upper Saddle River, NJ: Prentice Hall.

Stapleton, H.M., Dodder, N.G., Offenberg, J.H. et al. (2005). Polybrominated diphenyl ethers in house dust and clothes dryer lint. *Environ. Sci. Technol.* **39** (4): 925–931.

Suzuki, T., Nakagawa, Y., Takano, I. et al. (2004). Environmental fate of bisphenol A and its biological metabolites in river water and their xeno-estrogenic activity. *Environ. Sci. Technol.* **38** (8): 2389–2396.

Thacker, P.D. (2004). Livestock flood the environment with estrogen. *Environ. Sci. Technol.* **38** (13): 241A–242A.

(2002). *The New Encyclopedia Britannica*, 15e, vol. 18. Chicago: Encyclopedia Britannica, Inc.

Tomy, G.T., Budakowski, W., Halldorson, T. et al. (2004). Fluorinated organic compounds in an eastern arctic marine food web. *Environ. Sci. Technol.* **38** (24): 6475–6481.

Tuerck, K.S.J., Kucklick, J.R., Becker, P.R. et al. (2005). Persistent organic pollutants in two dolphin species with focus on toxaphene and polybrominated diphenyl ethers. *Environ. Sci. Technol.* **39** (3): 692–698.

Unites States Department of Health and Human Services, (2002). Public Health Service. Toxicological Profile For DDT, DDE, and DDD. Agency for Toxic Substances And Disease Registry. Prepared by Syracuse Research Corporation Under Contract No. 205-1999-00024. September, 2002.

USGA Fact Sheet FS-095-01 (2001). https://pubs.usgs.gov/fs/fs095-01/fs095-01.pdf (accessed 21 August 2018).

USGS Fact Sheet FS-095-01 (2001). Mercury in Coal—Abundance, distribution, and modes of occurrence, U.S. Department of the Interior, USGS (September, 2001).

Vives, I., Grimalt, J.O., Catalan, J. et al. (2004). Influence of altitude and age in the accumulation of organochlorine compounds in fish and high mountain lakes. *Environ. Sci. Technol.* **38** (3): 690–698.

Whittelsey, F.C (2003). Hazards of hydration: choose your plastic water bottles carefully. *Sierra Magazine* (16–18 November/December, 2003).

Wright, K. (2005). Our preferred poison. *Discover* (Tuesday 31 March).

www.epa.gov (accessed 21 August 2018).

www.usgs.gov (accessed 21 August 2018).

Yamamoto, T. and Yasuhara, A. (1999). Quantities of bisphenol a leached from plastic waste samples. *Chemosphere* **38** (11): 2569–2576.

Part VII

Supporting Laboratory Experiments

Tell me and I will listen. Show me and I will watch. Let me experience, and I will learn

Lao Tzu, 500 BC

Pollutant Fate and Transport in Environmental Multimedia, First Edition. Frank M. Dunnivant and Elliot Anders.
© 2019 John Wiley & Sons, Inc. Published 2019 by John Wiley & Sons, Inc.
Companion website: www.wiley.com/go/Dunnivant/pollutant_fate&Transp_EnvironMultimedia

16

Laboratory Experiments

16.1 Introduction

This chapter contains laboratory procedures, with student results, for a variety of experiments that are complimentary to the text materials. These have been used for years and include (i) proper laboratory notebook protocol, (ii) long-term lab demonstrations, (iii) standard wet chemistry experiments, and (iv) experiments supporting modeling efforts.

16.2 Keeping a Legally Defensible Laboratory Notebook

Chemistry is a laboratory science, and an important part of this course is to familiarize the student with proper laboratory techniques and recording of their work in laboratory notebooks. The practices we use here may seem rather detailed or "picky," but these are minor compared to the practices used in some industrial sectors.

You must keep an elaborate and highly organized account of each lab exercise in your notebook. At the beginning of each lab, I will look for a "Things to do" list, possibly a detailed procedure, a dilutions table, etc. in your lab notebook. This tells me you came prepared to lab.

Organization of your notebook:
Sections for every lab:

Title (in designated area)
Purpose
Chemicals and solution used (brand, lot number, who made them)
Calculations
Cited SOP with modifications and/or detailed procedures
Data, hand written, spreadsheets, plots to be completed during each lab (spreadsheets and plots should be neatly taped in your lab notebook)
Conclusions, what you found in this lab.

A few widely accepted details concerning the environmental laboratory are as follows:

(1) Leave a few pages at the beginning for a table of contents (by experiment), a list of commonly used tables, and a list of commonly used figures, and appendices. Our current notebooks have a place for this. Note that at the end of the day you must initial and date each page.
(2) If someone else collects the data and you need to recopy it in your notebook, note where and when you obtained the data.
(3) Neatness is important, but you must write in procedures as you plan them (in the lab) and enter data *as you collect it* (not on note book paper or napkins for later transfer to your official record). EVERY LAB BOOK MUST CONTAIN ALL DATA COLLECTED IN THE LAB, NO MATTER WHO IN YOUR GROUP COLLECTED THE DATA.
(4) Record every detail in your notebook for future reference. Some items that may not be immediately evident, but should be included, are the type, model number, serial number of equipment (serial number only if more than one piece of the equipment is available), type of glassware used in dilutions, standard dilutions, and brand name and lot number of all chemicals used.
(5) No skipped pages are allowed in official (legally defensible) notebooks without proper procedures. Any page that contains more than 1/4 of a blank page at the bottom or top or one side of a page must have a line drawn through it and be signed and dated by you.
(6) All entries must be made with a ballpoint pen (not the alcohol-based, water soluble pens that are more commonly available). If you make a mistake, draw a single line through the mistake and write in the correction. Do not obliterate a number that you think is in error as you may find that you actually needed that number.

Pollutant Fate and Transport in Environmental Multimedia, First Edition. Frank M. Dunnivant and Elliot Anders.
© 2019 John Wiley & Sons, Inc. Published 2019 by John Wiley & Sons, Inc.
Companion website: www.wiley.com/go/Dunnivant/pollutant_fate&Transp_EnvironMultimedia

(7) Remember that the goal of this entire process is it to produce a document that you, or one of your peers, can pick up (even years in the future) and exactly reproduce your procedure and hopefully your results.
(8) Everything that you do in the lab should be recorded in your lab notebook as you do it, including lengthy procedures and calculations as you do them.
(9) Do not remove any pages from your lab notebook.

16.3 Quarter- and Semester-Long Experiments

16.3.1 Creation of Natural Organic Matter (NOM)

Natural organic matter (NOM) is a major chemical component of natural waters directly and indirectly involved in chemical reactions with pollutants and affecting pollutant toxicology, and regulating pollutant fate and transport in aqueous environments. The creation of NOM from yard waste is an easy way to illustrate the creation and presence of NOM.

- Add 900 ml of tap or DI water to a 1000 ml graduated cylinder.
- Collect most any yard waste (leaves, grass, etc.) and add approximately 100 ml volume, wet or dry, to the top of 1000 ml graduated cylinder and gently immerse in the water.
- Observe the solution of NOM over the following days and weeks.
- *Class observation:* Note the influx of NOM to systems during the fall as dead plant materials are washed into aquatic systems.

16.3.2 Winogradsky Column: A Microcosm of Aquatic Environments

This experiment qualitatively illustrated the variety of microorganisms and chemical reactions occurring in aquatic systems that can affect pollutant chemical reactions and fate and transport. The column can be easily set up a day and observed throughout the school quarter or semester as it evolves and changes. Detailed instructions, illustrations, photograms, and information can be found at:

https://en.wikipedia.org/wiki/Winogradsky_column.
http://archive.bio.ed.ac.uk/jdeacon/microbes/winograd.htm.
https://www.scientificamerican.com/article/bring-science-home-soil-column/.

16.4 Pollutant Fate and Transport Experiments for the Last Two Dispersion Experiments

16.4.1 The Determination of Alkalinity in Water Samples

Purpose:
To determine the alkalinity of a natural water sample by titration.
Prelab preparation:
This experiment is simulated in the Flash software package EnviroLab, specifically Alkalinity, in English and Spanish, available at https://sites.google.com/a/whitman.edu/frank-dunnivant-webpage/environmentalsoftware.
Background:
Alkalinity is an expression of water's ability to neutralize acids. Therefore, alkalinity measurements are important for considering the fate of acidic industrial pollution emitted directly into our waterways and as acid precipitation. Alkalinity is also a measure of a water's buffering capacity or its ability to resist changes in pH upon the addition of acids or bases. Alkalinity in natural waters is due primarily to the presence of weak acid salts, although strong bases (e.g. OH^-) may also contribute in industrial waters. Bicarbonates represent the major form of alkalinity in natural waters and are derived from the dissolution of CO_2 from the atmosphere and the weathering of carbonate minerals in rocks and soil. Other salts of weak acids, such as borate, silicates, ammonia, phosphates, and organic bases from NOM, may be present in small amounts. Alkalinity, by convention, is reported as mg/l $CaCO_3$, since most alkalinity is derived from the weathering of carbonate minerals, rather than from CO_2 partitioning with the atmosphere. Alkalinity for natural water (in molar units) is typically defined as the sum of the carbonate, bicarbonate, hydroxide, and hydronium concentrations, such that

$$[\text{alkalinity}] = 2[CO_3^{2-}] + [HCO_3^-] + [OH^-] - [H_3O^+] \quad (16.1)$$

Alkalinity values can range from zero in acid rain–impacted areas, where the pH of the water will be below 5.4, to less than 20 mg/l (as $CaCO_3$) for waters in contact with noncarbonate bearing soils, to 2000–4000 mg/l (as $CaCO_3$) for waters from the anaerobic digesters of domestic wastewater treatment plants.

Neither alkalinity nor its converse, acidity, has any known adverse health effects, although highly acidic or alkaline waters are frequently considered unpalatable. However, alkalinity can be affected by or affect other parameters. Below are some of the most important effects of alkalinity.

(1) The alkalinity of a body of water determines how sensitive that water body is to acidic inputs such as acid rain. A water with high alkalinity better resists changes in pH upon the addition of acid (from acid rain or from an industrial input).

(2) Turbidity is frequently removed from drinking water by the addition of alum, $Al_2(SO_4)_3$, to the incoming water followed by coagulation, flocculation, and settling in a clarifier. This process releases H^+ into the water through the reaction

$$Al^{3+} + 3H_2O \rightarrow Al(OH)_3 + 3H^+ \quad (16.2)$$

In order for effective and complete coagulation to occur, alkalinity must be present in excess of the H^+ released. Usually, additional alkalinity, in the form of $Ca(HCO_3)_2$, $Ca(OH)_2$, or Na_2CO_3 (soda ash), is added to ensure optimum treatment conditions.

(3) Hard waters (with excessive metal ion concentrations) are frequently softened by precipitation using CaO (lime), Na_2CO_3 (soda ash), or NaOH. The alkalinity of the water must be known in order to calculate the lime, soda ash, or sodium hydroxide requirements for precipitation.

(4) Maintaining alkalinity is important to corrosion control in piping systems. Corrosion is of little concern in modern domestic systems, but many main water distribution lines and industrial pipes are made of iron. Low pH waters cause corrosion in metal pipe systems, which are costly to replace.

(5) Bicarbonate (HCO_3^-) and carbonate (CO_3^{2-}) can complex other elements and compounds, altering their toxicity, transport, and fate in the environment. In general, the most toxic form of a metal is its uncomplexed hydrated metal ion. Complexation of this free ion by carbonate species can reduce toxicity.

Safety hazards:

- As in all laboratory exercises, safety glasses must be worn at all times.
- Avoid skin and eye contact with NaOH and HCl solutions. If contact occurs, rinse your hands and/or flush your eyes for several minutes. Seek immediate medical advice for eye contact.
- Use concentrated HCl in the fume hood and avoid breathing its vapor.

For the student:

Samples:

You will have several water samples to titrate. These will include one or more of the following: distilled water, deionized water, tap water, river water, and groundwater. You will determine the alkalinity of one or all of these samples (in triplicate), as per your instructor's instructions.

Procedures:

Characterizing the alkalinity of a water sample requires titrating to two different end points, each with its own significance. Titrating to pH 8.3, HCl yields the total amount of hydroxide and carbonate alkalinity, while titrating farther to pH 4.5 yields the total alkalinity.

1. First, an adequate sample volume for titration must be determined. This is accomplished by performing a test titration. Select a volume of your sample, such as 100 ml, and titrate it to pH 8.3 with standardized 0.02 HCl solution to estimate the carbonate alkalinity of your sample. This will tell you how much of your acid titrant is required to neutralize all the base (alkalinity) in the volume of sample that you chose. For best accuracy in adding the titrant, you should use at least 10 ml but not more than 50 ml from a 50.0 ml buret. Because this is the case for each of the titration end points, the volume required for titration to 4.5 should be at the high end of this range.
   To titrate to pH 4.5, add bromocresol green or the mixed bromocresol green–methyl red indicator solution. Slowly add 0.020 M HCl and note the color change indicating a pH value of 4.5. Alternatively, a pH meter can be used to determine the inflection point (pH 4.5).
   Adjust your sample size so that it requires ~40–50 ml to titrate.
2. *Titration to pH* 8.3: Titrate the determined volume of sample with standardized 0.02 M HCl solution. Add phenolphthalein or metacresol purple indicator solution, and then add HCl slowly from a buret. The equivalence point is marked by a color change around pH 8.3. Alternatively, a pH meter can be used to determine the inflection point. This measurement will be a combination of the hydroxide and carbonate alkalinity.
3. Continue the titration to the ~4.5 end point (as described in 1 above) with the same HCl solution, either with the same sample aliquot or a new one. Better results will be obtained by titrating a new sample to the ~4.5 end point. This will avoid potential color interferences between the 8.3 and 4.5 pH indicators.
4. Repeat the titrations of the sample, Steps 2 and 3, at least three times.
   - Calculate the hydroxide and carbonate (phenolphthalein end point described in the following) alkalinity and total alkalinities for your samples. Report your values in mg $CaCO_3$/l. Show all calculations in your notebook.

Calculation:

Phenolphthalein alkalinity

$$= \frac{(\text{l of acid to pH 8.3})(\text{M of acid})\left(\frac{100\text{ g}}{\text{mol}}\text{ CaCO}_3\right)\left(1\text{ mol}\frac{\text{CaCO}_3}{2}/2\text{ mol alkalinity}\right)}{\text{l of sample}}$$

Total alkalinity

$$= \frac{(\text{l of acid to pH 4.5})(\text{M of acid})\left(\frac{100\text{ g}}{\text{mol}}\text{ CaCO}_3\right)\left(1\text{ mol}\frac{\text{CaCO}_3}{2}/2\text{ mol alkalinity}\right)}{\text{l of sample}}$$

Waste disposal:

After neutralization, all solutions can be disposed of down the drain with water.

Student assignment:

Report the alkalinity (as $CaCO_3$) of each sample that you titrate. Summarize your data and turn in a hard copy to your instructor.

For the instructor (to be prepared prior to the lab period):

Chemicals and solutions:

Limits of the method: Typically 20 mg $CaCO_3$/l. Lower detection limits can be achieved by using a 10-ml microburet.

Laboratory time requirements: One 3–4-hour laboratory period.

Level of difficulty: Easy

Glassware:

- *Standard laboratory glassware:* 50-ml buret, 250-ml Erlenmeyer flasks, 50-ml beakers, Pasteur pipets.
- *Sample handling:* Alkalinity is a function of the dissolved CO_2 in solution. Thus, any chemical or physical manipulation of the sample that will affect the CO_2 concentration should be avoided. This includes filtering, diluting, concentrating, or altering the sample in any way. Nor should the sampling temperature be exceeded, as this will cause dissolved CO_2 to be released. Samples containing oil and grease should be avoided. Sampling and storage vessels can be plastic or glass without headspace.
- *Sodium carbonate solution (~0.025 M):* Primary standard grade Na_2CO_3 must be dried for three to four hours at 250 °C and allowed to cool in a desiccator. Weigh 0.25 g to the nearest 0.001 g and quantitatively transfer all the solid to a 100-ml volumetric flask. Dilute to the mark with distilled or deionized water. Calculate the exact molarity of the solution in the 100-ml flask.
- *Standardized hydrochloric acid (~0.02 M):* Add 8.3 ml of concentrated (12 M) HCl to a 1000-ml volumetric flask and dilute to the mark with deionized (DI) or distilled water. This solution has a molarity of approximately 0.10 M. Transfer 200 ml of this solution to another 1000-ml volumetric flask to prepare the 0.020 M solution. Standardize the dilute HCl solution (~0.020 M) against the Na_2CO_3 primary standard solution. This is done by pipeting 10.00 ml of the ~0.025 M Na_2CO_3 solution into a 250-ml Erlenmeyer flask and diluting to about 50 ml with distilled or deionized water. Add 3–5 drops of the bromocresol green indicator (more if needed) to the Erlenmeyer flask and titrate with ~0.02 M HCl solution. Bromocresol green changes from blue to yellow as it is acidified. The indicator end point is intermediate between blue and yellow and appears as a distinct green color. From the titration volume, determine the molarity of the HCl solution. Remember to wash down any droplets of solution from the walls of the flask. Label the container with the measured molarity for the students.
- Bromocresol green indicator solution, ~0.10%, pH 4.5 indicator. Dissolve 0.100 g of the sodium salt into 100 ml distilled or deionized water. Colors: yellow in acidic solution, blue in basic solution.
- Phenolphthalein solution, alcoholic, pH 8.3 indicator. Colors: colorless in acidic solution, red in basic solution.
- Metacresol purple indicator solution, pH 8.3 indicator. Dissolve 100 mg metacresol purple in 100 ml water. Colors: yellow in acidic solution, purple in basic solution.
- Mixed bromocresol green–methyl red indicator solution. You may use either the water- or alcohol-based indicator solution. Water solution: dissolve 100 mg bromocresol green sodium salt and 20 mg methyl red sodium salt in 100 ml distilled or deionized water. Ethyl or isopropyl alcohol solution: dissolve 100 mg bromocresol green and 20 mg methyl red in 100 ml 95% alcohol (see Table 16.1).

16.4.2 Total Suspended and Dissolved Solids in Water Samples

Purpose:

To determine the amount of suspended and/or dissolved solids in water samples.

Prelab Preparation:

This experiment is simulated in the Flash software package *EnviroLab*, specifically *total suspended solids (TSS)*, in English and Spanish, available at https://sites.google.com/a/whitman.edu/frank-dunnivant-webpage/environmentalsoftware.

Table 16.1 Total suspended solid and total dissolved solid.

	Total suspended solids				
	Massed filter paper and watch glass				
	Initial mass (g)				
Sample No.	1	2	Average	Final mass (g)	TSS (mg/l)
1	14.5947	14.5914	14.5931	14.6801	870.50
2	19.5827	19.5796	19.5812	19.6652	840.50
3	23.043	23.0410	23.0420	23.1200	780.00

	Total dissolved solids			
	Massed 150-ml beaker			
Sample No.	Initial beaker mass (g)	Sample volume (l)	Dried beaker mass (g)	TDS (mg/l)
1	73.2604	0.096	73.3238	660.42
2	61.0671	0.099	61.1266	601.01
3	68.4542	0.100	68.5124	582.00

Source: Data from Braden Hussey, Environmental Chemistry and Engineering Class, spring, 2018.

Background:

Environmental waters may contain a variety of solid or dissolved impurities. In quantifying levels of these impurities, "suspended solids" is the term used to describe particles in the water column. Practically, they are defined as particles large enough to not pass through the filter used to separate them from the water. Smaller particles, along with ionic species, are referred to as dissolved solids. In considering waters for human consumption or other uses, it is important to know the concentrations of both suspended and dissolved solids.

First, let us consider some implications of TSS.

- High concentrations of suspended solids may settle out onto a streambed or lake bottom and cover aquatic organisms, eggs, or macroinvertebrate larva. This coating can prevent sufficient oxygen transfer and result in the death of buried organisms.
- High concentrations of suspended solids decrease the effectiveness of drinking water disinfection agents by allowing microorganisms to "hide" from disinfectants within solid aggregates. This is one of the reasons the TSS, or turbidity, is removed in drinking water treatment facilities.
- Many organic and inorganic pollutants sorb to soils, so that the pollutant concentrations on the solids are high. Thus, sorbed pollutants (and solids) can be transported elsewhere in river and lake systems, resulting in the exposure of organisms to pollutants away from the point source.

Second, consider the importance of monitoring total dissolved solids (TDS).

- TDS of potable waters ranges from 20 to 1000 mg/l. In general, waters with a TDS less than 500 mg/l are most desirable for domestic use.
- Waters with TDS greater than 500 mg/l may cause diarrhea or constipation in some people.
- Water with a high TDS is frequently hard (i.e. has a high Ca^{2+} and/or Mg^{2+} concentration) and requires softening (the removal of hardness cations) by precipitation. The TDS of a water sample can be used to determine the most appropriate method of water softening because precipitation reduces TDS, while some ion exchange processes may increase TDS.
- Waters with high TDS may result in clogged pipes and industrial equipment through the formation of scale (Ca and Mg solids precipitated in the pipes).

Procedures for the students:

Note: Each of these procedures will require you to come in at unusual times during the next week. You must observe safety procedures set up by your school pertaining to working in the laboratory alone or in pairs.

TSS and TDS measurements:

Overview: In this procedure, you will first use 100 ml of your sample to perform the most commonly used solid measurement, of the TSS. This requires you to filter a known volume of sample through a preheated and pretared glass fiber filter, to remove the suspended solids. The difference between the initial (just filter) and final (filter with solids) weights, divided by the volume of sample, will yield the TSS. The TSS measurement accounts for all solids that do not pass through the filter (typically 0.2–0.45 μm in size). The filter is weighed after drying at 104 °C, both before and after the filtering procedure. The filtrate (the remaining sample liquid) is then used in the dissolved solids determination (TDS, the next procedure). The TDS is a simple experiment where the filtrate is placed in a tared (preweighed) beaker and evaporated to dryness.

Student procedures:
Preparing your filters:

1. Rinse three filters with 20–30 ml DI to remove any solids that may remain from the manufacturing process. Place the filters in separate, labeled aluminum weight pans, dry them in a 104 °C oven for 30 minutes, place them (filter and pan) in a desiccator, and obtain a constant weight by repeating the oven and desiccation steps.

Obtaining the TSS measurement:

2. Filter 100. ml of sample through each tared filter.
3. Place each paper in its aluminum weight pan in the 104 °C oven for one hour. Cool the filter and pan in a desiccator and obtain a constant weight by repeating the drying and desiccation steps. (This step will be completed after your normal lab meeting time.)

Calculation:

$$\text{TSS in}\frac{\text{mg}}{\text{l}} = \frac{(\text{average weight from Step 3 in g} - \text{average initial weight from Step 1 in g})\left(\frac{1000\ \text{mg}}{\text{l}}\right)}{\text{sample volume in l}}$$

TDS measurement:

4. Obtain a constant weight for three 150-ml beakers, using the same procedure used for the filters in Step 1.
5. Add 100. ml of your filtered sample to each beaker and evaporate it in the 98 °C oven overnight.
6. The next day, place the beakers in the 104 °C oven and heat them for one hour.
7. Place the beakers in a desiccator until cool and obtain a constant weight.
8. Repeat Steps 5 and 6 until you obtain a constant weight (within 0.5 mg of each other). The differences between this weight and the weight of the beakers originally will be the total masses of dissolved solids in the 100-ml samples.

Calculation:

$$\text{TDS in}\frac{\text{mg}}{\text{l}} = \frac{(\text{average crucible weight from Step 7 in g} - \text{average crucible weight empty in g})\left(\frac{1000\ \text{mg}}{\text{g}}\right)}{\text{sample volume in l}}$$

Hints for success:

- Always, ALWAYS completely mix your sample before removing any solution/suspension. The soil/sediment particles will settle and bias your results if you do not completely mix the sample every time you remove an aliquot.
- Perform all measurements in triplicate.
- Carefully clean all containers and prewash all filters with DI water prior to use. As the procedures notes, you must heat filters to the maximum temperature that you will use experimentally, before filtering. Also as noted in the procedures, you must obtain a constant weight (generally within 0.5 mg) before you end each experiment. (Fingerprints and dust weigh enough to significantly affect your results.)
- Your balances have been calibrated, but for best results, you should still use the same balance for every measurement. Even if the calibration on a balance is slightly off, the change in weight will probably be accurate.

Your assignment:
Turn in a table showing your measurements for each sample, your calculations TSS and TDS, and your final averages.

For the instructor:
Laboratory time requirements:
One 3–4-hour laboratory period plus a few short visits to the lab during the following week.
Level of difficulty: Easy
Equipment needed:

500 ml of unknown sample per student
8–10 150-ml beakers (predried and desiccated; three per student)
10–15 glass fiber filters (0.2–0.45 μm nominal pore size, without organic binders) and filtration setups
Five filtration flasks (500 ml capacity; several can be shared between students)
Three graduated cylinders (100 ml capacity; one per student)
98 °C oven
104 °C oven
Desiccators (one large desiccator for the group or one small desiccator per student).

Sample student data:

16.4.3 The Determination of Hardness in a Water Sample

In the past, water hardness was defined as a measure of the capacity of water to precipitate soap. However, current laboratory practices define total hardness as the sum of divalent ion concentrations, especially those of calcium and magnesium, expressed in terms of mg $CaCO_3$/l. There are no known adverse health effects of hard or soft water, but the presence of hard waters results in two economic considerations: (i) hard waters require considerably larger amounts of soap to foam and clean materials and (ii) hard waters readily precipitate carbonates (known as scale) in piping systems at high temperatures. Calcium and magnesium carbonates are two of the few common salts whose solubility decreases with increasing temperature. This is due to the removal of dissolved CO_2 as temperature increases due to the changing Henry's law constant. The advent of synthetic detergents has significantly reduced the problems associated with hard water and the "lack of foaming." However, scale formation continues to be a problem in domestic hot water heaters and for industry.

The source of a water sample usually determines its hardness. For example surface waters usually contain less hardness than ground waters, which spend longer in contact with geologic materials that are sources of divalent cations. Carbonates in surface soils and sediments also increase the hardness of surface waters, and subsurface limestone formations also increase the hardness of ground waters. Hardness values can range from a few to hundreds of mg $CaCO_3$/l. Therefore, depending on your water's source, some modifications to the following procedure may be necessary.

The divalent metal cations responsible for hardness can react with soap to form precipitates in hot water pipes (the version known as scale occurs where particular anions are present). The major hardness-causing cations are calcium and magnesium, while strontium, ferrous iron, and manganese can also contribute. It is common to compare the alkalinity values of a water sample to the hardness values, with both expressed in mg $CaCO_3$/l. Hardness up to the value of the alkalinity is referred to as the "carbonate hardness." Thus, when the hardness is equal to or less than the total alkalinity, all hardness is carbonate hardness. When the hardness value is greater than the total alkalinity, the amount in excess is referred to as the "noncarbonate hardness."

The following method described relies on the competitive complexation of divalent metal ions by ethylenediaminetetraacetic acid (EDTA) or an indicator. The chemical structure for the disodium salt of EDTA is in Figure 16.1. Note the lone pairs of electrons on the two nitrogen atoms. These, combined with the dissociated carboxyl groups, enable formation of a one-to-one hexadentate complex (one cation bonded to six sites on one EDTA molecule) with each divalent ion in solution. Although the complexation constant (describing the

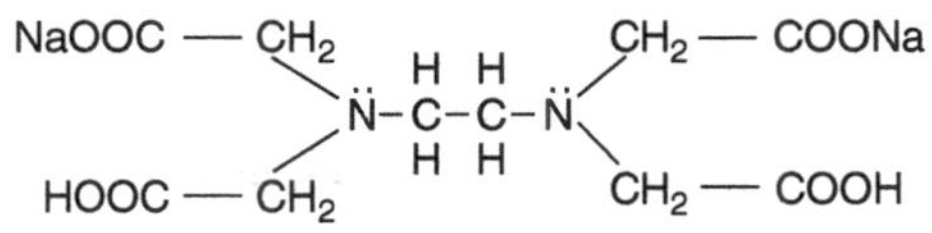

Figure 16.1 EDTA structural formula.

location of the complexation equilibrium) is a function of pH, virtually all common divalent ions will be complexed at pH values greater than 10, the pH used in this titration experiment, and in most hardness tests. Thus, the value for hardness calculated from the experimental results includes all divalent ions in a water sample.

Three indicators are commonly used in the EDTA titration: Eriochrome Black T (Erio T), Calcon, and Calmagite. The use of Eriochrome Black T requires that a small amount of Mg^{2+} ion be present at the beginning of the titration. Calmagite is used in this experiment because its end point is sharper than that of Eriochrome Blank T.

Prelab preparation:

This experiment is simulated in the Flash software package *EnviroLab*, specifically *Hardness*, in English and Spanish, available at https://sites.google.com/a/whitman.edu/frank-dunnivant-webpage/environmentalsoftware.

Safety hazards:

- As in all laboratory exercises, safety glasses must be worn at all times.
- Avoid skin and eye contact with pH 10 buffer. In case of skin contact, rinse the area for several minutes. For eye contact, flush eyes with water and seek immediate medical advice.

In the laboratory:

Two methods are available for determining the hardness of a water sample. The one described and used here is based on a titration method using a chelating agent. The basis for this technique is that at specific pH values, EDTA binds with divalent cations to form a strong complex. Thus, by titrating a sample of known volume with a standardized (known molarity) solution of EDTA and a complexing indicator that competes with EDTA, you can measure the amount of divalent metals in solution. The end point of the titration is observed using a colorimetric indicator, in our case Calmagite. When a small amount of indicator is added to a solution containing hardness (at pH = 10.0), it combines with a few of the hardness ions and forms a weak wine red complex. During the titration, EDTA complexes more and more of the hardness ions until it has complexed all the free ions and "out-competes" the weaker indicator complex for hardness ions. At this point, the indicator returns to its uncomplexed color (blue for Calmagite), indicating the end point of the titration, where only EDTA-complexed hardness ions are present.

Student procedures:

- Pipet an aliquot of your sample into a 250-ml Erlenmeyer flask. The initial titration will only be a trial, and you will probably need to adjust your sample volume to obtain the maximum precision from your pipeting technique (obtained by adding more than 10 ml but less than 50 ml of EDTA titrant). Increase or decrease your sample size as needed depending on the results of your initial titration.
- Add 3 ml of the pH 10 buffer solution and ~1 ml of the Calmagite indicator. Check to ensure that the pH of your sample is at or above a pH value of 10. Add additional buffer solution if needed.
- Titrate with EDTA solution and note the color change as you reach the end point. Continue adding EDTA until you obtain a stable blue color with no reddish tinge. (*Note:* Incandescent light can produce a reddish tinge at and past the end point.)
- Repeat until you have at least three titrations that are in close agreement.
- Titrate at least two types of water.
- Calculate the hardness for each of your samples. Express your results in mg $CaCO_3$/l. If the stockroom gave you a 0.0100 M solution of EDTA, 1.00 ml of EDTA solution is equivalent to 1.00 mg $CaCO_3$.

Calculation:

Hardness is usually expressed in terms of $CaCO_3$ (mg $CaCO_3$/l). We calculate hardness from your titration results by

$$\frac{\text{mg}}{\text{l}}\text{CaCO}_3 = \left(0.0100\ \text{mol}\frac{\text{EDTA}}{\text{l}}\right)(\text{l of EDTA used}) \times \left(\frac{1\ \text{mol divalent metal}}{1\ \text{mol EDTA}}\right) \times \left(\frac{1\ \text{mol CaCO}_3}{2\ \text{mol divalent metal}}\right)\left(\frac{100\ \text{g}}{\text{mol}}\text{CaCO}_3\right) \times \left(\frac{1000\ \text{mg}}{\text{g}}\right)\left(\frac{1}{\text{l}}\text{of sample}\right)$$

Waste disposal:

After neutralization, all solutions can be disposed of down the drain with rinsing.

Your assignment:

Prepare a summary table showing all your results with the averages for each water tested.

For the instructor (to be prepared prior to the lab period):

Limits of the method:

Detection limits are dependent on the volume of sample titrated.

Laboratory time requirements:

One 3–4-hour laboratory period.

Level of difficulty: Easy

Chemicals and Solutions (to be prepared by the stockroom before the lab period):

- *Sampling handling:* Plastic or glass sample containers can be used. A minimum of 100 ml is needed, but for replicate analysis of low hardness water, a liter of sample is suggested. If you are titrating the sample on the day of collection, no preservation is needed. If longer holding times are anticipated, the sample can be preserved by adding nitric or sulfuric acid to a pH value of less than 2.0. Note that acidified samples must be adjusted to a pH above 10 before the titration.
- Students should be provided with several samples that may include tap water, river water, ground water, and a synthetically prepared sample containing a known amount of hardness.
- *pH 10 buffer:* In a 250-ml volumetric flask, add 140 ml of a 28% by weight NH_3 solution to 17.5 g of NH_4Cl and dilute to the mark.
- *Calmagite (1-(1-hydroxy-4-methyl-2-phenylazo)-2-naphthol-4-sulfonic acid):* Dissolve 0.10 g of Calmagite in 100 ml distilled or deionized water. Use ~1 ml per 50 ml sample to be titrated.
- Dry ~5 g analytical reagent-grade Na_2EDTA (FW 372.25) at 80 °C for one hour and cool it in a desiccator. Accurately weigh 3.723 g (or a mass accurate to 0.001 g), dissolve in 500 ml deionized water with heating, cool to room temperature, quantitatively transfer to a 1-l volumetric flask, and fill to mark. Since EDTA will extract hardness-producing cations out of most glass containers, store the EDTA solution in a plastic container. This procedure produces a 0.0100 M solution.

Glassware:

- *Standard laboratory glassware:* 50-ml buret, 250-ml Erlenmeyer flasks, 50-ml beakers, Pasteur pipets.

16.4.4 The Determination of Dissolved Oxygen in Water using the Winkler Method (Iodometric Titration Method)

Purpose:

To determine the dissolved oxygen (DO) concentration in a water sample.

To learn the chemical reactions involved in the Winkler DO method.

Prelab Preparation:
This experiment is simulated in the Flash software package *EnviroLab*, specifically *Dissolved Oxygen – BOD* Simulator, in English and Spanish, available at https://sites.google.com/a/whitman.edu/frank-dunnivant-webpage/environmentalsoftware.

Background:
It is a common perception that all life is dependent upon the presence of oxygen, either in the atmosphere or in the water. However, this is anything but true. The first life forms to evolve on Earth are thought to have been anaerobic, requiring an oxygen-free environment to grow. In fact, free oxygen is toxic to anaerobic organisms' biochemical machinery. Oxygen was actually a waste product from these organisms, and their emission of oxygen over hundreds of millions of years enabled the evolution of aerobic organisms. Even today, there are many types of respiration (and organisms) that do not require the presence of oxygen as their terminal electron acceptor (TEA). Every life form needs a TEA, to accept the excess electrons from their reduced food sources. For example look at how we oxidize glucose with atmospheric oxygen to yield energy. Electrons on glucose are removed and added to diatomic oxygen, and in this process, oxygen is reduced from an oxidation state of zero to −2, while carbon is oxidized to +4. The net result is a generation of 2863 kJ of energy per mole of glucose oxidized, a higher energy yield than that achieved with more primitive TEAs (i.e. NO_3^-, SO_4^{2-}, etc.). Therefore, the first life forms only yielded small amounts of energy from their oxidation of food substrates. This is one reason why organisms that use oxygen as their TEA out-compete other life forms.

Oxygen is considered poorly soluble in water. It is interesting to note that air-breathing organisms have available around 19% oxygen (19 000 ppm on a volume per volume basis or 262 mg/l from $PV = nRT$) in air for consumption, while organisms respiring in water have only a maximum of ~0.15% oxygen (14.6 mg/l). As temperature increases or as salt content increases, the dissolved oxygen concentration decreases. The range of dissolved oxygen concentrations in water under normal conditions is shown in Table 16.2. Note that the range of dissolved oxygen (DO) in pure water (no salt content) is from 7.6 mg/l at 30 °C to 14.6 mg/l at 0 °C. Although this may seem like a narrow range, many organisms have become specialized so that they can live only in a small portion of this range. Important examples are mountain trout and several species of invertebrate insect larva that require very cold waters with the highest concentrations of dissolved oxygen.

Table 16.2 Solubility of dissolved oxygen for water in contact with the dry atmosphere (at 1.0 atmosphere containing 20.9% oxygen)[a] (increasing the salt content of water decreases the solubility of any dissolved gas).

Temperature (°C)	Chloride concentration (mg/l)				
	0	5 000	10 000	15 000	20 000
0	14.6	13.8	13.0	12.1	11.3
1	14.2	13.4	12.6	11.8	11.0
2	13.8	13.1	12.3	11.5	10.8
3	13.5	12.7	12.0	11.2	10.5
4	13.1	12.4	11.7	11.0	10.3
5	12.8	12.1	11.4	10.7	10.0
6	12.5	11.8	11.1	10.5	9.8
7	12.2	11.5	10.9	10.2	9.6
8	11.9	11.2	10.6	10.0	9.4
9	11.6	11.0	10.4	9.8	9.2
10	11.3	10.7	10.1	9.6	9.0
11	11.1	10.5	9.9	9.4	8.8
12	10.8	10.3	9.7	9.2	8.6
13	10.6	10.1	9.5	9.0	8.5
14	10.4	9.9	9.3	8.8	8.3
15	10.2	9.7	9.1	8.6	8.1
16	10.0	9.5	9.0	8.5	8.0
17	9.7	9.3	8.8	8.3	7.8
18	9.5	9.1	8.6	8.2	7.7
19	9.4	8.9	8.5	8.0	7.6
20	9.2	8.7	8.3	7.9	7.4
21	9.0	8.6	8.1	7.7	7.3
22	8.8	8.4	8.0	7.6	7.1
23	8.7	8.3	7.9	7.4	7.0
24	8.5	8.1	7.7	7.3	6.9
25	8.4	8.0	7.6	7.2	6.7
26	8.2	7.8	7.4	7.0	6.6
27	8.1	7.7	7.3	6.9	6.5
28	7.9	7.5	7.1	6.8	6.4
29	7.8	7.4	7.0	6.6	6.3
30	7.6	7.3	6.9	6.5	6.1

a) After Wipple and Wipple (1911).

Theory:
Two methods are commonly used to determine the concentration of dissolved oxygen in water samples: the Winkler or iodometric method and the membrane electrode technique. Details on each of these methods can be found in Standard Methods (1998) and Sawyer and McCarty (1978). The iodometric method will be discussed first and is the focus of this laboratory procedure. A more recent development is the use of $Ru(bipy)_3$ optical sensor for O_2, described in the

environmental chemistry marketing literature and on the Internet.

The iodometric method, the more accurate of the two standard methods, determines the dissolved oxygen concentration through a series of oxidation-reduction reactions and is the reaction we will use here. First, Mn^{2+} (as $MnSO_4$) is added to a 250 or 300 ml sample. Next, the alkali-iodide reagent (KI in NaOH) is added. Under these caustic conditions, if oxygen is present in the water sample, the Mn^{2+} will be oxidized to Mn^{4+}, which precipitates as a brown hydrated oxide. This reaction is relatively slow, and the solution must be shaken several times to complete the reaction. This reaction can be represented by the following expressions:

$$2Mn^{2+} + 4OH^- + O_2 \rightarrow 2\,MnO_2(s) + 2\,H_2O$$

or

$$2\,Mn(OH)_2 + O_2 \rightarrow 2\,MnO_2(s) + 2\,H_2O$$

After the MnO_2 precipitate settles to the bottom of the flask, sulfuric acid is added to make the solution acidic. Under these low pH conditions, the manganese is dissolved as MnO_2 oxidizes the iodide (I^-) to free iodine (I_2) through the following reaction

$$MnO_2 + 2I^- + 4\,H^+ \rightarrow Mn^{2+} + I_2 + 2H_2O$$

Now, the sample is ready for titration with standardized sodium thiosulfate ($Na_2S_2O_3{\cdot}5H_2O$). In this step, thiosulfate ion is quantitatively added in order to convert the I_2 back to I^-. The reaction can be represented by

$$2S_2O_3^{2-} + I_2 \rightarrow S_4O_6^{2-} + 2I^-$$

The amount of I_2 present at this stage in the procedure is directly related to the amount of O_2 present in the original sample. The titration is complete when all of the I_2 has been converted to I^-. The end point of this titration can be determined through potentiometry or by using calorimetric indicators. The most common indicator is starch, which turns from deep blue to clear.

The dissolved oxygen concentration can be determined using the following equation, which also reflects the series of redox reactions in the following equations:

$$\frac{mg}{l}O_2 = \frac{(L\ S_2O_3^{2-})(M\ S_2O_3^{2-})\left(\frac{I_2}{2\ S_2O_3^{2-}}\right)\left(\frac{MnO_2}{I_2}\right)\left(\frac{O_2}{2\ MnO_2}\right)\left(\frac{32\ g}{mol\ O_2}\right)\frac{1000\ mg}{g\ O_2}}{l\ sample}$$

Several modifications of the Winkler method have been developed to overcome interferences. The azide modification, the most common, effectively removes interference from nitrite, which is commonly present in water samples from biologically treated wastewater effluents and incubated biochemical oxygen demand (BOD) samples. Nitrite interferes by converting I^- to I_2, resulting in overestimation of the dissolved oxygen in the sample. The nitrite is then regenerated by oxidation. This is illustrated in the following equations:

$$2\,NO_2^- + 2\,I^- + 4\,H^+ \rightarrow I_2 + N_2O_2 + 2\,H_2O$$

and

$$N_2O_2 + 1/2\,O_2 + H_2O \rightarrow 2NO_2^- + 2H^+$$

Note that N_2O_2 is oxidized by oxygen, which enters the sample during the titration procedure, and is converted to NO_2^- again, establishing a cyclic reaction that can lead to erroneously high results. This interference therefore yields apparent oxygen concentrations that are far in excess of the amounts that would be normally expected.

Nitrite interference can be easily overcome through the addition of sodium azide (NaN_3). Azide is usually added with the alkali-KI reagent, and when the sulfuric acid is added, the following reactions result in the removal of NO_2^-

$$NaN_3 + H^+ \rightarrow HN_3 + Na^+$$

$$HN_3 + NO_2^- + H^+ \rightarrow N_2 + N_2O + H_2O$$

Other methods can also be used to remove ferrous iron (the permanganate modification), ferric iron (the potassium fluoride modification), and suspended solids (the alum flocculation modification). We will only be using the azide modification in this laboratory experiment.

The electrode method offers several advantages over the titration method, including speed, elimination or minimization of interferences, field compatibility, continuous monitoring, and in situ measurement. However, it is also associated with some loss in accuracy. Modern electrodes rely on a selectively permeable membrane that allows only dissolved oxygen to enter the measurement cell, thus eliminating most interferences. A detailed description of the operation of this electrode can be found in Sawyer and McCarty (1978). The calibration and measurement is relatively simple, and a direct readout of the oxygen concentration (in mg/l) is given.

In the laboratory:

You will be given one or more samples by your instructor for titration using the Winkler method. We suggest using cold water that has been equilibrated for 24 hours open to the atmosphere. For this laboratory exercise, you do not have to be concerned with preservation of the sample or sample handling practices, but in the real world, there are many precautions that need to be taken. Most important is the preservation of field samples that need to be analyzed in the laboratory. The easiest way to avoid this is to use a field meter to determine the concentration of DO. This method is quick and relatively reliable. However, DO meters are expensive, and some monitoring programs may specify the use of the Winkler titration method because of its greater accuracy.

Two approaches are used to preserve samples for later DO determination. First, you can "fix" your samples using the following described procedures and then perform the titration when the samples are brought to the laboratory. Samples should be stored in the dark and on ice until titration. This preservation technique will allow you to delay the titration for up to six hours. However, this procedure may give low results for samples with a high iodine demand. In this case, it is advisable to use the second option which is to add 0.7 ml of concentrated sulfuric acid and 0.02 g of sodium azide. When this approach is used, it is necessary to add 3 ml of alkali-iodide reagent (below) rather than the usual 2 ml. In addition, avoid any sample treatment or handling that will alter the concentration of DO, including increases in temperature and the presence of atmospheric headspace in your sample container.

You will titrate your samples using the following described procedures. As in all titration experiments, you should do a quick titration to determine the approximate volume of titrant needed. Follow this first titration with at least three careful titrations. Average your values for each sample. Students should work independently or in pairs since each student will be using the Winkler titration in the next experiment (the BOD determination) for which lab technique must be comparable for all students in the class.

Safety and hazards:

- As in all laboratory exercises, safety glasses must be worn at all times.
- Avoid skin and eye contact with caustic and acidic solutions. If contact occurs, rinse your hands and/or flush your eyes for several minutes. Seek immediate medical advice for eye contact.
- Use concentrated acids in the fume hood and avoid breathing their vapors.
- Sodium azide is a toxin and should be treated as such.

Student procedures:

1. To 250 or 300 ml sample bottle full of sample, add 1 ml $MnSO_4$ solution, followed by 1 ml alkali-iodide-azide reagent. If your pipets are dipped into the sample (as they should be), rinse them before returning them to the reagent bottles. If the solution turns white, no dissolved oxygen is present.
2. Place stoppers on the sample bottles in a manner to exclude air bubbles and mix by rapidly inverting the bottle a few times. When the precipitate has settled to half the bottle volume, repeat the mixing and allow the precipitate to resettle.
3. Add 1.0 ml concentrated H_2SO_4.
4. Replace the stoppers and mix by rapidly inverting the bottles to dissolve the precipitate. You may open the bottle and pour the sample at this point since the DO and reagents have been "fixed" and will not react any further.
5. Titrate 200. ml of the fixed sample (delivered by a 200. ml pipet or with a graduated cylinder) with your standardized thiosulfate solution. First, titrate to a pale straw color, add starch indicator, and titrate to a clear end point.
6. Repeat the titration for two more samples and average your results.

Assignment:

1. Create a flow chart showing all the oxidation–reduction reactions involved in the Winkler titration method. Explain each reaction.
2. Calculate an average and standard deviation for each type of water sample.

Waste disposal:

After neutralization, all solutions can be disposed of down the drain with water.

For the instructor (to be prepared prior to the lab period):

Time requirements:

One 3–4-hour laboratory period

Level of difficulty: Moderate

Chemicals and Solutions (to be prepared by the stock room prior to lab period):

- *Manganese sulfate:* Dissolve 480 g $MnSO_4{\cdot}4H_2O$, 400 g $MnSO_4{\cdot}2H_2O$, or 364 g $MnSO_4{\cdot}H_2O$ in ~600 ml DI water. Filter the solution and dilute to 1.0 l.
- Alkali-iodide-azide reagent. Dissolve 500 g NaOH (or 700 g KOH) and 135 g NaI (or 150 g KI) in DI water and dilute to 1.0 l. Add 10 g NaN_3 dissolved in 40 ml DI water.
- Concentrated H_2SO_4 (1.0 ml of this solution is equivalent to approximately 3 ml alkali-iodide-azide solution).
- Starch solution. Dissolve 2 g laboratory-grade soluble starch and 0.2 g salicylic acid (as a preservative) in 100 ml hot distilled water. Allow to cool before use.
- Sodium thiosulfate titrant, 0.0250 M. Dissolve 6.205 g $Na_2S_2O_3{\cdot}5H_2O$ in DI. Add 1.5 ml 6 M NaOH or 0.4 g solid NaOH and dilute to 1.0 l. Standardize with bi-iodate solution.
- Standard potassium hydrogen bi-iodate solution, 0.002 10 M. Dissolve 0.8124 g $KH(IO_3)_2$ in DI and dilute to 1.000 l in a volumetric flask.
- Standardization of sodium thiosulfate titrant (to be standardized by the stockroom prior to lab period):

1. Dissolve a known mass of KI (approximately 2 g, free of iodate) in an Erlenmeyer flask containing 100–150 ml DI.
2. Add 1 ml 6 M H_2SO_4 or a few drops of concentrated H_2SO_4 and pipet 20.00 ml standard bi-iodate solution (the primary standard) into the flask. The bi-iodate will react according to

$$2IO_3^- + 10I^- + 12H^+ \rightarrow 6I_2 + 6H_2O$$

3. Titrate the liberated I_2 with thiosulfate titrant until a pale straw (yellow) color is reached. Add a few drops of starch indicator, which will result in a blue color, and continue the titration to the end point, when the solution turns clear.
4. If all solutions were made properly, 20.00 ml of the bi-iodate solution will require 20.00 ml thiosulfate titrant. If this result is not achieved, calculate the exact molar concentration of your titrant.

Glassware (per student group):

- Four Erlenmeyer flasks per student
- A 25 ml buret per student
- A 20.00 ml pipet per student
- Pasteur pipets
- Three, 1.00 ml pipets. At least one of these should be a wide-bore pipet for the viscous azide reagent.

16.4.5 The Determination of the Biochemical Oxygen Demand (BOD) of Sewage Influent: BOD_5 and/or BOD_{20}

Purpose:
To determine the BOD in a domestic wastewater sample.
Prelab Preparation:
This experiment is simulated in the Flash software package *EnviroLab*, specifically the *Dissolved Oxygen –BOD* Simulator, in English and Spanish, available at https://sites.google.com/a/whitman.edu/frank-dunnivant-web page/environmentalsoftware.
Background:
The focus of this laboratory exercise will be to determine the amount of oxidizable organic matter (sewage) in a wastewater sample. As we discussed in the dissolved oxygen chapter, the term DO refers to the chemical measurement of how much dissolved oxygen is present in a water sample, expressed in mg/l. The BOD is an estimate of how much total DO is required to oxidize the organic matter in a water sample. Thus, we will actually be measuring the change in DO in our experiments to estimate the BOD originally present in the water. But before we discuss the details of this experiment, it is important to gain an appreciation for the extent of the global sewage problem and environmental issues surrounding wastewater.

Our standard of living in the United States is a direct result of having adequate water and wastewater treatment, which are distinguishing features of developed countries. As early as 1700 BCE, people began to obtain the luxury of running water and then to deal with the disposal of associated wastes. Though there is evidence of plumbing and sewage systems at many age-old sites, including the *Cloaca Maxima*, or great sewer, of the ancient Roman empire, the common use of sewer and plumbing systems did not become widespread until modern times. Along with providing drinking water and disposing of sewage comes the challenge of preventing the rapid spread of disease within populations that utilize a common water source and treatment facilities.

The focus of modern sewage treatment is to remove turbidity, readily oxidizable organic matter, and pathogenic organisms. These three goals can easily be achieved at a minimal cost. Turbidity is removed in primary and secondary clarifiers and in sand bed filters. Organic matter is removed in biological contact units such as trickling filters and activated sludge lagoons. Most pathogens are naturally removed in the various treatment process, but removal is ensured with the use of sand bed filtration, chlorination, and ozonation. One of the major design criteria for a wastewater treatment plant, and in fact a daily monitoring parameter, is the BOD of the incoming and outgoing waste, a measure of the effectiveness of organic matter removal. In this laboratory exercise, we will be measuring the five-day BOD (BOD_5) or the ultimate BOD (BOD_L).

Theory:
In general, the utilization of oxygen by microorganisms is considered to be a pseudo-first-order process which for a closed system (no reaeration) is commonly described by

$$\text{BOD at time } t = \text{BOD}_{\text{Ultimate}} e^{-kt}$$

where

L is the concentration of oxygen at time t,
L_o is the original concentration of oxygen in a sample,
k is the first-order rate constant, the traditional value of 0.17 d^{-1} for sewage waste, and
t is time.

A similar expression can be used to describe the amount of remaining BOD in the sample, since it is the inverse of the oxygen consumption,

$$\text{BOD remaining} = \text{BOD}_{\text{Ultimate}} - \text{BOD}_{\text{Ultimate}} e^{-kt}$$

where

L is the concentration of biodegradable organic matter at time, t,
L_o is the original or ultimate concentration of biodegradable organic matter,

k is the rate constant, generally around $0.17\,d^{-1}$ for sewage waste, and
t is time.

Traditionally, we are concerned with the amount of oxygen required to oxidize the BOD over a five-day period. This time period was established years ago in England and results from the fact that it requires five days for the water in most English streams to reach the ocean. The BOD continues to exert an oxygen demand on the stream after this time, and the ultimate BOD, determined over a 20- to 60-day period, is becoming commonly used in the United States.

In the laboratory:

A BOD determination is made by taking a sample and incubating it over a 5- or 20-day period (we will be determining the five-day BOD), and monitoring the dissolved oxygen concentration at intervals of 12 or more hours. Alternatively, and to save lab time, one can also only measure the single five-day BOD measurement. For high concentrations of BOD, the sample must be diluted in order to avoid depleting all the original oxygen present in the water sample. There are several requirements for the dilution water. For example pure distilled water should not be used since microorganisms require certain salts for proper metabolism. Thus, potassium, sodium, calcium, magnesium, iron, and ammonium salts are added to the dilution water. Also, the water's pH should be buffered between 6.5 and 8.5 with phosphate buffers. Some water samples require a "seed" of viable microorganisms to complete the degradation process.

A general rule-of-thumb has been developed to provide sufficient accuracy in determining BOD values. This states that at least 2 mg/l of oxygen must be used over the course of the experiment (5 or 20 days), but at least 0.5 mg/l must remain in the final sample. The oxygen concentration can be measured by one of two methods described in the previous laboratory experiment, "The Determination of Dissolved Oxygen in Water using the Winkler Method."

Safety and hazards:

- As in all laboratory exercises, safety glasses must be worn at all times.
- Avoid skin and eye contact with caustic and acidic solutions. If contact occurs, rinse your hands and/or flush your eyes for several minutes. Seek immediate medical advice for eye contact.
- Use concentrated acids in the fume hood and avoid breathing their vapors.

Student procedures:

Dilution of your waste sample: A rule of thumb for estimating the dilution factor of your wastewater can be determined from Table 16.3 (from Standard Methods for the Examination of Water and Wastewater, 1998 or more recent edition). The best way to determine the appropriate dilution factor is to consult the operator of the sewage treatment plant where you obtain your wastewater sample. Your lab instructor will help you with this calculation.

Table 16.3 BOD_5 measurable from various dilutions of sample (Standard Methods for the Examination of Water and Wastewater 1998).

Using percent mixtures	By direct pipeting into 300-ml bottles		
% mixture	Range of BOD_5	ml sample	Range of BOD_5
0.01	20 000–70 000	0.02	30 000–105 000
0.02	10 000–35 000	0.05	12 000–42 000
0.05	4 000–14 000	0.10	6 000–21 000
0.1	2 000–7 000	0.20	3 000–10 500
0.2	1 000–3 500	0.50	1 200–4 200
0.5	400–1 400	1.0	600–2 100
1.0	200–700	2.0	300–1 050
2.0	100–350	5.0	120–420
5.0	40–140	10.0	60–210
10.0	20–70	20.0	30–105
20.0	10–35	50.0	12–42
50.0	4–14	100	6–21
100	0–7	300	0–7

Seeding: A "seed" is needed when your sample does not have sufficient microbial community to immediately support exponential microbial growth. A seed usually consist of a small amount of sewage added to your samples. If you are using domestic wastewater, then you will probably not have to seed your water, since viable microbial communities are already present. For the purposes of this experiment, we will assume that you do not need to seed your samples, but keep in mind that river, lake, and groundwater samples often need to have a seed added for their BOD determination. When you do use a seed, you must also run a blank for your BOD determination, since the seed will consume a small amount of the DO.

pH adjustment: Some domestic wastewater samples have industrial inputs to the sewer system and as a result may have extreme pH values (very high or low). In these cases, it will be necessary to adjust the pH of your original wastewater sample prior to making dilutions according to Table 16.3. Use 1 M HCl or 1 M NaOH for these adjustments.

Chlorine removal: Some samples may contain residual chlorine compounds that will inhibit the growth of microorganisms and will interfere with the BOD determination. If your sample contains residual chlorine compounds, these can be removed with sodium sulfite. Domestic sewage samples rarely have residual chlorine compounds, and we will not use sulfite in this procedure, but be aware that this is not always the case.

Student procedures:
Setup and titration of BOD samples:

1. Determine the appropriate dilution of your wastewater based on Table 16.3 and/or data from the wastewater treatment plant operator. It is best to have three dilutions, one 20–30% less dilute than suggested, one as suggested, and one 20–30% more concentrated than suggested. This approach should allow the determination of the BOD_5 or BOD_{20}.
2. Before you make your dilutions, homogenize your wastewater sample by blending it in a blender or food processor at high speed for five minutes. Also, adjust the temperature to 20 °C.
3. Add the desired volume of sample wastewater to each BOD bottle and fill with equilibrated, nutrient-added, 20 °C dilution water. (Alternatively, you may mix a larger volume of diluted wastewater and fill your BOD bottles.)
4. Make sure the bottles are filled to the top with dilution water. Insert the tapered cap in a manner to exclude any air bubbles from the BOD bottle.
5. Incubate the dilutions at 20 °C for five days, taking bottles from each dilution at each 12-hour increment and titrating them using the Winkler method to obtain a plot of BOD versus time. The necessary sampling times are dependent on the microbial oxidation rate, k, but we will use 12-hour sample intervals. After you have all of your samples set up, collect the members of the class together and assign sampling and titration times to cover the next five days.
6. Refer to the Winkler method (in the previous chapter) for the fixing and DO titration procedures.
7. Analyze your data and determine the BOD_5 or BOD_{20}.

Waste disposal:
After neutralization, all solutions can be disposed of down the drain with water.

Your assignment:
Make a table summarizing the class data. What is the BOD_5 of your sample?

For the instructor (to be prepared prior to the lab period):
Time requirements: One 3–4-hour laboratory period to set up the experiment to be monitored during the following week for a BOD_5 determination. For a BOD_{20} determination, three weeks of monitoring will be needed.

Level of difficulty: Moderate

Chemicals and Solutions (to be prepared by the stockroom prior to lab period):

- *Phosphate buffer solution:* Dissolve 8.5 g KH_2PO_4, 21.75 g K_2HPO_4, 33.4 g $Na_2HPO_4{\cdot}7H_2O$, and 1.7 g NH_4Cl in approximately 500 ml DI and dilute to 1.0 l. The pH of this solution should be 7.2.
- *Magnesium sulfate solution:* Dissolve 22.5 g $MgSO_4{\cdot}7H_2O$ in DI and dilute to 1.0 l.
- *Calcium chloride solution:* Dissolve 27.5 g $CaCl_2$ in DI and dilute to 1.0 l.
- *Ferric chloride solution:* Dissolve 0.25 g $FeCl_3{\cdot}6H_2O$ in DI and dilute to 1.0 l.
- Acid and alkali solutions for adjusting wastewater samples with extreme pH values: 1 M HCl and 1 M NaOH.
- Sodium sulfite solution for removal of residual chlorine: Dissolve 1.575 g Na_2SO_4 in 1.0 l DI. This solution is not stable and should be prepared daily.
- All of the reagents used in Laboratory Exercise 13.8.

Preparation of dilution water (to be prepared by the stockroom):
You will be determining the BOD of a domestic wastewater (sewage) sample, and you will need to dilute the wastewater in order to monitor the consumption of DO over a 5- or 20-day period. Since you will be diluting the wastewater, it is necessary to add inorganic nutrients to the dilution water. The nutrients include the phosphate buffer, calcium chloride, and ferric chloride solutions described earlier in the Chemicals and Solutions (to be prepared by the stockroom prior to lab period). First, estimate how much dilution water you will need. In doing this, you must determine the dilution factor of your wastewater and the number of replicates you will be taking.

After you have estimated the volume of dilution water needed (using Table 16.2), add 1.0 ml of each nutrient solution (phosphate buffer, $MgSO_4$, $CaCl_2$, and $FeCl_3$) per liter of dilution water needed. It is best if you store the dilution water at 20 °C overnight to allow O_2 equilibration with the atmosphere.

Glassware and equipment:

- BOD incubator or a 20 °C water bath that can be kept in the dark.
- 250- or 300-ml BOD bottles.
- 1-ml wide bore pipets for pipeting the blended sewage mixture.
- Several 1.00-ml pipets.
- All of the glassware needed in the Winkler titration experiment.

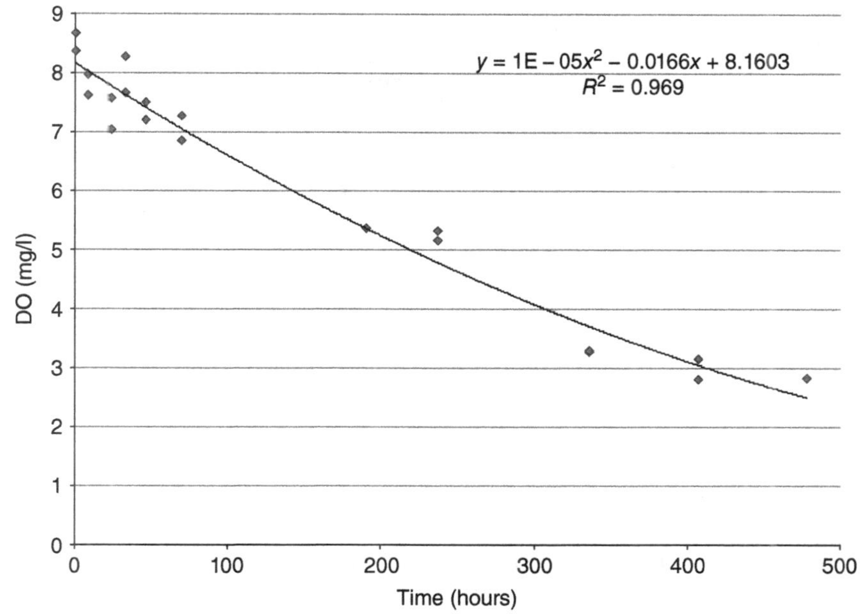

Figure 16.2 Dissolved oxygen remaining in diluted BOD sample.

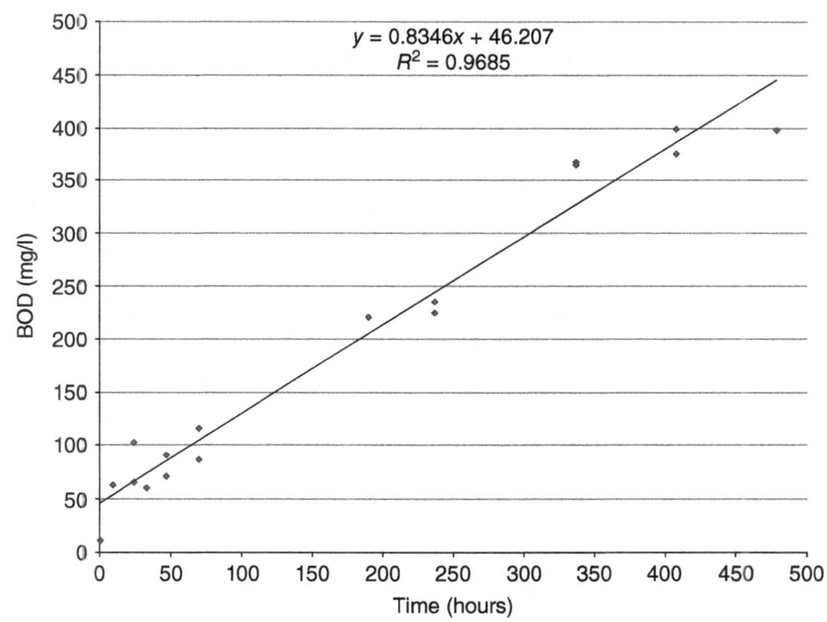

Figure 16.3 BOD as a function of time.

Student Results, Figures 16.2–16.4: Data from Environmental Chemistry and Engineering, Adele Thomson and F.M. Dunnivant, Spring, 2011.

16.4.6 Determination of a Clay-Water Distribution Coefficient for Copper

Purpose:

To determine the distribution coefficient of a metal on a characterized soil.

Background:

Perhaps the most important fate and transport parameter is the distribution coefficient, K_d. The distribution coefficient is a measure of adsorption phenomena between the aqueous and solid phases and is fundamental to understanding processes responsible for the distribution of pollutants in aquatic systems. For applications of the distribution coefficient to fate and transport modeling of groundwater, lakes, and riverine systems, refer to the modeling Chapters 5, 6, and 8. Mathematically, it can be represented as the ratio of the equilibrium pollutant concentration in the solid (sediment or soil) phase to the equilibrium pollutant concentration in the dissolved (aqueous) phase:

$$K_d = \frac{\text{soil conc.} \frac{\text{mg}}{\text{kg}}}{\text{water conc.} \frac{\text{mg}}{\text{l}}}$$

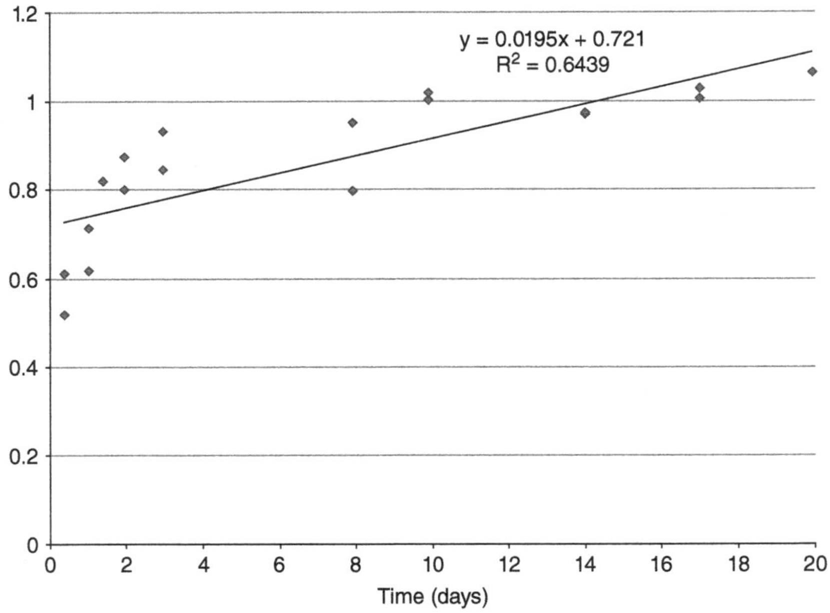

Figure 16.4 Thomas slope analysis of BOD data.

The purpose of the distribution coefficient is to quantify the pollutant's relative preferences for the two phases (solid and aqueous) and thus to determine the mass of pollutant present in each phase. The distribution coefficient is used in virtually every fate and transport model for the estimation of pollutant concentrations in aqueous systems. The aqueous phase concentration is important because the free aqueous phase concentration is usually the most toxic form of pollutants, especially in the case of dissolved metal cations. Inorganic and organic colloids and suspended solids in natural waters will increase the apparent water phase concentration, but pollutants adsorbed to these particles are usually not available for biological uptake. These particles can eventually settle out in quiescent regions of the natural water body or in estuaries and remove sufficient amounts of pollutant from the aquatic system.

Distribution coefficients are relatively easy to determine by allowing a pollutant-soil-water mixture of known composition to equilibrate, separating the mixture into solid and aqueous phases, and determining the pollutant concentration in each phase. This technique can be simplified by measuring (or knowing) the total mass of pollutant added to each sample (as determined from a blank sample), measuring the pollutant in the dissolved phase after equilibration, and estimating the mass of pollutant on the solid phase by difference (total mass of pollutant in blank minus aqueous phase mass). The distribution coefficient is then calculated using the aforementioned equation.

The major problem with designing K_d experiments for the laboratory is the variability (and unpredictability) of results that are obtained given the variety of solid phases available, the nature of the pollutant used (ionic metals or hydrophobic organic compounds), and the experimental aqueous conditions used (pH values, ionic strengths, solids concentrations, and pollutant concentrations). Aqueous conditions are especially important when measuring the K_d for ionic pollutants. Without conducting the experiment under the exact experimental conditions to be used in lab, it is difficult to tell whether aqueous solutions will contain sufficient pollutant in the aqueous phase to be measured, or whether all of the pollutant will be present in the aqueous phase. Given these experimental design problems, it is not surprising that this vital experimental parameter (K_d) is not typically taught in environmental chemistry lab courses but is usually covered in lecture material. This chapter contains a procedure, using standardized materials and conditions, for the determination of a distribution coefficient for copper. The procedure is also environmentally friendly since no (or limited) hazardous waste is generated.

In the laboratory:
Safety and hazards:

- As in all laboratory exercises, safety glasses must be worn at all times.
- Avoid skin and eye contact with NaOH, HCl, and HNO_3 solutions. If contact occurs, rinse your hands and/or flush your eyes for several minutes. Seek immediate medical advice for eye contact.

Student procedures:

Week 1: Prerinse all plastic vials and caps with DI water. "Sling" the vials several times to remove excess water.

Team 1: Refer to Table 16.4.

K_d as a function of Cu concentration (kaolinite):

1. The mineral phase to be used as your adsorbent is kaolinite (KGa-1b).

Table 16.4 Cu solutions table for Teams 1 and 2.

Desired Cu solution concentration in a vial (ppm)	Additional volume (ml) of the Cu solution to the right to yield the desired Cu concentration to the left (to the left)	Standard Cu solution[a)](mg/l)
50.0	2.00	1000.
25.0	1.00	1000.
10.0	4.00	100.
5.00	2.00	100.
1.00	4.00	10.0
0.500	2.00	10.0

a) Instructions for these solutions are given in the instructor procedures.

2. Your overall goal is to measure the K_d as a function of Cu concentration for a kaolinite clay. In order to compare our various K_d results, we need to prepare solutions in which the mass of solid phase, the ionic strength, and the pH are as close to identical as possible. There are probably several ways that we can do this, but we will use the following approach.
   Prepare two vials for each Cu concentration. You will use a TSS concentration of 5000 mg/l. You will be using a total volume in each sample vial of 40.0 ml. Weigh 0.200 g of kaolinite (for the 5000 mg/l TSS) into each vial (except your blank vials). Be as close as you can to this mass, and record your significant figures to four decimal places. You will also need to have two blanks for each Cu concentration. These blank vials will contain ionic strength adjustor, Cu, and water (see Step 4), but no solid phase. Label each with masking tape and a number (for example "T1-1" means "Team one, vial 1"; "T1-B1" means "Team one, blank 1"; ""T1-C1" means "Team one, Cu concentration 1"). Although the nonblank vials are identical at this point, labeling them prepares them for the addition of solutions of different copper concentrations.
3. Next, prepare the solutions to add to the clay-containing vials and blanks. Each vial will have a different amount of copper solution added but the same ionic strength. Make the following solutions in 100-ml (or better yet 50-ml) graduated cylinders, using Cu^{2+} solutions of 1000, 100, and 10 ppm, provided by your lab instructor:
   Combine 2.00 ml of $Ca(NO_3)_2{\cdot}4H_2O$ stock solution (ionic strength adjustor) with the appropriate amount of Cu solution for each concentration (Table 16.3) and fill to 40.0 ml with DI water.
4. Mix the solutions from Step 3 immediately prior to adding them to the kaolinite-containing and blank vials and cap. Again, be sure to prepare 2 blanks for each Cu concentration (containing everything, including Cu standard, but no solid phase). These will be necessary to determine if any Cu adsorbs to the container walls.
5. Place the vials on the mixer for at least three days. (You may leave these vials on the mixer for seven days, until your next lab period, but it is best to remove them one day prior to performing the filtration step that occurs during week 2. This will allow the filtration to proceed faster.)

Team 2: Refer to Table 16.4.

K_d as a function of Cu concentration (montmorillonite (STx-1 or SAz-1):

1. The mineral phase to be used as your adsorbent is montmorillonite (STx-1).
2. Your overall goal is to measure the K_d as a function of Cu concentration for a montmorillonite clay. In order to compare our various K_d results, we need to prepare solutions in which the mass of solid phase, the ionic strength, and the pH are as close to identical as possible. There are probably several ways that we can do this, but we will use the following approach.
   Prepare two vials for each Cu concentration. You will use a TSS concentration of 5000 mg/l. You will be using a total volume in each sample vial of 40.0 ml. Weigh 0.200 g of montmorillonite (for the 5000 mg/l TSS) into each vial (except your blank vials). Be as close as you can to this mass, and record your significant figures to four decimal places. You will also need to have two blanks for each Cu concentration. These blank vials will contain ionic strength adjustor, Cu, and water (see Step 4), but no solid phase. Label each with masking tape and a number (for example "T2-1" means "Team two, vial 1"; "T2-B1" means "Team two, blank 1"; "T2-C1" means "Team two, Cu concentration 1"). Although the nonblank vials are identical at this point, labeling them prepares them for the addition of solutions of different copper concentrations.
3. Next, prepare the solutions to add to the clay-containing vials and blanks. Each vial will have a different amount of copper solution added, but the same ionic strength. Make the following solutions in 100-ml (or better yet ~~a~~ 50-ml) graduated cylinders, using Cu^{2+} solutions of 1000, 100, and 10 ppm, provided by your lab instructor:
   Combine 2.00 ml of $Ca(NO_3)_2{\cdot}4H_2O$ stock solution (ionic strength adjustor) with the appropriate amount of Cu solution for each concentration (Table 16.3) and fill to 40.0 ml with DI water.
4. Mix the solutions from Step 3 immediately prior to adding them to the montmorillonite-containing and blank vials, and cap. Again, be sure to prepare 2 blanks for each Cu concentration (containing everything, including Cu standard, but no solid phase). These will

be necessary to determine if any Cu adsorbs to the container walls.

5. Place the vials on the mixer for at least three days. (You may leave these vials on the mixer for seven days, until your next lab period, but it is best to remove them one day prior to performing the filtration step that occurs during week 2. This will allow the filtration to proceed faster.)

Week 2: There will be several demonstrations at the beginning of lab to illustrate the use of the filter apparatus and mixing system.

1. Turn on the AAS to warm up the lamp.
2. Prepare calibration standards at concentrations of 0.100, 0.500, 1.00, 5.00, 10.0, 25.0, and 50.0 ppm Cu^{2+} (the instructor should check the linear range on your particular instrument). Prepare these in 1% HCl.
3. Filter the solutions that you prepared last week as illustrated by your lab instructor. First, filter them through the Gelman type A/E glass fiber filter, then through a 0.2-μm HPLC nylon filter with a syringe. Filter both the blanks and the actual samples. *Note:* You do not need to filter the entire 40.0 ml suspension, nor should you mix the vial, since it is already at equilibrium after the seven days of mixing. Mixing the vial again will only slow the filtration step and delay your leaving lab. You only need a sufficient volume of the filtrate to run on the atomic absorption unit. Typically, 10–20 ml will suffice, but check with you lab instructor to see how much you will need for your particular instrument.
4. Analyze the samples using AAS as demonstrated.
5. Calculate the K_d as illustrated in tables in Chapter 3.

Waste disposal:

Check with your laboratory instructor for disposal instructions for the Cu waste.

Assignment:

Prepare plots of Cu concentration in the dissolved phase (measured mg/l; placed on the x-axis) versus the calculated Cu concentration on the solid phase (mg/kg; placed on the y-axis). Calculate the slope of the line, which will be the K_d.

For the instructor (to be prepared prior to the lab period):

Time requirements:

Two 3–4-hour laboratory periods.

Level of difficulty: Moderate to difficult.

There are a number of ways to conduct this laboratory exercise; you should decide which is best for your class. One option is to divide the class into two groups, one group for the determination of K_d as a function of Cu concentration for kaolinite clay suspensions and another group for the determination of K_d as a function of Cu concentration for montmorillonite clay. Or, only one clay may be used by the entire class.

Chemicals and Solutions (to be prepared by the stockroom):

- *Sorbents*:
  1. Ca-montmorillonite (may be obtained from The Clay Minerals Society, Source Clays Repository [http://www.clays.org/sourceclays_history.html]. Product STx-1 or SAz-1. The origin of the clay, chemical composition, cation exchange capacity, and surface area are also given. Ordering information is also available at this website. Mass requirements are ~0.100 + −0.001 g per vial).
  2. K-Kaolinite (may be obtained from The Clay Minerals Society, Source Clays Repository [http://www.clays.org/sourceclays_history.html]. Product KGa-1b. The origin of the clay, chemical composition, cation exchange capacity, and surface area are also given. Ordering information is also available at this website. Mass requirements are ~0.100 + −0.001 g per vial).
- Preparation of stock copper solution:
  1. Make a 1000 mg/l solution by adding 2.683 g of $CuCl_2 \cdot 2H_2O$ to a 1-l volumetric flask and filling to the mark with DI water. (Do not add acid.) Make a 100:1000 dilution of this solution to obtain a 100.0 mg/l solution of Cu^{2+}. Make a 100 : 1000 dilution of this solution to obtain a 10.0 mg/l solution of Cu^{2+}. These Cu solutions will be referred to in Table 16.3.
  2. Prepare the Cu calibration standards for the FAAS unit at concentrations of 0.100, 0.500, 1.00, 5.00, 10.0, 25.0, and 50.0 ppm Cu^{2+}. Prepare these in 1% HCl.
- Preparation of calcium nitrate solution:
  Two 100-ml portions of 2.00 M $Ca(NO_3)_2 \cdot 4H_2O$
- Preparation of $CaNO_3$ ionic strength adjustor
  $Ca(NO_3)_2 \cdot 4H_2O$ stock solution – transfer 29.54 g of $Ca(NO_3)_2 \cdot 4H_2O$ (GMW 236.16 g/mol) to a 250 ml volumetric flask and fill to mark. This will yield a 0.500 M solution.

Glassware and supplies:

50-ml plastic, sealable vials (24 vials per solid evaluated) (Blue Max Disposable Centrifuge Tubes, Polystyrene, Conical-Bottom, Sterile, Falcon, VWR Scientific Products Number 21008-939).

100- or 50-ml graduated cylinders.

Plastic filter holders and filters. (Polycarbonate Filter Holder, 25-mm filter,12 per pack, VWR Scientific Product Number 22001-800.)

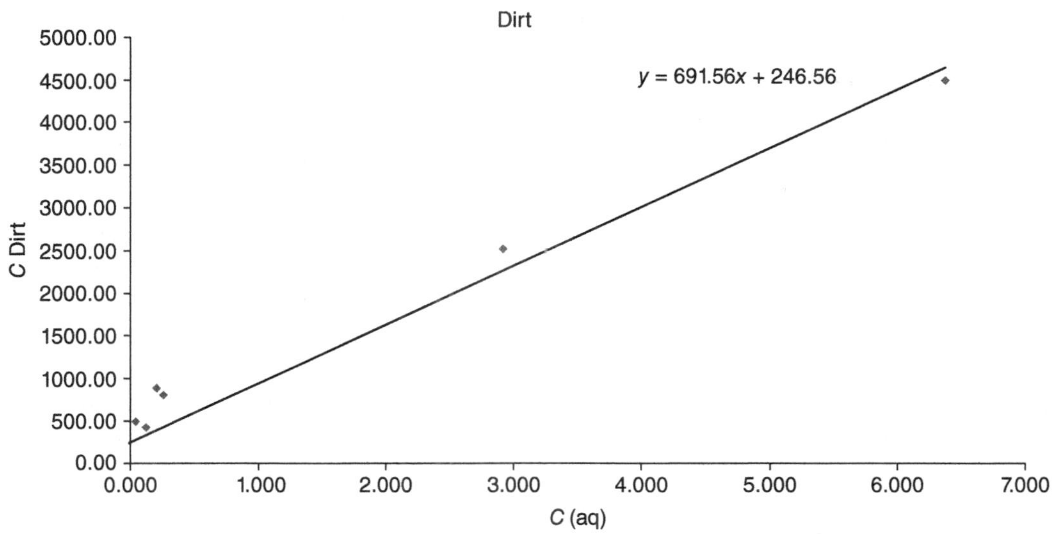

Figure 16.5 Cu K_d on backyard soil as a function of Cu concentration (Plot).

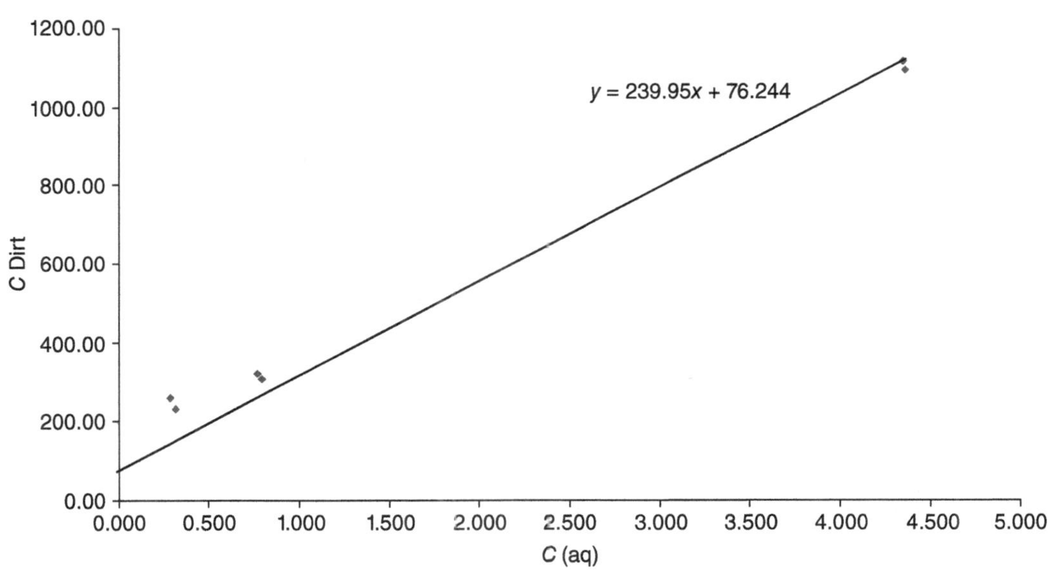

Figure 16.6 Plot of Cu K_d on montmorillonite sorbent as a function of Cu concentration; environmental chemistry and engineering, spring, 2016.

25-ml plastic syringes.

25-mm Gelman type A/E glass fiber filter, Glass Fiber Filters, Type A/E, Pall Gelman, VWR Scientific Products Number 28150-178.

Spartan-13 0.2-μm membrane filter or similar brand. (Agilent Technologies, HP-5061-3366.)

50-ml plastic beakers for holding filtered samples.

Test tube rotator, Glas-Col Mini-Rotator, 120V, VWR Scientific Products Number 33725-042. (Test tube rockers will probably work just as well.)

-Cu flame atomic absorption spectroscopy (FAAS) lamp.

-FAAS unit.

Sample student data:

(see Figures 16.5 and 16.6) and (Tables 16.5 and 16.6).

16.5 The Measurement of Dispersion in a Simulated River System

In Chapter 6, we presented two basic models for the transport of pollutants in river systems. We made a simple assumption for both models – that the river system was completely mixed upon the addition of

Table 16.5 Cu K_d on backyard soil as a function of Cu concentration (data); Environmental Chemistry and Engineering, spring, 2016.

Concentration (ppm)	Abs	Expected abs	Difference	Aqueous concentration (ppm)	Dirt (kg)	Concentration dirt
0.0000	0.003	0.0044	0.0014	−0.009	3.98E−05	8.56
0.0000	0.0035	0.0044	0.0009	−0.005	3.97E−04	0.55
0.0110	0.0032	0.0062	0.0030	−0.007	4.07E−05	17.97
0.0110	0.0034	0.0062	0.0028	−0.006	4.16E−05	16.39
0.0550	0.0037	0.0133	0.0096	−0.004	3.98E−05	59.51
0.0550	0.0035	0.0133	0.0098	−0.005	4.01E−05	60.29
0.5502	0.0111	0.0935	0.0824	0.041	4.10E−05	496.27
0.5502	0.0242	0.0935	0.0693	0.122	4.03E−05	424.61
1.1004	0.0381	0.1826	0.1445	0.208	4.02E−05	887.70
5.5018	0.4771	0.8954	0.4183	2.919	4.11E−05	2513.80
11.0035	1.0368	1.7865	0.7497	6.375	4.11E−05	4504.92

Table 16.6 Cu K_d on montmorillonite sorbent as a function of Cu concentration (data); Environmental Chemistry and Engineering, spring, 2016.

Concentration (ppm)	Abs	Expected abs	Difference	Aqueous concentration (ppm)	Dirt (kg)	Concentration dirt
0	0.0029	0.0044	0.0015	−0.009	3.98E−05	9.18
0	0.0026	0.0044	0.0018	−0.011	3.97E−04	1.11
0.0110	0.0026	0.0062	0.0036	−0.011	4.07E−05	21.61
0.0110	0.0026	0.0062	0.0036	−0.011	4.16E−05	21.14
0.0550	0.0027	0.0133	0.0106	−0.010	3.98E−05	65.71
0.0550	0.0027	0.0133	0.0106	−0.010	4.01E−05	65.22
0.550	0.0504	$	#VALUE!	0.284	4.10E−05	259.53
0.550	0.056	0.0935	0.0375	0.319	4.03E−05	229.72
1.100	0.133	0.1826	0.0496	0.794	4.02E−05	304.65
1.100	0.1292	0.1826	0.0534	0.771	4.11E−05	320.82
5.502	0.7094	0.8954	0.1860	4.353	4.11E−05	1117.85
5.502	0.711	0.8954	0.1844	4.363	4.16E−05	1094.91

our pollutant and that dispersion only occurred in the longitudinal direction (the direction of water flow downhill). This is essentially true in most cases, but the degree of longitudinal mixing changes as you move from calm regions of the river to riffle areas or eddies. In this laboratory exercise, we will look at several different "physical model" river systems, each containing different degrees of mixing. Our system is simple, but it will clearly illustrate how increasing the number of riffle areas and pools can be affect dispersion.

In the laboratory:
Safety and hazards:

- As in all laboratory exercises, safety glasses must be worn at all times.
- All the chemicals used in this laboratory are relatively safe, but standard laboratory cautions and safety should be practiced.
- You will spill and spread dilute fluorescein solution everywhere in this lab. Dilute concentrations are safe when spilled on your skin but may turn your skin yellow for a brief time. We recommend wearing latex gloves during this laboratory exercise.

Student procedures:

Your laboratory instructor will divide the class into several groups, corresponding to the number of different river systems have been prepared. Each group will have a specially prepared PVC model river channel (refer to Figure 16.5), a fluorescein dye for addition to your model river, a set of test tubes or beakers for sampling the effluent from river, and access to a spectrophotometer. Each group will have a model river system with a different number of weirs, producing distinct mixing patterns, so that the results from the three groups can be compared.

Note: This procedure does not result in exact science. Your sampling times and volume of sample may be considerable off from ideal values, but this will be fine. We are not conducting a detailed chemistry experiment in this laboratory, but only illustrating the concept of dispersion.

(1) Turn on the spectrophotometer and allow it to warm up. Zero the spectrophotometer with DI. Complete step two as the spectrophotometer is warming up.
(2) Adjust the slope of your river system as illustrated by your instructor and allow it to directly drain into a sink.
(3) Turn on the water source to your river system and adjust the flow to 1.0 l/min (adjust the flow as close to 1.0 l/min as possible; ±5%).
(4) As one student adds 5.0 ml of the dye solution to the top of the river system, other students should be ready to rapidly sample the outlet of the river. For the river systems with no weirs, the sampling must be very fast. We suggest having test tubes lined up in a holder and taking a sample at the specified times below. For river systems with weirs, you will sample at slightly longer intervals. The sampling times are (i) for no weirs, sample every two seconds, (ii) for one-weir systems, sample every five seconds, (iii) for two-weir systems, sample every 10 seconds, and (iv) for four-weir systems, sample every 10 seconds.
(5) Sample the effluent from your system until you do not see any dye color in the test tubes. The more dams present in the river systems, the longer it will take to flush all the dye from your model.
(6) Analyze the samples with respect to absorbance using the spectrophotometer.
(7) Repeat your experiment at least twice.
(8) Plot all the class data on one plot (x-axis: sample time in seconds; y-axis: absorbance at time "t" divided by absorbance of original dye solution) and answer the following questions:
   a. Was the tracer (dye) input a step or pulse input?
   b. Estimate the dispersion (in time units) of your river systems by taking the width of the dispersion peak at one-half the maximum absorbance value for each dye concentration. This will allow you to compare each river system.
   c. Which system showed the least dispersion?
   d. Which system showed the highest dispersion?
   e. Why are the dye concentrations (as measured by absorbance) highest for the system with no weirs?
   f. Why does the width of the dye plot increase with increasing number of weirs?

For the instructor (to be prepared prior to the lab period):

Time requirements:

One 3–4-hour laboratory period.

Level of difficulty: Easy

Chemicals and Solutions (to be prepared by the stockroom):

Fluorescence dye solution:

Dissolve ~0.30 g of fluorescein dye (EMD brand, part number FX0325-5, $C_{20}H_{10}Na_2O_5$; CAS number 518-47-8) dye (10^{-5}–10^{-4} M range) in 500. ml of water. The professor or stockroom technician should ensure the stock solution yields and acceptable absorbance (within Beer's law) on the spectrophotometer before use.

Glassware and supplies:

- A photograph of one of the PVC river system is shown in Figure 16.7. It was made by taking 3-in PVC tubing and 135° (45°) connection elbows and cutting them in half, lengthwise, and cutting half-circle weirs at the PVC joint intersections. Ten-inch sections of PVC tubing were connected with the elbows using PVC glue. The entire assembly was mounted on a 1-in by 8-in by 100-in board. Weirs were placed at

Figure 16.7 Photogram of a constructed PVC river.

the desired connection elbows with aquarium seal to allow removal and replacement. The entire assembly is placed at a slope of 12 in over the entire 92 in PVC river and directly drained into a sink. Other slopes can be used, but it is important that the water does not back up and overflow behind the weirs. Water is delivered to the top of the PVC river with tubing. Note that maintaining the consistency of sampling the water at the end of the PVC river is slightly difficult since water pressure (and thus flow) fluctuations occur in most laboratory plumbing. This will result in error in sampling times and volumes, but this error has not interfered with illustrating the concept of dispersion. Again, all systems should have the same slope and flow rate in order to compare the class data.

- A set of Spectronic-20 test tubes (10–15 per group).
- A spectrophotometer set to 494 nm.
- A stopwatch or clock with a second hand.

Student Results: Table 16.7 contains student data for a simulated PVC river with an increasing number of weirs (dams). The data are plotted in Figure 16.8.

Table 16.7 Student data for Figure 16.8; simulated river with weirs.

Time (s)	Free	One weir	Two weir	Three weir	Four weir	Five weir
0	100	100	100	100	98	97
2	98					
4	15					
5	12	71				
6	42					
8	89					
10	100	54	91.5	96		
15		75			93	
20		88	56.5			98
25		98				
30		100	71	78	75.5	
40			82.5	79		90
45					78	
50			88.5	84		
60			97	85.5	92	84
70			100	90		
75					90	
80				94		90
90				100	88	
100						92
105					95	
120					94	95
135					97	
140						98
150					94	
160						98
165						
180						100

16.6 The Measurement of Dispersion and Sorption in a Simulated Groundwater System

In Chapter 8, we studied the fate and transport of pollutants in groundwater systems. A few scientists have had the opportunity to experiment and study with contaminated real-world groundwater systems, but this process can be very prohibitively expensive. To reduce the expense of studying these sites, many scientists use microscale systems. This involves taking soil from the contaminated site, packing it into a small column, studying the movement of pollutants through the column, and then relating the results to the large-scale system. In this experiment, we will study the movement of cadmium ion through a sand column.

Before we conduct the experiment, it is useful to review the concept of distribution coefficient, K_d, and the retardation factor, R. Again, the distribution coefficient can be mathematically defined as

$$K_d = \frac{\text{pollutant conc. in the solid phase}}{\text{pollutant conc. in the aqueous phase}} = \frac{\text{mg/kg}}{\text{mg/l}}$$

and represents the relative affinity of the pollutant for the solid or liquid phases. K_d values greater than 1.0 show that the pollutant prefers the solid phase to the water phase. In order to incorporate this into the general transport equation for groundwater system, we must relate K_d to the degree to which the velocity of the pollutant in the system will be slowed down due to its preference for the solid phase. This is accomplished by

$$R = 1 + \frac{\rho_b K_d}{n}$$

where ρ_b is the bulk density (of the sand) and n is the volumetric water content of the soil column. R is directly used in step and pulse pollutant transport equations for groundwater systems. R can also be defined by a ratio of the rate of pollutant movement over the rate of water movement. Our "measure" of water and pollutant movement will be the volume of water needed to push the water or pollutant through the column. The volume of water needed for these measurements will be defined in units of column pore volumes. A pore volume is the volume of water contained in the soil column at any

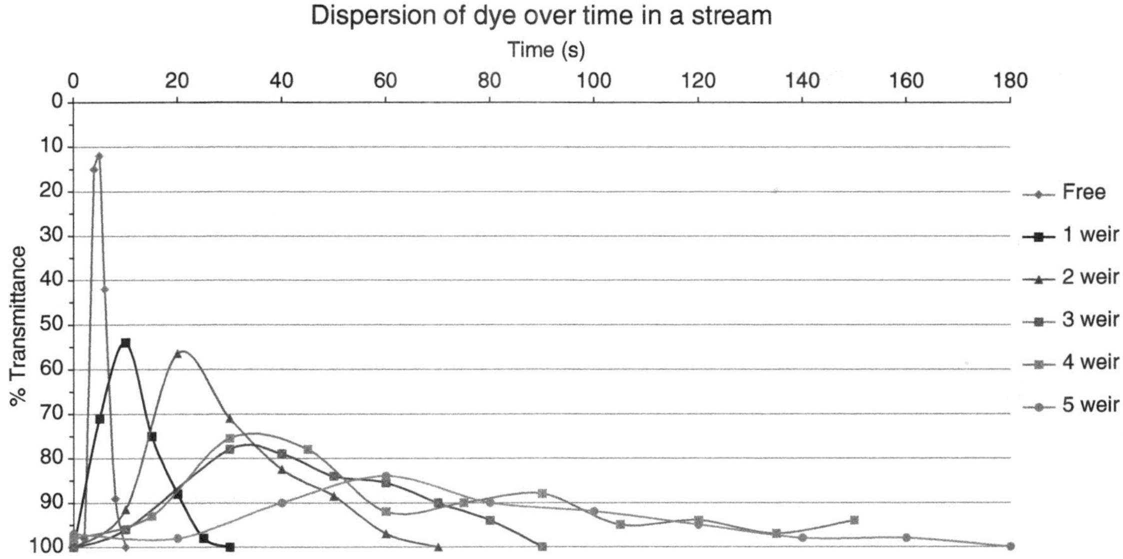

Figure 16.8 A plot of simulated river results for the environmental chemistry and engineering class of 2014.

given time. This can easily be measured, as described in the experimental procedures section, by weighing the soil column with and without water in it. Measuring the water velocity is not easy, so we place a chemical tracer in the water that is not retained by the soil and moves at a velocity at or very near that of the water. In our experiment, we will use fluorescein dye as this conservative chemical tracer. We will measure the velocity of the dye (volume of water needed to push out the dye) and the velocity of the cadmium ions (volume of water needed to push out the cadmium).

In the laboratory:
Safety and hazards:

- As in all laboratory exercises, safety glasses must be worn at all times.
- Dilute concentrations of fluorescein dye are safe when spilled on your skin but may turn your skin yellow for a brief time. In order to avoid this, we recommend wearing latex gloves during this laboratory exercise.
- Cadmium is a toxic heavy metal and should be handled with care. You will be using dilute concentrations. Wash any exposed skin with soap and water.
- Nitric acid will be used and should be handled with caution since it is a strong acid.

Chemicals and equipment to be supplied by your instructor:

- The experimental setup shown in Figures 16.9 and 16.10.
- A 1–2 mg/l solution of Cd^{2+} depending on the sensitivity of your FAAS.
- Fluorescein dye solution, instructions below (10^{-5}–10^{-4} M range).

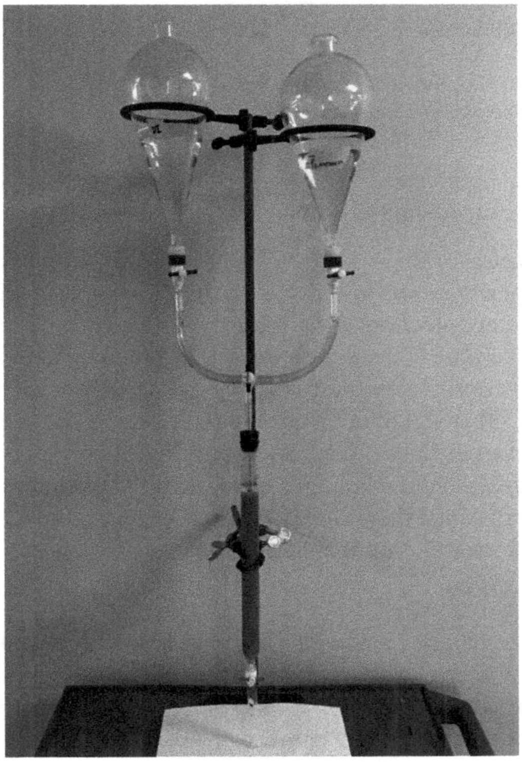

Figure 16.9 Overview photograph of the step groundwater experimental apparatus.

- Up to 100 test tubes for collection of samples (your instructor will estimate how many you will need for your experiment).
- A 10-ml graduated cylinder.
- A Spectronic-20 spectrophotometer for measuring fluorescein concentrations.

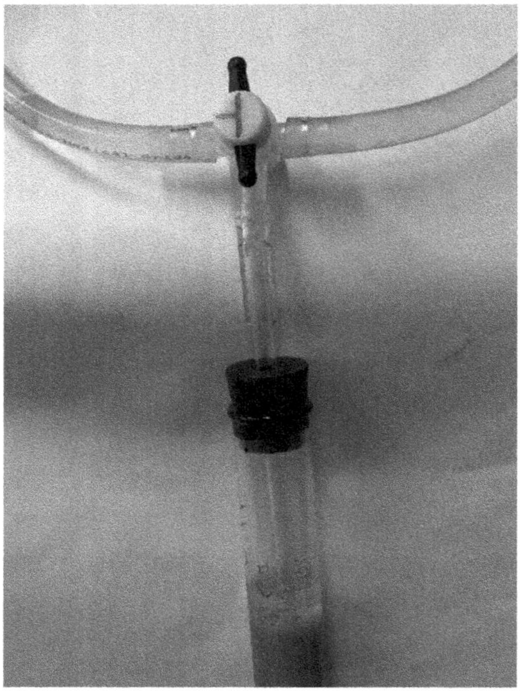

Figure 16.10 Close-up image of the step groundwater "T" joint and the transfer tube–sand interface.

- A flame atomic absorption spectrophotometer (FAAS unit) for measuring Cd concentrations.

Student procedures:

This experiment will last for two to three weeks and will involve a step injection of cadmium solution during the first week, EDTA-Cd solution during the second week, and an injection of conservative tracer dye during the second week. This will allow you to determine the relative velocity of each tracer. We have found that it is important to do the dye tracer experiment after the cadmium experiment. *Note:* This is a complicated experiment and requires close attention to detail.

Week 1:

1. Observe your experimental setup and identify all the parts and how they are put together as you read over the procedure.
2. Disconnect the glass column from the setup and place a very small ball of class wool in the narrow base of the column. Weigh the column. Fill the column between 70% and 80% full with dry sand. Weight the column again to determine the weight of dry sand in your column. Close the stopcock and completely fill the column with water. Place your hand over the top of the column while avoiding creating any air bubbles to seal the column. Invert the column to displace all air pockets in the sand. Add more water to remove the air and repeat the process until all the air has been removed from the soil column. (*Note:* Be careful not to dislodge the glass wool ball at the bottom of the column. If you do, you will need to repeat the entire Step 2.) Hold the column upright and tap the column to settle the sand.
3. Wash the sand column by opening the stopcock and adding water to the top of the column. *Note:* Never let the water level fall below the top of the sand packing or you will introduce new air pockets to the sand. If you accidently do allow the water level to fall below the top of the sand, repeat the last part of Step 2 to remove the air pockets. After you washed the column with 100 ml of DI, drain the excess water from the column, stopping just above the top of the sand (to within ~2 mm). Weigh the sand column again to determine the volume of water in the column. This mass minus the weight of the sand and glass column, when converted to volume (assume 1.00 ml water = 1.00 g water), is your column pore volume.
4. Reassemble the experimental setup keeping all stopcocks closed (the *y*-junction stopcock should be at a diagonal to the DI reservoir) and completely sealing the stopper in the top of the column.
5. Fill one separatory funnel with DI.
6. Slowly open the stopcock on the DI separatory funnel reservoir, then the T-junction stopcock (towards the DI only) allowing water to flow into the column. The water level will only slightly rise if you have a good seal throughout the apparatus.
7. Slowly open the stopcock at the effluent of the sand column to obtain a flow rate of 0.25 ml/s. Measure the flow rate by placing a 10-ml graduated cylinder under the column and timing how long it takes to collect 5.00 ml of water. Adjust the stopcock as needed.
8. Allow the column to continue to flush until no suspended particles are present in the effluent while you prepare the remainder of the experiment.
9. Fill the second separatory funnel with cadmium solution from the instructor after ensuring that the separatory funnel stopcock is closed.
10. Number each test tube and record its dry weight.
11. Decide who will (i) open the cadmium (or Cd-EDTA or dye tracer) stopcock, (ii) collect the samples in test tubes, (iii) weigh the test tube samples once collected, (iv) analyze the samples on the FAAS to determine the Cd^{2+} concentration, and (v) record all data from the experiment.
12. Before you start the experiment, you should (i) recheck the column flow and adjust it as necessary and (ii) estimate how many test tube volumes (about 7–8 ml for this experiment) it takes to equal one pore volume.

13. Start the experiment by opening the cadmium stopcock and turning the *y*-junction stopcock away from the DI reservoir and toward the cadmium solution reservoir. *Note:* take care to turn the stopcock so that it does *NOT* connect the two reservoirs to the column at the same time. Immediately start sampling. Collect approximately 7–8 ml in each test tube.
14. As you finish collecting each sample, weigh the test tube to determine the mass of water in each sample.
15. Measure the absorbance of Cd^{2+} in the original stock solution as well as each sample using the FAAS.
16. Collect samples until the Cd^{2+} in the effluent is equal to that in the original stock solution. Check with your instructor to make sure you can stop the experiment.
17. To stop the experiment, close the stopcocks starting at the bottom and working your way up: close the column effluent stopcock, the T-junction stopcock, and the cadmium reservoir stopcock.
18. While you work on the data analysis, flush the column with DI water. After ~5–10 pore water volumes have passed through, check the effluent for residual Cd^{2+}.
19. Create a table with your data and calculate two final columns: fraction pore water volume and C/C_0. Create a plot of C/C_0 versus fraction pore water volume.

Week 2:

1. Repeat the experimental procedure from Week 1, but with a cadmium-EDTA (1 : 2 molar) solution.

Week 3:

1. Repeat the experimental procedure from Week 1, but with a fluorescein solution and monitor fluorescein absorbance by UV–vis spectrophotometer.

Assignment:

1. Create a combined plot of C/C_0 versus fraction pore water volume for all three experiments.
2. Determine the retardation factor, R, for cadmium and cadmium-EDTA.
3. Determine the K_d for cadmium and cadmium-EDTA based on your column experiment.
4. Explain the differences in the cadmium and cadmium-EDTA plots.

Note: It has been found that conducting the fluorescein experiment first, alters the surface of the sand, and Cd in the next experiment does not properly adsorb to the sand.

Determine the retardation factor for cadmium.

Determine the K_d for cadmium based on your column experiment.

Why is the dye plot narrower than the cadmium plot?

Why is the height of the cadmium plot lower than the dye plot?

For the instructor (to be prepared prior to the lab period):

Time requirements:
Three 3–4-hour laboratory periods

Level of difficulty: Moderate

Chemicals and Solutions (to be prepared by the stockroom):

Fluorescence dye solution: Dissolve ~0.30 g of fluorescein dye (EMD brand, part number FX0325-5, $C_{20}H_{10}Na_2O_5$; CAS number 518-47-8) dye (10^{-5}–10^{-4} M range) in 500. ml of water. Check the absorbance on your spectrophotometer to ensure that the concentration is near the top of your linear range (less than or equal to 1.0 absorption units).

Cadmium solution: 5–10 mg Cd^{2+}/l (in a plastic bottle; no acid; to be prepared fresh each day) depending on the sensitivity of your FAAS.

Glassware and supplies:

- A photograph and illustration of the setup is shown in Figures 16.9 and 16.10. It consists of a two 1-l separatory funnel with a stopcock, T-junction connecting the two separatory funnels, a narrow glass tube to deliver the water flow to the top of the sand, and a glass chromatography column (1.5–2.0 cm in diameter by ~30 cm long) equipped with a stopcock. A fraction collector is optional, but a manual student version of this works very well.
- Up to 100 test tubes for collection of samples (a preliminary experiment by the stock room will help you determine exactly how many test tubes will be needed by each student group). *Note:* 20–30 for the tracer dye and Cd-EDTA experiments, but considerable more for the Cd^{2+} experiment.
- A 10-ml graduated cylinder.
- A Spectronic-20 spectrophotometer.
- A flame atomic absorption spectrophotometer (FAAS).

If you do not use the recommended sand, it is strongly recommend conducting a K_d determination experiment with your specific batch of sand to determine the K_d (low K_d is highly desirable in order to have a short, time-manageable experiment). Details on how to conduct this experiment were given in Experiment 16.10. From this experiment, you can determine R and estimate the volume of water that needs to be passed through the column to observe the pulse of cadmium. The preparatory experiment will be instrumental to the success of the students' experiment since different batches of sand will adsorb cadmium to a different degree.

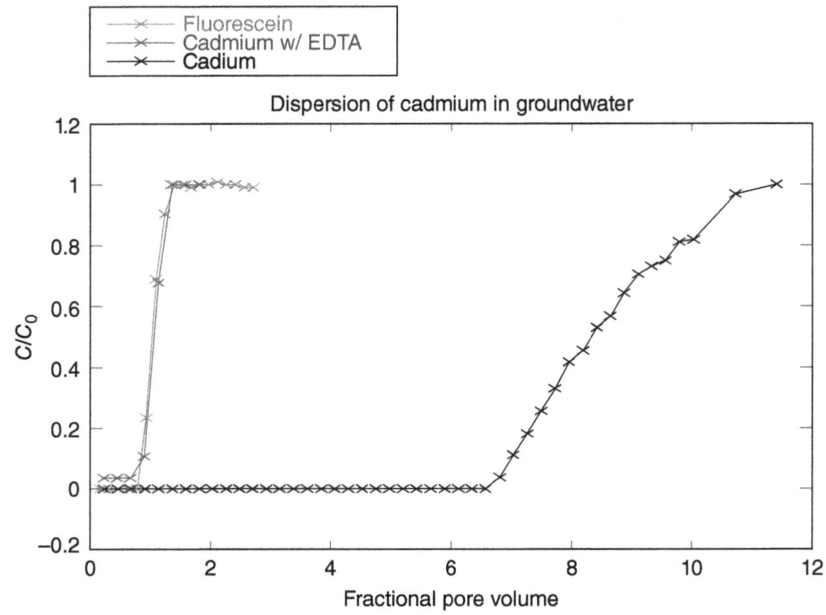

Figure 16.11 Sample student-generated experimental data for the conservative tracer, cadmium pollutant, and cadmium–EDTA mixture.

Results from our experiments:

Flow rate = 0.25 ml/s
130 g of sand placed in column
Pore volume = 30. ml
Volumetric water content = $n = 0.298$
Bulk density = $\rho_b = 1.132$.

A data summary and plot of student results are shown below: (Environmental Chemistry and Engineering, spring, 2011) (Table 16.8 and Figure 16.11).

Your laboratory instructor will give you plans and procedures for this exercise, since these procedures will vary greatly depending on his or her particular plans. You will be visiting a local stream and conducting several physical, chemical, and biological measurements to characterize the water quality. While playing outside and in the water is fun, keep in mind the goal at hand: evaluating the quality of the stream.

For the instructor (to be prepared prior to the lab period):

This exercise will be the most memorable of the entire class and has inspired numerous students to become science majors. It is a common exercise for high school students but can effectively be used in college. There are a variety of ways to conduct this exercise, and your approach will most likely be determined by the enrollment in your class. For large classes, we suggest following the exercise described in Dunnivant et al. (1999). For smaller classes (less than 15–20 students), we suggest taking the students on an overnight sampling and monitoring trip of a local stream. Most streams have a diurnal cycle with respect to dissolved oxygen, pH, temperature, and for most nutrients. Most of the needed equipment is given in the article previously cited, but the addition of a HACH water testing kit adds considerably to the project. We usually collect data every two to three hours for 24 hours and have the students compile the data into plots.

Time requirements:

This ranges from one 3–4-hour laboratory period up to an overnight camping trip.

Level of difficulty: Easy

Chemicals and Solutions (to be prepared by the stockroom):

Suggested equipment for river survey	• Water current meter (optional) • Tennis ball (for water velocity measurements) • A 10-ft section of rope (for water velocity measurements) • A rope marked at 0.5- or 1-ft intervals (depending on the size of the stream) and long enough to reach across the stream; alternatively a long measuring tape • Stop watch • Yardstick or meterstick (for water depth measurements) • Laboratory notebook and ball point pens
Suggested equipment for chemical analysis	• Portable laboratory kit with dissolved oxygen probe, spectrophotometer, conductivity meter, pH meter • Sampling bottles • Laboratory notebook and ball point pens
Suggested equipment for macroinvertebrate Sampling	• Four sticks for staking off the sampling area and a fine mesh net or a Surber macroinvertebrate sampler • Dissecting microscope (optional) • White buckets, white sorting trays, and forceps • Macroinvertebrate key (referenced below) • Laboratory notebook and ball point pens

Table 16.8 Student data for Experiment 16.6.

(a) Fluorescein

Sample	Abs.	Empty mass (g)	Full mass (g)	Mass of sample (g)	26.5848 Total pore volume (ml)	Pore volume (ml) Fractional pore volume	C/C_0
Initial conc. (mg/l)						33.65	
1	0.00	8.56	14.1	5.54	5.54	0.16	0.000
2	0.00	8.62	13.66	5.04	10.58	0.31	0.000
3	0.00	8.56	13.67	5.11	15.69	0.47	0.000
4	0.00	8.51	13.93	5.42	21.11	0.63	0.000
5	0.00	8.61	13.73	5.12	26.23	0.78	0.000
6	0.26	8.53	13.59	5.06	31.29	0.93	0.236
7	0.75	8.56	13.7	5.14	36.43	1.08	0.690
8	0.99	8.63	13.72	5.09	41.52	1.23	0.904
9	1.09	8.58	13.79	5.21	46.73	1.39	1.000
10	1.09	8.55	13.58	5.03	51.76	1.54	1.000
11	1.08	8.56	13.21	4.65	56.41	1.68	0.991
12	1.09	8.53	13.52	4.99	61.4	1.82	1.000
13	1.09	8.57	13.45	4.88	66.28	1.97	1.000
14	1.10	8.66	13.59	4.93	71.21	2.12	1.009
15	1.09	8.58	13.51	4.93	76.14	2.26	1.000
16	1.09	8.64	13.69	5.05	81.19	2.41	1.000
17	1.08	8.61	13.64	5.03	86.22	2.56	0.991
18	1.08	8.7	13.67	4.97	91.19	2.71	0.991

(b) Cadmium

Sample	C (mg/l)	Empty mass (g)	Full mass (g)	Mass of sample (g)	Total pore volume (ml)	Fractional pore volume	C/C_0
1	0	8.57	16.32	7.75	7.75	0.230	0.00
2	0	8.63	16.23	7.6	15.35	0.456	0.00
3	0	8.57	16.11	7.54	22.89	0.680	0.00
4	0	8.52	16.18	7.66	30.55	0.908	0.00
5	0	8.61	16.32	7.71	38.26	1.137	0.00
6	0	8.54	16.24	7.7	45.96	1.366	0.00
7	0	8.57	16.27	7.7	53.66	1.595	0.00
8	0	8.63	16.05	7.42	61.08	1.815	0.00
9	0	8.59	16.12	7.53	68.61	2.039	0.00
10	0	8.74	16.44	7.7	76.31	2.268	0.00
11	0	8.57	15.84	7.27	83.58	2.484	0.00
12	0	8.54	15.82	7.28	90.86	2.700	0.00
13	0	8.57	16.16	7.59	98.45	2.926	0.00
14	0	8.66	16.23	7.57	106.02	3.151	0.00
15	0	8.58	16.23	7.65	113.67	3.378	0.00
16	0	8.64	16.46	7.82	121.49	3.610	0.00
17	0	8.61	16.52	7.91	129.4	3.845	0.00
18	0	8.71	16.17	7.46	136.86	4.067	0.00
19	0	8.61	16.31	7.7	144.56	4.296	0.00
20	0	8.53	16.11	7.58	152.14	4.521	0.00
21	0	8.53	16.21	7.68	159.82	4.749	0.00

(Continued)

Table 16.8 (Continued)

(b) Cadmium							
Sample	C (mg/l)	Empty mass (g)	Full mass (g)	Mass of sample (g)	Total pore volume (ml)	Fractional pore volume	C/C_0
22	0	8.59	16.18	7.59	167.41	4.975	0.00
23	0	8.55	16.16	7.61	175.02	5.201	0.00
24	0	8.66	16.32	7.66	182.68	5.429	0.00
25	0	8.56	16.42	7.86	190.54	5.662	0.00
26	0	8.54	16.29	7.75	198.29	5.893	0.00
27	0	8.62	16.36	7.74	206.03	6.123	0.00
28	0	8.57	16.2	7.63	213.66	6.349	0.00
29	0	8.6	16.36	7.76	221.42	6.580	0.00
30	0.006	8.62	16.34	7.72	229.14	6.810	0.04
31	0.018	8.69	16.32	7.63	236.77	7.036	0.11
32	0.029	8.57	16.32	7.75	244.52	7.267	0.18
33	0.041	8.57	16.27	7.7	252.22	7.495	0.26
34	0.053	8.58	16.4	7.82	260.04	7.728	0.33
35	0.067	8.59	16.33	7.74	267.78	7.958	0.42
36	0.073	8.67	16.5	7.83	275.61	8.190	0.46
37	0.085	8.57	16.34	7.77	283.38	8.421	0.53
38	0.091	8.57	16.35	7.78	291.16	8.653	0.57
39	0.103	8.51	16.22	7.71	298.87	8.882	0.64
40	0.113	8.57	16.26	7.69	306.56	9.110	0.71
41	0.117	8.49	16.05	7.56	314.12	9.335	0.73
42	0.12	8.59	16.3	7.71	321.83	9.564	0.75
43	0.13	8.59	16.37	7.78	329.61	9.795	0.81
44	0.131	8.59	16.31	7.72	337.33	10.025	0.82
45		8.66	16.56	7.9	345.23	10.259	
46		8.62	16.34	7.72	352.95	10.489	
47	0.155	8.62	16.44	7.82	360.77	10.721	0.97
48		8.56	16.31	7.75	368.52	10.952	
49		8.62	16.48	7.86	376.38	11.185	
50	0.16	8.62	16.42	7.8	384.18	11.417	1.00
(c) Cadmium and EDTA							
1	0.01	8.61	16.19	7.58	7.58	0.225	0.0357
2	0.01	8.57	15.97	7.4	14.98	0.445	0.0357
3	0.01	8.58	16.01	7.43	22.41	0.666	0.0357
4	0.03	8.62	16.26	7.64	30.05	0.893	0.1071
5	0.19	8.55	16.33	7.78	37.83	1.124	0.6786
6	0.28	8.53	16.2	7.67	45.5	1.352	1.0000
7	0.28	8.56	16.29	7.73	53.23	1.582	1.0000
8	0.28	8.59	16.31	7.72	60.95	1.811	1.0000

Bibliography

American Water Works Association (1998). *Standard Methods for the Examination of Water and Wastewater*, 20e. American Water Works Association, or more recent edition.

Dunnivant, F.M., Newman, M.E., Brzenk, R. et al. (1999). A comprehensive stream study designed for an undergraduate non-majors course in earth science. *JGE* **47**: 158–165.

Sawyer, C.N. and McCarty, P.L. (1978). *Chemistry for Environmental Engineering*, 3e. New York: McGraw-Hill Book Company.

Wipple, G.C. and Wipple, M.C. (1911). Solubility of oxygen in sea water. *J. Amer. Chem. Soc.* **33**: 362.

Index

Pollutant Fate and Transport in Environmental Multimedia, First Edition. Frank M. Dunnivant and Elliot Anders.
© 2019 John Wiley & Sons, Inc. Published 2019 by John Wiley & Sons, Inc.
Companion website: www.wiley.com/go/Dunnivant/pollutant_fate&Transp_EnvironMultimedia

Periodic Table of the Elements

Carbon 6 C 12.01	Element Element # Symbol Atomic Mass

	IA	IIA	IIIB	IVB	VB	VIB	VIIB	VIII	VIII	VIIII	IB	IIB	IIIA	IVA	VA	VIA	VIIA	VIIIA
	Hydrogen 1 H 1.01																	Helium 2 He 4.00
2	Lithium 3 Li 6.94	Beryllium 4 Be 9.01											Boron 5 B 10.81	Carbon 6 C 12.01	Nitrogen 7 N 14.01	Oxygen 8 O 16.00	Fluorine 9 F 19.00	Neon 10 Ne 20.18
3	Sodium 11 Na 22.99	Magnesium 12 Mg 24.31											Aluminum 13 Al 26.98	Silicon 14 Si 28.09	Phosphorous 15 P 30.97	Sulfur 16 S 32.07	Chlorine 17 Cl 35.45	Argon 18 Ar 39.95
4	Potassium 19 K 39.10	Calcium 20 Ca 40.08	Scandium 21 Sc 44.96	Titanium 22 Ti 47.88	Vanadium 23 V 50.94	Chromium 24 Cr 52.00	Manganese 25 Mn 54.94	Iron 26 Fe 55.85	Cobalt 27 Co 58.93	Nickel 28 Ni 58.69	Copper 29 Cu 63.55	Zinc 30 Zn 65.39	Gallium 31 Ga 69.72	Germanium 32 Ge 72.61	Arsenic 33 As 74.92	Selenium 34 Se 78.96	Bromine 35 Br 79.90	Krypton 36 Kr 83.80
5	Rubidium 37 Rb 85.47	Strontium 38 Sr 87.62	Yttrium 39 Y 88.91	Zirconium 40 Zr 91.22	Niobium 41 Nb 92.91	Molybdenum 42 Mo 95.94	Technetium 43 Tc 99	Ruthenium 44 Ru 101.07	Rhodium 45 Rh 102.91	Palladium 46 Pd 106.42	Silver 47 Ag 107.87	Cadmium 48 Cd 112.41	Indium 49 In 114.82	Tin 50 Sn 118.71	Antimony 51 Sb 121.75	Tellurium 52 Te 127.60	Iodine 53 I 126.90	Xenon 54 Xe 131.29
6	Cesium 55 Cs 132.91	Barium 56 Ba 137.33	Lanthanum 57 La 138.91	Hafnium 72 Hf 178.49	Tantalum 73 Ta 180.95	Tungsten 74 W 183.85	Rhenium 75 Re 186.21	Osmium 76 Os 190.2	Iridium 77 Ir 192.22	Platinum 78 Pt 195.08	Gold 79 Au 196.97	Mercury 80 Hg 200.59	Thallium 81 Tl 204.38	Lead 82 Pb 207.2	Bismuth 83 Bi 208.98	Polonium 84 Po 209	Astatine 85 At 210	Radon 86 Rn 222
7	Francium 87 Fr 223	Radium 88 Ra 226	Actinium 89 Ac 227	Rutherfordium 104 Rf 261	Dubnium 105 Db 262	Seaborgium 106 Sg 263	Bohrium 107 Bh 262	Hassium 108 Hs 265	Meitnerium 109 Mt 266	Ununnilium 110 Uun 269	Unununium 111 Uuu 272	Ununbium 112 Uub 277						

Lanthanide Series	Cerium 58 Ce 140.12	Praseodymium 59 Pr 140.91	Neodymium 60 Nd 144.24	Promethium 61 Pm 147	Samarium 62 Sm 150.36	Europium 63 Eu 151.97	Gadolinium 64 Gd 157.25	Terbium 65 Tb 158.93	Dysprosium 66 Dy 162.50	Holmium 67 Ho 164.93	Erbium 68 Er 167.26	Thulium 69 Tm 168.93	Ytterbium 70 Yb 173.04	Lutetium 71 Lu 174.97
Actinide Series	Thorium 90 Th 232.04	Protactinium 91 Pa 231	Uranium 92 U 238.03	Neptunium 93 Np 237	Plutonium 94 Pu 244	Americium 95 Am 243	Curium 96 Cm 247	Berkelium 97 Bk 247	Californium 98 Cf 251	Einsteinium 99 Es 252	Fermium 100 Fm 257	Mendelevium 101 Md 258	Nobelium 102 No 259	Lawrencium 103 Lr 260